W0051388

Research Reports ESPRIT

Project 818/2252 · Delta-4 · Vol. 1

Edited in cooperation with
the Commission of the European Communities

D. Powell (Ed.)

Delta-4:
A Generic Architecture
for Dependable
Distributed Computing

Springer-Verlag

Berlin Heidelberg New York London Paris
Tokyo Hong Kong Barcelona Budapest

Editor

David Powell
LAAS-CNRS
7, avenue du Colonel Roche
F-31077 Toulouse, France

ESPRIT Project 818/2252 "Definition and Design of an Open Dependable Distributed Computer System Architecture (Delta-4)" belongs to the Subprogramme "Advanced Information Processing" of ESPRIT, the European Strategic Programme for Research and Development in Information Technology supported by the Commission of the European Communities.

The Delta-4 objectives are to formulate, develop and demonstrate an open, fault-tolerant, distributed computing architecture. The key features of the architecture are: a distributed object-oriented application support environment, built-in support for user-transparent fault-tolerance, use of multicast or group communication protocols, and use of standard off-the-shelf processors and standard local area network technology with minimum specialized hardware.

CR Subject Classification (1991): C.2, D.4.5, D.4.7, B.6.2

ISBN-13:978-3-540-54985-7   e-ISBN-13:978-3-642-84696-0
DOI: 10.1007/978-3-642-84696-0

Publication No. EUR 14025 EN of the
Commission of the European Communities,
Scientific and Technical Communication Unit,
Directorate-General Telecommunications, Information Industries and Innovation,
Luxembourg
Neither the Commission of the European Communities nor any person acting on behalf of the Commission is responsible for the use which might be made of the following information.

Typesetting: Camera ready by author

45/3140 – 543210 – Printed on acid-free paper

FOREWORD

Delta-4 is a 5-nation, 13-partner project that has been investigating the achievement of dependability in open distributed systems, including real-time systems.

This book describes the design and validation of the distributed fault-tolerant architecture developed within this project. The key features of the Delta-4 architecture are: (a) a distributed object-oriented application support environment; (b) built-in support for user-transparent fault-tolerance; (c) use of multicast or group communication protocols; and (d) use of standard off-the-shelf processors and standard local area network technology with minimum specialized hardware.

The book is organized as follows:

The first 3 chapters give an overview of the architecture's objectives and of the architecture itself, and compare the proposed solutions with other approaches.

Chapters 4 to 12 give a more detailed insight into the Delta-4 architectural concepts. Chapters 4 and 5 are devoted to providing a firm set of general concepts and terminology regarding dependable and real-time computing. Chapter 6 is centred on fault-tolerance techniques based on distribution. The description of the architecture itself commences with a description of the Delta-4 application support environment (Deltase) in chapter 7. Two variants of the architecture — the Delta-4 *Open System Architecture* (OSA) and the Delta-4 *Extra Performance Architecture* (XPA) — are described respectively in chapters 8 and 9. Both variants of the architecture have a common underlying basis for dependable multicasting, i.e., an atomic multicast protocol and fail-silent hardware; these are described respectively in chapters 10 and 11. The important topic of input to, and output from, the fault-tolerant Delta-4 architecture is tackled in chapter 12.

Chapters 13 and 14 describe the work that has been carried out in the fields of security, by intrusion-tolerance, and the tolerance of software design faults by diversified design.

Finally, chapter 15 gives an extensive overview of the validation activities carried out on various sub-systems of the architecture.

Several annexes that detail particular points are included together with a glossary of abbreviations that are used throughout the book.

ACKNOWLEDGEMENTS

The authors wish to acknowledge the strong support given to the members of the project by their respective organizations: *Bull SA, Crédit Agricole, Ferranti International plc, IEI-CNR, IITB-Fraunhofer, INESC, LAAS-CNRS, LGI-IMAG, MARI Group, SRD-AEA Technology, Renault Automobile, SEMA Group* and the *University of Newcastle-upon-Tyne*. The enthusiastic and close collaboration between partners, which this backing has encouraged from the outset, was an important factor in achieving the project objectives.

In particular, the authors acknowledge the work and the support of: *Martine Aguera, Marc Alciato, Susanne Angele, John Baldwin, Yada Banomyong, Mario Baptista, Gilles Bellaton, Patrick Béné, David Briggs, Derek Brown, Errol Burke, Gérard Carry, Jean Catala, John Clarke, Alain Costes, Michael Dean, Ken Doyle, John Etherington, Paul Ezhilchelvan, Joseph Faure, Jean-Michel Fray, Jacques Goze, Frédéric Grardel, Georges Grunberg, Christian Guérin, Jerôme Hamel, Rainer Helfinger, Alain Hennebelle, Peter Hinde, Wolfgang Hinderer, Alex Hogarth, Marie-Thérèse Ippolito, Ross Kavanagh, Ivor Kelly, Philippe Lebourg, Marc Liochon, Gilles Lorenzi, Pascal Martin, Navnit Mehta, Denis Michaud, Stephen Mitchell, Luís Nacamura, Christopher Nichols, Gilles Pellerin, Joëlle Penavayre, Eliane Penny, Laurence Plancher, Bernard Quiquemelle, Pierre-Guy Ranéa, Alan Reece, John Reed, Philippe Reynier, René Ribot, David Riley, Bruno Robert, Bethan Roberts, Carlos Rodriguez, Paul Rubenstein, Heike Schwingel-Horner, Bruno Séhabiague, Philippe Sommet, Isabelle Theiller, Touradj Vassigh, David Vaughan, François Waeselynck* and *Jiannong Zhou*.

We also wish to acknowledge the contribution made in previous phases of this project by our earlier partners: *BASF, GMD-First, Jeumont Schneider, Telettra* and the *University of Bologna*.

✳ ✳ ✳

The Delta-4 project was partially financed by the *Commission of the European Communities* as projects 818 and 2252 of the European Strategic Programme for Information Technology (ESPRIT).

LAAS-CNRS would like to acknowledge additional sponsorship of the Delta-4 project from the *Conseil Régional de Midi-Pyrénées*.

✳ ✳ ✳

TABLE OF CONTENTS

Chapter 1

Requirements and Objectives

The requirements and objectives of the Delta-4 system architecture are naturally and properly determined by the demands of the targetted application domains. After nominating these application domains, this chapter contains a brief discussion of the major technical and commercial influences that arise; these influences affect both the architectural design and its implementation. Although Delta-4 is a precompetitive applied research project, commercial influences have particular significance since early exploitation of results is intended.

1.1. Fields of Application

Delta-4 aims to provide an on-line infra-structure to support certain large-scale information processing systems in both commerce and industry. The following applications characterize the distributed application systems addressed by Delta-4:

- Computer Integrated Manufacturing and Engineering (CIME) Systems for aerospace and automotive industries
- Industrial Systems e. g. process control, power generation and distribution systems
- Commercial Systems, e.g., banking and office systems
- Integrated Information Processing (IIP) Systems, e.g., command, control and communications systems for police, fire and defence systems
- Distributed Data Base Management (DBDM) Systems

It is recognised that satisfying the communication and computation needs of such diverse application areas is a major commitment. However, these application areas do have important common characteristics that are summarised in the following paragraphs and which constrain the nature of architectural solutions.

In classifying the application areas of interest as large-scale information processing systems, the term "large-scale" is not a reference to the scale of individual computers used to build such systems — systems may be built from a range of computer sizes from PCs through Workstations to mainframes. The significance of the term large-scale is rather that we are addressing systems that supply a wide range of integrated services, as opposed to "stand-alone" applications.

1.2. Architectural Attributes

1.2.1. Distribution

Each of the application domains requires distributed system solutions. This tends to be seen as a self-evident requirement, so it is sufficient for the present purposes to list a few of the reasons why the Delta-4 application domains in particular need distributed system solutions:

- In an industrial production environment, to minimize cabling costs in feeding many thousands of analogue and digital signals to and from primary processing elements.

- To take advantage of intelligent display capabilities and avoid I/O bottle-necks in passing information to and from a centralized system.

- To take advantage of low cost, mass produced, specialized computing sub-systems, e.g., workstations, programmable logic controllers, array processors, PCs, etc. Integration of these heterogeneous sub-systems, inevitably leads to a distributed system solution.

- To allow systems to grow incrementally in functionality and performance.

- To reduce communication costs by performing processing close to the physical location of an operator while maintaining access to shared resources.

- To provide opportunities for concurrency and parallelism in both closely-coupled (multiprocessor) systems and loosely-coupled (networked) systems. As a corollary, use of an object-oriented computational model and associated configuration tools encourages the user to exploit, fully and efficiently, opportunities for parallelism in either of these environments.

The distributed systems designed for the chosen application areas are usually based on the use of Local Area Networks (LANs); such networks may connect systems over distances ranging from a few kilometres with commonly-used technology to several tens of kilometres for emerging technologies such as the Fiber-optic Distributed Data Interface, FDDI. Delta-4 aims to meet the need of large-scale processing systems using such Local Area Network protocols. Delta-4 systems may interact with other systems by way of Wide Area Network (WAN) connections but Delta-4 services are provided only in the LAN environment.

1.2.2. Dependability

In practice, users come to depend on many of the services offered by an information processing system. The fact is that service failures may be very costly in terms of:

- loss of production (e.g., manufacturing applications),
- loss of clients (due to lack of user confidence),
- loss of secrecy (systems with sensitive data), or even,
- loss of human life (life-critical applications).

Systems are susceptible to accidental hardware and software faults and to intentional faults (intrusions). The main requirement of the applications domains from the dependability viewpoint is the tolerance of (accidental) hardware faults. Cost-effectiveness rather than extremely high levels of dependability is sought. Nevertheless, the Delta-4 architecture should be usable in safety-related sub-systems where the overall system, comprising the process plant and its control system, is required to exhibit levels of safety appropriate to life-critical applications.

Techniques for tolerating intentional interaction faults (intrusions) and accidental software design faults are also of potential long-term interest.

It follows that the considered application domains not only require distributed system solutions but also require dependable system solutions. However, distributed systems, and particularly open distributed systems using off-the-shelf hardware, need to possess specific architectural attributes if dependable operation is to be achieved. Important considerations here are that:

- the typically large number of hardware resources employed in large-scale information processing systems naturally increases the probability of, at least partial, system failure.

- off-the-shelf hardware fails in arbitrary ways and has no special built-in mechanisms either to reliably detect errors when they occur or to contain the effect of errors so that they do not subsequently spread in an uncontrolled and unpredictable way through the system. The fault assumptions of such a basic dependability model cannot be justified.

Proprietary fault-tolerant products, although widely available, tend to be in the form of stand-alone systems primarily applicable to centralized solutions. Using such products in the considered application domains entails either that:

- they are applied to all nodes of the distributed system where dependable service provision is required with a consequent rapid increase of cost and lack of flexibility, or that

- the solution provided becomes centralized with the consequent loss of the advantages mentioned in §1.2.1 above.

Since many industrial and commercial applications do not require dependable availability of all services, and cannot justify the costs involved, the Delta-4 architecture is required to offer the user the option to obtain specifiable levels of dependability on a service-by-service basis.

Delta-4 aims to overcome the cost escalation implicit in the use of current proprietary fault-tolerant products in distributed systems, by allowing the use of controlled levels of redundancy of off-the-shelf hardware components, so minimizing overall system costs. Just as most proprietary products aim to provide fault-tolerance based on valid fault assumptions, so, in Delta-4, the objective is to provide a complementary valid basis for distributed system solutions. It then becomes possible to build on either the inherent (or supplementary) redundancy in distributed systems to provide dependable operation. It must, however, be emphasised that dependability is not a natural property of distributed systems (this point is discussed further in chapter 3).

The requirement for dependable system operation outlined above can be met by the provision of redundancy through *software* component replication *on* distinct hosts within an open distributed system. This, in turn, requires support for logical component replication (e.g., coherent dissemination (multicasting) of information to multiple replicates) and majority voting, or comparison, of the interactions between replicated components. This, in turn, involves communication between replicates. Furthermore, advanced system administration mechanisms are required to implement fault diagnosis, fault passivation and system reconfiguration, *and extensive validation activities need to be carried out to justify user reliance on the services provided.*

1.2.3. Real-Time

Many (although not all) of the target application domains require service provision with a real-time response. It is not necessary, at this stage, to offer a rigorous definition of the term "real-time"; that definition follows in chapter 5. For the purposes of summarising architectural requirements, it is sufficient to offer three aspects which characterize systems addressed by Delta-4:

- Certain application domains are concerned with hard real-time in the sense that failure to provide a service by a deadline means service failure.

- The time scales of interest are typical of those encountered in industrial processes — deadlines of the order of 10 milliseconds must be met.

- Not all services are required to be provided with hard real-time constraints. Some may be required to meet soft deadlines and some may have no effective deadline. A soft deadline is one where it is necessary to make a best endeavour to meet the deadline and provide an expectation that it will be met on some specifiable percentage of occasions.

The consequence of this last point is that Delta-4 must be able to offer real-time support in systems that have a dynamic workload and that are sized in the knowledge that, while all hard deadlines can be met, short term overloads will nevertheless occur as regards the "best-endeavour" behaviour. It is neither commercially realistic, nor technically feasible, to size a large-scale industrial system to have sufficient processing power to meet all deadlines under all load conditions — especially where such systems are required to respond to uncontrolled external events. Costs would be prohibitive.

A consequential objective is to introduce mechanisms that explicitly support the requirements of real-time systems concerning both throughput and response. Delta-4 must be able to offer explicit support for the priorities and deadlines of real-time services; these must be respected both by the remote invocation of computation and by the communication protocols.

1.2.4. Performance

1.2.4.1. The Requirement. One of the principal requirements of the application domains is that of specifiable performance on a service-by-service basis. This attribute is, naturally, of particular significance to real-time systems. Real-time applications are required to satisfy performance criteria from the viewpoints of both responsiveness (computational results delivered as soon as possible) and timeliness (computation results delivered in respect of real-time deadlines).

1.2.4.2. Performance Requirements and Metrics. There are at least two alternative scenarios in which communications performance is of interest; the different aims of the architectures lead, in each case, to the choice of different performance metrics.

The first underpins OSI-related conceptual work and most current implementations and applications. Here, the major design concern is with the transport of large quantities of data from end-point to end-point; performance is therefore measured in terms of throughput.

The second concerns the support of distributed computations *where messages are used to support the passage of computational control between nodes; such interactions are typically represented by Remote Procedure Call (RPC).* Performance is then determined by the total round-trip delays, including communications and processing overheads, in support of the RPC paradigm. Birrell and Nelson [Birrell and Nelson 1984], who pioneered the concept of RPC, comment that when these overheads are too high, "applications that might otherwise get developed would be distorted by (the designer's) desire to avoid communicating".

Production-quality, fully open implementations of the ISO/OSI communications protocols offer a wide variety of services appropriate to the broad spectrum of applications envisaged in the definition of the ISO/OSI model. Precisely because of this generality and openness, however, they cannot offer the performance attributes or characteristics needed for the demanding real-time applications that are included in the nominated Delta-4 application domains.

In addition to the Delta-4 Open System Architecture (D4-OSA), work has, therefore, progressed on an Extra Performance Architecture (D4-XPA) in which openness is traded for performance. One objective of this work is to minimize and bound the time for replicated null-parameter RPC calls. It remains to be seen, in the light of application experience, whether Birrell and Nelson's objective can be achieved in the context of the Delta-4 approach to dependability.

1.2.5. Heterogeneity

The typical customer for large-scale distributed systems does not wish to be restricted to a single vendor and seeks an architecture having an evident basis for long-term support. This leads both to a requirement for heterogeneity of hardware and software and to the need for an OPEN architecture (see §1.2.5 below). It is neither desirable (nor interesting) to specify and design new hardware or software (operating systems, languages, etc.) where components of adequate functionality already exist, nor to make the architecture specific to, or dependent on, any such particular hardware or software component, or communications environment.

Rather, it is desirable to accelerate exploitation of the architecture by encouraging implementations based on commercially available components. A consequence of this independence to the underlying base technology is to make the architecture both future-proof and attractive to multiple vendors.

Since it is highly desirable, for commercial reasons, to build systems from off-the-shelf components, and since the ability to construct open, distributed, dependable systems in a predictable way is not yet a commercial reality, one of the objectives of Delta-4 is to achieve truly dependable operation with such components. Indeed, Delta-4 was founded on the concept that the property of dependability could be conferred on commercial processor systems if these systems were linked through a dependable communication system. How we achieve the necessary error detection and containment is central to each of the dependability models described in chapter 2.

1.2.6. Openness

The term "open" has many interpretations all of which are relevant to Delta-4.

1) The architecture is published and made freely available to interested vendors and users with a view to widespread adoption of the architecture.

2) Implementations of the architecture are possible on a wide variety of processors, operating systems, languages and communications media. The architecture is not dependent on any proprietary base components.

3) The architecture makes use of existing standards where possible and otherwise contributes to the evolution of relevant new standards.

However, in relation to item (3) above, ISO/OSI standards do not yet address:

- Support for computational dependability as opposed to dependability of the communication process itself, which is well addressed in various reliable communication protocols and services.

- The achievement of real-time deadlines in a distributed environment — neither computation nor communication issues are addressed here.

- Certain computational issues — applications program models for transparent distribution (as opposed to communicating computation), and the issue of program structure and its language representation.

Precisely because existing standards do not support the facilities required by Delta-4 objectives, it is not possible to develop the architecture purely from existing products and standards. While it is clearly possible to engineer large scale, dependable, real-time systems without recourse to such standards, the proprietary nature of such systems leads to unacceptably high development costs and long-duration, high risk, developments. Hence the strong commercial incentive to *seek an extension to the scope of standards to provide an open, standard architecture for the considered application domain*.

The requirement is therefore to create an architecture which:

- Minimizes the need for new or extended standards, and

- Is an *open* architecture to encourage multiple vendors to support it and to encourage the standardization process. A non-open architecture would quite simply return us to our starting point.

It is also recognised that the enhanced performance required for the real-time application domains considered can only be achieved at the cost of some loss of generality in host connectivity. In response to this relaxation in the openness requirement, D4-XPA (introduced in §1.2.4.2 above) has evolved. This architecture may be seen as a complement to the open system architecture. The loss of generality is required to be minimal; D4-XPA subsystems are required to be able to co-exist, and inter-work, with standard D4-OSA systems. D4-XPA may only be closed in the sense that standards that cannot be adapted to meet performance requirements will not be used; however, this must be on as limited a scale as possible and, as with other Delta-4 work, the rationale and results will be published.

1.2.7. Applications Software

Software for the target application areas may be of three types:

1) Conformant with ISO/OSI application layer standards.

Computer-integrated manufacturing is an area in which much recent progress has been made in the definition of open communication standards; the Manufacturing Automation Protocol (MAP) with its Manufacturing Message Service (MMS) is of particular relevance to Delta-4. MMS itself confers the desirable property on distributed applications that participating applications may cooperate without prior knowledge of either application component language or the supporting operating systems used. *Portability of MAP- and other ISO-conformant applications to Delta-4 systems is an essential requirement*.

Inter-working and co-existence with MAP *and other ISO standard applications protocols* are considered as major functional requirements of the Delta-4 architecture; this places constraints on communications sub-system design as well as the applications support environment.

2) Targetted towards the emerging ISO/IEC Reference Model for Open Distributed Processing (RM-ODP) and the corresponding Support Environment (SE-ODP) standards.

This is of particular interest in the context of large-scale system development. Such developments typically involve multiple development teams, with large numbers of software components being developed by each team; these components must interact both with each other and with components developed by the other teams. Such interfacing issues are well-addressed for single large programs by programming languages with separate compilation of modules and with strong type checking applied across their interfaces in languages such as Ada. ODP extends this facility to the inter-working of software in distributed systems. The services of a distributed application are provided by a set of cooperating, logically separate components, which may be sited at physically remote hosts or co-located on a single host.

The support is such that application designers can quickly specify, develop and deliver dependable, high-performance, distributed applications. Systems and subsystems can readily be installed, changed and maintained.

3) Third party software that may or may not have been designed with distributed application in mind.

Where possible, the architecture should be able to confer dependability on such software without modification. For example, there is a strong commercial incentive to continue to use certain relational databases that have established themselves as de-facto standards.

1.2.8. Communications

Multicasting has been recognised as an important communication paradigm not only for handling replication but also for facilitating the definition and implementation of distributed applications. Delta-4 proposes extensive multicasting communication services based on a set of specific protocols. Again, inter-working and co-existence with existing (ISO/OSI) communication protocols are seen as major requirements as is convergence with current standardization activities on multicasting (see, for example, [ISO N2031]).

1.2.9. Portability

Delta-4 aims to achieve portability in at least two ways:

1) The architecture itself should be portable in the sense that implementations can be achieved on a wide variety of underlying proprietary hardware and software systems. An important consequence is that the architecture is then, to a high degree, future-proof to underlying technological developments.

2) Applications should be portable from one Delta-4 implementation to another. Here, portability means the ability to transfer either complete software systems or software "building-blocks" to different environments and/or to different hardware. Portability thus allows re-use of software with its attendant advantages in terms of economy and dependability (increased confidence in the correctness of the software).

Both aspects of portability are assisted by the requirement that Delta-4 support a high level of abstraction, such that the specifications and designs of the interactive aspects of dependable, distributed applications are independent of networks, operating systems and programming languages.

Chapter 2

Overview of the Architecture

Despite the enormous improvements in the quality of computer technology that have been made during the last few decades, failures of computer system components cannot be ruled out completely. Some application areas therefore need computer systems that continue to provide their specified service *despite* component failures — such computer systems are said to be *fault-tolerant*. Distributed computing systems — in which multiple computers interact by means of an underlying communication network to achieve some common purpose — offer attractive opportunities for providing fault-tolerance since their multiplicity of interconnected resources can be exploited to implement the redundancy that is necessary to survive failures.

The Delta-4 fault-tolerant architecture aims to provide a computational and communication infrastructure for application domains that require *distributed* system solutions with various *dependability* and *real-time* constraints (e.g., computer integrated manufacturing and engineering, process control, office systems, etc.). The scale of distribution in the targetted application domains is commensurate with the distances that can be covered by local area networks, i.e., from a few metres up to several kilometres. The Delta-4 architecture also seeks to be an *open* architecture, i.e., one that can use off-the-shelf computers, accommodate *heterogeneity* of the underlying hardware and system software and provide portability of application software. However, in those application domains where *real-time* response is paramount, the heterogeneity and openness may need to be sacrificed to provide the appropriate assurance of timeliness. To this end, the Delta-4 architecture offers two variants (both based on sub-systems that present a high degree of commonality):

- the Delta-4 *Open System Architecture* (D4-OSA) which, as its name suggests, is an *open* architecture able to accommodate heterogeneity,
- the Delta-4 *Extra Performance Architecture* (D4-XPA) which provides explicit support for assuring timeliness.

This chapter sketches out the major characteristics of the Delta-4 architecture and its variants. The key attributes of the architecture — dependability and real-time — are discussed in the section 2.1 and section 2.2 outlines the Delta-4 approach to attaining these attributes. Section 2.3 presents the hardware architecture of Delta-4 and discusses the communication issues. The application software support environment and administration system are presented in section 2.4. Finally, section 2.5 describes the validation of the architecture.

2.1. Dependability and Real-Time Concepts

This section introduces some basic concepts and terminology regarding dependable and real-time computing (for a more detailed exposition of these concepts, see chapters 4 and 5).

2.1.1. Dependability

Dependability can be defined as the "trustworthiness of a computer system such that reliance can justifiably be placed on the service it delivers" [Carter 1982]. Dependability is thus a *global* concept that subsumes the usual attributes of *reliability* (continuity of service), *availability* (readiness for usage), *safety* (avoidance of catastrophes) and *security* (prevention of unauthorized handling of information).

When designing a dependable computing system, it important to have very precise definitions of the notions of faults, errors, failures and other related concepts. A system *failure* is defined to occur when the service delivered by the system no longer complies with its *specification*, the latter being an agreed description of the system's expected function and/or service. An *error* is that part of the system state that is liable to lead to failure: an error affecting the service, i.e., becoming visible to the user, is an indication that a failure occurs or has occurred. The adjudged or hypothesized cause of an error is a *fault*. An error is thus the manifestation of a fault *in the system*, and a failure is the effect of an error *on the service*.

Faults are thus the potential source of system undependability. Faults may either be due to some physical phenomenon (inside or outside the system) or caused (accidentally or intentionally) by human beings. They may either occur during the operational life of the system or be created during the design process. *Fault tolerance* — i.e., the ability to provide service complying with the specification in spite of faults — may be seen as complementary to *fault prevention* techniques aimed at improving the quality of components and procedures to decrease the frequency at which faults occur or are introduced into the system. Fault tolerance is achieved by *error processing* and by *fault treatment* [Anderson and Lee 1981]: error processing is aimed at removing errors from the computational state, if possible before a failure occurs; fault treatment is aimed at preventing faults from being activated again[1].

Error-processing may be carried out either by error-detection-and-recovery or by error-compensation. In error-detection-and-recovery, the fact that the system is in an erroneous state must first be (urgently) ascertained. An error-free state is then substituted for the erroneous one: this error-free state may be some past state of the system (backward recovery) or some entirely new state (forward recovery). In error-compensation, the erroneous state contains enough redundancy for the system to be able to deliver an error-free service from the erroneous (internal) state. Classic examples of error-detection-and-recovery and error-compensation are provided respectively by atomic transactions (see, for instance, [Lampson 1981]) and triple-modular redundancy or voting techniques (see, for instance, [Wensley et al. 1978]).

Fault treatment is a sequel to error processing; whereas error processing is aimed at preventing errors from becoming visible to the user, fault treatment is necessary to prevent faults from causing further errors. Fault treatment entails fault diagnosis (determination of the cause of observed errors), fault passivation (preventing diagnosed faults from being activated again) and, if possible, system reconfiguration to restore the level of redundancy so that the system is able to tolerate further faults.

When designing a fault-tolerant system, it is important to define clearly what types of faults the system is intended to tolerate and the assumed behaviour (or failure modes) of faulty components. If faulty components behave differently from what the system's error processing and fault treatment facilities can cope with, then the system will fail. In distributed systems, the behaviour of a node can be defined in terms of the messages that it sends over the network. The assumed failure modes of nodes are thus defined in terms of the messages that faulty nodes send or do not send. The simplest and most common assumption about node failures is that nodes are *fail-silent* [Powell et al. 1988], i.e., that they function correctly until the point of

---

[1] In anthropomorphic terms, error processing can be viewed as "symptom relief" and fault treatment as "curing the illness".

failure when they "crash" and then remain forever silent (until they are repaired). The most severe failure mode that can be imagined is that of *fail-uncontrolled* nodes that fail in quite *arbitrary* or "Byzantine" ways [Lamport et al. 1982]. Such nodes can fail by producing messages with erroneous content, messages that arrive too early or too late, or indeed "impromptu" messages that should never have been sent at all. In between these two extremes it is possible to define failure modes of intermediate severity [Cristian et al. 1985, Powell 1991]. Generally, when the assumed failure modes of system components become more severe, then more redundancy (and complexity) must be introduced into the system to tolerate a given number of simultaneously active faults.

2.1.2. Real-Time

Real-time systems are those which are able to offer an assurance of *timeliness* of service provision. The very notion of *timeliness* of service provision results from the fact that real-time services have associated with them not only a functional specification of "what" needs to be done but also a timing specification of "when" it should be done. Failure to meet the timing specification of a real-time service can be as severe as failure to meet its functional specification. According to the application, and to the particular service being considered, the assurance of timeliness provided by a real-time system may range from a *high expectation* to a *firm guarantee* of service provision within a defined time interval[2].

At least three sorts of times can enter into the timing specification of a real-time service:

- a *liveline* and a *deadline* indicating respectively the beginning and the end of the time interval in which service must be provided if the service is to be considered timely,

- a *targetline* (or "soft" deadline) specifying the time at which it would somehow be "best" for the service to be delivered, i.e., the point at which maximum benefit or minimum cost is accrued.

The essential difference between a service deadline and a service targetline is that the former *must* be met if the service is to be timely whereas missing a targetline, although undesirable, is not considered as a failure. An alternative way to specify the timing constraints of real-time services is that of value-time or worth-time functions that indicate the benefit of service delivery as a function of the instant of service delivery (see, for example, [Jensen et al. 1985, Jensen and Northcutt 1990]).

In systems that are "real-time" in the sense defined above, the available resources (finite in any practical system) must be allocated to computation and communication activities so that, as long as specified "worst-case" environmental constraints (e.g., event occurrence rate < specified maximum) are satisfied, it is guaranteed that all critical deadlines are met or that the cost of not meeting targetlines is minimized. In some applications, even if the "worst-case" conditions are exceeded, it may be required that the system make a best effort to ensure that the number or cost of missed deadlines or targetlines is minimized.

Scheduling concerns the allocation of resources that must be time-shared; the resources in question could be processing power, memory, communication bandwidth, etc. In real-time systems, scheduling algorithms are concerned more with the respect of livelines, deadlines and targetlines than with ensuring fairness of resource usage. Consequently, most scheduling decisions need to take account of time explicitly when determining what activity should be scheduled next on a particular resource. Schedules can be calculated off-line (during the system design phase). This is often the case when it must be guaranteed that all critical deadlines be met under specified worst-case conditions. A set of such pre-calculated static schedules may be

---

2   Note that this view of real-time systems means more than just "high performance" or the ability to react "quickly" in response to asynchronous external events.

necessary if the system is to operate in several different modes (e.g., startup, production, normal shut-down, emergency shut-down, etc.). The performance penalties incurred by such a static approach can be partially mitigated if it can be arranged for non-critical computation or communication to make use of the "holes" that are left in the pre-calculated schedules. In complex real-time applications, on-line (dynamic) scheduling may be preferable to off-line (static) scheduling since the complexity of the application may be such that no *a priori* "worst-case" can be defined. Even if the worst-case is known, it may be desired that the system degrade gracefully whenever the "worst" case is exceeded (consider a radar-tracking system designed to track, at most, 100 aircraft — if 101 aircraft happen to be within range, it may be better to track as many as possible rather than none at all).

If consistent time-dependent decisions (see figure 1) must be made by different nodes of a *distributed* real-time system, then, if the expense of an agreement protocol is not to be incurred for every such time-dependent decision, each node must have a *consistent* view of time. Consequently, the local clocks of each node in a distributed real-time system must be synchronized to a specified *precision* so that all (non-faulty) nodes agree on an *approximate global time*. Furthermore, many applications require this global time to be synchronized to a known *accuracy* of some external standard of *physical time*. The precision and accuracy of clock synchronization determine the granularity at which different nodes can make consistent time-dependent decisions.

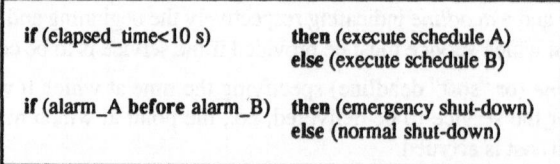

Fig. 1 - Examples of Time-Dependent Decisions

2.2. Dependability and Real-Time in Delta-4

The Delta-4 architecture is a distributed architecture that seeks to be both open and dependable, and capable of supporting real-time applications. To be able to satisfy a large range of application requirements in a cost-effective manner, the Delta-4 architecture can provide various degrees of dependability and real-time functionality. This section outlines the dependability and real-time features of the architecture.

2.2.1. Dependability in Delta-4

The Delta-4 architecture provides mechanisms for achieving dependability on a service-by-service basis in "money-critical" applications for which the relevant dependability attributes are reliability, availability and security. The architecture is *not* aimed at life-critical applications for which safety is the major concern. Note however, that reliability, availability and security are of very real interest in safety-*related* applications, e.g., systems that, if they fail, cause emergency shutdown and thus induce not only unavailability but also wear-out of the primary, safety-critical protection system (see annex A).

Delta-4 is concerned essentially with accidental physical faults, i.e., faults in the hardware components and, to a lesser extent, with accidental design faults in software.

The basic paradigm for tolerating hardware faults in Delta-4 is that of *replicated computations* executed by distinct nodes of a distributed system. The units of replication are *software components* — logical run-time units of computation and data encapsulation that communicate with each other by means of messages (only). Software components may be replicated to different degrees according to their degree of criticality and the assumptions made about the failure modes of the underlying node hardware (cf. section §2.1.1). For a given service, if a fail-silence assumption for the underlying hardware can be justified to a degree commensurate with the dependability objectives of that service, then the service can be made single-fault tolerant by duplication. Indeed, duplication of the software components providing the service is sufficient to allow recovery from a fault that causes a node to stop sending messages over the communication system. However, if the fail-silent assumption cannot be justified to a sufficient degree for a particular service, then triplication becomes necessary so that errors due to fail-uncontrolled behaviour can be masked by voting techniques.

It is of course assumed that hardware faults will occur independently in different nodes. In fact, any faults (of hardware or software origin) that manifest themselves *independently* in different nodes can be tolerated by straightforward replication techniques. Some design faults in system software at each node and, to a lesser degree, design faults in application software, can be expected to have such independence in their manifestations since the execution environments of replicas on distinct nodes are essentially different (due to loose synchronization and differing workloads). Such faults are commonly called "Heisenbugs" [Gray 1986]. In addition to this possibility of tolerating certain software design faults, distributed fault-tolerance techniques have the following advantages over tightly-coupled "stand-alone" fault-tolerance techniques:

- specialized hardware is kept to a minimum since distributed fault-tolerance is implemented primarily in software; the cost of specialized hardware design — or re-design when a technology update is required — is therefore minimized,
- geographical separation of resources does not have to be "added on" if disaster recovery is to be provided; the same distributed fault-tolerance techniques can be used regardless of whether the replicas are close to or distant from each other,
- loose-synchronization of replicas (through message-passing) leads to improved tolerance of transient faults that could otherwise simultaneously affect all redundant computations at the same point of execution [Kopetz et al. 1990].

Software design faults that do not manifest themselves in the independent "Heisenbug" fashion cannot be tolerated by replication. Diverse designs must be used to define *multiple variants*. These multiple variants may be encapsulated within a software component that can then be replicated for hardware fault-tolerance. Alternatively, the variants could be executed on distinct nodes in such a way as to tolerate simultaneously both hardware and software faults. The latter approach has been investigated in Delta-4 although it has not yet been implemented (see chapter 14).

In some multi-user application domains where data of a sensitive nature is manipulated, human faults of an intentional nature become a concern of considerable importance. There are two varieties of intentional faults: malicious logic and *intrusions*. The latter variety of fault has been explicitly addressed by the Delta-4 project in the form of *intrusion-tolerance* mechanisms aimed at supplementing conventional intrusion prevention schemes. Replication is of little help when trying to tolerate intrusions. From the confidentiality viewpoint, replication is in fact detrimental to security since the number of different places where sensitive information can be found is greater than in a system without replication. This therefore leads to more potential loopholes for an intruder to penetrate. The project has therefore investigated techniques based on *fragmentation-scattering*; sensitive information is split into fragments and scattered over different nodes so that intrusions only lead to access to partial information (see chapter 13).

The remainder of this sub-section is devoted to Delta-4 replication techniques for tolerating hardware faults and "Heisenbug" software faults.

2.2.1.1. Error Processing. In the context of replicated computation, error processing consists of those techniques for coordinating replicated computation that allow communication and computation to proceed despite the fact that some of the replicas may reside on faulty nodes (figure 2).

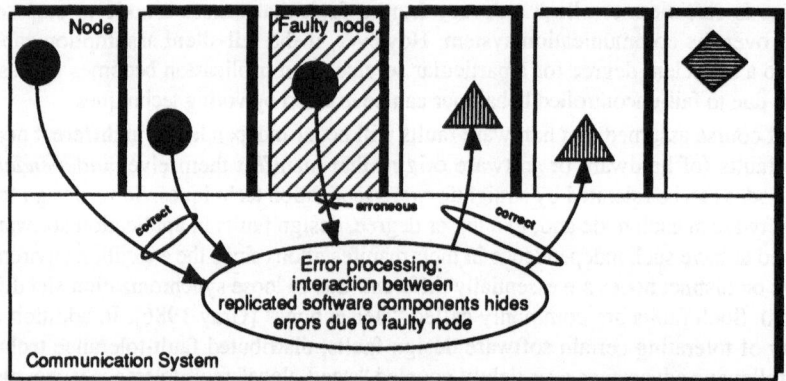

Fig. 2 - Illustration of the Principle of Error Processing in the Context of Replicated Computation

Delta-4 provides three different — but complementary — techniques for coordinating replicated computation: active, passive and semi-active replication.

Active replication is a technique in which all replicas process all input messages concurrently so that their internal states are closely synchronized — in the absence of faults, outputs can be taken from any replica. The active replication approach allows quasi-simultaneous recovery from a node failure. Furthermore, it is adapted to both the fail-silent and fail-uncontrolled node assumptions (cf. §2.1.1) since messages produced by different (active) replicas can be cross-checked (in value and time). However, active replication requires that all replicas can be guaranteed to be *deterministic* in the absence of faults, i.e., it must be guaranteed that if non-faulty replicas process identical input message streams, they will produce identical output message streams.

Passive replication is a technique in which only one of the replicas (the primary replica) processes the input messages and provides output messages. In the absence of faults, the other replicas (the standby replicas) do not process input messages and do not produce output messages; their internal states are however regularly updated by means of checkpoints from the primary replica. Passive replication can only be envisaged if it is assumed that nodes are fail-silent. Unlike active replication, this technique does not require computation to be deterministic [Speirs and Barrett 1989]. However, the performance overheads of transferring checkpoints and rolling-back for recovery may not be acceptable in certain applications — especially in real-time applications.

Semi-active replication can be viewed as a hybrid of both active and passive replication. Only one of the replicas (the leader replica) processes all input messages and provides output messages. In the absence of faults, the other replicas (the follower replicas) do not produce output messages; their internal state is updated either by direct processing of input messages or, where appropriate, by means of "notifications" or "mini-checkpoints" from the leader replica. Semi-active replication seeks to achieve the low recovery overheads of active replication while relaxing the constraints on computation determinism. A notifications can be used to force the followers to obey all non-deterministic decisions made by the leader replica [Barrett et al. 1990]. This possibility is particularly relevant for allowing replica-consistent preemption decisions. Like passive replication, this technique resides on the assumption that nodes are fail-silent. This technique is particularly suitable for replicating large "off-the-shelf" software components about which no assumption can be made on replica determinism and internal states (e.g., commercially available database management software).

Table 1 summarises the relative merits of each technique; further details are given in chapter 6.

Table 1 - Comparison of Replication Techniques

Replication technique	Recovery overhead	Non-determinism	Accommodates fail-uncontrolled behaviour
Active	*Lowest*	*Forbidden*	*Yes*
Passive	*Highest*	*Allowed*	*No*
Semi-active	*Low*	*Resolved*	*No[3]*

2.2.1.2. Fault Treatment. In the context of replicated computation, fault treatment consists essentially of three activities:

1) diagnosing the cause of error — i.e. finding out what entity (node or replica) is at fault,

2) if necessary, "passivating" the entity that is judged to be faulty (so that it will not cause further errors),

3) if possible, creating new replicas on fault-free nodes to restore the level of redundancy and thus be able to tolerate further faults (figure 3).

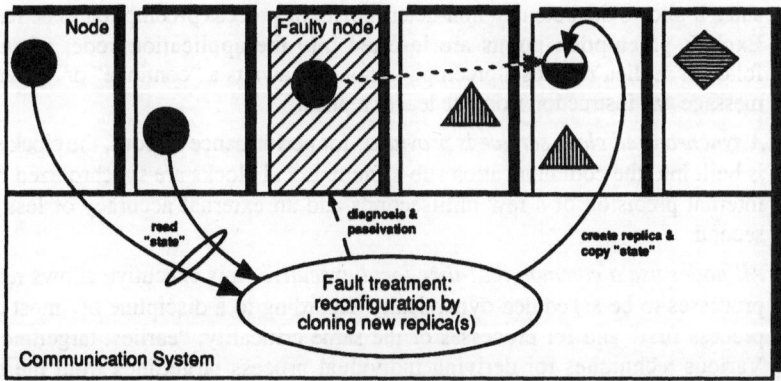

Fig. 3 - Illustration of the Principle of Fault Treatment in the Context of Replicated Computation

---

3   An extension of the semi-active replication technique to accommodate fail-uncontrolled behaviour is presently being investigated.

The set of sites on which replicas of a given software component can be located is termed the component's *replication domain*. The creation of a new replica on a fault-free node is called *cloning*. The location at which a new replica is cloned may be:

a) the same site at which the original replica was located; this could be the case:

  - either because the fault was considered to a be soft fault (e.g., a Heisenbug) such that re-initialization of the state of the failed replica is sufficient to bring it back on line,

  - or when corrective maintenance of the faulty node has been carried out,

b) on any other site in the component's replication domain thus allowing corrective maintenance to be deferred.

Fault treatment is further detailed in section 2.4.3.

2.2.2. Real-Time in Delta-4

Support for heterogeneity and openness and support for real-time are antagonistic aims; in heterogeneous distributed systems, the local schedulers at each node may not even be the same, let alone time-dependent. This is one of the motivations for the homogeneous XPA (Extra Performance Architecture) variant of the Delta-4 architecture [Barrett et al. 1990] in which heterogeneity and openness are sacrificed to provide assurance of timeliness.

The XPA real-time variant of the Delta-4 architecture differs from the OSA (Open System Architecture) variant by the following features:

- *Nodes are homogeneous:* this eliminates the need, and therefore the performance penalties, of the translation of data representations during node interactions.

- *Nodes are fail-silent:* since openness has been sacrificed in order to achieve real-time performance, nodes are purpose-designed with built-in self-checking to support the fail-silence assumption. This eliminates the need, and therefore the performance penalties, of voting on the contents of transmitted messages.

- *Exclusive use of semi-active replication for fault-tolerance:* the semi-active replication technique (cf. §2.2.1.1) was pioneered during the development of XPA since it allows the potential non-determinism of process preemption to be resolved. Explicit preemption points are inserted into the application code; whenever a follower replica reaches a preemption point, it awaits a "continue" or "preempt by message #*n*" instruction from the leader replica.

- *A synchronized clock service is provided:* for performance reasons, the clock service is built into the communication sub-system. Local clocks are synchronized with an internal precision of a few milliseconds and an external accuracy of less than a second.

- *All nodes use a common real-time local executive:* this executive allows real-time processes to be scheduled dynamically according to a discipline of "most critical process first" and for processes of the same criticality, "earliest targetline first". Various techniques for deriving individual process targetlines from the overall targetline/deadline of a distributed service are being investigated as is the possibility of mapping an off-line pre-calculated schedule onto the dynamic run-time scheduling infrastructure.

- *The communication system is optimized for real-time performance:* in particular, a collapsed-layering philosophy is followed based on an atomic multicast protocol providing multiple "qualities of service" (see §2.3.2.3 below).

Note, however, that not all the Delta-4 target application domains require real-time response. As long as "sufficient" performance is attainable, the heterogeneity and openness criteria may be more important — in which case the OSA variant of the architecture is to be preferred. Some application domains may need features of both variants of the architecture. In this case, the overall system may contain sub-systems constructed according to either the OSA or XPA variants.

2.3. Architectural Sub-Systems

This section identifies the main hardware components of the Delta-4 architecture and outlines the communication facilities provided in the OSA and XPA variants of the architecture.

2.3.1. Hardware

The Delta-4 distributed system architecture has been designed so that it is capable of accommodating the arbitrary modes of failure of *fail-uncontrolled* hosts (cf. §2.1.1) by means of *active replication* techniques (cf. §2.2.1.1). To be able to use standard local area networks instead of resorting to the costly interconnection topologies normally required to accommodate arbitrary failures (see chapter 6), each node consists of two distinct parts: a *host* computer and a *network attachment controller* (NAC) (figure 4).

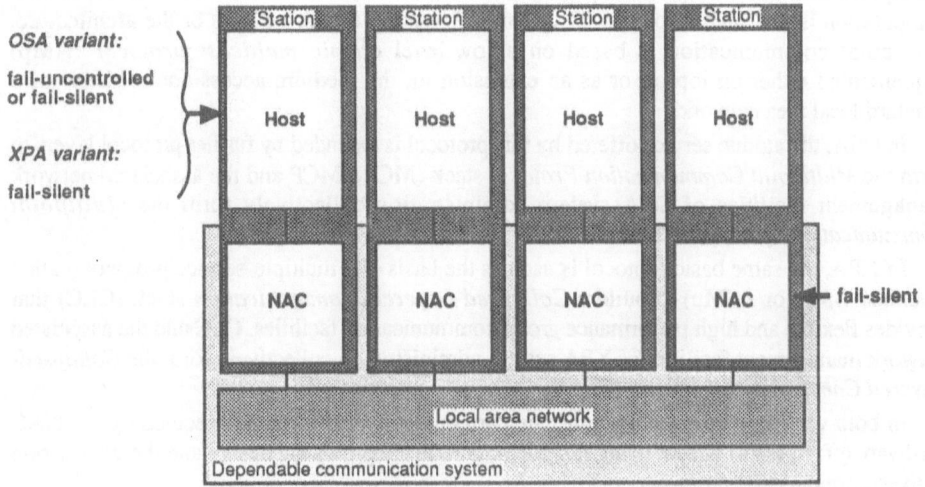

Fig. 4 - Delta-4 Hardware Architecture

Host computers (on which replicas are executed) may be fail-uncontrolled or fail-silent; however, NACs are assumed to be fail-silent. This assumption for the NAC is substantiated by the use of hardware self-checking techniques. The important consequence of this split in failure mode assumptions between host and NAC is that, even when a (fail-uncontrolled) host forwards erroneous data to its NAC, the latter will either process this data in a consistent manner or simply remain silent. The possibility of forwarding the data *inconsistently* to multiple destinations is effectively removed.

As mentioned in §2.2.1.1, active replication has the potential disadvantage of requiring computation to be deterministic. However, when either the *passive* or *semi-active replication* techniques are preferred, it is necessary to assume that hosts are fail-silent. This is possible when the coverage of the host self-checking mechanisms[4] is commensurate with the dependability objectives of the supported application. In the XPA variant of the architecture, where openness and heterogeneity are sacrificed to support real-time applications, all hosts are purpose-designed with built-in self-checking and are therefore assumed to be fail-silent.

To summarise, the OSA variant of the architecture can accommodate both fail-silent hosts and fail-uncontrolled hosts whereas the XPA variant accommodates only fail-silent hosts. Fail-silent NACs are used in both variants.

As for the local area network itself, implementations exist for token bus [ISO 8802-4], token ring [ISO 8802-5] and FDDI [ISO 9314]. For performance reasons, FDDI is preferred in the XPA variant of the architecture. FDDI is also of very real interest in the OSA context since it is able to accommodate a high number of interconnected sites spread out over quite long distances (up to 200 km). Since the local area network must also be dependable if the complete architecture is to be dependable, dual communication media can be employed in all LAN implementations.

2.3.2. Communications

An essential feature of Delta-4 communications is the provision of *multipoint* services. Multipoint communication is necessary in some form or another as soon as replicated computation is considered for providing fault-tolerance. In both variants of the architecture, multipoint communication is based on a low-level *atomic multicast protocol* (AMp) implemented either on top of, or as an extension to, the medium access control layer of a standard local area network.

In OSA, the atomic service offered by this protocol is extended by further protocol layers to form the *Multipoint Communication Protocol* stack (MCP). MCP and the associated network management facilities of OSA system administration collectively form the *Multipoint Communication System* (MCS).

In XPA, the same basic protocol is used as the basis of a multiple-service protocol (called *extended AMp*, or *xAMp*) to build a *Collapsed-Layered Communication* stack (CLC) that provides flexible and high performance group communication facilities. CLC and the associated network management facilities of XPA system administration collectively form the *Collapsed-Layered Communication System* (CLCS).

In both variants of the architecture, the communication software is executed by the NAC hardware (cf. figure 4) which, being fail-silent, greatly simplifies the design and the verification of the communication protocols.

2.3.2.1. Atomic Multicast Protocol. The Delta-4 atomic multicast protocol (AMp) allows data frames to be delivered to a group of logically-designated *gates*. The protocol ensures, for each frame, that either all addressed gates on non-faulty nodes receive the frame or none[5]. The protocol also ensures that frames are delivered to all addressed gates in a consistent order and that any changes in the membership of a gate group (due to node failure or re-insertion) are notified consistently to all members of that group.

The protocol is based on a centralized, two-phase accept protocol. In the first phase, the data frame is transmitted to all members of the group. The latter then inform the sender whether

---

4   Or, equivalently, the conditional probability that, when a host fails, it fails by going silent.
5   This occurs, for example, if any recipient cannot accept a frame due to lack of receive credit.

or not they are able to receive that frame. If all intended recipients can accept the data frame, the sender transmits an *accept* frame (if a participant perceives the data frame, but cannot accept it because of buffer limitations, a *reject* frame is sent). The protocol tolerates faults of both the sending and receiving nodes as well as transmission faults; it has been implemented in such a way that it can be ported to different underlying networks. It has been successfully implemented *on top of* the Medium Access Control layer of token bus [ISO 8802-4] and token ring [ISO 8802-5].

Another protocol, providing the same service, has been implemented as a *modification of* the Medium Access Control (MAC) layer of the 8802/5 token ring [Guérin et al. 1985]. This extension decreases the number of frames that need to be exchanged by making use of the "on-the-fly" bit flipping possibility of the token ring and is implemented partially in hardware.

Atomic multicasting in Delta-4 is described in more detail in chapter 10.

2.3.2.2. OSA Multipoint Communication Protocol Stack. The OSA multipoint communication protocol stack (MCP) provides two major innovative features:

- the ability to coordinate communication to and from *replicated endpoints*,
- the provision of *multipoint associations* for connection-oriented communication between groups of peer entities.

The *replicated endpoint* paradigm is used to implement the error processing associated with the *active replication* model. Replicated endpoints are provided by an *inter-replica protocol* (IRp) situated at the bottom of the session layer. This protocol coordinates the flow of information sent and received by the different replicas of a replicated software component. The information sent by each replica can be cross-checked for validity (in both the value and time domains) with that sent by the other replicas in the set. This is carried out in the (fail-silent) network attachment controllers (NACs) before sending the actual information over the network. The cross-check itself is based on the exchange of checksums of the data to be sent (for value domain errors) and on the comparison of inter-replica desynchronization (for time domain errors). The flow of information to a replicated software component is also controlled by this protocol; a buffer-status voting mechanism is used to ensure that flow control towards a replicated destination is ensured despite the fact that a faulty replica could choose to refuse messages from the network. The protocol can be configured for either fail-uncontrolled hosts or for fail-silent hosts.

Whereas the replicated endpoint facility provides for the transparent (or *invisible*) multicasting of information to a replicated destination, the MCP *multipoint association* facility allows *visible* multicasting between groups of peer software components (which may or may not be individually replicated). This service is delivered by the MCP multipoint session layer protocol. Facilities are provided that allow software components to join and leave multipoint associations and for information to be multicasted to all or some members of the association. A sending entity may choose to include or exclude itself from the set of destinations for a particular message. All information transferred over such a multipoint association is delivered in a consistent order to all overlapping destinations.

The MCP stack also provides a fully-conforming ISO session layer service so that applications conforming to ISO standards can be used. The replicated endpoint facility is offered both for the multipoint session-layer service and for the ISO-compatible bi-point service.

The MCP stack is further detailed in section §8.1.

2.3.2.3. XPA Collapsed-Layered Communications. The XPA communication subsystem provides the high performance that is a necessary (but not sufficient) condition for stringent real-time applications. To this end, a collapsed-layering philosophy is followed. Of course, removing layers also implies removing services so the functionality of the remaining layers needs to be increased to palliate this. In XPA, only four layers are defined — from the top down: the *group management layer*, the *group communication layer* and the (standard) LAN *medium access control* and *physical* layers.

The group management layer is responsible for choosing the group communication "quality of service" (see below) that is appropriate for the model of replication that is used within a particular group. It also implements an appropriate inter-replica protocol to ensure replica consistency and error-processing (in the semi-active replication or leader-follower model).

The group communication layer is based on an extension of the basic AMp protocol (xAMp) that provides multiple *qualities of service* (QOS) that can be selected according to the specific group communication requirements (see table 2).

Table 2 - Summary of xAMp Service Properties

Quality of service	Agreement	Total order	Causal order per clabel	Rx queue re-ordering
bestEffortN	*best effort to N*	*no*	*FIFO*	*no*
bestEffortTo	*best effort to list*	*no*	*FIFO*	*no*
atLeastN	*assured to N*	*no*	*FIFO*	*no*
atLeastTo	*assured to list*	*no*	*FIFO*	*no*
Reliable	*all*	*no*	*FIFO*	*no*
Atomic	*all or none*	*yes (same gate)*	*yes*	*no*
Tight	*all or none*	*yes (same gate)*	*yes*	*yes*

The *bestEffortN* and *bestEffortTo* services ensure that the frame is received by a number N, or a specified list, of the addressed recipients, *if* the sender does not fail. The *atLeastN* and *atLeastTo* qualities of service enforce stronger agreement, assuring that a given number or sub-set of the participants will receive the frame, even if the sender fails. The *Reliable* service ensures that, if any recipient received the frame, *all* addressed recipients receive it, even if the sender fails. In all of these first three services, there is no guarantee of order at the receivers and no control of receive credit; frames are forwarded directly to the service users as soon as they are received. The *Atomic* service is the same as that provided by the basic AMp protocol: it guarantees, even if the sender fails, that all recipients, or none of them, receive the frame. This service ensures both consistent ordering (across multiple receive queues) and causal ordering (within a receive queue). The *Tight* service extends on the *Atomic* service by providing the possibility to allow more urgent frames to overtake less urgent ones in the receive queues. Of course, this can only be done if the resulting frame deliveries to users are still consistently and causally ordered.

2.4. Software Environment and Management Issues

Figure 5 provides an abstract view of the overall Delta-4 architecture. The left-hand "slice" of the diagram recapitulates the hardware architecture discussed in the previous section.

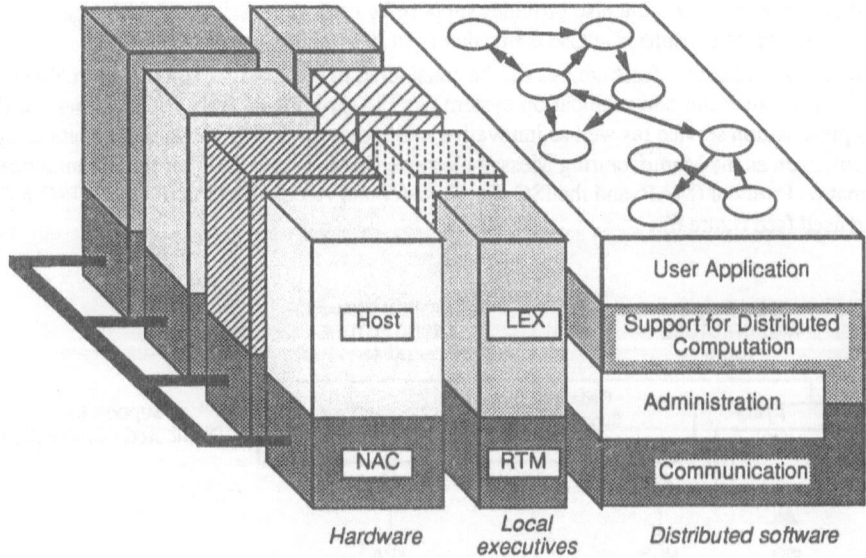

Fig. 5 - Abstract View of the Delta-4 Architecture

The middle slice of the figure represents the local executives residing on the host and NAC hardware. The local execution environments (LEXes) of the hosts are shown shaded differently (like the hosts) to underline that, in the OSA variant of the architecture, heterogeneous host hardware and executive software may be accommodated. In practice, the present implementations all use different flavours of UNIX. In principle, however, the design philosophy of the OSA variant of the architecture would allow an implementation in which both UNIX and non-UNIX systems could co-exist. In the XPA variant of the architecture, LEXes (and hosts) are homogeneous; today, a specially-developed real-time version of UNIX (RT-UNIX [SVC200]) is used but future implementations may adopt some other real-time operating system. In both variants of the architecture, the NACs use a real-time monitor (RTM) that may be homogeneous across all NACs (although this is not of course mandatory).

The right-hand slice of the figure represents the *distributed Delta-4 software* that can be represented in four parts:

- the distributed user application software represented as a set of "software components" (logical units of distribution) that communicate by messages (only),
- the host-resident infrastructure for support of distributed computation,
- the computation and communication administration software (executing partly on the host computers and partly on the NACs),
- the communication protocol software (executing on the NACs).

A particular host-resident infrastructure for supporting open *object-oriented* distributed computation has been developed for the Delta-4 architecture: the *Delta-4 Application Support Environment* (Deltase). According to the philosophy of "open" distributed processing, Deltase facilitates the use of heterogeneous languages for implementing the various objects of a distributed application and allows the differences in underlying LEXes to be hidden (see section §2.4.1 below). Deltase provides the means for generating software components called "capsules" (executable representations of objects). Capsules are coordinated by the Deltase execution support system that also provides support for error processing and fault treatment by means of replicated capsules. Deltase is mandatory in XPA and is optional in OSA.

Other host-resident infrastructures can be considered in the OSA variant of the architecture. In particular, since the communication system in OSA provides a fully ISO-compatible (bi-point) presentation service (as well as innovative multipoint services), standard application layer protocols such as the Manufacturing Message Service (MMS) [ISO 9506] of the Manufacturing Automation Protocol (MAP) and the ISO File Transfer and Access Method (FTAM) [ISO 8571] can be used (see figure 6).

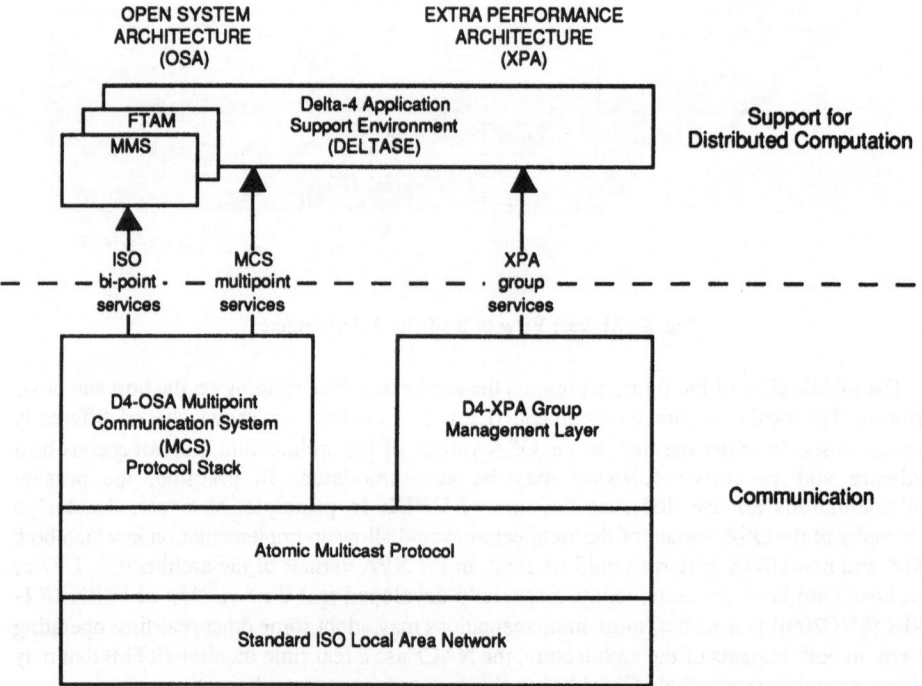

Fig. 6 - Computation and Communication Support in OSA and XPA

In addition to the communication software (already outlined in the previous section), a further important sub-system in the distributed software slice of figure 5 is that of *system administration*. Delta-4 system administration provides the mechanisms for managing a Delta-4 system: it consists of both support for *network management* in the classic sense as well as support for managing the computation system (see section 2.4.2 below).

In conjunction with Deltase/XEQ, the Delta-4 administration system carries out the automatic *fault treatment* functions mentioned in §2.2.1.2: this is detailed in section 2.4.3 below.

2.4.1. Application Support Environment

The purpose of *Deltase*, the *Delta-4 Application Support Environment*, is to provide a *single virtual machine* for the support of modular applications on Delta-4 systems. This single virtual machine conceals the (possibly-heterogeneous) underlying hardware and software, consisting of computers, local operating systems, language systems and the communication system.

The *computational model* provides the model to be used for application programs that are to be supported by Deltase. This model has its origins in work on *Open Distributed Processing* (ODP) within the ANSA (later ISA) project [ANSA 1989] and related work on standardisation of a *Support Environment for ODP (SE-ODP)* within ECMA [ECMA TR/49 1990]. ODP is concerned with the definition of a generic architecture for distributed processing systems. The current implementation of Deltase is a prototype Support Environment for Open Distributed Processing extended to include the Delta-4 approach to fault-tolerance and real-time.

The decomposition of an application into a number of *language-level* program modules, which interact with one another by means of procedure calls, is a widely-used method for applications that are implemented as a single program and executed on a single machine. Deltase enables this application structure to be supported on Delta-4 systems; that is, to extend, to distributed fault-tolerant systems, a widely-used application structure.

Since fault-tolerance and distribution are made *transparent* to the application programmer, an application written for use with Deltase can be used, without change to the application code, on a small-scale system, consisting of a single host machine with a suitable implementation of Deltase.

Deltase supports this computational model through a combination of:

- Generation Support — referred to as Deltase/GEN.
- Run-Time Support — referred to as Deltase/XEQ.

The language-level program modules are independent of the underlying computing environment, and can be ported between different computing environments where that language is supported. Thus Deltase provides vendors, and users, with a natural and economic way to migrate their software investment to new technology.

2.4.1.1. Concepts

2.4.1.1.1. Computational Objects and Services. For use with Deltase, the program modules (which together constitute an application) must be *computational objects*. Each one encapsulates a private *state*, and provides one (or more) clearly-defined *service interface(s)* for operations on that state; each service interface offers a set of *services*. These services provide the only way for one computational object (the one invoking the service) to read or modify the private state of another computational object (the one offering the service). The specification of a service interface is independent of its many possible implementations; each service is defined by its parameters, results, effects and constraints.

2.4.1.1.2. Remote Procedure Call. The *Remote Procedure Call (or RPC)* mechanism is a direct extension, to distributed systems, of the widely-used procedure call mechanism provided by many conventional programming languages. RPC provides a means of programming the interactions between objects, using existing language constructs, with each service treated as a separate procedure. The same language construct is therefore used both for the invocation of a service that is provided internally (by a local procedure), and for the invocation of a service provided externally (by another object). RPC is an abstraction away from messages, offering a familiar language-level construct for programming the interactions between computational objects. With language-based type-checking mechanisms at both the calling and the called objects, the use of RPC provides an assurance that code at both ends conforms to the same interface specification.

2.4.1.1.3. Interface Trading. Before two computational objects can interact by way of RPC, a *logical path* between them must be established, through *interface trading*. An object (service provider) may *offer* its services for use by other objects, by *exporting* that service interface; this offer will be added to a catalogue of such offers. An object requiring the use of a particular set of services (service user) may find a service provider by invoking a search of that catalogue for offers of service, based on the specified interface type name. The effect of *importing* a service interface is to establish a logical interaction path; such a path would enable the service user to invoke the services provided by the service provider, through that particular interface.

2.4.1.1.4. Threads. To allow a service provider to process a number of service requests concurrently, Deltase supports the use of multiple threads of control within an object. Each thread handles one service request at a time; such threads are called *server* threads, and are managed by Deltase. Server threads are not directly visible to the applications programmer, but the mutual exclusion of access to shared internal data remains the responsibility of the application programmer. Within an object, a server thread may create distinct subsidiary (*forked*) threads that disappear when their activity is complete. Typically, such forked threads are used to allow local processing to continue in parallel with service requests invoked by RPC (whose threads are blocked); computational progress may then occur on several hosts on behalf of a single original service request.

2.4.1.1.5. Transformers. To interact with alien computational worlds, Deltase uses the concept of a *transformer*. This is a "half-object"; it appears from one side as a conventional Deltase object, but from the other side, interacts according to the rules and conventions of some other computational world. A transformer object can be used to provide access from the Deltase world to an existing (non-Deltase) software package. Typically, the software accessed in this way would be proprietary and commercially important. Conversely, a transformer object may provide access, from the non-Deltase world, to a service provided in the Deltase world (with the consequent benefits in terms of representational transparency and Delta-4 dependability), whilst preserving non-Deltase interface conventions and standards. More generally, transformer objects provide a means of interworking with other proprietary or standard computational worlds, presenting these to the Deltase world in a consistent, and preferably generic, manner. An important use for transformer objects is in interfacing to input-output devices, either directly or by way of existing drivers.

2.4.1.2. Implementation. Deltase/GEN is used to generate executable software components from the language-level computational objects; the term *capsule* is used for a software component that is generated from a computational object, using Deltase/GEN. A capsule is directly executable on a particular type of host and under the chosen local operating system; for example, for operation under UNIX, a capsule would be generated (and then handled) as a UNIX process. A capsule consists of the compiled source code of one or more computational objects, together with additional code that provides the environment to map the computational object(s) onto the local operating system. This additional software is referred to as the *envelope*; each capsule contains one envelope, which represents all the code necessary to support all the computational objects within that capsule. The envelope is generated automatically by Deltase/GEN and includes the following:

- *interface modules* corresponding to the services exported or imported by the computational objects; interface modules are generated automatically from the service interface specifications,
- a set of environment-dependent library procedures,
- a thread scheduler, for scheduling of threads local to that capsule,
- object-dependent support for error-processing and fault-treatment (see below).

2.4.1.3. Impact of Fault-Tolerance. Deltase combines ODP concepts with the Delta-4 approach to fault-tolerance based on user-transparent replicated computation. From the application programmer's viewpoint, fault-tolerance is transparent. Deltase/GEN can automatically generate capsules and provide them with the additional run-time code necessary to support the Delta-4 replication models.

The error processing associated with the *active replication* model is carried out by the underlying communication system. However, this model requires that the observed behaviour of each replica be identical in the absence of faults (cf. §2.2.1.1). From the viewpoint of Deltase/XEQ, this means that each replica of a capsule must process requests *deterministically* despite any possible internal parallelism. One means of ensuring this determinism is to structure each capsule as a *state machine* [Schneider 1990] such that capsule replicas process requests and supply responses in a strictly identical order. Adherence to the state machine model means that the object envelope must schedule threads in the same sequence and that the transfer of control from one thread to another must be at the same point in computation at all replicas.

In the case of the *passive replication* model, Deltase/GEN generates an object envelope that allows a capsule replica to act either as the *primary* replica or as a *back-up* replica. It is this special software in the object envelope that is responsible for sending and receiving checkpoints at appropriate times and updating the state of passive replicas. In the *semi-active replication* model, Deltase/GEN must again generate an object envelope that allows a capsule replica to act either in *leader* mode or *follower* mode.

Deltase also provides support for the *cloning* mechanism associated with fault treatment (see section 2.4.3 below).

2.4.2. System Administration

System administration provides the mechanisms for managing a Delta-4 system: it consists of both support for *network management* in the classic sense as well as support for *computation management*.

2.4.2.1. Management Functions. The term "administration" covers the set of management functions concerned with planning, organising, supervising and controlling a Delta-4 system. In particular, these management functions are concerned with maintaining a specified level of service and enabling the system to evolve by providing the means to add new facilities. Management of distributed systems is a complex and often ill-defined topic, for the following reasons [Sloman 1987]:

- distributed systems are large and complex,
- their components, very diverse in nature, all need to interact and be managed,
- there are many different facets to management; management has to deal not only with configuration, designation, performance, faults, security and accounting, but also with people and computers,

- there is a floating boundary between management and the normal functionality of the system.

The management of distributed systems is currently under intense discussion in the academic field as well as in the immense ISO standardisation work, international multi-vendor initiatives (MAP, CNMA, OSI/Network Management Forum) and network management product developments (IBM's Netview, HP's Open View, DEC's EMA, etc.). An important consequence of the sophisticated Delta-4 fault-tolerance facilities is that system administration must cover not only the management of communication resources but also those necessary for computation. Presently, as is the case in MAP, three categories of management functions are supported [MAP]:

- management of system configuration and naming,
- performance management,
- fault and maintenance management.

The latter category of functions is particularly important in Delta-4 because it includes all aspects of fault treatment mentioned earlier. A further category of management functions concerned with security is also being investigated (see chapter 13).

2.4.2.2. Management Model and Design Principles. Delta-4 management is based on the ISO/OSI concept of "managed objects", which has been consistently extended such that it can be applied to objects outside the OSI scope, for instance hardware and software components. Apart from their normal functionality, managed objects are characterized by:

- *attributes* such as their version identification, their state and their operational parameters, information relative to their error and performance statistics, their dependencies with respect to other managed objects, etc.,
- specially-defined management *operations* to allow access to, and manipulation of, object attributes, e.g., create, clone, delete, re-initialize, set_value, get_value, etc.,
- *events,* which are a means by which a managed object delivers management information asynchronously (events are a special form of attribute).

In accordance with the current state-of-the-art in distributed systems management [ANSA 1989, Sloman 1989], management specific to a *set* of managed objects is termed *domain management* and the set of objects, a *domain*. Examples of domains are sets of nodes, sets of replicas or sets of software components. Since replicas of a given software component may only be located on certain nodes (i.e., those possessing the resources necessary for their execution), the set of such nodes is termed the software component "replication domain".

Each management domain in Delta-4 is assigned an architectural component called a *domain manager*. A domain manager may consist of a single *domain manager process* (which may be replicated) or a set of *peer domain manager processes* that cooperate within the domain boundary to carry out domain-specific management tasks. Domain managers of different domains cooperate to fulfil common management policies.

Domain managers make use of two sorts of management information: a) management information *about* managed objects that is integrated within the domain manager, and b) management information integrated within the managed objects themselves that is closely related to their normal functionality. Both kinds of management information are conceptually summarised under the term *Management Information Base* or *MIB*, which is therefore, by its very principle, distributed.

The Delta-4 implementation presently comprises three types of domain managers:

- a manager of communication objects within a communication domain — in accordance with the MAP terminology, a domain manager process of the communication domain manager is termed a "Systems Management Application Process" or SMAP,

- managers of application objects within replication domains, or "Replication Domain Manager" or RDM,
- a manager of a particular object within a replication domain, that contains management information, called the Global-MIB ("MIB Domain Manager").

Communication domain managers in Delta-4 manage types of communication objects beyond those defined in present ISO, MAP and IEEE standards. Furthermore, these communication objects may be replicated. By essence, the SMAPs cannot be replicated themselves. Consequently, SMAPs are executed by self-checking hardware — the fail-silent network attachment controllers (cf. §2.3.1). A special protocol (M-CMIP: Multipoint — Common Management Information Protocol) is used for the exchange of information between SMAPs. To manage replicated communication objects, SMAPs on different Delta-4 nodes must have a consistent view of the non-local features of such objects. To achieve this, the corresponding communication management information is stored in a separate global management data base, the "Global-MIB". As this is critical information, the Global-MIB is replicated and thus managed by a (replication) domain manager of its own (the MIB Domain Manager).

The architectural approach has also been applied to objects on the application level (see section §2.4.3):

- *Processes:* They form the basis on which (existing) applications are built. In the present Delta-4 implementation, a Deltase capsule is represented by a UNIX process.
- *Files:* Global files may be useful in certain applications, so a server for managing replicated global files also been implemented.

These kernel management components, which support the Delta-4 fault-tolerance approach, are supplemented by a set of management application tools with graphical human interfaces to allow the control and the visualisation of the Delta-4 system behaviour. Among these are a configuration toolbox for the Global-MIB and various status, utilization and performance monitoring tools.

2.4.3. Cloning

Cloning techniques have been investigated and implemented for both Deltase capsules and files. Cloning of higher level objects, for instance file servers or databases, may be based on the cloning mechanisms of these basic managed objects.

The implemented cloning machinery for both capsules and files comprises:

- a) A *replication domain manager* (RDM), which applies a reconfiguration strategy to a set of replicated processes or files that have the same replication domain. The reconfiguration strategy defines when and where new replicas are to be instantiated, for instance:
  - restoration of the replication degree of objects which have lost a replica due to node failure by cloning them to nodes offering spare redundancy, or
  - migration of all objects from a node, e.g., due for maintenance.
- b) *Object manager entities* (OMEs), local to the managed application objects, that perform the cloning protocol on request of the replication domain manager.

The cloning of Deltase capsules is carried out by three generic components: a capsule RDM, capsule OMEs and *factories*. A factory exists on each node and is responsible for instantiation of capsule replicas on that node. To instantiate a capsule replica at a particular node, the replication domain manager makes use of the services offered by the factory resident at that node. Each capsule includes an OME as part of its envelope that is responsible for carrying out those operations that can only be done from within the capsule. The capsule OME consists of a set of library procedures, which are included in the envelope as part of the capsule generation process.

The initial state of a new capsule replica is obtained from the file generated by the capsule generation process. The current state of the replicated capsule is obtained from the OMEs of the existing replica(s) by what is essentially the same activity as the checkpointing activity associated with the passive replication model. Further local state information is handled locally by the OME at the node where the new replica is created. This information consists of data specific to the local environment of the new replica such as references to local resources.

✛ ✛ ✛

Further details on Deltase can be found in chapter 7, and on system administration and cloning in section 8.2 for OSA and in section 9.5 for XPA.

2.5. Validation

For an architecture to be worthy of the epithet "dependable", it is necessary that users of the architecture may *justifiably* place their confidence in the architecture. Consequently, such an architecture must undergo extensive validation both from the verification viewpoint (removal of faults in the specification, design and implementation) and the evaluation viewpoint (quantification of the provided dependability and performance).

Validation can and should be carried out at each step in the process of producing a system. At the specification stage, validation consists essentially of verifying that the specifications of the future architecture are mutually consistent with each other and with the requirements of the intended application domains. This "informal" verification is carried out in Delta-4 by a "peer review process" either explicitly during scientific and technical committee meetings or implicitly, as in the production of this book describing the architecture's concepts.

More tangible validation activities are carried out during the design and implementation phases. Ideally, all components of a system should be extensively validated. However, for the money-critical (as opposed to life-critical) applications for which Delta-4 is intended, it was decided to restrict the validation to the most important (or the most critical) sub-systems.

2.5.1. Design Validation

Design validation is centred on descriptions or models of the future implementation. Its purpose is: a) to verify that these models are consistent with the specifications or b) to evaluate (predict) some characteristics (e.g., performance, dependability) of the future implementation. Two design validation activities have been carried out.

- *Protocol verification* aimed at removing faults in the protocol design has been carried out on various versions of the essential atomic multicasting protocol, AMp. This work employed temporal logic specifications of the required properties of AMp and verification that a formal description of the protocol, in Estelle/R, satisfied these properties [Baptista et al. 1990]. Present protocol verification work is centred on the inter-replica protocol (IRp) of the session layer of the OSA variant of the architecture.

- *Dependability evaluation* work is being carried out with a view to quantifying the dependability actually achievable by the Delta-4 architecture. The work carried out to date has centred on the communications infrastructure and has shown the importance of coverage of the network attachment controller self-checking mechanisms (to substantiate the fail-silence assumption) and identified the conditions under which redundant communication media should be employed [Kanoun and Powell 1991]. Present dependability evaluation work is aimed at evaluating the availability and reliability of applications making use of the various Delta-4 fault-tolerance models.

2.5.2. Implementation Validation

Implementation validation is centred on *testing* actual prototype versions of the architecture instead of on models. Like design validation, its purpose is twofold: a) to verify that the implementation provides the specified functionality and b) to evaluate (measure) some characteristics of the actual implementation. Implementation validation has centred on two aspects.

- *Fault injection* (into the prototype hardware) has been used as a means for validating: a) the self-checking mechanisms of the Delta-4 network attachment controllers (NACs), and b) the implementation, on these NACs, of the atomic multicasting protocol (AMp) [Arlat et al. 1990]. This activity contributes to the verification of the absence of implementation faults (and residual design faults) — in particular, it verifies that the system works as intended *in the presence of the very faults it is meant to tolerate*. Fault-injection also enables the measurement of the effectiveness of the built-in error detection and fault-tolerance mechanisms by means of coverage, dormancy and latency estimations.

- *Software reliability evaluation* is being carried out on many of the major software subsystems of the architecture. Static testing tools are being used to identify important characteristics of the implemented software. In addition, failure data is being collected during the software development and testing phase to predict the rate at which it can be expected that residual design or implementation faults will cause the system to fail when in operational use.

✧ ✧ ✧

Further details on the validation of the Delta-4 architecture can be found in chapter 15.

Chapter 3
Comparison with other Approaches

The purpose of this chapter is to position the Delta-4 architecture with respect to commercially recognised and available techniques. The comparison is made with respect to the Delta-4 application domains, characterized as large-scale, distributed information processing systems, and summarised in 1. It would not be appropriate here to make comparisons or draw conclusions about the respective merits of these architectures in other domains, e.g., transaction processing. Rather, the objective is to compare Delta-4 with alternative approaches to the design of distributed systems that are also required to be dependable, or exhibit real-time characteristics, or both. Before making this comparison, a summary of currently used commercial fault-tolerance techniques is included for reference. No similar analysis of real-time systems is offered since conventional stand-alone real-time operating systems are well-known and there are no known actual examples or de facto standards for distributed real-time systems.

3.1. Commercial Fault-Tolerance Techniques

The concepts behind the achievement of fault-tolerance are described in section §4.5.2, together with different techniques that may be applied. These techniques provide a structure for the general discussion below of commercial solutions based on their combination.

The general issue of non-disruptive on-line repair is referred to in all cases examined.

The assumption of *fail-silent* components, i.e., components that fail only by stopping delivery of outputs, is necessary in some approaches. The *assumption* of fail-silence is part of the model; vendors must provide mechanisms to *support* this assumption to the degree demanded by the market for their product. Thus, these mechanisms vary widely, from highly justifiable techniques such as sophisticated lock-step synchronized hardware with self-checking voting logic to address the most critical markets, to use of self-analysis or watchdog software running in largely fail-uncontrolled hardware.

3.1.1. Backward Error Recovery

In this approach, faults are tolerated by use of regular state-capture to mirrored recovery caches commonly accessible to two fail-silent subsystems, one of which is active until failure. The combination of active fail-silent subsystem and mirrored recovery cache permits the other fail-silent subsystem to discover a correct state from which to continue execution in the event of failure of the first, even during state capture.

One issue of importance is the degree of coupling which permits the cache accessibility with minimal overhead, whilst avoiding execution disruption during on-line repair activity.

3.1.2. Error Compensation Techniques

3.1.2.1. Error Compensation by Majority Voting. Practical commercial systems using majority voting techniques are based on *triple modular redundancy*; faults are tolerated by locally triplicating subsystems. Thus, one node contains tripled CPUs, memories, and so on. To permit redundancy to be restored on-line, arrangements are made for physical replacement of faulty components and to cause the state of the restored components to converge with the state of the remaining system without disrupting service provision. The system MTBF is improved by reducing the granularity of containment of failure; to this end, a dependable power-supply can be constructed as a separate subsystem. Several proprietary means of doing this and allowing on-line repair exist.

3.1.2.2. Error Compensation by Redundant Fail-silent components. If a component is arranged to fail only by stopping delivery of outputs, then the error can be compensated by providing more than one such component. A mechanism is required which transforms the several (necessarily equivalent) outputs from all non-failed components so that they are seen as a single output. Sometimes this is done by arranging one component to be a "Master" until it fails, when another takes over.

A typical implementation of this approach is to tolerate faults by locally duplicating subsystems that are themselves constructed, through duplicating cross-checked components, to exhibit fail-silence. Thus, one node contains quadrupled CPUs, memories, and so on. In the event of failure, mechanisms are provided to restore redundancy on-line without disrupting execution of the remaining system. Again, a dependable power-supply is provided.

3.1.2.3. Replica Group Determinism. With either of the above approaches, replica group determinism must be assured, a property which products based on these approaches confer by use of one of the following techniques:

3.1.2.3.1. Microsynchronous Execution. In this approach, replica hardware may be sufficiently coupled to assure replica-determinism without any special software provision, by executing the entire ensemble in a lock-step manner by synchronising all replicated components to a common dependable clock subsystem. Proprietary solutions are used to resolve such close-coupling issues with the need to be sufficiently decoupled to allow redundancy to be restored on-line.

3.1.2.3.2. Macrosynchronous Execution. An alternative to the lock-step synchronization solution to replica group determinism without special software provision is the use of deterministic resynchronization. Each CPU may have an independent and therefore asynchronous clock and hence avoid the need for a dependable clock subsystem. The equivalent of lock-step synchronism can be achieved by resynchronising whenever necessary, for voting and other events. Necessary resynchronization events include interrupts, and voting events such as reads or writes to an external address space. Other resynchronization events may also form part of a proprietary solution. For example, use of a maximum time elapsing since the previous resynchronization event offers a simple resynchronization mechanism where the asynchronous clocks are sufficiently close in frequency to only drift slowly past each other.

3.2. Comparison of Approaches by Concept/Technique

In comparing these architectures with Delta-4, the lowest level of interest is the basis on which dependability is achieved.

3.2.1. Stand-Alone vs. Distributed

A first observation is that, in all the cases examined above, the intention has been to construct "stand-alone" equipment whose external appearance is of a dependable single node. Delta-4, however, uses a distributed software solution, with assistance from a minimum amount of specialized hardware, the fail-silent NACs. At this level, the above solutions may be viewed as competing with that offered by Delta-4. Ignoring any benefits *per se* arising from inherent distribution, Delta-4 provides a more economic solution when based on standard mass-produced nodes, and a more flexible solution arising from both the incremental nature of its dependability and its ability to accommodate nodes with special properties, such as array processors or other mathematical engines, workstations or other nodes supporting specialized I/O equipment. There is even an advantage in accommodating nodes with their own proprietary mechanisms for achieving node-level dependability; see section §3.3.1 below for a discussion of such configurations.

These benefits carry a cost, that of the performance overheads of inter-replica communication protocols. There are, however, some valuable natural advantages to the geographical separation of redundancy underlying distributed fault-tolerance:

a) Replicas execute in physically distinct nodes that may be separately located within limits dictated only by the LAN technology used and any node input-output requirements; the nodes concerned may be located in different rooms, floors or even buildings. For this reason alone replicas can be rendered less susceptible to the common-mode effects of power surges, electromagnetic interference, floods, fires and so on.

b) Such physical separation requires loose coupling between replicas. The mechanisms that permit this also offer natural tolerance of a notorious and quite common class of software design errors that have become known as "Heisenbugs". This term was initially coined to refer to software (or hardware) design errors that lead to faults that can be tolerated by simple re-execution, the definition of the class being that a slight difference in execution circumstances is enough for the fault not to re-manifest itself. Such slight differences in execution circumstances occur between the support environments of loosely-coupled replicas. Heisenbugs are discussed further in chapter 6.

3.2.2. Fundamental Concept

Unlike some products that have briefly competed in this market, all *successful* products have been based on recognisably viable techniques for achieving dependability. This fact is, we believe, of great significance and lends support to the thesis that success for a product offering dependability requires that it be very evident to customers *how* that dependability has been achieved. In Delta-4, as with all the above approaches, a structured approach is taken in which a comprehensive set of identified problems is encapsulated and addressed by clearly defined techniques. A positive consequence of this approach is that customer doubts are localized; such doubts generally concern engineering details ("how do you make your common clock dependable?"), rather than fundamental principles ("on what basis can you be sure to detect faults of *this* variety?"). In consequence, these doubts are more effectively reassured under such a structured approach.

The achievement and justification of node-level dependability do require extensive attention to be paid to the low-level mechanisms that discover faults and isolate sub-components both to avoid common-mode failure and to permit on-line repair. In the Delta-4 *distributed* approach to fault-tolerance, equally extensive low-level analysis is important with respect to particular sub-components such as the fail-silence of the NAC and, where this is required, of the host. Assuring fail-silence by low-level mechanisms is, however, a lesser goal than providing a complete solution for fault-tolerance within a node as well as providing for local on-line repair. In Delta-4, the distributed basis for fault-independence, fault-tolerance and the isolation needed to permit repair is handled rigorously. However, whereas in any dependable-node approach, details of node hardware raise important design issues (how power is supplied, what bus structures exist inside the node, how to carry out physical component replacement on-line, and so on), in the Delta-4 approach, the mechanisms are implemented at a level of abstraction in which such details are unimportant.

3.2.3. Real-Time Mechanisms

Whereas there are many commercial products that provide mechanisms to support some form of real-time behaviour, such products do not offer dependability support mechanisms. Conversely, products offering node-level dependability do not offer any mechanisms to support real-time behaviour, even in the relaxed sense of handling events and priorities (rather than the strong sense discussed in chapters 5 and 9).

The explanation for this is partly cultural, and partly a question of historical timing of the order in which problems have been tackled. Researchers and manufacturers active in the field of dependability achieved success addressing the issues of a large system market entirely devoid of real-time requirements, e.g., transaction processing. Meanwhile, a quite distinct "real-time culture" addressed the issues of a more specialized market in which dependability, when required, was traditionally addressed at the application level, not always with great success. Although several research projects have shown that the two fields largely overlap in issues and solutions [Dertouzos and Mok 1989, Jensen et al. 1985, Kopetz et al. 1988], this has yet to be reflected in the marketplace.

Section 3.3.3 below discusses Delta-4 as a product in relation to specialized real-time systems.

3.3. Comparison of Approaches by Target System Characteristics

The products discussed above provide or contribute to the platforms on which target systems are constructed. We now consider various system characteristics and their relationship to our chosen commercial examples and Delta-4.

3.3.1. Inherently Distributed Systems

Chapter 1 identified a broad range of application areas (CIME, Industrial, Commercial, IIP, DDBM) which are addressed by systems that are *necessarily* distributed. Most of these have been subject to rapid evolution within the last decade. This evolution has been stepwise, subject to the availability of appropriate technology to solve the most important perceived problem with sufficient economy.

3.3.1.1. Dependability in Distributed Systems. Inevitably, the need for services to exhibit varying degrees of dependability has been discovered by this market. This is currently addressed by arranging to provide node-level dependability and various mechanisms whereby nodes that do not possess enhanced dependability properties may access services provided by these.

To examine a single example, the danger of file loss is, unless special arrangements are made, a problem on specialized workstations linked to centralized file services. The consequences of file loss are more severe when critical information is on a single node rather than distributed. The probability of loss of a particular file is not changed significantly and when loss occurs, the result is more likely to be catastrophic (affecting all files) than unfortunate (affecting some files).

Although other directions have been taken in the marketplace, one identifiable evolutionary step in response to this has been to provide systems where such a central file service is based upon fault-tolerant nodes. This approach is taken in several systems that address the Automated Office market; it is perceived to be a cost-effective means of significantly reducing the probability of loss of files.

3.3.1.2. Distributed Consistency — the Need. The history of such evolution is one of "the next important problem" becoming recognised, as solutions to previous problems are accepted as the "State of the Art", based always on the changing economics of technology and cost. As noted in section §3.2.1 above, a file server based on a single fault-tolerant node remains susceptible to the common-mode effects of power surges, electromagnetic interference, floods, fires, etc. One approach might be to include several such nodes, separating them physically to the necessary degree (e.g., different buildings and electrical supplies). It is then necessary to perform data manipulations across several separately located files. Such manipulations must be made according to a model of distributed consistency acceptable within the context of the application. Indeed, distributed consistency seems to be a valuable mechanism; once it has been constructed, it may be applied, beyond its original purpose, to many applications (see chapter 6). However, such mechanisms are not native to conventional communications systems, nor efficient when built above them.

3.3.1.3. Distributed Consistency — the Delta-4 Solution. Unlike other approaches, distribution is part of the fabric of the Delta-4 solution to dependability. The problem of distributed consistency has been confronted, and the particular solution developed is based upon low-level mechanisms that support multicasting with atomicity (see chapter 10). This is not the case for the "dependable node" approaches described above, where vendors typically provide conventional OSI point-to-point communications support. When the need for distributed consistency arises, for instance when committing a distributed transaction, additional protocols must be implemented on top of the point-to-point protocols. In Delta-4, since distributed consistency of communications is a pre-requisite for ensuring the distributed consistency of replicated computation, extensive use is made of the explicit atomic multicasting protocols built into the Delta-4 communication system. This basic atomic multicasting protocol has allowed the design and implementation of a range of multi-partner protocols that greatly reduce the amount of application-specific code that must be written whenever the need for distributed consistency arises.

Moreover, Delta-4 is in the position of permitting the next step to be taken by present approaches. The Delta-4 solution to distributed consistency may be inherited by existing commercial fault-tolerant nodes by adding Delta-4 communication mechanisms. Distributed-system-level dependability could then properly be claimed for node-based approaches, and in addition, "mix-and-match" strategies become possible; these dependability technologies may interwork transparently with components that use Delta-4 distributed replication models, to enable a tradeoff of cost against performance requirements. The facility is often provided on commercial fault-tolerant nodes to include externally supplied VME-bus based equipment; this would allow direct use of fail-silent NACs based on the prototypes developed by Delta-4, executing either MCS or XPA communications software. Otherwise, a fail-silent NAC interfaced to a proprietary bus could be developed. In either case, a port of the relevant host driver software is needed, or alternatively, MCS or XPA communications software could be ported in place of the OSI stack running on the host. Since the latter is fault-tolerant, the need for a specialized fail-silent NAC is removed.

3.3.2. "Bespoke-Tailored" Systems

Where the specification of a particular site requirement cannot be fulfilled by a standard product, a system must be specially engineered to fulfil the purpose. Such engineering is called "bespoke-tailoring" by analogy with the business of crafting a well-fitting suit of clothes from a chosen cloth, as opposed to the alternative of offering a choice from a range of "standard" off-the-peg suits.

Vendors of such systems remain competitive by optimising the design along many dimensions; time-scales, costs, quality, maintainability, etc. Advantage is taken of the fact that, even if the overall system is unique, it can be built using methods and from components that are not unique. Vendors therefore invest in *enabling technology*; that is, construction facilities, tools and philosophies that simplify the design process and allow systems and system components to be constructed in a modular and flexible way; to be maintained, reused and ported to new technology as it arises. An enabling technology should offer *transparency* with respect to a number of issues; in particular, transparency to distribution, to the underlying technology, and to dependability.

The ideal is to minimize any new development needed, by allowing system construction to be as much as possible the project engineering task of selecting and combining preexisting components, and as little as possible the programming task of constructing new components or changing the facilities of old components. To protect what can easily become a massive investment in applications software, a vendor must find means to maximize software portability across a range of underlying technologies, so that such software can be reused.

Usually, but by no means always, such systems are large. Methods developed to be applicable *in the small* do not usually scale well. In contrast, methods applicable *in the large* can, if due attention is paid to means of excluding any unnecessary "baggage" of support environment functionality from the system, also be applied in the small. It is highly desirable to a vendor that a spectrum of systems can take advantage of a single enabling technology, and that components can be developed and reused across this spectrum.

The ability to accommodate off-the-shelf components into systems developed under such a regime is also desirable, since it serves both to minimize new-development costs and to address issues of "fashion", such as where a particular relational database is specified by the customer.

Where the complexity of the requirement demands that the system itself is complex, distribution is generally an inherent physical requirement, and dependability becomes essential. If different services can be configured with degrees of dependability related to their role within the system, significant cost optimizations can be achieved. Incremental fault-tolerance is, of course, one of the objectives of Delta-4 (see chapter 1).

For the suppliers of bespoke-tailored systems, Delta-4 is in the position of supporting (rather than competing with) the above approaches. The requirements identified here are addressed by the Delta-4 Application Support Environment (Deltase, see chapter 7). Deltase is intended to be simple to port to any host and local execution environment (LEX), and this will be trivial for any approach based on 680X0 and UNIX, as are the Delta-4 prototypes. This is the case with several commercial offerings falling into the categories discussed above.

The result of such a port would enhance these approaches through giving access to an ODP world. This would remove the dependency of application programs on what nodes they run on or across. Programmers of dependable nodes typically have dependability transparency, but "see" their own flavour of LEX. They explicitly take account of distribution, with all the well-known difficulties of assuring and maintaining the mutual correctness of separate units of remote interfacing code (typically constructed manually and by different programming teams), compounding the issue discussed above of assuring distributed consistency. Several manufacturers have addressed this by supporting a proprietary RPC across their own flavour of LEX and host, usually built above native OSI. As standards continue to evolve, those intended to support distributed computation will no doubt be accommodated by many manufacturers. Deltase programmers already live in an ODP world; even with the further compounding effect of heterogeneous machines, operating systems and languages, the difficulties outlined above disappear and the investment in code is protected against technology change.

With that transparency, many of the approaches described can be viewed as providing a "high-end" solution to dependability. Engineers could use different builds of system to meet different customer requirements, whilst preserving the software base.

3.3.3. Real-Time Systems

The meaning of the term "real-time" in Delta-4 is discussed in chapter 5. We repeat here our view that *reasoning about run-time timeliness behaviour is crucial to real-time system design.*

To allow such reasoning with respect to applications, run-time support-environment properties must be assumed and therefore must be supported. However, the features offered by some vendors of such environments sometimes do not allow whole classes of timeliness requirements to be assured, or lead to inordinate complexity in the design in order to achieve that assurance.

An environment claiming to be real-time must, at the least, provide some sort of support for handling events, timed activation and timed notification, with bounded latencies and predictable scheduling properties. Support environments that only offer mechanisms to handle events, or to promote such attributes as "a short average preemption latency", do not provide a good basis to address real-time support. (See [Le Lann 1989] for an analysis of the value of such mechanisms.)

A run-time support environment must explicitly support the design process with run-time mechanisms whose properties are based on sound timeliness modelling. To our knowledge, only Delta-4 has considered dependability requirements in the context of large-scale real-time system design. The XPA instance of the Delta-4 architecture is unique in offering both building blocks that allow software to be constructed according to the above perspective, and transparent support for the replicated execution of such software under a high-performance and criticality-respecting execution regime. This is discussed at length in chapter 9, and many of the relevant dependable input/output issues (also commonly neglected) are discussed in chapter 12.

Such systems are really a specialized subset of the "bespoke-tailored" category. Few offerings based on node-level dependability provide any support for this application area; if they were to be included as hosts into a Delta-4 real-time system, they would not be able to support real-time activity. However, they would not necessarily compromise other nodes offering this support. The XPA version of the Delta-4 architecture provides the necessary properties of a communication system that could accommodate such hosts; an appropriate port would only offer low-precedence communications access.

3.3.4. "Shrink-Wrap" Software

Manufacturers are concerned to maximize the return on effort; small-scale application porting exercises are seen as dissipative (in that such effort can generally be deployed more productively elsewhere) and lead to licensing complications, lack of clarity of maintenance issues, unwanted visibility of the inside of the application and so on.

What has become colloquially known as "shrink-wrap" software is often represented as part of the growing recognition that well-defined standard interfaces and clear structural separation of issues and responsibilities are valuable. In fact, shrink-wrap software really represents recognition of the financial rewards of providing the solution accepted by the market to problems common to large numbers of systems. Evidence for this view is the proprietary nature of many offerings (utility interfaces, language dialects, etc.), which runs exactly counter to the above ideal.

Shrink-wrap software is nevertheless emerging as the form of supply which customers, as well as manufacturers, prefer. Customers want to buy software off-the-shelf and run it immediately, whilst being as little constrained as possible over their choice of platform. There is little evidence that they are at present concerned by the consequences of becoming locked-in to a particular manufacturer.

This particular market is moving away from any need for software to be supplied in source or even unlinked-binary form. To achieve this end, major suppliers in the UNIX community are defining a series of binary standards. These must, of course, be specific to the underlying microprocessor; a number of these are recognised as worth supporting. There is strong pressure to avoid diversification of microprocessor types; the list can be expected to remain small. Most "stand-alone" fault-tolerant products based on the techniques described above claim to offer a solution to the problem of conferring dependability on "black-box" software. To guarantee that dependability will be conferred on the "shrink-wrap" market-driven variety, the technology concerned must also be aligned with this movement.

For this particular class of software, distributed fault-tolerance techniques encounter a somewhat different set of difficulties. Can we construct or make use of a host type that is able to accept such software and render it dependable using Delta-4 models?

3.3.4.1. Passive Replication.

Passive replication relies on mechanisms that can efficiently take, transmit and install checkpoints. With black box software, checkpoints are difficult to collect and equally difficult to install. There is an additional complexity associated with correctly activating a passive replica after the active replica fails, unless the checkpoint is of a suspended state and the LEX data structures themselves represent this correctly. This would seem to rule out use of Delta-4 passive replica techniques.

3.3.4.2. Active and Semi-Active Replication.

To arrange for "shrink-wrap" software to be executed in a distributed replica-group deterministic manner, some assistance must be provided by the LEX. Some recent work aimed at identifying how such a LEX might be constructed is given in appendix B. As with other prototypes developed within the Delta-4 project, this work is based upon UNIX as a representative, popular and open LEX, but the techniques developed are not dependent on any of its characteristics and therefore can be applied elsewhere. Engineering an accepted open version of UNIX in order to achieve the properties described is by no means a trivial exercise. Nevertheless, a possible mechanism does appear to exist that "shrink-wrap" software could be rendered dependable under Delta-4 support.

3.4. Case Studies

This section presents as two case studies the two applications that were used as pilot applications during the present phase of the Delta-4 project. One is a banking application for electronic payment authentication, the other is a production application for car manufacturing.

3.4.1. Pilot Site One

One partner of the Delta-4 project provides a good example of requirements for "bespoke-tailored" dependable distributed systems.

The partner runs an electronic payment business, using two separately located sites, each housing a large fault-tolerant host (dual configuration to provide continuous system availability as well as data integrity).

Because falling profit margins accompanied market growth, the partner faced new constraints and found it necessary to redefine its position with respect to operating costs and performance. The issues included: the overall cost of hardware and software, the cost of upgrading, operating difficulties linked to changes of scale, the inconsistency of batch processing with services intended to support on-line transaction processing,...

The two main sets of constraints that were identified match well the characteristics of Delta-4:

- To decrease the high cost of the fault-tolerant equipment, whilst increasing system capacity to keep pace with market growth, without impairing with system dependability (protection against file corruption, fire hazards and the like).

  One way of doing this is to replace large-scale expensive fault-tolerant hosts by a dependable distributed system that uses sets of smaller stations, linked through a LAN, but physically separated to the necessary degree. This has additional advantages in operational terms. It allows the necessary growth whilst enabling optimal system use, by supporting modular dependability and load-balancing. For example, the batch programs could run on stations other than those dedicated to the on-line services.

- To decrease the high cost of software maintenance, without reducing the span and frequency of the changes in application software that must take place because of the need to maintain position in an increasingly competitive and rapidly-evolving market.

  This can be done by:

  - modifying software production methods, in particular by shifting to the object-oriented modes advocated in the ODP model and implemented in the Delta-4 Application Support Environment;

  - switching from proprietary to standard operating systems and languages, to decrease training costs at a time of high turnover among information systems staff.

If the partner were to adopt generally the Delta-4 pilot site solution, additional advantages would accrue, such as the possibility of achieving very high levels of hardware fault tolerance and of using the fragmentation and scattering technique to ensure the extensive protection of sensitive files (see chapter 13).

3.4.2. Pilot Site Two

This pilot is based on a real case that has been implemented and is now operational in a car assembly factory. The CIME (Computer Integrated Manufacturing and Engineering) subset that has been selected for the Delta-4 project is one of the most critical of the factory since a failure of this subset would lead to a halt in production. This subset comprises several application components distributed on different levels of the factory (shop, cell, operator's level) thus representing a meaningful part of a CIME system:

1) At the shop floor level:

- The *production management application* identifies the necessary components and the operations (manual, mechanical) to be carried out to build mechanical parts. It works on data produced by the production planning activities and stored in the shop database.

- The *shop database* stores information about the sequence of cars to be produced with the associated references of mechanical parts, about the current state of production and about the list of components to use and operations to carry out per mechanical part reference.

- The *message scheduling application*, under the control of production management, schedules messages to mechanical or human operators at the cell level. This scheduling is driven by the state of the work in progress (product tracking), and by the nature of data stored in the database.

2) At the cell level

- The *message dispatching application* sends messages received from the shop floor level to the right destination within the cell and at the right time.

- The *automation application* controls the real production process and ensures that this process executes the right operations; this component communicates with PLCs (e.g., for controlling movements of automatic guided vehicles), with identification posts and with human operators through asynchronous terminals.

The complexity of such a system is very high, and is increased by strong distribution and dependability requirements. The risk is therefore real of loosing the mastership of the overall system if the basic technology does not allow simplification and integration. In its initial implementation, the system was implemented over an infrastructure of heterogeneous standard computers communicating over LANs with standard OSI protocols, thus offering a message-passing paradigm with no built-in mechanisms for fault-tolerance. To decrease complexity and to achieve dependability, the application designers and programmers therefore had to fill the gap that leads to a higher level of abstraction by including some additional mechanisms within the application itself. This leads inevitably to mechanisms specific to the application that generally solve the problems of dependability and distributed consistency in an imperfect way.

The use of a Delta-4 platform for such an application places the application designer and programmer at the level of abstraction that enables him to concentrate on the resolution of the application-specific problems, and not of those which are implied by its distributed nature or by the dependability requirements. The use of RPC paradigms associated with multi-threading facilities offered by Deltase has for example much simplified the design of the message scheduling application, as opposed to the use of message-passing support. The distributed consistency of production data and events is transparently offered by the Delta-4 infrastructure, and does not have to be reconsidered in the application (as opposed to what was necessary in the initial implementation).

This application was therefore a good candidate for Delta-4 as an enabling technology for building large-scale complex distributed applications.

3.5. Conclusion

As is proper in the context of an open project such as Delta-4, with both academic and commercial aspects to the development, the above comparison is unusually frank and unbiased by comparison with what would be found in, let us say, the sales literature of a new commercial product. It should be evident that, as is common in all engineering situations, ideal solutions do not exist. We can approach, but not achieve "the best of all possible worlds", and the approach taken is inevitably coloured by the intended market.

For this market and, we believe, for many others, Delta-4 does offer a "shopping list" of significant advantages and advances. Taken in isolation, the list given below, whilst impressive, would be open to a criticism of bias from vendors operating in any market that did not recognise the value of the attributes listed; the above discussion is intended to redress the balance. We nevertheless contend that the contents of the present book amply justify our claiming the following list of benefits to our architecture:

- extension of the concept of *openness* to dependable systems:
  - removal of the traditional restrictions to single vendor;
  - support for hardware and software from many sources;
  - contributions to the evolution of standards;
  - public availability of Implementation Guides;
- selectable levels of dependability at the granularity of individual services;
- accommodation of variants:
  - off-the-shelf hardware which does *not* possess native mechanisms for fault-tolerance;
  - specialized non-Delta-4 hardware which *does* possess native mechanisms for fault-tolerance;
  - new technology as it arises;
- mix-and-match capability allowing cost/performance tradeoffs between any of the above variants:
  - maximization of software portability;
  - uniform set of technology-independent support mechanisms;
  - mapping to existing languages and development tools;
  - minimal restrictions on choice of language;
- accommodation of software not corresponding to Delta-4 principles:
  - preexisting popular applications such as SQL databases;
  - MMS applications software;
- support for specialized application domains such as *distributed real-time* in a much more coherent manner than exhibited by current commercial products.

Chapter 4

Dependability Concepts[1]

This chapter does not deal with Delta-4 *per se* but is aimed at giving informal but precise definitions characterizing the various attributes of computing systems dependability. It is a contribution to the work undertaken within the "Reliable and Fault Tolerant Computing" scientific and technical community [Anderson and Lee 1981, Avizienis 1978, Avizienis and Laprie 1986, Carter 1979, Cristian et al. 1985, FTCS12, Jessep 1977, Laprie 1985, Laprie 1989, Melliar-Smith and Randell 1977, Randell et al. 1978, Siewiorek and Johnson 1982] in order to propose clear and widely acceptable definitions for some basic concepts. Readers already familiar with this terminology may wish to skip this chapter.

Dependability is first introduced as a global concept that subsumes the usual attributes of reliability, availability, safety, security. The basic definitions given in the first section are then commented, and supplemented by additional definitions, in the subsequent sections. A glossary is given in annex, which recapitulates the definitions given throughout the chapter. The presentation has been structured so as to avoid forward referencing. Underlining is used when a term is defined, italic characters being an invitation to focus the reader's attention. The guidelines that have governed this presentation can be summed up as follows:

- search for the minimum number of concepts enabling the dependability attributes to be expressed;
- use of terms which are identical to — whenever possible — or as close as possible to those generally used; as a rule, a term which has not been defined retains its ordinary sense (as given by any dictionary);
- emphasis on integration [Goldberg 1982, Randell and Dobson 1986] (as opposed to specialization) through the independence of the given definitions with respect to the classes of faults.

This contents of this chapter can be seen as a minimum consensus within the community in order to facilitate fruitful interactions; in addition the material presented is hoped to be suitable a) for being used by other bodies (including standards organizations), and b) for educational purposes. In this view, the associated terminology effort is not an end in itself: words are only of interest in so far as they transmit ideas, subject them to criticism, and enable viewpoints to be shared. There is no pretension of this chapter representing *the* state-of-the-art or "Tablets of Stone": the presented concepts have to evolve with technology, and with our progress in understanding and mastering the design and the assessment of dependable computer systems.

---

[1] This chapter is a result of work partially financed by the Esprit basic research action project PDCS (Predictably Dependable Computing Systems). It is also the basis of pre-standardization work being carried out by the IFIP 10.4 working group on Dependable Computing and Fault-Tolerance. It has been included in this book about the Delta-4 architecture in order to provide a well-defined terminological and conceptual framework.

4.1. Basic Definitions

Dependability is defined as the trustworthiness of a computer system such that reliance can justifiably be placed the service it delivers [Carter 1982]. The service delivered by a system is its behavior *as it is perceived* by its user(s); a user is another system (human or physical) which *interacts* with the former.

Depending on the application(s) intended for the computer system under consideration, different emphasis may be put on different facets of dependability, i.e., dependability may be viewed according to different, but complementary, *properties*, which enable the *attributes* of dependability to be defined:

- with respect to the *readiness for usage*, dependable means available;
- with respect to the *continuity of service*, dependable means reliable;
- with respect to the *avoidance of catastrophic consequences on the environment*, dependable means safe;
- with respect to the *prevention of unauthorized handling of information*, dependable means secure.

A system failure occurs when the delivered service no longer complies with the specification, the latter being an *agreed* description of the system's expected function and/or service. An error is that part of the system state that is liable to lead to failure: an error affecting the service, i.e., becoming user-visible, is an indication that a failure occurs or has occurred. The *adjudged or hypothesised* cause of an error is a fault.

The development of a dependable computing system calls for the *combined* utilization of a set of methods that can be classed into:

- fault prevention: how to prevent fault occurrence or introduction;
- fault tolerance: how to provide a service complying with the specification in spite of faults;
- fault removal: how to reduce the presence (number, seriousness) of faults;
- fault forecasting: how to estimate the present number, the future incidence, and the consequences of faults.

Fault prevention and fault tolerance may be seen as constituting dependability procurement: how to *provide* the system with the ability to deliver a service complying with the system specification; fault removal and fault forecasting may be seen as constituting dependability validation: how to *reach confidence* in the system's ability to deliver a service complying with the system specification.

Reliance on the system's service, and justification for reliance, are based on the *assessment* of the system, conducted primarily with respect to the *attributes* of dependability.

The notions introduced up to now can be grouped into three classes (figure 1):

- the impairments to dependability: faults, errors, failures; they are undesired — but not in principle unexpected — circumstances causing or resulting from un-dependability (whose definition is very simply derived from the definition of dependability: reliance cannot, or will not any longer, be placed on the service);
- the means for dependability: fault prevention, fault tolerance, fault removal, fault forecasting; these are the methods, tools, and solutions enabling one a) to provide the ability to deliver a service on which reliance can be placed, and b) to reach confidence in this ability.

- the <u>attributes</u> of dependability: reliability, availability, safety, security; these enable a) which properties are expected from the system to be expressed, and b) the system quality resulting from the impairments and the means opposing to them to be assessed.

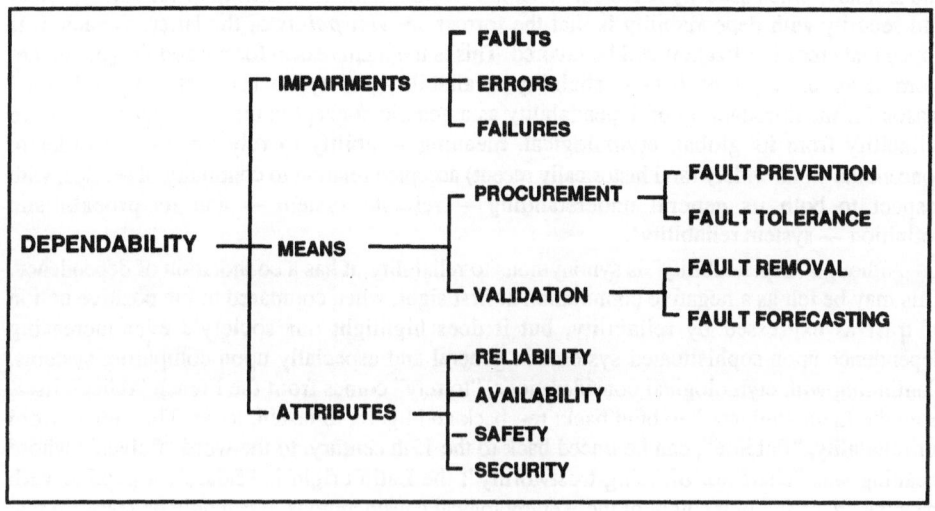

Fig. 1 - The Dependability Tree

4.2. On the Introduction of Dependability as a Generic Concept

A natural tendency of any emerging scientific or technical discipline is, in a first step, to restrict its field of investigation in order to make — rapid — progress in solving the associated problems. Then comes a time when its interactions with other disciplines can no longer be ignored. A great temptation is then to declare those other disciplines as being "special cases" of the considered discipline. This usually results in large debates, often conducted by the adherents of each discipline in their own jargon. This was the case for reliability, safety, and security of computing systems. Initially, the main concern was to have computing systems work: *reliability*. The utilization of computing systems in critical applications brought in the concern for *safety*. The safest system is often the one that does not do anything, which is not very helpful; so, people concerned with safety tend to consider reliability a subset of safety. The advent of distributed systems has exacerbated the *security* issues. Again, a secure system is intended to fulfil functionalities; in addition, security violations can be catastrophic; so, people concerned with security tend to consider safety and reliability as subsets of security.

However, the relations between reliability, safety and security are more complex than a simple dependence. Let us consider the example of the so-called "softbombs", i.e., faults deliberately introduced in a computing system in order to provoke, at a moment chosen by the "terrorist" — and under his/her control — a system failure, of consequences preferably felt by the user as non-catastrophic (until he becomes aware of the failure causes). This example clearly involves reliability, safety, and security, in a very intricate and varying manner depending upon the viewpoint considered[2]. What is certain is that the user cannot, or should not, place reliance on the service delivered by such a system, which is not *dependable*, in the primary sense of the word.

The preceding discussion clearly shows that it is *not* this chapter's intention to contribute to the controversy concerning whether reliability is a broader concept than safety or vice-versa, and similarly when security is added. What is essential in the relationship of reliability, safety and security with dependability is that the former are *viewpoints* of the latter: as such, it is hoped that cross-fertilization will be favored. This is the main reason for the addition of another word to an already long list — reliability, availability, safety, security, etc. An additional reason for the introduction of dependability as a generic concept is the willingness to relieve reliability from its global, etymological, meaning — ability to rely upon — in order to concentrate on its widely (and historically recent) accepted relation to continuity of service, with respect to both its general understanding — reliable system — and its probabilistic definition — system reliability[3].

Although "dependability" is synonymous to reliability, it has a connotation of dependence. This may be felt as a negative connotation at first sight, when compared to the positive notion of trust as expressed by reliability, but it does highlight our society's ever increasing dependence upon sophisticated systems in general and especially upon computing systems. Continuing with etymological considerations, "to rely" comes from the French "relier", itself from the Latin "religare", to bind back: re-, back and ligare, to fasten, to tie. The French word for reliability, "fiabilité", can be traced back to the 12th century, to the word "fiableté" whose meaning was "character of being trustworthy"; the Latin origin is "fidare", a popular verb meaning "to trust". In the light of these etymological considerations, it can only be regretted that the definition of reliability currently employed in many engineering fields[4] has substituted the notion of "ability" for the notion of "trust", for (at least) the two following reasons:

a) from the viewpoint of the system user, what is of real interest is not so much the *ability* to provide functionalities, as the *service* which is *actually* delivered to the user;

b) from the viewpoint of the system producer who is willing to admit the possible existence of faults in his design, the interpretation of the term "ability" must be questionable, despite the fact that it has been adopted in software engineering glossaries (see, e.g., [IEEE 729]).

4.3. On System Function, Behavior, Structure and Specification

Up to now, a <u>system</u> has been — implicitly — considered as a whole, emphasizing its externally perceived behavior. A definition complying with this "black box" view is: an entity having interacted or interfered, interacting or interfering, or likely to interact or interfere with other entities, i.e., with other systems. These other systems have been, are, or will constitute

---

2   It is also noteworthy that the events reported in the section on "Risk to the Public in Computer Systems" of the ACM Software Engineering Notes relate to reliability, to safety, and to security.

3   It is interesting to note that:

a) most books having the word "reliability" in their title actually deal with how to evaluate, measure, predict the reliability of systems, not really with how to build reliable systems;

b) viewing dependability as a more general concept than reliability, availability, etc., and embodying the latter terms, has already been attempted in the past (see, e.g., [Hosford 1960]); this was however attempted with less generality than here, since the goal was to define a measure embodying availability and reliability, and security was not of concern.

4   For example, "Reliability: The ability of an item to perform a required function under given conditions for a given time interval" [IEC 191].

the environment of the considered system[56]. A system user is that part of the environment that
interacts with the considered system: the user provides inputs to and receives outputs from the
system, its distinguishing feature being to *use the service* delivered by the system.

The function of a system is what the system *is intended for* [Kuipers 1985]. The behavior
of a system is what the system *does*. What *makes it do what it does* is the structure of the
system [Ziegler 1976]. Adopting the spirit of [Anderson and Lee 1981], a system, from a
structural ("glassbox") viewpoint, is a set of components bound together in order to interact; a
component is another system, etc. The recursion stops when a system is considered as being
atomic: any further internal structure cannot be discerned, or is not of interest and can be
ignored. The term "component" has to be understood in a broad sense: layers[7] of a system as
well as intra-layer components; in addition, a component being itself a system, it embodies the
interrelation(s) of the components of which it is composed. A more classical definition of
system structure is what a system *is*. Such a definition fits in perfectly when representing a
system without accounting explicitly for any impairments to dependability, and thus in the case
where the structure is considered as *fixed*. We do not want to restrict ourselves to systems
whose structure is fixed. In particular, we need to allow for structural changes caused by, or
resulting from, dependability impairments. It thus appears that a structure may have states[8,9].
Hence a definition for the notion of state: a condition of being with respect to a set of
circumstances, *whether of behavior or of structure*.

From its very definition (the user-perceived behavior), the service delivered by a system is
clearly an *abstraction* of its behavior. It is noteworthy that this abstraction is highly dependent
on the application that the computer system supports. An example of this dependence is the
important role played in this abstraction by time: the time granularities of the system and of its
user(s) are generally different, and the difference varies from one application to another one.

The specification of a system may describe the system's expectations in terms of either or
both its expected function and its expected service; there is usually not a single specification,
but several ones, according to:

- varying degrees of detail: requirement specification, design specification, realization
  specification, etc.;
- different viewpoints [Anderson and Lee 1981]: functional relationship between
  inputs and outputs, performance criteria (e.g., limits on response time), attributes of
  dependability (reliability, availability, safety, security).

Clearly, a system may fail with respect to some of these multiple specifications, and still
comply with other ones, leading to the notion of *degraded* mode of operation.

It is essential that a specification be *agreed upon* by two persons or corporate bodies — in
fact, legal personnae: the system supplier (in a broad sense of the term: designer, builder,

---

5   Giving recursive definitions is not for recursion's sake. The aim is to emphasise relativity with respect to
    the adopted viewpoint. So is it for the notion of system: a given system's boundaries may vary depending
    on whether it is viewed by its designer(s), by its user(s), by its maintenance crew, etc.

6   The passive, present and future forms are employed to stress that a system's environment may vary with
    time, especially with respect to the phases of its life-cycle. For instance, the notion of "programming
    environment" fits into the given definition, as well as the physical environment a system is confronted with
    during operation.

7   In the sense of protocols, i.e., a given layer using the services provided by lower layer(s), including
    hardware, and delivering services to the upper layer(s).

8   It could therefore be said that a "structure" has also a "behavior", especially with respect to the dependability
    impairments, even if the considered velocities of evolution with respect a) to the user's request on one hand,
    and b) to the impairments on the other, are —hopefully— different.

9   The given definition enables other types of systems with varying structures to be embodied, e.g., adaptive
    —especially knowledge-based— systems.

vendor, etc.) and its human user(s)[10]. The agreement is necessary so that the specification can serve as a basis for adjudicating whether the delivered service is acceptable or not, or, equivalently, whether a failure has occurred or not. What can be judged as an acceptable service with respect to a specification at a given level of detail may not comply with the specification at a less detailed level, because of mistakes occurred when detailing the specification, resulting in fact in *specification faults*. Specification faults may in turn affect any of the various specifications. More generally, a specification cannot be claimed to be immutable once established. This would be simple ignorance of the facts of life, which imply *change*. The changes may be motivated by modifying the system requirements: modification of the expected function and/or service, or correction of some faults[11]. Once more, what is important is that the specification is (again) agreed upon. Finally, it is noteworthy that such matters as environment, exposure time, observability, etc., can — and should — be captured by an appropriately stated specification.

Based on the preceding view of system structure, the notions of function, of service and of their specification apply equally naturally to the components. This is especially interesting in the design process, when off-the-shelf components, either hardware or software are used: what is more of interest to the designer is the function and/or the service they are able to provide, rather than their detailed (internal) behavior.

4.4. The Impairments to Dependability

4.4.1. Faults

Faults and their sources are extremely diverse. They can be classified according to three main viewpoints that are their nature, their origin and their persistence.

The *nature* of faults leads one to distinguish:

- accidental faults, which occur or are created fortuitously;
- intentional faults, which occur or are created deliberately.

The *origin* of faults may itself be decomposed into three viewpoints:

- the *phenomenological causes*, which lead one to distinguish [Avizienis 1978]:
    - physical faults, which are due to adverse physical phenomena,
    - human-made faults, which result from human imperfections;
- the *system boundaries*, which lead one to distinguish:
    - internal faults, which are those parts of the state of a system which, when invoked by the computation activity, will produce an error,
    - external faults, which result from interference to the system from its physical environment (electromagnetic perturbations, radiation, temperature, vibration, etc.), or from interaction with its human environment;
- the *phase of creation* with respect to the system's life, which leads one to distinguish:

---

[10] The agreement may be implicit, as when purchasing a system that comes with its specification and user's manual, or when using off-the-shelf systems.

[11] We are thus faced with a circular problem: a reference is needed for adjudicating whether a delivered service is acceptable or not, and this reference may be faulty. Improving specifications has long been devoted a significant amount of attention, including proposals for life-cycle models aimed at this objective [Boehm 1988].

- design faults, which result from imperfections arising either a) during the initial design of the system (broadly speaking, from requirement specification to implementation) or during subsequent modifications, or b) during the establishment of the procedures for operating or maintaining the system;

- operational faults, which occur during the system's exploitation.

A distinction can also be made with respect to temporal *persistence* of faults, leading to:

- permanent faults, whose *presence* is not related to pointwise conditions whether they be internal (computation activity) or external (environment),

- temporary faults, whose presence is related to such conditions, and are thus present for a limited amount of time.

Security issues are dominated by — but not restricted to — intentional faults, which are clearly *human-made* faults. Intentional faults can be either internal or external; typical examples are:

- concerning internal faults, the incorporation of malicious logic (e.g., the so-called "Trojan horses"), which is an intentional *design* fault;

- concerning external faults, an intrusion that is an intentional *operational external* fault.

To be "successful", intentional faults may take advantage of accidental faults, e.g., an intrusion exploiting a security breach due to an accidental design fault; there are interesting and obvious similarities between this example and an accidental temporary external fault "exploiting" a lack of shielding.

It could be argued that introducing the *phenomenological causes* in the classification criteria of faults may lead recursively "a long way back", e.g., why do programmers make mistakes? why do integrated circuits fail? The very notion of fault is *arbitrary*, and is in fact a facility provided for stopping the recursion. Hence the definition given: *adjudged or hypothesised* cause of an error. This cause may vary depending upon the chosen viewpoint: fault tolerance mechanisms, maintenance engineers, repair shop, designer, semiconductor physicist, etc. In our view, recursion stops at *the cause that is intended to be avoided or tolerated*. This view provides consistency with the distinction between human-made and physical faults: a computing system is a human artifact and as such any fault in it or affecting it is ultimately human-made since it represents human inability to master all the phenomena that govern the behavior of a system. In an absolute sense, a distinction between physical faults and human-made faults (especially design faults) may be considered unnecessary; however, this distinction is of importance when considering the (current) methods and techniques for procuring and validating dependability. If the recursion mentioned above is not stopped, then *a fault is nothing other than the consequence of a failure of some other system* (including the designer) *that has delivered or is now delivering a service to the given system*.

Examples of the preceding discussion follow:

- a design fault results from a designer failure;

- a physical internal fault is due to a hardware component failure, which is itself the consequence of (an) error(s) at the electrical or electronic level (the "physics reliability" community rarely characterizes failures as "sudden and unpredictable"), in turn originating from physico-chemical disorders, again originating from the hardware production, or from — the limits of — our knowledge of semiconductor physics;

- a physical or human-made external fault is in fact a design fault: the inability to foresee all the situations the system will be faced with during its operational life, or the refusal to consider some of them (e.g., for economic reasons); for instance:

- in the case of an electromagnetic perturbation: is it an external fault or a design fault, i.e., the lack of adequate shielding?

- in the case of a failure caused by an operator typing a single inappropriate character: is it an interaction fault or a design fault, i.e., the lack of confirmation asked by the system [Norman 1983]?

The temporal persistence viewpoint deserves the following comments:

1) Temporary external faults originating from the physical environment are often termed <u>transient faults</u>.

2) Temporary internal faults are often termed <u>intermittent faults</u>; such faults result from the presence of rarely occurring combinations of conditions; examples are a) "pattern sensitive" faults in semiconductor memories, changes in the parameters of a hardware component (effect of temperature variation, delay in timing due to parasitic capacitance, etc.), or b) situations — affecting either hardware or software — occurring when system load goes beyond a certain level, such as marginal timing and synchronization. In fact, the term "fault" in such cases is actually an abstraction for *fault conditions*. The very notion of intermittent faults is in an absolute sense arbitrary: such faults are nothing other than (permanent) faults whose conditions of activation cannot be reproduced or which occur rarely enough; however, as already pointed out for the distinction between physical faults and design faults, their consideration is a useful facility. Permanent faults whose conditions of activation can be reproduced are often termed <u>recurrent faults</u>.

From the above discussions, it appears that *any fault is a permanent design fault*. This is indeed true in an absolute sense, but is not very helpful for the designers and assessors of a system.

Figure 2 summarises the various classes of faults that have been dealt with, with respect to the various viewpoints that have been considered.

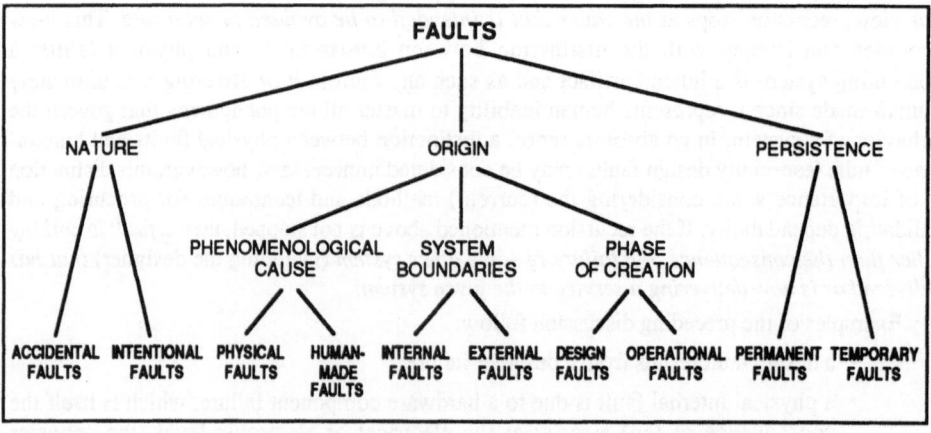

Fig. 2 - The Classes of Faults according to Various Viewpoints

If all the combinations of fault classes according to the 5 viewpoints of figure 2 were possible, there would be 32 different fault classes. In fact, the number of likely combinations is more restricted: 10 combinations are indicated by the rows of table 1, which also gives the usual labelling of these combinations — not their definition.

Table 1 - The Classes of Faults resulting from Combinations according to the Various Viewpoints

Nature		Origin						Persistence		Usual Labelling
		Phenomenological cause		System Boundaries		Phase of creation				
Accidental Faults	Intentional Faults	Physical Faults	Human-made Faults	Internal Faults	External Faults	Design Faults	Operational Faults	Permanent Faults	Temporary Faults	
✓		✓		✓			✓	✓		Physical faults
✓		✓			✓		✓		✓	Transient faults
✓		✓		✓			✓		✓	Intermittent faults
✓			✓	✓		✓			✓	
✓			✓	✓		✓		✓		Design faults
✓			✓		✓		✓		✓	Interaction faults
	✓		✓	✓		✓		✓		Malicious Logic
	✓		✓	✓		✓			✓	
	✓		✓		✓		✓	✓		Intrusions
	✓		✓		✓		✓		✓	

4.4.2. Errors

An error was defined as being *liable* to lead to failure. Whether or not an error will actually lead to a failure depends on three major factors:

1) The system composition, and especially the nature of the existing redundancy:
   - *intentional* redundancy (introduced to provide fault tolerance) which is explicitly intended to prevent an error from leading to failure,
   - *unintentional* redundancy (it is practically difficult if not impossible to build a system without any form of redundancy[12]) which may have the same — unexpected — result as intentional redundancy.

2) The system activity: an error may be overwritten before creating damage.

3) The definition of a failure from the user's viewpoint: what is a failure for a given user may be a bearable nuisance for another one. Examples are a) accounting for the user's time granularity: an error which "passes through" the system-user interface may or may not be viewed as a failure depending on the user's time granularity, b) the notion of "acceptable error rate" — implicitly before considering that a failure has occurred — in data transmission. This discussion explains why it is often desirable to explicitly mention in the specification such conditions as the maximum outage time (related to the user time granularity).

4.4.3. Failures

A system may not, and generally does not, always fail in the same way. The ways a system can fail are its failure *modes*, which may be characterized according to three viewpoints: domain, perception by the system users, and consequences on the environment.

The *failure domain* viewpoint leads one to distinguish:
   - <u>value failures</u>: the value of the delivered service does not comply with the specification;

---

[12] A classical problem in hardware testing is the removal of such "false redundancies", whose effect may be to mask faults, and as such to make the task of test pattern generation more complicated.

- **timing failures**: the timing of the service delivery does not comply with the specification.

Such general definitions (non compliance with the specification) apply to **arbitrary failures**. Refined modes of failures can be distinguished. For instance, the notion of timing failure may be refined into *early* timing failure or *late* timing failure, depending on whether the service is delivered too early or too late. A class of failures relating to both value and timing is the **stopping failures**: system activity, if any, is not any more perceptible to the users, and a constant value service is delivered; the constant value delivered may vary according to the application, and thus the specification, e.g., last correct value, predetermined value, etc. A special case of stopping failures is constituted by **omission failures** [Cristian et al. 1985, Ezhilchelvan and Shrivastava 1986]: no service is delivered. Such a failure can be seen as a common limiting case for both value failures (null value) and timing failures (infinitely late failure); a *persistent* omission failure is a **crash failure**. A system whose failures can only be — or more generally are to an acceptable extent — stopping failures, is a **fail-stop system**[13], and a system whose failures can only be — are to an acceptable extent — crash failures, is a **fail-silent system** [Powell et al. 1988]; on the opposite, a system whose failures may be arbitrary is a **fail-uncontrolled** system [Powell et al. 1988].

When a system has several users, the *failure perception* viewpoint leads one to distinguish:

- **consistent failures**: all system users have the same perception of the failures;

- **inconsistent failures**: the system users may have different perceptions of a given failure; inconsistent failures are usually termed, after [Lamport et al. 1982], *Byzantine failures*.

The failure **severities** result from grading the *consequences of the failures* upon the system environment. They thus enable the failure modes to be ordered. A special case of great interest is that of systems whose failure modes can be grouped into two classes whose severities differ considerably:

- **benign failures**, where the consequences are of the same order of magnitude (generally in terms of cost) as the benefit provided by service delivery in the absence of failure;

- **catastrophic failures**, where the consequences are incommensurably greater than the benefit provided by service delivery in the absence of failure.

A system whose failures can only be — or more generally are to an acceptable extent — benign failures is a **fail-safe system**. The notion of failure severity enables the notion of criticality to be defined: the **criticality** of a system is the highest severity of its (possible) failure modes[14].

Based on the given definition of a system's structure, the discussion of whether "failure" applies to a system or a component is simply irrelevant, since a component is itself a system. When atomic systems are dealt with, the notion of an "elementary" failure comes naturally.

---

[13] The concept of fail-stop processor has been defined in [Schlichting and Schneider 1983] in the context of distributed systems. The definition of fail-stop system we give, when interpreted in the context of distributed systems where information is exchanged by messages, is consistent with the definition of fail-stop processor.

[14] As an example, the criticality levels accepted by the aviation community are defined as follows [RTCA 178A]:
- **critical**: functions for which the occurrence of any failure would prevent the continued safe flight and landing of the aircraft;
- **essential**: functions for which the occurrence of any failure would reduce the capability of the aircraft or the ability of the crew to cope with adverse operating conditions;
- **non essential**: functions for which failure could not significantly degrade aircraft capability or crew ability.

4.4.4. Fault Pathology

The creation and manifestation mechanisms of faults, errors, and failures may be summarised as follows:

1) A fault is <u>active</u> when it produces an error. An active fault is either a) an internal fault that was previously <u>dormant</u> and that has been activated by the computation process (including the simultaneous existence of the fault conditions for an intermittent fault), or b) an external fault. An internal fault may cycle between its dormant and active states. Physical faults can directly affect the hardware components only, whereas human-made faults may affect any component.

2) An error may be latent or detected. An error is <u>latent</u> when it has not been recognised as such; an error is <u>detected</u> by a detection algorithm or mechanism. An error may disappear before being detected. An error may, and in general does, propagate; by propagating, an error creates other — new — error(s).

3) A failure occurs when an error "passes through" the system-user interface and affects the service delivered by the system. A component failure results in a fault a) for the system that contains the component, and b) as viewed by the other component(s) with which it interacts; the failure modes of the failed component then become fault types for the components interacting with it.

These mechanisms enable the "fundamental chain" to be completed:

$$\dots \rightarrow \text{failure} \rightarrow \text{fault} \rightarrow \text{error} \rightarrow \text{failure} \rightarrow \text{fault} \rightarrow \dots$$

Some illustrative examples of fault pathology:

- the result of a programmer's *error* is a *(dormant) fault* in the written software (faulty instruction(s) or data); upon activation (invoking the component where the fault resides and triggering the faulty instruction, instruction sequence or data by an appropriate input pattern) the fault becomes *active* and produces an error; if and when the erroneous data affect the delivered service (in value and/or in the timing of their delivery), a *failure* occurs;

- a short circuit occurring in an integrated circuit is a *failure* (with respect to the service specification of the circuit); the consequence (connection stuck at a Boolean value, modification of the circuit function, etc.) is a *fault* which will remain dormant as long as it is not activated, the continuation of the process being identical to the previous example;

- an electromagnetic perturbation of sufficient energy is a *fault;* this fault may
  a) directly create an *error*, e.g., by electromagnetic interference with the electrical charges circulating along wires,
  b) create another, (internal) fault; for instance, if the perturbation acts on a memory's inputs in the write position in changing some bit values, these errors will subsequently remain as faults in the memory; the latter faults will remain dormant until the particular memory location(s) are read; the error-failure sequence from the external transient fault to the internal fault still exists, at the electronic level;

- an inappropriate man-machine interaction performed by an operator during the operation of the system is a *fault* (from the system viewpoint); the resulting altered processed data is an *error;* etc.

- a maintenance or operating manual writer's error may result in a fault in the corresponding manual (faulty directives) which will remain dormant as long as the directives are not acted upon in order to deal with a given situation, etc.

From the above examples, it is easily understood that the fault dormancy may vary considerably, depending upon the fault, the given system's utilization, etc.

The man-made faults can be either accidental or intentional. The previous example relating to a programmer's error and its consequences may be rephrased as follows: a logic bomb is created by a malicious programmer; it will remain dormant up to being activated (e.g., at some predetermined date); it then produces an error which may lead to a storage overflow or to slowing down the program execution; as a consequence, service delivery will suffer from a so-called denial-of-service, a special type of failure.

These examples were deliberately kept simple. Real life is, as usual, much more complicated; four examples:

a) a given fault in a given component may result from different possible sources; for instance, a permanent fault in a physical component — e.g., stuck at ground voltage — may result from:

- a physical failure (e.g., caused by a threshold change),
- an error caused by a design fault — e.g., faulty microinstruction — propagating "top-down" through the layers and causing a short between two circuit outputs for a duration long enough to provoke a short-circuit having the same consequence as the threshold change;

b) a fault of a given class may, through propagation, create a fault of another class; for instance, the fault having led to an error during the execution of the microinstruction in the preceding example could have been a transient fault;

c) some viewpoints may become — temporarily at least — of lesser importance during the propagation process; for instance, when dealing with external faults producing input errors during execution of a software component (thus, invoking it in its so-called exceptional input domain [Cristian 1980]), the fact that the fault is physical or human-made may not be of importance for the failure behavior of the given component;

d) a failure often results from the combined action of several faults; this is especially true when considering security issues: a trap-door (i.e., some way to bypass access control) which is inserted into a computer system, either accidentally or intentionally, is a design fault; this fault may remain dormant until some malicious human makes use of it to enter the system; the intruder login is an intentional interaction fault; when the intruder is logged in (while he or she should not be), he or she may deliberately create an error, e.g., in modifying some file (integrity attack); when this file is used by an authorized user, the service will be affected, and a failure will occur.

Two additional comments, relative to the words, or labels, "fault", "error", and "failure":

a) their exclusive use in this paper does not preclude the use in special situations of words which designate, briefly and unambiguously, a specific class of impairment; this is especially applicable to faults (e.g., bug[15], defect, deficiency) and to failures (e.g., breakdown, malfunction, denial-of-service);

---

[15] Including specialization of the term "bug", as in [Gray 1986], which distinguishes "Heisenbugs" (for intermittent software faults, from the Heisenberg uncertainty principle) from "Bohrbugs" (for permanent software faults, "like the Bohr atom, solid, easily detected by standard techniques, and hence boring").

b) the assignment made of the particular terms fault, error, failure simply takes into account current usage: i) fault avoidance, tolerance, and diagnosis, ii) error detection and correction, iii) failure rate.

Finally, it has to be stressed that the definitions given in this section are *syntactic*; accordingly, the criteria for the various classifications performed have been emphasised, and are in our view more important than the classes themselves.

4.5. The Means for Dependability

4.5.1. Dependencies between the Means for Dependability

All the "how to's" that appear in the basic definitions given in section 4.1 are in fact goals that cannot be fully reached, as all the corresponding activities are human activities, and thus imperfect. These imperfections bring in *dependencies* that explain why it is only the *combined* utilization of the above methods — preferably at each step of the design and implementation process — that can best lead to a dependable computing system. These dependencies can be sketched as follows: in spite of fault prevention by means of design methodologies and construction rules (imperfect in order to be workable), faults occur. Hence the need for fault removal. Fault removal is itself imperfect, as are the off-the-shelf components — hardware or software — of the system, hence the importance of fault forecasting. Our increasing dependence on computing systems brings in the requirement for fault tolerance, which is in turns based on construction rules; hence fault removal, fault forecasting, etc. It must be noted that the process is even more recursive than it appears from the above: current computer systems are so complex that their design and implementation need computerized tools in order to be cost-effective (in a broad sense, including the capability of succeeding within an acceptable time scale). These tools have themselves to be dependable, and so on.

The preceding reasoning illustrates the close interactions between fault removal and fault forecasting, and motivates their gathering into the single term *validation*. This is despite the fact that validation is often limited to fault removal, and associated with one of the main activities involved in fault removal, verification: e.g., in "V and V" [Boehm 1979]; in such a case the distinction is related to the difference between "building the system right" (related to verification) and "building the right system" (related to validation)[16]. What is proposed here is simply an extension of this concept: the answer to the question "am I building the right system?" (fault removal) being complemented by "for how long will it be right?" (fault forecasting)[17]. In addition, fault removal is usually closely associated with fault prevention, forming together fault avoidance, i.e., how to *aim at* a fault-free system. Besides highlighting the need for validating the procedures and mechanisms of fault tolerance, considering fault removal and fault forecasting as two constituents of the same activity — validation — is of great interest in that it enables a better understanding of the notion of coverage, and thus of an important problem introduced by the above recursion: *the validation of the validation,* or how to reach confidence in the methods and tools used in building confidence in the system. Coverage refers here to a measure of the representativity of the situations to which the system is submitted during its validation compared to the actual situations it will be confronted with during its

---

[16] It is noteworthy that these assignments are sometimes reversed, as in the domain of communication protocols (see, e.g., [Rudin 1985]).

[17] Validation stems from "validity", which encapsulates two notions:
- validity at a given moment, which relates to fault removal;
- validity for a given duration, which relates to fault forecasting.

operational life[18]. Imperfect coverage strengthens the relation between fault removal and fault forecasting, as it can be considered that the need for fault forecasting stems from imperfect coverage of fault removal.

In the remainder of this section, we examine in turns fault tolerance, fault removal and fault forecasting; fault prevention is not dealt with as it clearly relates to "general" system engineering.

4.5.2. Fault Tolerance

Fault tolerance is carried out by error processing and by fault treatment [Anderson and Lee 1981]. Error processing is aimed at removing errors from the computational state, if possible before failure occurrence; fault treatment is aimed at preventing faults from being activated — again.

Error processing can be carried out in two ways:

- error recovery, where an error-free state is substituted for the erroneous state; this substitution may take on two forms [Anderson and Lee 1981]:

  - backward recovery, where the erroneous state transformation consists of bringing the system back to a state already occupied prior to error occurrence; this involves the establishment of recovery points, which are points in time during the execution of a process for which the then current state may subsequently need to be restored;

  - forward recovery, where the erroneous state transformation consists of finding a new state, from which the system can operate (frequently in a degraded mode);

- error compensation, where the erroneous state contains enough redundancy to enable the delivery of an error-free service from the erroneous (internal) state.

When error recovery is employed, the erroneous state needs to be (urgently) identified as being erroneous prior to being transformed; this is the purpose of error detection, hence the term of *error detection-and-recovery* that is usually employed. The association into a component of its functional processing capability together with error detection mechanisms leads to the notion of self-checking component, either in hardware [Carter and Schneider 1968, Nicolaidis et al. 1989, Wakerly 1978] or in software [Yau and Cheung 1975]; one of the important benefits of the self-checking component approach is the ability to give a clear definition of *error confinement areas* [Siewiorek and Johnson 1982]. When error compensation is performed in a system made up of self-checking components partitioned into classes executing the same tasks (the so-called "active redundancy"), then state transformation is nothing else than switching within a class from a failed component to a non-failed one, hence the corresponding approach to fault tolerance: *error detection-and-compensation*[19]. On the other hand, compensation may be applied systematically, even in the absence of errors, then providing error masking (e.g., in majority vote). However, this can at the same time correspond to an unknown decrease in

---

[18] The notion of coverage as defined here is very general; it may be made more precise by indicating its field of application, e.g.:
- coverage of a software test with respect to its text, control graph, etc.
- coverage of an integrated circuit test with respect to a fault model,
- coverage of fault tolerance with respect to a class of faults,
- coverage of a design assumption with respect to reality.

[19] Error detection and compensation may be seen as a limiting case of error detection-and-recovery, where recovery is performed using the present (erroneous) state of the system instead of substituting an error-free state to the erroneous state.

redundancy. So, practical implementations of masking generally involve error detection, which may then be performed *after* the state transformation.

Backward and forward error recovery are not exclusive: backward recovery may be attempted first; if the error persists, forward recovery may then be attempted. In forward recovery, it is necessary to *assess the damage* caused by the detected error, or by errors propagated before detection; damage assessment can — in principle — be ignored in the case of backward recovery, provided that the mechanisms enabling the transformation of the erroneous state into an error-free state have not been affected [Anderson and Lee 1981].

The operational time overhead necessary for error processing is radically different according to the adopted error processing form:

- in error recovery, the time overhead is longer upon error occurrence than before; especially, in backward recovery it is related to the provision of recovery points, thus in fact to preparing for error processing;

- in error compensation, the time overhead required by compensation is the same, or almost the same, whether errors are present or not.

In addition, the duration of error compensation is much shorter than the duration of error recovery, due to the larger amount of (structural) redundancy.

This remark

a) is of high practical importance in that it often conditions the choice of the adopted fault tolerance strategy with respect to the user time granularity;

b) has introduced a relation between operational time overhead and structural redundancy; more generally, a redundant system always provides redundant behavior, incurring at least some operational time overhead; the time overhead may be small enough not to be perceived by the user, which means only that the *service* is not redundant; an extreme opposite form is "time redundancy" (redundant *behavior* obtained by repetition) which needs to be at least initialized by a structural redundancy, limited but existing; roughly speaking, the more the structural redundancy, the less the time overhead incurred.

The first step in *fault treatment* is fault diagnosis, which consists of determining the cause(s) of error(s), in terms of both location and nature. Then come the actions aimed at fulfilling the main purpose of fault treatment: preventing the fault(s) from being activated again, thus aimed at making it(them) passive, i.e., fault passivation. This is carried out by removing the component(s) identified as being faulty from further executions. If the system is no longer capable of delivering the same service as before, then a *reconfiguration* may take place.

If it is estimated that error processing could directly remove the fault, or if its likelihood of recurring is low enough, then fault passivation need not be undertaken. As long as fault passivation is not undertaken, the fault is regarded as a soft fault; undertaking it implies that the fault is considered as hard, or solid. At first sight, the notions of soft and hard faults may seem to be respectively synonymous to the previously introduced notions of temporary and permanent faults. Indeed, tolerance of temporary faults does not require fault treatment, since error recovery should in this case directly remove the effects of the fault, which has itself vanished, provided that a permanent fault has not been created in the propagation process. In fact, the notions of soft and hard faults are useful due to the following reasons:

- distinguishing a permanent fault from a temporary fault is a difficult and complex task, since a) a temporary fault vanishes after a certain amount of time, usually before fault diagnosis is undertaken, and b) faults from different classes may lead to very similar errors; so, the notion of soft or hard fault in fact incorporates the subjectivity associated with these difficulties, including the fact that a fault may be declared as a soft fault when the fault diagnosis is unsuccessful;

- the ability of those notions to incorporate subtleties of the modes of action of some transient faults; for instance, can it be said that the dormant internal fault resulting from the action of alpha particles (due to the residual ionization of circuit packages), or of heavy ions in space, on memory elements (in the broad sense of the term, including flip-flops) is a *temporary* fault? Such a dormant fault is however a *soft* fault.

The preceding definitions apply to physical faults as well as to design faults: the class(es) of faults that can actually be tolerated depend(s) on the fault hypothesis that is being considered in the design process, and thus relies on the *independence* of redundancies with respect to the process of fault creation and activation. An example is provided by considering tolerance of physical faults and tolerance of design faults. A (widely-used) method to attain fault tolerance is to perform multiple computations through multiple channels. When tolerance of physical faults is foreseen, the channels may be identical, based on the assumption that hardware components fail independently; such an approach is not suitable for the tolerance to design faults where the channels have to provide *identical services* through *separate designs and implementations* [Avizienis 1978, Elmendorf 1972, Randell 1975], i.e., through design diversity [Avizienis and Kelly 1984].

An important aspect in the coordination of the activity of multiple components is that of preventing errors to propagate and to affect the operation of non-failed components. This aspect becomes particularly important when a given component needs to communicate some information to other components that is private to that component. Typical examples of such *single-source information* are local sensor data, the value of a local clock, the local view of the status of other components, etc. The consequence of this need to communicate single-source information from one component to other components is that non-failed components must reach an *agreement* as to how the information they obtain should be employed in a mutually consistent way. Specific attention has been devoted to this problem in the field of distributed systems (see, e.g., clock synchronization [Kopetz and Ochsenreiter 1987, Lamport and Melliar-Smith 1985] or membership protocols [Cristian 1988]). It is important to realize, however, that the inevitable presence of structural redundancy in any fault-tolerant system implies distribution at one level or another, and that the agreement problem therefore subsists. Geographically localized fault-tolerant systems may employ solutions to the agreement problem that would be deemed too costly in a "classical" distributed system of components communicating by messages (e.g., inter-stages [Lala 1986], multiple stages for interactive consistency [Frison and Wensley 1982]).

The knowledge of some system properties may limit the necessary amount of redundancy, leading to the so-called "low-cost fault tolerance". Examples of these properties are regularities of a structural nature: error detecting and correcting codes [Peterson and Weldon 1972], robust data structures [Taylor et al. 1980], multiprocessors and computer networks [Pradhan 1986, Rennels 1986], algorithm-based fault tolerance [Huang and Abraham 1982]. The faults that are tolerated are then dependent upon the properties accounted for, as they intervene directly in the fault hypotheses.

Of importance is the signalling of a component failure to its users. This may be accounted for within the framework of *exceptions* [Anderson and Lee 1981, Cristian 1980, Melliar-Smith and Randell 1977]. *Exception handling* facilities provided in some languages may constitute a convenient way for implementing error recovery, especially forward recovery[20].

---

[20] The use of the term "exception", due to its origin of coping with exceptional situations —not only errors— is to be carefully used in the framework of fault tolerance: it could appear as contradicting the view that fault tolerance is a natural attribute of computing systems, considered from the very initial design phases, and not an "exceptional" attribute.

Fault tolerance is (also) a recursive concept: it is essential that the mechanisms aimed at implementing fault tolerance be protected against the faults that can affect them. Examples are voter replication, self-checking checkers [Carter and Schneider 1968], "stable" memory for recovery programs and data [Lampson 1981].

Fault tolerance is not restricted to accidental faults. Protection against intrusions traditionally involves cryptography [Denning 1982]. Some mechanisms of error detection are directed towards both intentional and accidental faults (e.g., memory access protection techniques) and schemes have been proposed for the tolerance to both intrusions and physical faults [Fraga and Powell 1985, Rabin 1989], as well as for tolerance to malicious logic [Joseph and Avizienis 1988].

4.5.3. Fault Removal

Fault removal is composed of three steps: verification, diagnosis, correction. Verification is the process of checking whether the system adheres to properties, termed the *verification conditions* [Cheheyl et al. 1981]; if it does not, the other two steps have to be undertaken: diagnosing the fault(s) that prevented the verification conditions from being fulfilled, and then performing the necessary corrections. After correction, the process has to be resumed to check that fault removal had no undesired consequences; the verification performed at this stage is usually termed (non-)regression verification. The verification conditions can take two forms:

- general conditions, which apply to a given class of systems, and are therefore — relatively — independent of the specification, e.g., absence of deadlock, conformance to design and realization rules;

- conditions specific to the considered system, directly deduced from its specification.

The verification techniques can be classed according to whether or not they involve exercising the system. Verifying a system without actual execution is static verification. The verification can be conducted:

- on the system itself, in the form of a) *static analysis* (e.g., inspections or walk-through [Myers 1979], data flow analysis [Osterweil and Fodsick 1976], complexity analysis [McCabe 1976], compiler checks, etc.) or b) *proof-of-correctness* (inductive assertions [Craigen 1987, Hoare 1969]);

- on a model of the system behavior (e.g., Petri nets, finite state automata), leading to *behavior analysis* [Diaz 1982].

Verifying a system through exercising it constitutes dynamic verification; the inputs supplied to the system can be either symbolic in the case of symbolic execution, or valued in the case of testing.

Testing exhaustively a system with respect to all its possible inputs is generally impractical. The methods for the determination of the test patterns can be classed according to two viewpoints:

- *criteria* for selecting the test inputs, which may relate to either the function or the structure of the system, leading respectively to functional testing and structural testing; in both cases, the criteria may relate to

  - the system's ability to deliver service (e.g., path sensitization [Rapps and Weyuker 1985], input boundary values in software [Ntafos 1988]),

  - revealing specific classes of faults (e.g., stuck-at-faults in hardware production [Roth et al. 1967], physical faults affecting the instruction set of a microprocessor [Thatte and Abraham 1978], design faults in software [DeMillo et al. 1978, Goodenough and Gerhart 1975, Howden 1987]);

- *generation* of the test inputs, which may be deterministic or probabilistic:
    - in deterministic testing, test patterns are predetermined by a selective choice according to the adopted criteria,
    - in random, or statistical, testing, test patterns are selected according to a defined probability distribution on the input domain; the distribution and the number of input data are determined according to the adopted criteria [David and Thévenod-Fosse 1981, Duran and Ntafos 1984].

Observing the test outputs and deciding whether they satisfy or not the verification conditions is known as the *oracle* problem [Adrion et al. 1982]. The verification conditions may apply to the whole set of outputs or to a compact function of the latter (e.g., a system signature when testing for physical faults in hardware [David 1986], or a "partial oracle" when testing for design faults of software [Weyuker 1982]). When testing for physical faults, the results — compact or not — anticipated from the system under test for a given input sequence are determined by simulation [Levendel 1986] or from a reference system ("golden unit"). For design faults, the reference is generally the specification; it may also be a prototype, or another implementation of the same specification in the case of design diversity ("back-to-back testing", see, e.g., [Bishop 1988]).

Some verification methods may be used in conjunction, e.g., symbolic execution may be used a) for facilitating the determination of the testing patterns [Adrion et al. 1982], or b) as a proof-of-correctness method [Carter et al. 1978].

As verification has to be performed all along a system's development, the above techniques apply naturally to the various forms taken by a system during its development: prototype, component, etc. Verifying that the system cannot do *more* than what is specified is especially important with respect to intentional faults [Gasser 1988].

Designing a system in such a way as to facilitate its verification is the design for verifiability. This is especially developed for hardware with respect to physical faults, where the corresponding techniques are then termed *design for testability* [McCluskey 1986, Williams 1983].

Fault removal during the operational phase of a system's life is corrective maintenance, aimed at preserving or improving the system's ability to deliver a service complying with the specification[21]. Corrective maintenance can take two forms:

- curative maintenance, aimed at removing faults which have produced one or more errors and have been reported;
- preventive maintenance, aimed at removing faults before they produce errors; these faults can be
    - physical faults having occurred since the last preventive maintenance actions,
    - design faults having led to errors in other similar systems [Adams 1984].

These definitions[22] apply to non-fault-tolerant systems as well as to fault-tolerant systems, which can be maintainable on-line (without interrupting service delivery) or off-line. It is finally

---

[21] The other forms of maintenance usually distinguished are [Ramamoorthy et al. 1984]:
- *adaptive maintenance*, which adjusts the system to environmental changes (e.g., change of operating systems or system data-bases);
- *perfective maintenance*, which improves the system's function by responding to customer — and designer — defined changes, which may involve removal of specification faults.

[22] It is noteworthy that current discussions about the irrelevance of the use of the term "maintenance" when applied to software simply forget the etymology of the word: in the Middle Ages, maintenance designated the actions performed in order to keep an army battleworthy, thus including the corrective, adaptive and perfective forms of maintenance. The association of maintenance with repairing hardware is actually a

noteworthy that the frontier between corrective maintenance and fault treatment is relatively arbitrary; especially, curative maintenance may be considered as an — ultimate — means of achieving fault tolerance.

4.5.4. Fault Forecasting

Fault forecasting is conducted by performing an *evaluation* of the system behavior with respect to fault occurrence or activation. Evaluation has two aspects:

- *non-probabilistic*, e.g., determining the minimal cutset or pathset of a fault tree, conducting a failure mode and effect analysis;
- *probabilistic*, aimed at determining the conformance of the system to dependability objectives expressed in terms of probabilities associated to some of the attributes of dependability, which may then be defined as *measures* of dependability.

The life of a system is perceived by its user(s) as an alternation between two states of the delivered service with respect to the specification:

- <u>correct service</u>, where the delivered service *complies with* the specification[23];
- <u>incorrect service</u>, where the delivered service *does not comply with* the specification.

A failure is thus a transition from correct to incorrect service, and the transition from incorrect to correct service is a <u>restoration</u>. Quantifying the alternation of correct-incorrect service delivery enables reliability and availability to be defined as measures of dependability:

- <u>reliability</u>: a measure of the *continuous* delivery of correct service — or, equivalently, of the *time to* failure;
- <u>availability</u>: a measure of the delivery of correct service *with respect to the alternation* of correct and incorrect service.

A third measure, <u>maintainability</u>, is usually considered, which may be defined as a measure of the time to restoration from the last experienced failure, or equivalently, of the continuous delivery of incorrect service.

As a measure, safety can be seen as an extension of reliability. Let us group the state of correct service together with the state of incorrect service subsequent to benign failures into a safe state (in the sense of being free from catastrophic damage, not from danger); <u>safety</u> is then a measure of continuous "safeness", or equivalently, of the time to catastrophic failure. Safety can thus be considered as reliability with respect to the catastrophic failures. A direct extension of availability, i.e., a measure of safeness with respect to the alternation of safeness and incorrect service after catastrophic failure, would not provide a significant measure. When a catastrophic failure has occurred, the consequences are generally so important that service restoration is not of prime importance for — at least — the two following reasons:

- it comes second to repairing (in the broad sense of the term, including legal aspects) the consequences of the catastrophe;
- the lengthy period before being allowed to operate the system again (commissions of enquiry, etc.) would lead to insignificant numerical values.

A "hybrid" reliability-availability-type measure can however be defined: a measure of correct service delivery with respect to the alternation of correct service and incorrect service

---

(recent) deviation; associating "to maintain" with the notion of *service* would enable this etymological meaning to be revived, while at the same time removing the very source of discussion.

23 We deliberately restrict the use of "correct" to the service delivered by a system, and do not use it for the system itself: in our view, non-faulty systems do not exist, there are only systems that may have not yet failed.

after benign failure. This measure is of interest in that it provides indeed a quantification of the system availability *before* occurrence of a catastrophic failure, and as such enables quantification of the so-called "reliability- (or availability-) safety tradeoff".

In the case of multi-performing systems, several modes of service delivery can be distinguished, ranging from full capacity to complete disruption, which can be seen as distinguishing less and less correct service deliveries. Performance-related measures of dependability for such systems are usually termed performability [Meyer 1978, Smith et al. 1988].

When performing a probabilistic evaluation, the approaches differ significantly according to whether the system is considered as being in stable reliability or in reliability growth, which may be defined as follows [Laprie et al. 1990]:

- stable reliability: the system's ability to deliver correct service is *preserved* (stochastic identity of the successive times to failure);
- reliability growth: the system's ability to deliver correct service is *improved* (stochastic increase of the successive times to failure)[24];

Practical interpretations of stable reliability and of reliability growth are as follows:

- stable reliability: at a given restoration, the system is identical to what it was at the previous restoration; this corresponds to the following situations:
    - in the case of a hardware failure, the failed part is changed for another one, identical and non-failed,
    - in the case of software failure, the system is restarted with an input pattern different from the one having led to failure;
- reliability growth: the fault whose activation has led to failure is diagnosed as a design fault (in software or in hardware) and is removed.

Evaluation of the dependability of systems in stable reliability is usually composed of two main phases:

- *construction* of the model of the system from the elementary stochastic processes which model the behavior of the components of the system and their interactions;
- *processing* the model in order to obtain the expressions and the values of the dependability measures of the system.

Evaluation can be conducted with respect to a) physical faults [Trivedi 1984], b) design faults [Arlat et al. 1988, Littlewood 1979], or c) a combination of both [Laprie 1984, Pignal 1988]. The dependability of a system is highly dependent on its environment, either in the broad sense of the term [Hecht and Fiorentino 1987], or more specifically its load [Castillo and Siewiorek 1981, Iyer et al. 1982]. When evaluating fault-tolerant systems, the coverage of error processing and fault treatment mechanisms has a very significant influence [Arnold 1973, Bouricius et al. 1969]; its evaluation can be performed either through modeling [Dugan and Trivedi 1989] or fault-injection [Arlat et al. 1989].

Many models of reliability growth have been proposed, devoted to hardware [Duane 1964], software, or both [Laprie et al. 1990]. Most of them are devoted to software, and they are aimed at evaluating either the reliability [Miller 1986a, Yamada and Osaki 1985], or the number of (remaining) faults [Goel and Okumoto 1979, Tohma et al. 1989]; as these models

---

[24] *Reliability decrease* (the system's ability to deliver correct service is degraded, and there is thus a stochastic decrease of the successive times to failure) is theoretically, and practically, possible, e.g., upon introduction of new faults during corrective actions, whose probability of activation is greater than for the removed fault(s). In such a case, it has to be hoped that the decrease is limited in time, and that reliability is globally growing over a long observation period of time.

are aimed at predicting the future reliability from the failure data accumulated in the past, particular attention has been devoted to the prediction problem [Littlewood 1988].

4.6. The Attributes of Dependability

The attributes have been defined in §4.1 according to different properties, which may be more or less emphasised depending on the application intended for the computer system under consideration. This may be refined as follows:

- readiness for usage is always required as a property, although to a varying degree depending on the application;
- continuity of service, avoidance of catastrophic consequences on the environment, preservation of confidentiality may or may not be required according to the application.

An additional property, which may be viewed as a prerequisite for the other properties to be fulfilled, is integrity, i.e., the condition of being unimpaired, in the broad sense of the term: a) for either data or programs, and b) with respect to either accidental or intentional faults[25].

The variations in the emphasis to be put on the attributes of dependability have a direct influence on the appropriate balance of the techniques addressed in the previous section to be employed so that the resulting system be dependable. This problem is all the more difficult since some of the attributes are antagonistic (e.g., availability and safety, availability and security), necessitating tradeoffs to be performed. Considering the three main design dimensions of a computer system, i.e., cost, performance and dependability, the problem is still exacerbated by the fact that the dependability dimension is less understood than the cost-performance design space [Siewiorek and Johnson 1982].

The *assessment* of whether a system is truly dependable — justified reliance on the delivered service — or not thus goes beyond the validation techniques as they have been addressed in the previous section for, at least, the three following reasons:

- checking with certainty the coverage of the design or validation assumptions with respect to reality (e.g., relevance to actual faults of the criteria used for determining test inputs, fault hypotheses in the design of fault tolerance mechanisms) would imply a knowledge and a mastering of the technology used, of the intended utilization of the system, etc. which are by far superior to what is generally achievable;
- performing an evaluation of a system according to some attributes of dependability with respect to some classes of faults is currently considered as non feasible or yielding non-significant results: probability-theoretic bases do not exist or are not widely accepted; examples are safety with respect to accidental design faults[26], security with respect to intentional faults;
- the specifications "against" which validation is performed are not generally non-faulty — as any system.

The consequence is an emphasis put on the development and production process when assessing a system: methods and techniques utilized and how they are employed; in some

---

[25] This definition of integrity incorporates the notion of fault secureness as defined in the theory of self-checking circuits [Anderson and Metze 1972; Wakerly 1978].

[26] The following sentence is excerpted from [RTCA 178A]: "During the preparation of this document, techniques for estimating the post-verification probabilities of software errors were examined. The objective was to develop numerical requirements for such probabilities for digital computer-based equipment and systems certification. The conclusion reached, however, was that currently available methods do not yield results in which confidence can be placed to the level required for this purpose."

cases, a *grade* is assigned and delivered to the system according to a) the nature of the latter and to b) an assessment of their utilization[27].

---

[27] For instance:

- systems are ranked from the security viewpoint [DoD 5200.28] from A1 ("verified design") to D ("minimal protection");
- software for civil transportation airplanes is classed as Level 1, 2 or 3 according to the criticality of the function to be accomplished by the software: critical, essential, non-essential.

GLOSSARY

Warning: this glossary is provided as an aid for reading the chapter. *Do not* consider it independently of the chapter.

Accidental fault..................Fault occurring or created fortuitously.

Active fault ....Fault producing an error.

Arbitrary failure.................see Failure

Atomic system.............. System whose internal structure cannot be discerned, or is not of interest and can be ignored.

Attributes of dependability.Attributes enabling the system quality resulting from the impairments and the means opposing to them to be assessed. Reliability, availability, maintainability, safety, security.

Availability ....Readiness for usage. Measure of correct service delivery with respect to the alternation of correct and incorrect service.

Avoidance (fault ~) ....Methods and techniques aimed at producing a fault-free system. Fault prevention and fault removal.

Backward recovery......... Form of error recovery where the erroneous state transformation consists of bringing the system back to previously occupied state.

Behavior (system ~) ....What a system does.

Benign failure..................Failure whose penalties are of the same order of magnitude as the benefit provided by correct service delivery.

Catastrophic failure...........Failure whose consequences are incommensurably greater than the benefit provided by correct service delivery.

Compensation (error ~)......Form of error processing when erroneous state contains enough information to enable correct service delivery.

Component (system ~).......Another system.

Consistent failure..............Failure perceived similarly by all system users.

Corrective maintenance.... Preservation or improvement during its operational life of a system's ability to deliver a service complying with the specification. Fault removal during the operational life of a system.

Coverage.........................Measure of the representativity of the situations to which a system is submitted during its validation compared to the actual situations it will be confronted with during its operational life.

Crash failure ....Persistent omission failure

Criticality (system ~).........Highest severity of failure modes.

Curative maintenance ...... Corrective maintenance aimed at removing faults that have produced errors and which have been reported.

Dependability ....trustworthiness of the delivered service such that reliance can justifiably be placed on this service.

Design diversity............ An approach to the production of systems, involving the provision of identical services from separate designs and implementations.

Design fault.....................Human-made internal fault.

Design for verifiability......Methods and techniques when designing a system that facilitate its verification.

Detection (error ~)............The action of identifying that a system state is erroneous.

Detected error..................Error recognised as such by a detection algorithm or mechanism.

Deterministic testing.........Form of testing where the test patterns are predetermined by a selective choice.

Diagnosis (fault ~)............The action of determining the cause of an error in location and nature.

Dormant fault ....Internal fault not activated by the computation process.

Dynamic verification..........Verification involving exercizing the system.

Environment (system ~).....The other systems interacting or interfering with the given system.

Error .... Part of system state that is liable to lead to failure.
Manifestation of a fault in a system.

External fault...................Fault resulting from environmental interference.

Fail-safe system...........System whose failures can only be, or are to an acceptable extent, benign failures.

Fail-silent system..........System whose failures can only be, or are to an acceptable extent, crash failures.

Fail-stop system...............System whose failures can only be, or are to an acceptable extent, stopping failures.

Fail-uncontrolled system....System whose failures may be arbitrary.

Failure .... Deviation of the delivered service from compliance with the system specification.
Transition from correct service delivery to incorrect service delivery.

Fault.............................Adjudged or hypothesised cause of an error.
Error cause that is intended to be avoided or tolerated.
Consequence for a system of the failure of another system that has interacted or is interacting with the considered system.

Forecasting (fault ~).........Methods and techniques aimed at estimating the present number, the future incidence, and the consequences of faults.

Forward recovery.............Form of error recovery where the erroneous state transformation consists of finding a new state.

Function (system ~) ....What a system is intended for.

Functional testing............Form of testing where the testing inputs are selected according to criteria relating to the system's function.

Hard fault or **solid fault**......Fault requiring passivation.

Human-made fault.............Consequence of human imperfection.

Impairments to dependability   Undesired, but not unexpected, circumstances causing or resulting from un-dependability.
Faults, errors, and failures.

Incorrect service .... Service delivered not in compliance with the system specification.

Inconsistent failure...........Failure such that system users may have different perceptions of it.

Integrity...........................Condition of being unimpaired.

Intentional fault...............Fault occurring or created deliberately.

Intermittent fault..............Temporary internal fault.

...................................Faults whose conditions of activation cannot be reproduced or which occur rarely enough.

Internal fault ....Fault inside a system.

Intrusion........................Intentional operational external fault.

Latent error.....................Error not recognised as such.

Maintainability .... Measure of continuous incorrect service delivery.
Measure of the time to restoration from the last experienced failure.

Malicious logic................Intentional design fault.

Masking (fault ~)..............The result of applying error compensation systematically, even in the absence of error.

Means for dependability.....Methods and techniques enabling a) to provide a system with the ability to deliver a service on which reliance can be placed, and b) to reach confidence in this ability.
Fault prevention, fault tolerance, fault removal, fault forecasting.

Measures of dependability..Attributes enabling the service quality resulting from the impairments and the means opposing to them to be appraised.
Reliability, availability, maintainability, safety.

Omission failure................Failure such that no service is delivered.

Operational fault ....Faults that occur during the system's exploitation.

Passivation (fault ~)..........The actions taken in order that a fault cannot be activated.

Performability ....Performance-related measure of dependability.

Permanent fault.................Fault whose presence is not related to pointwise conditions of the system, either internal or external.

Physical fault ....Fault resulting from adverse physical phenomena.

Preventive maintenance.... Corrective maintenance aimed at removing faults before they are activated.

Prevention (fault ~)..........Methods and techniques aimed at preventing fault occurrence or introduction.

Processing (error ~)..........The actions taken in order to eliminate errors from a system.

Procurement of dependability   Methods and techniques intended to provide a system with the ability to deliver a service complying with the system specification.
Fault prevention and fault tolerance.

Correct service .... Service delivered in compliance with the system specification.

Recovery (error ~)............Form of error processing where an error-free state is substituted for an erroneous state.

Recovery point .... Point in time during the execution of a process for which the then current state may subsequently need to be restored.

Recurrent fault .... Permanent fault whose conditions of activation can be reproduced.

Regression verification......Verification performed after a correction, in order to check that the correction has no undesired consequences.

Reliability................... Dependability with respect to the continuity of service.
Measure of continuous correct service delivery.
Measure of the time to failure.

Reliability growth.............The system's ability to deliver correct service is improved (stochastic increase of the successive times to failure).

Removal (fault ~).............Methods and techniques aimed at reducing the presence (number, seriousness) of faults.

Restoration (service ~).......Transition from incorrect to correct service delivery.

Random testing.................See Statistical testing

Safety .... Dependability with respect to the non occurrence of catastrophic failures.
Measure of continuous delivery of either correct service or incorrect service after benign failure.
Measure of the time to catastrophic failure.

Security.................... Dependability with respect to the preservation of confidentiality and integrity.

Self-checking component...Component comprising error detection mechanisms associated with its functional part.

Service..................... System behavior as perceived by the system user.

Severity (failure ~)...........Grade of the failure consequences upon the system environment.

Soft fault ....Fault for which fault passivation is not undertaken.

Specification (system ~) ....Agreed description of the system's requirements.

State (system ~)................A condition of being with respect to a set of circumstances.

Stable reliability ....The system's ability to deliver correct service is preserved (stochastic identity of the successive times to failure).

Static verification..............Verification conducted without exercising the system.

Statistical testing ....Form of testing where the test patterns are selected according to a defined probability distribution on the input domain.

Structure (system ~)..........What makes a system do what it does.

Structural testing ....Form of testing where the testing inputs are selected according to criteria relating to the system's structure.

Symbolic execution ....Dynamic verification performed with symbolic inputs.

System .... Entity having interacted, interacting, or able to interact with other entities.
Set of components bound together in order to interact.

Temporary fault................Fault that is present for a limited amount of time.

Testing ....Dynamic verification performed with valued inputs.

Timing failure..................Failure such that the timing of service delivery does not comply with the specification.

Tolerance (fault ~) ....Methods and techniques aimed at providing a service complying with the specification in spite of faults.

Transient fault.................Temporary physical external fault.

Treatment (fault ~)...........The actions taken in order to prevent a fault from being re-activated.

Un-dependability ....Property of a computing system such that reliance cannot, or will not, any more be justifiably placed on the service it delivers.

User (system ~) ....Another system (physical, human) interacting with the considered system.

Validation .... Methods and techniques intended to enable confidence to be reached in a system's ability to deliver a service complying with the system specification.
Fault removal and fault forecasting.

Value failure ....Failure such that the value of the delivered service does not comply with the specification.

Verification ....The process of determining whether a system adheres to properties (the verification conditions) which can be a) general, independent of the specification, or b) specific, deduced from the specification.

Chapter 5

Real-Time Concepts

Real-time systems are those which are able to offer some assurance of the timeliness of service provision. This assurance may range from a firm guarantee of service provision within a defined time interval to a high expectation of service provision before a defined time elapses. In order to meet the timing constraints of the more important real-time services, a graceful degradation in the timeliness of less important services may be necessary.

This chapter is concerned with the definition of real-time concepts, rather than with the Delta-4 techniques to support real-time systems. A detailed account of these techniques is given in chapter 9.

5.1. Concepts and Definitions

5.1.1. Definitions

It is common practice to begin a discussion of real-time with a set of definitions of terms, in particular, the term "real-time" itself. As an example, consider: "If a real-time system misses a deadline, it has failed". Definitions like this exclude from analysis many issues that clearly involve the timeliness of computation. Consider, for example, control systems based on periodic sampling; here it is cumulative rather than individual service untimeliness that leads to risk to control stability, so deadlines are not an appropriate basis to distinguish correct from incorrect control. One conventional practice is to notice that a parameter other than time has exceeded a defined bound.

It is not the purpose of this chapter (or of the architecture) to reject all but a convenient sub-domain of these real-time issues; instead, we seek inclusive models. All-inclusive models are unachievable, so we shall instead examine models for their generality, looking for evident opportunities to solve a large class of commonly-encountered problems. Under such an intent, definitions are entities to be examined, rather than representations of truth.

The following definitions are derived from [PDCS 1990], since they are perhaps more inclusive than most. They are offered as a starting point for the evolution of the necessary concepts.

A *real-time service* is a service that is required to be delivered within time intervals dictated by the environment. (Note that to deliver a service may mean to sample an input rather than to produce an output.)

A *real-time system* is a system that delivers at least one real-time service.

A distinction is normally made between hard and soft real-time. Whilst it is universally agreed that this distinction exists, its meaning is subject to subtle differences in interpretation. The following are typical definitions; note that they express service requirements rather than properties of an implementation:

A *hard* real-time system is a real-time system in which at least one of the timing failure modes is costly or damaging to the environment.

A *soft* real-time system is a real-time system for which all possible timing failures are benign, i.e., system service at the wrong time may be useless, but it is not catastrophic.

Similar, but not equivalent, definitions are to be found in [Jensen 1991]:

A *real-time service* is one that produces results having time constraints that are part of the results' correctness, rather than performance, criteria.

A *hard* real-time service is one that has zero or negative utility if it is delivered outside a certain time interval.

A *soft* real-time service has positive but sub-optimal value if delivered outside a certain time interval, and maybe zero or negative value if delivered outside a wider time interval.

These definitions reflect a change in perspective: Jensen is concerned to address what are termed "mission-critical systems", where a condition of overload exists when services are needed most and timeliness assurances are in general impossible. The notion of the value or utility of a service is introduced to develop the concept of dynamically maximising the overall utility of system behaviour. The design goal is to achieve a particular form of behaviour optimization, rather than the behaviour assurance that is implied by the earlier definitions.

XPA is intended to support hard and soft real-time services according to either perspective and within the same distributed system. This leads directly to one important restriction in the applicability of the architecture; such support is incompatible with the constraints necessary to address safety-critical systems. This is discussed further in annexe A.

Whereas definitions such as these are in common use, they must be used carefully as they appear to draw a sharp distinction among three classes of service: *hard-*, *soft-* and *non-real-time*. The boundary between classes is indistinct; in practice there are many compounding factors. To accept the definitions too literally results in polarization and inflexible architectures.

This point is well illustrated through use of a graphical representation due to [Jensen et al. 1985], which is able to capture quite subtle characteristics of services and service interactions. In particular, it reveals that definitions such as those given above represent only special cases amongst many others that it might prove equally useful to recognise. Figures 1 to 4 give examples based on Jensen's graphical representation of a utility function of service-delivery against time, i.e., the benefit or cost of providing a particular service at a particular instant in time[1].

In figure 1, the cost (or "negative utility") of delivering the service outside a window defined by the "liveline" and "deadline" is unacceptably large, so this example corresponds to both definitions of hard real-time. In figure 2, the cost just outside a similar window is small, so this corresponds to the first definition of soft real-time. Note, however, that if service is seriously delayed, the cost eventually becomes as unacceptably large as in figure 1. However, there is a difficulty in defining a window that allows this service to be classified as "hard" according to the first definition above, since service provision just *inside* any such window would be almost as costly.

If the above definitions were to be cast in terms of the rate of change in utility at significant points in time, the two cases illustrated would be distinguished, but other cases are then rendered anomalous. Is the service requirement of figure 3 harder (because greater rate-of-change of utility occurs) or softer (because no negative utility occurs) than that of figure 2?

---

[1]   Note that Jensen proposes that such utility functions can both define a service requirement and be internally represented and used by a dynamic scheduling mechanism. Such scheduling has not been implemented on Delta-4, but it should be noted that a local execution environment using such a scheme could be included under the architectural mechanisms described in chapter 9.

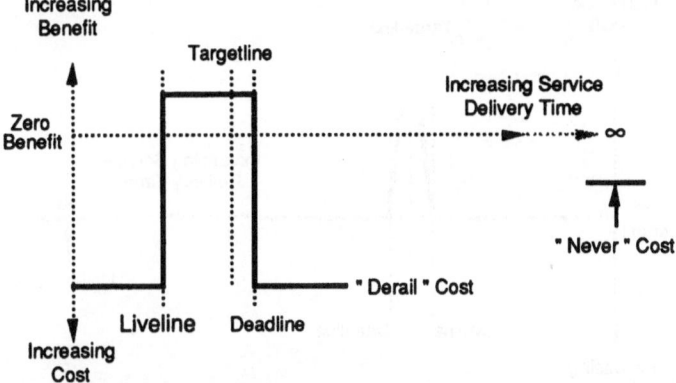

Fig. 1 - Benefit of a Hard Real-Time Service against Delivery Time

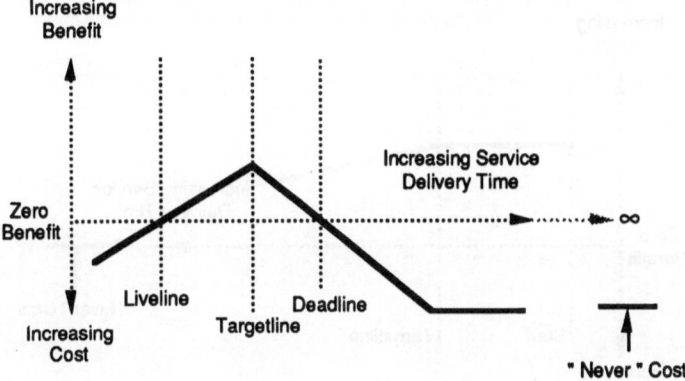

Fig. 2 - Benefit of a Soft Real-Time Service against Delivery Time

These issues are further discussed in [Burns 1990b], through use of the same graphical representation.

Many commercial Local Execution Environments (LEXes) claiming to offer real-time facilities provide a single mechanism (e.g., a priority- or deadline-based scheduler), and allow its shortcomings to be addressed through making various "escape" facilities available to the programmer (e.g., priority or deadline may be manipulated explicitly from the application code). In this chapter, we will assume the use of such off-the-shelf "real-time" nodes, and discuss the distributed aspects of linking these together under the Delta-4 model of dependability.

We therefore define *precedence* to be a generic representation of the necessary timeliness characteristics of the service as seen by the system designer. A system component is presumed to exist on each node to transform and interpret the precedence of each service, from the form chosen to represent the system requirements into the form required by the local execution environment. The distributed environment does not prescribe the scheduling philosophy that will be used.

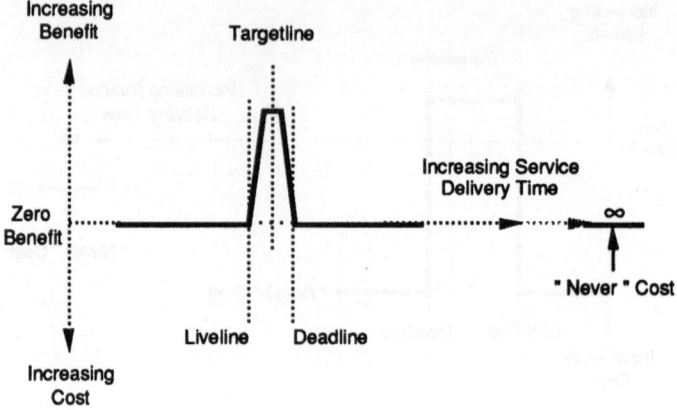

Fig. 3 - Benefit of Anti-Aircraft Gun Service against Delivery Time

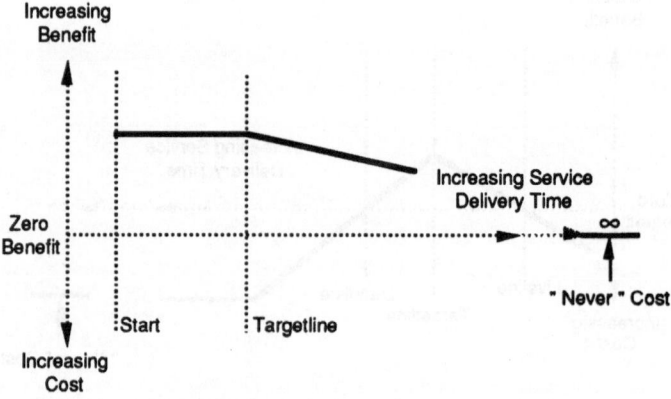

Fig. 4 - Benefit of a Non-Real-Time Service against Delivery Time

Since it is difficult to examine instances of real-time issues exclusively in terms of such a general abstraction as precedence without some prescription for its implementation, we will map this, where example is needed, onto a set of common concepts as defined below. The reader is reminded that alternative mappings are possible, through re-implementation of the precedence interpreter.

Notice that the graphs have been extended to show the cost of *never* delivering the service. In figure 1, which shows the benefit or cost of switching railway points at certain times, the cost of never switching the points is actually less than the cost of doing so when it will derail a train. Some of the graphs are further discussed in section 5.1.3.

The *priority* of a real-time service is a measure of the cost of missing its timing constraints. Hard real-time services are therefore given the highest priority values and non-real-time services the lowest priority.

The priority of a component that supports provision of a real-time service is naturally derived from that of the service. If the timing of component C can cause a timing failure of service S, and C can cause no more costly timing failures, C's priority is exactly the priority of

S. This is called "priority inheritance" (see [Sha et al. 1987]). The "ceiling protocol" [ibid] is a refinement of simple priority inheritance that bounds the number of times a high priority process is blocked by other components.

A *deadline* is the end of a time interval within which a real-time service is required to be delivered.

So if the timing of component *C* can cause service *S* to miss its deadline, *C* should be assigned a deadline earlier than or equal to the deadline of *S*. Alternative strategies for this "deadline inheritance" are discussed in chapter 9.

A *liveline* is the start of a time interval within which a real-time service is required to be delivered.

The *targetline* is the time at which the system designer aims to deliver the service. The targetline is normally chosen as the time of maximum benefit, if known. For a real-time component, it is chosen between the liveline and deadline. The system designer might equate the targetline of a periodic process to the start of the next period, or to an earlier time if required. The targetline of a non-real-time process might be equated to the time at which the user expects to collect the output.

The *precedence* of a real-time service will therefore be assumed in this chapter to be a combination of priority and targetline.

At or before a deadline, a software component may receive an *event* to notify it that it should take some special action, e.g., interact with the environment at once, before it misses the deadline, or abandon itself, because it has missed the deadline. In figure 3, for example, the event might instruct the gun controller to fire, and the event time might coincide with the targetline, if that was chosen as the optimum time to fire.

5.1.2. Systems and their Environments

A control system is acted upon by, and in turn acts upon, its enclosing environment. Events in the enclosing environment that are presented to the system are termed *input events*. Note that, by design, a system might not always change its internal state as a result of these. However, if it does so, these are then called *observed events*. Events initiated by the system on the enclosing environment are termed *output events*. Note that, by design, the environment might not always be affected by such events. However, if it does so, these are termed *control events*.

An *operational envelope* is a part of the universe of all possible behaviours of the enclosing environment, specified in such a way as to permit design-time assurances that the system will behave in a timely manner within the envelope.

Although a requirements specification is treated as if it captures facts about the environment with which a system interacts, this is not quite what it does. It captures assumptions, which in the judgement of the specifier resolve two conflicting issues:

- The assumptions must be useful. They establish a basis for or assist in the design process. They must, for example, establish sharp, finite bounds that permit a design to be constructed and design assurances to be made.

- The assumptions must be realistic. They do not misrepresent reality to the point where a system designed to those assumptions cannot give useful service when presented with reality.

For many requirements in many disciplines, the real world is convenient and permits satisfactory resolution of this conflict.

For some, however, the real world is inconvenient in that, although its parameters are finite, they are generally not sharply bounded, instead exhibiting probability distributions with extended tails of uncertain extent. On the surface of the Earth, what are the limits on temperature, humidity, wind-speed, earthquake-magnitude, lightning-strikes? How will an enemy behave in sending attack aircraft? How predictable are future alarm conditions?

We may not like it, but even where the extent is known, perhaps by appeal to physics, such limits may not be useful as design bounds, being orders-of-magnitude beyond what is practical or economic in the market concerned. A lesser, artificial limit must then be set.

Thus, bridge and skyscraper builders set boundaries for wind-speed and the magnitude of earthquakes. Electronic systems designers bound such parameters as temperature, humidity, supply voltage, duty cycle and so on. Command-and-control systems are based on assumed minima for the inter-arrival times of enemy aircraft. Electricity supply and other process control systems are based on bounded inter-arrival times for input events.

On what basis are artificial bounds plausible? A room containing electronic equipment may be air-conditioned, or the climate may be considered by the customer to be sufficiently well-behaved that he can accept responsibility for the consequences of abnormal weather. Probabilistic evidence may also be used in other cases where external control cannot be applied, such as the command-and-control example.

Generally, a commercial contract between supplier and customer is involved. This raises several issues that directly influence the choice of technology. Who takes the risk in specifying or making use of an artificial bound? What is the nature of the risk: is safety involved, or lost production, or merely inconvenience? What is the cost of not taking the risk? In a competitive world, an over-cautious supplier can lose an order completely with a marginally more expensive solution. What is the cost of taking the risk? A supplier, or customer, or insurance company, might carry the cost of lost production or worse.

The simplest systems-environment relationship is such that first, all output events are control events, and second, all input events result from previous control events; the design can therefore arrange to ensure that all input events are observed and that all necessary computation is completed in a timely manner.

Most systems-environment relationships are not so simple. In many, the environment can present the system with events that do not result from previous control events but from *autonomous behaviour*. The system-environment relationship is still simple if such input events are sufficiently predictable, e.g., with a minimum duration between consecutive events, such as might occur with manufactured parts delivered by a conveyor belt. The design can again ensure that these are observed and that all computation is timely. In this case, the operational envelope specifies the entire universe of environment behaviour.

If such is not the case, as with *aperiodic events* (i.e., events with no known minimum inter-arrival time) such as the attack of enemy aircraft, operator actions, the occurrence of alarm conditions in all their diverse forms, changes of operating mode, the failure of nodes, etc., then *there is a possible behaviour of the environment for which timely system behaviour cannot be assured by the design process*. This chapter will discuss this issue at some length, since it is the experience of the industrial partners of Delta-4 that it cannot be ignored in the target markets for the architecture.

Since aperiodic events are incompatible with design-time assurances, a designer may try to transform the system-environment relationship as follows. Instead of receiving aperiodic events, the system periodically samples environmental state information, and is sized so that it can always process each periodic sample before the next sample is received. Between samples, some aperiodic events will not be observed, but in some applications this is acceptable, e.g., because it only entails the loss of intermediate values of a continuous variable. Even when it is possible, however, the transformation does not always lead to practical and economic limits on the system resources required.

In the context of Delta-4, the failure of system hardware components is viewed as autonomous environment behaviour, for which specialized sensors and event management mechanisms are constructed within the Delta-4 support environment itself. Such events are only statistically predictable; it is possible (but of infinitesimal likelihood) for a cascade of failures to fall outside any defined operational envelope.

Within the operational envelope, a system is required to meet timeliness criteria needed for correct system operation. This assurance must be provided during the system design process. The designer must demonstrate that at run time, providing the enclosing environment complies with the specified operational envelope, timeliness requirements will be met.

If the resulting system subsequently encounters circumstances and events that fall outside this envelope, then no such absolute guarantee can be given. One may, however, still be able to prove that timeliness criteria will be met with very high probability, or that the timeliness of the higher priority system functions can be guaranteed. Systems offering such qualified assurances can still be useful in the target market areas.

5.1.3. Some Examples

To assist the discussion, we now consider a few examples of real-time problems. A hard real-time problem is described first.

5.1.3.1. Example 1. Imagine a process control plant that uses motorized rail-trolleys to transport materials from one part of the plant to another. One of these trolleys approaches a set of points; the time when the trolley will reach the points can be calculated. If the points are currently in the wrong state, they must change state before that time; this is a hard deadline in the real world (see figure 5).

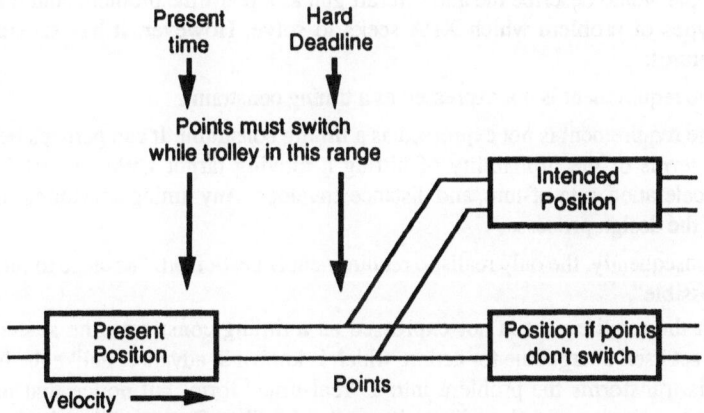

Fig. 5 - Rail-Trolley Example

The example is of course simplified. A wealth of practical details complicate the issue. For instance, the deadline might be estimated from position and velocity; velocity might be determined from the times at which front and rear axles pass fixed track-side position sensors; there might be a velocity-changing phenomenon such as friction that can only be estimated. The time when the points will be reached cannot be predicted exactly so that an earliest bound must be established. Also, the points take some time to move: this too must be bounded.

These bounds perhaps establish the actual deadline, the latest time by which the points must have started to move. If the points can jam, this must be detectable and it might be necessary to determine another deadline, the latest time at which the trolley brakes may be applied. Events could be sent to the controlling capsule at either or both these deadlines, instructing it to take the appropriate action.

None of these elaborations alter the requirement for the controlling component to complete some action by some point in time. If it fails to do so, the trolley can be derailed, which could have catastrophic costs; therefore by definition this is indeed hard real-time. As such, design time assurances of timeliness will normally be required.

This raises the question discussed in section 5.1.2. Can the rail-trolley system encounter environmental circumstances and events outside any definable operational envelope? If so, we cannot provide absolute design time assurances of timeliness, but should still seek to provide qualified assurances that timeliness will be maintained with a high quantified probability. If deadlines are missed, the design must include measures to limit the damage.

Thus, the rail-trolley example has a short-term and a long-term consequence to missing the deadline (see figure 1). In the short term, there is a deadline after which the component must no longer change the state of the points, or the trolley will derail, at considerable cost. In the long term, if the points are never switched, the trolley may be on the wrong track, which has lesser costs, but requires longer term management to get the trolley onto the right track by a longer route. In other words, *although the component controlling the points has a hard deadline, the containing system does not.*

5.1.3.2. Example 2. As a second example, consider the controller of an anti-aircraft gun that tracks the predicted trajectory of an enemy aircraft, i.e., points to where it is estimated the aircraft will be when the shell reaches it. The anti-aircraft gun is required to hit as many enemy aircraft as possible. Figure 3 shows the benefit or cost of firing the gun at different times.

Most people would describe the anti-aircraft gun as a real-time problem, and it is certainly one of the types of problem which XPA seeks to solve. However, it has several possibly surprising features:

- The requirement is not expressed as a timing constraint.
- The requirement is not expressed as a timing constraint. It can perhaps be expressed in terms of the probability of hitting a moving target within a certain velocity, acceleration, rate-of-turn, and distance envelope. Any timing constraints arise as part of the design process.
- Consequently, the only realistic requirement is best-effort: "as close to the aircraft as possible".

Although the requirement is not expressed as a timing constraint, the system can still calculate and set itself a targetline for action, which is known in advance, unlike the liveline and deadline. This transforms the problem into a "real-time" form, but notice that missing the targetline is not a criterion of failure, like missing the deadline. One possible criterion of failure would be a loss of service due to counterattack from the aircraft!

Having moved the gun to the required position, the gun controller might suspend itself waiting for the order to fire, which might take the form of an event at the targetline.

5.1.3.3. Example 3. Consider a 3-term digital control loop based on a periodic process calculating and responding to a so-called "error_term" so as to control an environmental variable within acceptable bounds. The requirement is to remain within a predefined range of the (changing) ideal value at each instant, otherwise there will be a costly consequence (in exactly the sense of section 5.1.1). The process implements a control theory prescription to minimize the absolute value of the "error_term", and can check that this remains less than some predefined amount (i.e., ABS(error_term) < δ). It is therefore able to discover any violation of this condition, and signal this as a failure. It is likely to fail if it suffers long-term delayed cycle activation or execution time instability.

Again the requirement is not expressed in terms of time and the failure is not a failure to meet a specified deadline, although conceivably it might have been predicted that another delayed activation would cause failure. In this case, however, the requirement is "hard" rather than "best-effort". Again we can transform the non-time requirement into a real-time form, e.g.:

"Each cycle activation time should be within an interval on either side of a targetline, and it will then follow that ABS(error_term) < δ."

This transformation has in fact strengthened the original requirement, but if such a targetline and interval can be defined, we can schedule the component in the same way as other periodic real-time tasks, e.g., by sending it an event at the targetline to activate another cycle. So we might take the design decision to adopt the transformed requirement as a basis for scheduling; but notice that the only criterion of failure is still the original requirement: "ABS(error_term) < δ".

If we try to express the transformed requirement as a graph, we get something like figure 2, where only a little cost is assigned to missing the deadline, because one missed deadline will normally not lead to failure. Notice we are here reaching the limitations of these graphs, which have no way of expressing the real failure criterion.

Such a 3-term controller is certainly required to operate in "real-time", even though its failure criterion does not mention time. It is of considerable interest to XPA, and its implementation requires components such as clocks, which are unarguably "real-time".

5.1.3.4. Example 4. Consider the updating of an operator's display; this might meet some desired delay requirement of the form: "the screen must be updated on average within 0.5 seconds over any 30 second period, except when the system is in emergency mode". The only cost of exceptional delay may be temporary inconvenience to the operator.

This is a typical soft-real-time timing constraint. Notice that the failure criterion is statistical: there are no precise deadlines. The designer might decide to schedule each update operation by assigning it a targetline 0.5 seconds after the start, but this does not mean it has failed if it occasionally takes 0.6 seconds.

5.2. Delta-4 Approach to Real-Time

5.2.1. Approaches to System Design

If a system is presented with circumstances outside the operational envelope, then it is not possible to assure that deadlines will be met. Two views can be taken on this, which roughly speaking identify two schools of thought on design, called here the "bounded-demand" school and the "unbounded-demand" school.

5.2.1.1. The Bounded-Demand School. This school considers that to take the possibility of failing to meet deadlines into account during design is to propose a method of tolerating an inadequate system size. A system in which some deadlines might not be met is not, in this view, a real-time system. Arrangements made to tolerate missed deadline faults are uninteresting since the design process is concerned to assure that this cannot happen.

In this view, an operational envelope is determined which bounds environmental behaviour, and then the system is sized so that deadlines will be met within this envelope. For all relevant input events, bounds must be determined which are commensurate with the control system technology. They must be *periodic* (occurring at regular intervals), or *sporadic* (occurring at irregular intervals, but in such a way as to impose a bounded system load in any finite duration [Burns 1990a]). The bounds must be absolutely assured, perhaps by the laws of physics, since events beyond the operational envelope will either be ignored or result in "undefined behaviour".

Bounds on internal behaviour must also be established. Unfortunately, combinatorial explosions (e.g., of state space) can, even in quite modest sized systems, exceed the practical limits of existing formal methods and design tools. Some techniques for reducing the state explosion are discussed in section 15.2, but despite these techniques there is a limit on the size of bounded-demand systems established by the state of the art.

5.2.1.2. The Unbounded-Demand School.

This school owes its existence to the fact that requirements exist where an "adequate" system size, in the above sense, is simply not able to be established in design. Although finite, the bounds are not discoverable, are too extreme to be reached or supported in practice, or for some other reason have not been specified. "Unbounded" is therefore used here as shorthand for "without applicable bounds" rather than for "without intrinsic bounds". Input events may be *aperiodic* (with no specified minimum inter-arrival time, so that no upper bound can be specified for the system load imposed by the environment). The consequences of output events may be too complex for complete analysis. System size calculations then depend on a large number of assumptions. Even where they are fully understood, these assumptions are often inseparably cross-dependent and resist formal analysis.

In practice, these assumptions are probable rather than certain, and if any of them prove invalid, events may move outside the operational envelope in which timeliness can be guaranteed. If so, the consequence must not be "undefined behaviour", which may in fact be the loss of the system's most important function. Some form of "graceful degradation", i.e., some effort to control and define behaviour outside the operational envelope, is preferable and is seen by this school as responsible defensive engineering.

A similar situation arises when an "adequate" system size *can* be calculated, but is found to imply a noncompetitive system price. A commercial decision may then be taken to use a smaller system size, assuming that the probability of this size proving inadequate is negligible in the lifetime of the system, or its mission time. This assumption may be justified by statistical arguments (see §5.2.2), but there is still a need for defensive engineering in case it is violated.

Even the execution time of a single component can be unbounded. For example, an algorithm for inverting a matrix may converge in a known duration for nearly all possible input values, but may exceed this limit, or never converge at all, for a few input values. Whatever the system size, we must therefore design for the possibility that the matrix inverter will miss its deadline.

5.2.1.3. Argument between the Schools.

A major criticism of the "unbounded-demand" concept from the perspective provided by the "bounded-demand" school of thought can be summarised as follows:

- To design for behaviour outside an operational envelope, a wider operational envelope (or several envelopes) must be defined, since one cannot design anything to work reliably in an unbounded environment. So an "unbounded-demand" system is in fact a "bounded-demand" system with several operational envelopes.

This is certainly true, and it follows that neither type of system is predictable outside the widest imaginable operational envelope, i.e., if all the hardware fails. However, as the environment grows more hostile, the two design approaches lead to different system behaviours:

- The "bounded-demand" designer may argue that, since an operational envelope is expected to bound the environment in all but the most drastic circumstances, all one can do outside this envelope is abandon all real-time services and perform a safe shutdown, provided one is still within the wider envelope assumed by the shutdown routine. The result is a clearly-defined but rather brittle degradation.

- The "unbounded-demand" designer expects the "inner" operational envelopes to be violated more often; so he abandons or postpones system services one by one, starting with the least important or urgent. The result is a graceful degradation that provides a service probabilistically even when it cannot be guaranteed. This is more compliant with a specification stating that the service is desirable but not essential.

The first approach certainly applies to hard real-time services; if they cannot be guaranteed, the whole system has failed, and should stop, preferably in a safe and consistent state. The second approach is more appropriate to soft real-time (or non-real-time) services that can be delivered belatedly without damaging the environment.

This suggests that both design approaches could be used in a system that provided both hard real-time and non-hard real-time services. This appears to be possible in the special case where the following conditions are met:

- The demand for hard real-time services is bounded, so that the system can be sized to assure them.

- The demand for the non-hard services is unboundable.

- The hard real-time subsystem is always able to preempt the resources it requires in a bounded time, so that it can be designed as if the whole system is available to it.

- In particular, the hard real-time subsystem must be able to preempt any input or output channel. It must be possible to ignore or lock out the (possibly unbounded) inputs to non-hard components without losing inputs to hard real-time components.

In such a system, the environment stays within a wide operational envelope in which the hard real-time services can be guaranteed. However, it sometimes violates a narrower operational envelope, outside which the non-hard services have to be progressively postponed or abandoned.

Since Delta-4 is concerned to support the whole range of real-time systems, with and without bounded environmental behaviour, it must make use of both bounded-demand and unbounded-demand approaches, as appropriate.

5.2.2. System Sizing

To determine the required system size, a bounded-demand designer must bound all execution and communication times for all combinations of components in all circumstances. For example, maximum task execution times can be calculated by source code analysis [Puscher and Koza 1989], provided the application obeys certain restrictions such as bounded loops.

An unbounded-demand designer has the less stringent requirement of determining bounds that will only be exceeded with a known probability, so that the risk of graceful degradation can be quantified. It should therefore be possible to size the system by arguing from observed probability distributions.

Distributions based not on theory but on observation are modelled by mathematics that is more able to reflect regions of high probability than low-probability tails. Such models typically display infinite tails even though these may be known to be unrealistic, and still be useful for many purposes. However, establishing a reasonable artificial bound requires low probability behaviour to be quantified; what is being said is something like the following:

- "It is believed that the probability of an event exceeding this bound is negligible in the lifetime of the system, (or mission, or critical period,....)."

A good mathematical model of high probability behaviour may exhibit order-of-magnitude error in quantifying low probability behaviour. Setting an artificial bound must be done on some additional basis. One traditional way is to allow an adequate "safety-margin"[2]; this leaves unanswered the question of how to establish what is adequate.

Instead, a contract may specify a bound, a maximum probability for exceeding the bound, and a behavioural requirement if the bound is exceeded. The behavioural requirement is typically that the contractual requirements for service timeliness no longer apply but that some form of graceful degradation must be exhibited by the technology concerned. In a competitive world, the form this takes is of some importance.

5.2.3. Types of Algorithm

Given a choice between algorithms exhibiting good average but poor worst-case execution times and those exhibiting medium average and medium worst-case execution times, which should be chosen? Note that such a choice is quite common, if we consider algorithms pertaining to all levels in the systems concerned. Memory access can be augmented normally, but slowed down on "misses", by cacheing algorithms. Preemption benefits the preemptor but disadvantages the preempted. Speeding up the sorting of disordered data can slow down the sorting of data in an unfortunate order. The benefits of recognising special cases incur costs to other cases. See, for many examples, [Knuth 1973].

This question is easy to answer in an exclusively hard real-time context. The choice of "medium average and medium worst case" execution times minimizes the system power required to provide a-priori assurance of timeliness, all else being equal. However, in a context in which hard real-time activity is a subset of the total activity, the issue is much less clear. As an example, consider two systems, C being based on "good average but poor worst-case" algorithms and D being based on "medium average and medium worst-case" algorithms. Clearly the timeliness of the hard real-time activities can be a-priori assured in D with less computing power than in C.

However, we are also interested in the expected timeliness of other, desired but soft real-time activities that are allowed execution resources when the hard real-time activities are satisfied. A posteriori, timely behaviour will normally be exhibited for a larger set of soft real-time activities in C than in D, assuming C and D have equal power. Given a sufficient proportion of soft real-time activity, the conclusion is therefore reversed. C is the preferred choice.

Moreover, the uncertainty concerning the proportion of the activity population that will exhibit timeliness depends on the size of this population and the independence of its members. A large, busy system is less variable than the execution-time variance of individual activities would suggest. The subset of activities exhibiting timeliness appears to vary less (in timing or set-membership) than its members, by a significant and useful factor. This is the same sort of mass-statistical effect as permits the hypotheses of large-scale physics or chemistry to be accepted as laws, despite the uncertainty that exists on the microscopic scale (the reasons for these observed effects are discussed in annexe C). These statistical effects are particularly significant in large and complex systems.

---

2   Even with a large safety margin, a bound can still be artificial in that it is less than that which theory can deliver.

The preference for good average but poor worst-case algorithms is a fortunate conclusion, since in practice we may choose such algorithms for other reasons. We may be obliged, because of their power, cost and commercial acceptability, to build systems with microprocessors that use pipelining, cacheing and busses with contention algorithms. Similar comments apply to operating systems and the more advanced language compilers and applications.

5.2.4. Types of Scheduler

We now examine the nature of run-time support for the assumptions taken above; for example, how to grant activities execution resource preference.

A *scheduler* is a resource allocator that affects the timing of a real-time component. The resource may be processing power, network bandwidth, memory, access to a monitor, etc.

An *off-line scheduler* is executed during the system design phase, when it works out fixed schedules for all anticipated component execution sets. The result is a timetable; "slots" are reserved for every process execution and message transmission, and collisions cannot occur. When an event occurs which requires a change of behaviour, this is discovered in a prearranged "slot" and an alternative predetermined schedule is installed. The fixed schedules take account of dependencies between components, which are assumed to be predictable at design time, so that they do not cause delay.

An *on-line scheduler* is executed at run-time, using heuristics that must not themselves impose a significant system load. Dependencies and deadlocks must either be avoided by component design or the resultant delays must be minimized by priority inheritance protocols [Sha et al. 1987] and distributed deadlock detection techniques (see chapter 6 in [Moss 1981]).

Note that both types of scheduler can be combined on one system. An off-line scheduler can calculate a fixed schedule at design time, and encode this schedule by assigning suitable precedence parameters to each component. An on-line scheduler can then reproduce the fixed schedule at run-time by interpreting the precedence parameters. In the free slots of the fixed schedule, the on-line scheduler can run additional components not anticipated by the off-line scheduler.

The requirement that hard real-time components have provable timeliness, despite the presence of lower precedence components, leads to a requirement for *preemption*. This means that when a higher precedence component becomes executable because of an external stimulus or internal event, and a lower precedence component is using the resources it requires, and an unbounded number of lower precedence components are waiting for the resources, the scheduler nevertheless gives the resources to the higher precedence component after a bounded delay.

An on-line scheduler preempts the lower precedence component by waiting for the next interruptible state in its execution — perhaps as soon as the end of the current machine instruction. An off-line scheduler may take longer to preempt, but still schedules the higher precedence component in a bounded time, typically at a pre-arranged point in a fixed cyclic schedule.

There is a large body of research into on-line schedulers that are guaranteed to find a feasible schedule dynamically, if one exists (e.g., [Halang 1986, Sha et al. 1988]). This research assumes that feasible schedules can separately be proved to exist, at least within some operational envelope.

Others, such as [Kopetz 1986], working with systems of the type characterized above as "bounded-demand", use off-line schedulers to generate static cyclic schedules for both computation and communication that assure the required timeliness of behaviour within each operational envelope. The construction of a static schedule can itself be regarded as a proof of the existence of a feasible schedule, again provided the real world does indeed stay within the operational envelope.

These approaches lead to different types of run-time delay: those of an on-line scheduler include schedule calculation times and preemption delays when components collide; those of an off-line scheduler include delaying the start times of components until their pre-arranged slots begin, when in practice they could often have started sooner. Off-line schedulers normally reserve one slot in the cycle for switching from one schedule to another; so the latency before such a switch may be a whole cycle.

However, if we neglect the difference between the run-time delays of on-line and off-line schedulers, then within an operational envelope, the schedules they produce will both exhibit acceptable (but not necessarily identical) behaviour.

A "bounded-demand" system may have several operational envelopes, sporadic external stimuli, variable execution times and different operating and failure modes. So an off-line scheduler may face a combinatorial explosion of scenarios leading to a similar number of different static schedules, or to one "worst-case" schedule that is vastly more pessimistic than the normal case. With the on-line approach, the scenarios can often be collapsed into a single set of dynamic control rules, or at worst one set of rules for each user-defined operational mode. (For example, a military command and control system may have "combat" and "training" modes in which the same component is assigned different priorities.)

When environmental demand is unbounded, outside the operational envelope, a system based on static schedules continues to sample inputs periodically, so that it may (nonselectively) ignore events. If execution times are unbounded, it may also miss deadlines. If the static slot and cycle times are exceeded, and there is inadequate defensive programming, the system may even crash. A dynamic system is less vulnerable outside the operational envelope; provided its on-line scheduling algorithm is carefully chosen, then the most valuable functions are the last to be placed at risk. The choice of on-line scheduling algorithms is discussed in chapter 9.

Again, a set of dynamic control rules apply in each of a relatively modest number of user-defined operational modes. These rules amount to a set of ordered lists of latent activities, a subset of which are likely to be demanded in each mode. The ordering represents preference between activities that cannot all be carried out, but for which it is desirable to carry out as many as possible. Since the normal object is to minimize the cost of timing faults, this will be a priority order in the sense of section 5.1.1.

5.3. Real-Time Communications Protocols

Given the characteristics of real-time systems, just described, communications protocols for these systems should ensure timeliness properties, apart from other useful engineering features. We shall focus on timeliness properties of real-time broadcast protocols, which are associated with their synchronism, as defined below. The emerging requirements, together with engineering features concerning real-time, will be detailed in the appropriate chapters (chapters 9 and 10).

Building blocks, to manage real-time replicated objects, are *reliable broadcast or multicast* protocols, *group management* protocols, and *measures of the passage of time*. Broadcast protocols reliably disseminate information; group management protocols help to determine to whom or by whom the information is disseminated, how participants cooperate, and how groups of replicas are managed; measures of time (like clocks and timers) assist in establishing timeliness properties. A complete solution has normally a flavour of each of these three building blocks.

There are essentially two classes of approaches to build reliable broadcast and group membership services: the *clock-driven* [Cristian et al. 1985] and the *clock-less* [Birman and

Joseph 1987] approach. The former rely on the existence of a global timebase, which constitutes an absolute time reference, whereas the latter do not, although they may rely on timers, i.e., relative time references. With regard to synchronism, clock-driven and clock-less protocols have often been classified as equivalent to "synchronous" and "asynchronous" protocols, respectively. That analogy is mostly due to the "synchronized" nature of clock-driven protocols, which progress at pre-defined times. The clock as an implementation tool is, however, not mandatory to achieve synchronism, as we discuss below.

The importance of this argument stems from the fact that while clock-less protocols are a useful alternative to clock-driven protocols in a number of situations, their suitability for real-time fault-tolerant applications has been questioned, on the grounds of their "asynchronism". The two attributes relevant to this question are: achieving *execution times with a known bound*, i.e., being able to fulfil not only liveness, but also timeliness properties, and respecting *temporal orderings of events*. It is established in [Veríssimo 1990] that clock-less protocols do indeed have these attributes, and are therefore capable of supporting real-time systems. Having clarified what "clock-less" means, let us now introduce a metric for "synchronism" that applies equally to clock-driven and clock-less protocols. This way, we extend the criteria of suitability for real-time and fault-tolerant applications, to clock-less protocols.

5.3.1. Synchronism

Let us start by making some definitions. Consider a reliable broadcast execution, invoked through a *send* primitive, whereby message m_i is delivered to recipients through a *receive* primitive. Let us call *Execution Time* (T_e), the time between $send(m_i)$ request and the last $receive(m_i)$ indication. Similarly, *Inconsistency Time* (T_i), will be the maximum time between $receive(m_i)$ indications at any two different recipients.

If these variables have a variation, two broadcasts that start at slightly different times, may deliver their messages in diverse orders, according to the relative magnitude of those variations. These variations of T_e and T_i from execution to execution are paramount to characterize protocol timeliness, and in consequence, we define the following two attributes:

• **Tightness** (τ) — *Tightness of a protocol, is the measure of* $\Delta T_i = T_{imx}\text{-}T_{imn}$.

• **Steadiness** (σ) — *Steadiness of a protocol, is the measure of* $\Delta T_e = T_{emx}\text{-}T_{emn}$.

The *synchronism* of a reliable broadcast protocol is a measure of its tightness and steadiness. An asynchronous protocol is neither tight nor steady. A protocol is synchronous when its steadiness is known and bounded[3]. Synchronism can have several grades, depending on how tight and steady a protocol is.

Clock-less and clock-driven methods normally yield different grades of protocol synchronism, but they can both be measured as defined above. This establishes that both of them are capable of real-time operation, if only for requirements up to their possibilities.

Any protocol implementation that can be translated to a sequence of events forming a chain in time (coarse though it may be), can yield a synchronous protocol, if the maximum length of that chain is known and bounded. This provides the foundation for building real-time clock-less protocols, and the approach taken in the Delta-4 LAN-based protocol, AMp:

> MAC entities[4] cooperate with each other to fulfil timeliness of LAN operation. In some networks, it is possible to build a timing equation representing a frame transmission delay, show that it is bounded, and determine the upper bound, for

---

3   Since $T_i < T_e$, this is a sufficient condition for tightness to be also bounded.

4   Medium Access Control layer entities, which are responsible for exchanging frames and keeping the network in operation.

assumed load and fault patterns. Similarly, it is possible to impose a performance specification on the hardware and software of the communication adapters (CPU, kernel, etc.) so that processing times of protocol actions are also bounded, and known for the worst case traffic pattern specified. We will show that the protocol architecture can be designed so that these requirements are respected (chapter 10). In conclusion, if a protocol displays an execution time with a known bound, it is synchronous.

5.3.2. Temporal Order

The second attribute of real-time capability is respecting the temporal order of events. This is an extension of the logical causal ordering requirements discussed for decentralized operation of distributed algorithms [Lamport 1978], to real-time systems, mainly those with input/output — and possibly replicated — actions, which must be correctly ordered in time. Note however that temporal order is not always required; some of the required ordering properties can instead be ensured by appropriate computational and replication models. An example of this is discussed in chapter 9.

One way to enforce it, is by using real time clocks, approximately synchronized with a precision ε. Running a clock-driven protocol, all participants timestamp their messages, so that they are delivered by timestamp order. The fault-tolerant synchronization method proposed in [Lamport 1984] achieves such an order, which is also guaranteed in the atomic broadcast of [Cristian et al. 1985]. A number of methods exist to maintain local clocks approximately synchronized [Cristian 1989, Schneider 1986, Srikanth and Toueg 1987]. Since these protocols deliver messages by order of their timestamps, this also implies that messages are delivered in the same order everywhere, i.e., a consistent order, one of the requirements to ensure replica group determinism (see chapter 6).

Again, it can be shown that temporal order is intimately related to the grade of synchronism. When τ and σ are not negligible, compared with T_e, we may say a protocol is *loosely-synchronous*, to distinguish from the tightly-synchronous ones, an example of which are clock-driven Δ-protocols such as those of [Cristian et al. 1985]. As a rule, synchronous clock-less protocols are not able to achieve the fine degree of steadiness and tightness, that clock-driven ones are, and this is also true of AMp. However, their ability to fulfil temporal order properties is definable, and in consequence, so are the classes of applications for which a given protocol provides a correct implementation, vis-a-vis ordering of events.

It is worthwhile noting that these concepts apply to general distributed processes. In fact, synchronism of distributed — and in particular replicated — computations, i.e., their steadiness and tightness, is an important issue in real-time systems. We recall that they are related to familiar concepts like lock-step, action rate, simultaneity of replicated I/O, etc.

The approach used in Delta-4, for real-time distribution, relies on *synchronous clock-less* protocols, on local area networks. Although clock-driven protocols can be built on top of the basic communication mechanism (AMp, see chapter 10), clock-less protocols achieve the best possible average execution time. So this approach optimizes the provision of different qualities of service to support complex systems, in which for example periodic hard real-time tasks may coexist with sporadic and possibly aperiodic soft real-time and non-real-time ones.

5.4. Summary of Real-Time System Requirements

From the basic real-time requirement to offer design-time assurances of timeliness, and the arguments above, there follow a number of real-time system requirements, which are summarised below.

- There must be a requirements specification for a particular real-time system, defining the timeliness requirements and the operational envelope(s) in which they are to be achieved. The assumptions made should be useful and realistic. Probabilities should be estimated where assumptions are not certain, and the action to be taken for each timing fault defined.

- The design philosophy should cater for both "bounded-demand" and "unbounded-demand" requirements. Outside an operational envelope, it should still ensure, with high probability, as many as possible of the services that are guaranteed within the operational envelope.

- To this end, designers should choose algorithms exhibiting good average behaviour, to maximize the total activity normally achieved.

- To ensure a graceful degradation as each operational envelope is exceeded, and each priority level loses its guaranteed timeliness, on-line schedulers should be used. To prove the existence of a feasible schedule, or to satisfy simple bounded-demand requirements, off-line schedulers may also be used.

- Bounded-demand systems should be sized so that the kernel of hard real-time activity is still achieved within the worst-case operational envelope. Unbounded-demand systems where even this envelope can be exceeded should be sized so that the probability of this is acceptably small; if it occurs, timing failures should be detected, and predefined actions triggered to minimize the resultant damage.

- There a need for tools to measure maximum execution times and system latencies, and assist in the sizing of systems and the verification of the timeliness specification. For unbounded requirements, the probability that the system size will prove inadequate must also be estimated.

- All system latencies and support mechanisms must be time-bounded, notably communication metrics and the precision of clock synchronization. There must be support for bounded preemption, so that hard real-time components can co-exist with lower priority components without loss of timeliness.

The consequences of these requirements are analysed in more detail in chapter 9, so as to derive a set of techniques for supporting real-time systems in Delta-4.

5.5. Conclusions

The theory underlying bounded-demand systems that are amenable to a complete worst-case analysis has been well-explored and many techniques of design are now mature. These can be used successfully for some classes of system, characterized by limited complexity and a well-behaved environment.

These design techniques are, however, inappropriate for various other possible system requirements, i.e.:

- The environment imposes unboundable demands.

- The demands of the environment can be bounded, but the bounds do not lead to an economically competitive system size. The processing power required may even exceed the bounds of technology.

- Worst-case analysis at design-time may face combinatorial explosions that render it unfeasible or impossibly costly. Design validation is then only possible on the basis of simplifying assumptions that sacrifice the possibility of an absolute assurance.

- Absolute assurances may not be required, especially for soft real-time services, or the customer may choose to forgo them because of the cost and time required to provide them.

These cases seem to require the approach of an older but less mature school of design, called here the "unbounded-demand" school. This school recognises the limitations on scale and complexity imposed by the present state of the art in achieving mathematically exact and complete design assurance. Instead it uses statistical method and probability theory, in which large scale and diversity are an advantage rather than a disadvantage.

Delta-4 aims to support a wide range of real-time systems, including large-scale complex applications, and therefore admits the use of both design philosophies. In one special case, where the demand for hard real-time services can be bounded but the demand for soft real-time services is unboundable, it may be possible to use both design approaches in one system.

In the Delta-4 Open Systems Architecture (OSA), because of its openness and generality, it is difficult to achieve real-time objectives such as the bounding of all system latencies and application execution times. This is a major motivation for the development of the Extra Performance Architecture (XPA) described in chapter 9.

Chapter 6

Distributed Fault-Tolerance

Distribution and fault-tolerance are tightly related. Should a single element of a distributed system fail, users expect at worst a slight degradation of the service that is offered; distributed systems must thus at least have *some* built-in fault-tolerance. On the other hand, most fault-tolerant systems can, at some level or another, be seen as a distributed system due to their redundant processing resources. Distributed fault-tolerance is used here to refer to that class of techniques suitable for ensuring fault-tolerance in an architecture consisting of a set of processing elements (called *nodes* or *stations*) interconnected by a message-passing communication network (figure 1). The distributed fault-tolerance techniques discussed here are focussed towards distributed systems in which the communication network consists of one or more local area networks. In particular, the existence of high-bandwidth broadcast channels allowing efficient multicast communication is assumed.

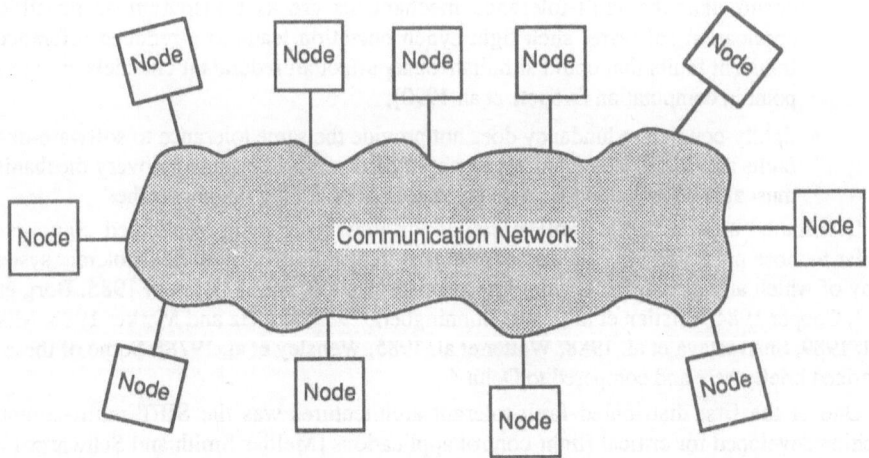

Fig. 1 - Distributed System

The chapter is organized as follows. After discussing some of the related work in this field, section §6.3 discusses the characteristics of the computational nodes of the considered class of distributed systems. In particular, this section details the assumed node failure modes and their impact on the design of distributed fault-tolerance techniques. The assumed models of distributed computation are sketched out in section 6.4, then section 6.5 considers various aspects relating to the replication of computation for the purposes of fault-tolerance.

The two issues of *error-processing* (preventing errors from affecting service delivery) and *fault treatment* (preventing faults from inducing further errors) are dealt with separately. Error-processing is described in sections 6.5 through to 6.7 which describe three classes of techniques for coordinating replicated computation, called active, passive and semi-active replication. Fault treatment is discussed in section 6.8.

Finally, section 6.9 is devoted to the definition of group communication facilities that the underlying communication system should offer to support the various distributed fault-tolerance techniques.

6.1. Related Work

Fault-tolerance is often implemented in single "stand-alone" machines rather than in the distributed fashion considered here. Such stand-alone fault-tolerant machines have several disadvantages:

- much special-purpose hardware must be designed — and re-designed when a technology update is required; our software-implemented distributed approach allows the use of standard, off-the-shelf machines;

- geographical separation of redundant resources has to be "added on" if disaster recovery is to be ensured; geographical separation of redundant resources is just a configuration parameter in our distributed approach — the same techniques are used whether the redundant resources are close to or distant from each other;

- stand-alone fault-tolerant machines often resort to tight synchronization in order to ensure that the fault-tolerance mechanisms are as transparent as possible to application software; such tight synchronization leads to a reduced tolerance of transient faults that could simultaneously affect all redundant channels at the same point in computation [Kopetz et al. 1990];

- tightly-coupled redundancy does not provide the same tolerance to software-design faults as a loosely-coupled approach; in particular, restart and recovery mechanisms must also be included to provide tolerance of operating system crashes.

Distributed approaches to fault-tolerance via message-passing replicated computation similar to those presented here have also been used in a number of other fault-tolerant systems, many of which are also aimed at supporting real-time applications [Birman 1985, Borg et al. 1983, Cooper 1984, Cristian et al. 1990, Gunningberg 1983, Kopetz and Merker 1985, Mishra et al. 1989, Shrivastava et al. 1988, Walter et al. 1985, Wensley et al. 1978]. Some of these are described briefly here and compared to Delta-4.

One of the first distributed fault-tolerant architectures was the SIFT multi-computer machine developed for critical flight-control applications [Melliar-Smith and Schwartz 1982, Wensley et al. 1978]. The MAFT system is a more recent example of the same type of architecture [Kieckhafer et al. 1988, Walter et al. 1985]. The SIFT architecture is based on general-purpose processors or nodes interconnected by a set of broadcast serial busses. Due to the criticality of the intended application domain (civil aircraft flight control), no restrictive assumption is made about node failure modes, i.e., nodes may be fail-uncontrolled (see section §6.2.2). To mask the arbitrary behaviour of such nodes, each node is connected to every other node by its own private broadcast bus. Single-source data is broadcast to all nodes by means of a clock-synchronous, phased Byzantine Agreement Protocol [Melliar-Smith and Schwartz 1982]. Such an interconnection structure was feasible in SIFT since the architecture was designed to accommodate at most eight nodes geographically located in the same equipment bay. It was required that the Delta-4 architecture be able to accommodate several tens of nodes spread out over quite considerable distances (i.e., a factory or several large neighbouring buildings) so such an interconnection structure is not economically viable. Error processing in SIFT is based on majority voting of the results of tasks that are replicated across several nodes. Tasks are executed according to a static frame-based cyclic schedule (calculated off-line). Delta-4 is intended to be an *open* architecture designed to accommodate heterogeneous nodes and local operating systems and serve a wider range of applications; systematic use of such synchronous frame-based scheduling was thus precluded. Finally, system reconfiguration after node failure is relatively simple in SIFT. Copies of the data necessary for the execution of all tasks can be maintained in every node of the system since the elements of computation (tasks) contain little or no internal state (persistent data). In Delta-4, the elements of computation may be as large as a complete database system. The reconfiguration mechanism must be able to create (or *clone*) new replicas whose internal state is initialized by copying the state of existing replicas on non-faulty nodes and transferring it across the network to the nodes on which new replicas are created.

MARS [Kopetz et al. 1988, Kopetz and Merker 1985] is an example of a distributed fault-tolerant system for real-time applications in which the geographical distance between nodes goes beyond that of a single equipment bay. The nodes in MARS are assumed to be fail-silent (nodes are designed to be self-checking, see section 6.2.1) and are interconnected by a serial baseband bus. The local clocks of each node are closely synchronized by means of a specially-designed clock synchronization chip [Kopetz and Ochsenreiter 1987] that achieves such a tight synchronization (less than $10\mu s$) that it can be used to control access to the serial bus by means of time-division multiplexing. Similarly to SIFT and MAFT, MARS uses cyclic scheduling of tasks based on a static (off-line) schedule taking into account the worst-case or peak-load application scenario. However, MARS does not need to resort to majority voting since nodes are assumed to be fail-silent; node failure only results in the absence of messages (detected in the time domain). Since all computation and communication in MARS are statically scheduled, the instants of communication and the quantity of transferred data are pre-established; therefore, the communication system does not have to worry about flow control to prevent buffer overflow. Furthermore, due to the fail-silent node assumption and the static communication schedule, reliable broadcasting can be achieved by systematic $k+1$ repetition of messages to mask k transmission errors. Like MARS, the real-time variant of the Delta-4 architecture (the XPA architecture, see chapter 9) also adopts a fail-silent node assumption to avoid the overheads of voting. However, unlike the static time-triggered approach of MARS, the XPA architecture adopts a dynamic event-triggered approach. This allows a more economical use of computation and communication resources for dynamic applications in which the load imposed on the system may suddenly vary due to the occurrence of asynchronous events. In many applications, no *a priori* worst-case load scenario can be determined; in such cases, the dynamic scheduling philosophy of XPA allows a best-effort approach to meeting application deadlines. Like MARS, the clocks of nodes in XPA are globally synchronized; however, for commercial reasons, clocks are synchronized without resorting to special-purpose hardware.

The ISIS system [Birman and Joseph 1987, Birman 1985, Birman and Joseph 1987] presents many similarities to Delta-4. Like Delta-4, ISIS is aimed at providing user-transparent fault-tolerance in a general-purpose distributed computing environment (as opposed to the SIFT and MARS systems that are tailored to clock-synchronous, real-time applications). The ISIS system provides a flexible *tool-kit* of basic primitives that allow an application programmer to build a distributed application that is made fault-tolerant by replication of code and data. ISIS assumes that nodes or processes fail only by crashing (i.e., that they are fail-silent) and provides a single mechanism for fault-tolerance at the process level based on a *coordinator-cohort* scheme. This scheme is similar in some respects to Delta-4's semi-active replication technique (see section 6.7). The ISIS coordinator-cohort scheme does not, however, address the issue of resolving replica non-determinism. The tools provided by ISIS could also allow the implementation — by the application programmer — of actively-replicated processes (restricted to a fail-silence assumption) and passively-replicated processes (by making coordinators systematically transfer their state to cohorts when a service request is completed). In Delta-4, many applications can be designed as if they were to run on a system that never fails. Since fault-tolerance is managed by built-in system facilities, the issues of replication can be entirely hidden from the application programmer and only specified at configuration time. ISIS also provides the basic *state transfer* mechanism [Birman and Joseph 1987] necessary to ensure the cloning of replicas for system reconfiguration during fault treatment. However, since ISIS assumes fail-silence, the state transfer tool does not provide a facility for error-detection during state transfer by cross-checking states copied from multiple source replicas. Like Delta-4, ISIS makes use of special communication facilities supporting *multicast* protocols based on a clock-asynchronous approach rather than the clock-synchronous techniques of SIFT and MARS. However, the ISIS multicast protocol suite is implemented on top of TCP/IP such that each multicast results in a number of point-to-point TCP/IP messages. In Delta-4, the basic atomic multicast protocol is implemented on top, or as extension of, the medium access control protocol of selected local area networks (see section 6.9 and chapter 10); this allows hardware broadcasting opportunities to be exploited for increased performance.

A system that resembles Delta-4 quite closely is IBM's Advanced Automation System (AAS) for the US Air Traffic Control network [Cristian et al. 1990]. The AAS concept of "server groups" is equivalent to that of a "replicated software component" developed here. In AAS, both active and passive replication techniques are available[1] but only for the case of server replicas that fail by responding late, by omitting to respond or by crashing. In our approach, the case of fail-uncontrolled (active) replicas — ones that can fail in quite arbitrary fashion — is also accommodated by means of a built-in voting mechanism. The Delta-4 atomic multicast protocol allows replicated entities to be logically addressed such that messages are delivered (with low overhead) to all replicas; this results in somewhat simpler management of active replicas than in AAS. Processors in AAS are organized in distinct *processor groups* that can each support replicas of a given set of servers. There is a *group service availability manager* (gSAM) for each processor group, with replicas on all processors of that group. The gSAM is responsible for ensuring that the number of replicas of all servers supported by the processor group is maintained according to the server group's replication policy (specifying the minimum number of required replicas). If a server group can no longer execute according to its replication policy then, in some cases, it may be moved to another processor group under the control of a *global service availability manager* (GSAM). Each processor group supports a group membership service and a group broadcast service that enables gSAM replicas to maintain a consistent view of the processor group's global state. A routed multicast facility is also provided for communication between server groups residing in the same or different processor groups. In Delta-4, management is based on the open distributed system concepts of "managed objects" and "management domains". The nearest equivalent to the AAS processor group and gSAM is that of a software component *replication domain* and its corresponding *replication domain manager* or RDM (see sections 6.4.3 and 8.2.3.3.1). However, since nodes in Delta-4 are not *a priori* split into groups, replication domains for different software components may overlap. Indeed, some software components may have a replication domain that spans all nodes in the system. A global processor membership service is provided over all nodes in a Delta-4 system and is built into the Delta-4 atomic multicast protocol.

6.2. Node Hardware Characteristics

This section identifies the different failure modes that can be assumed for nodes and sketches out the communication network topologies that are appropriate for achieving fault-tolerance

---

1 Referred to in AAS as *close* and *loose* synchronization of replicas.

(formal definitions of assumed failure modes are given in annexe E). A node consists of at least a processor, local memory and some sort of communication network interface; refinements of the internal node architecture are discussed later that enable simplifications of the distributed fault-tolerance techniques.

6.2.1. Fail-Silent Nodes

A fail-silent node is defined here to be a processing element that, viewed from the communication network, either operates in conformance with its specification or remains silent [Powell et al. 1988]. In particular, any message sent by a fail-silent node is a message that is correct in both value and time.

Some authors describe such nodes as being "fail-stop" (cf. chapter 4); the epithet "fail-silent" is preferred here since "fail-stop processors" have been previously defined to include not only the "halt-on-failure" characteristic implied here but also the fact that the other nodes are informed of the failure [Schlichting and Schneider 1983] or even that they are capable of accessing the node's stable storage [Schneider 1984]. Moreover, the term fail-silent makes explicit the necessary *external* perception of node activity and does not preclude the possibility of the node disconnecting itself from the network (i.e., going silent) but remaining active to carry out, for example, some local testing activity. In the terminology of [Cristian et al. 1985], a fail-silent node exhibits "crash" failure semantics.

Numerous implementations of fault-tolerance in distributed systems assume a "clean" node failure mode such as that embodied in the above definition of a fail-silent node. It is, in particular, an essential assumption in all previous work that we know of that uses transactions as a structuring concept for achieving fault-tolerance in distributed systems (see section §6.3.2).

There are several important simplifications that result from the fail-silent node assumption.

First, since fail-silent nodes never send any messages that are incorrect (thus eliminating any "two-faced" behaviour of a failed node), solutions to the distributed consensus problem are the simplest possible [Fischer 1983]. In particular, the minimum number of nodes necessary to achieve consensus in the presence of t faulty nodes, is given by $n \geq t+2$ [Lamport et al. 1982] (the problem is vacuous for $n \leq t+1$).

Second, since such a node either sends messages within specified time delays and with correct values or not at all, the other nodes of the distributed system can detect whether a node has failed by means of a simple interrogation and time-out mechanism or by timing out on regularly transmitted "I'm alive" messages (subject to the fact that communication delays are also bounded [Fischer et al. 1985]).

Third, data and/or code replication techniques for continued operation in the presence of t faulty nodes need only rely on $t+1$ replicas (see section 6.4).

Fourth, since such an assumption effectively precludes any possibility of the communication network being saturated by spontaneously-produced "garbage" messages, faults in the communication network can be dealt with independently from faults in the nodes. Thus, simple communication architectures using shared multipoint transmission channels can be envisaged (see figure 2).

However, a 100% guarantee that nodes are indeed fail-silent implies that the nodes are implemented with perfect, zero-latency self-checking mechanisms. Although many techniques are available for implementing self-checking hardware [Wakerly 1978], it is not easy to buy off-the-shelf general purpose computers that use these techniques extensively.

Consequently, although one can *assume* that an off-the-shelf computer is fail-silent, the coverage of this assumption may not be very high (see annex F). To attain a higher degree of dependability, a less stringent assumption with higher coverage can be adopted, e.g., that of "fail-uncontrolled" nodes.

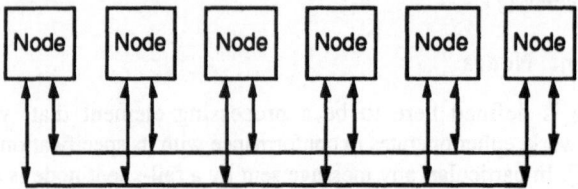

Fig. 2 - Interconnection Topology suitable for Fail-Silent Nodes

6.2.2. Fail-Uncontrolled Nodes

A fail-uncontrolled node represents the opposite extreme of the failure mode spectrum, i.e., a node that may fail in a quite arbitrary fashion. In particular, fail-uncontrolled nodes can:

   a) send messages that are late (including completely omitting to send messages);

   b) send messages sooner than expected;

   c) send messages with erroneous content;

   d) send unspecified or "impromptu" messages [Powell 1991].

Since this worst case assumption (see annexe E, expression 3) is — by essence — true, the probability of the assumption being satisfied in a real system (the assumption coverage, see annexe F) can be set equal to 1. The resulting ease with which it is possible to quantify the dependability achievable with such a worst-case assumption must however be weighed against several important disadvantages.

First, a fail-uncontrolled node can exhibit quite malicious behaviour[2]. For example, when relaying a message received from one processor, it could relay different copies to different destinations (the so-called "Byzantine" faults). It can be shown that the minimum number of nodes necessary to achieve consensus in the presence of t faulty nodes exhibiting such two-faced behaviour is given by $n \geq 3.t + 1$ (e.g., 4 nodes for 1 fault) [Lamport et al. 1982].

Second, it is impossible to use a simple interrogation and time-out mechanism to detect whether a fail-uncontrolled node has failed since a faulty node can reply correctly to an interrogation yet still send erroneous messages to other nodes. Node failures can only be revealed by comparing the activity of different nodes (and of course assuming that nodes fail independently).

Third, since node failure can manifest itself by messages being sent at the wrong time or with erroneous content, data and/or code replication techniques for continued operation in the presence of t faulty nodes must be based on at least $2.t+1$ replicas so that a minority of erroneous messages can be masked (see §6.4).

Fourth, since a faulty fail-uncontrolled node may generate an arbitrary number of "impromptu" messages, any (or all) transmission channel(s) to which it is attached may be saturated by garbage messages that prevent the channel(s) from being used by other nodes. This is a simple illustration of the fact that, from the viewpoint of error propagation, there is no built-in "error containment barrier" at the node interface like the one that exists for a fail-silent

---

2   Any assumption to the contrary would not be the worst case and would need to quantified by an appropriate less-than-unity assumption coverage.

node. Furthermore, since the content of erroneous messages may be arbitrary, messages may be sent with erroneous source address fields that would make it impossible for a receiving node to know where they came from. A faulty node could thus masquerade as an arbitrary number of non-faulty nodes and thus foil any attempt at a consensus by the latter. It is therefore necessary to rely on the node interconnection topology to identify the source of all messages or to use authenticated messages (which essentially defines a *different* failure mode assumption that is less than arbitrary and consequently has an assumption coverage less than 1, [Powell 1991]). Consequently, the only viable fault-tolerant architectures with such a node failure assumption are those in which nodes do not all share the same transmission channels ([Lamport et al. 1982, Melliar-Smith and Schwartz 1982]). Figure 3 gives two possible interconnection topologies suitable for constructing a fault-tolerant distributed system based on the fail-uncontrolled node assumption. The first of these shows each node connected to every other node by a unidirectional multi-drop bus (this is in fact the architecture of the SIFT multiprocessor [Melliar-Smith and Schwartz 1982]). The second shows a meshed network in which each node is connected to each of its neighbours by a unidirectional channel (such an architecture is implied in [Cristian et al. 1985]). In both cases, the immediate consequence of node failure is limited to the fault-containment domain constituted by the node itself and its associated channel(s) (examples are shown in heavy lines on figure 3).

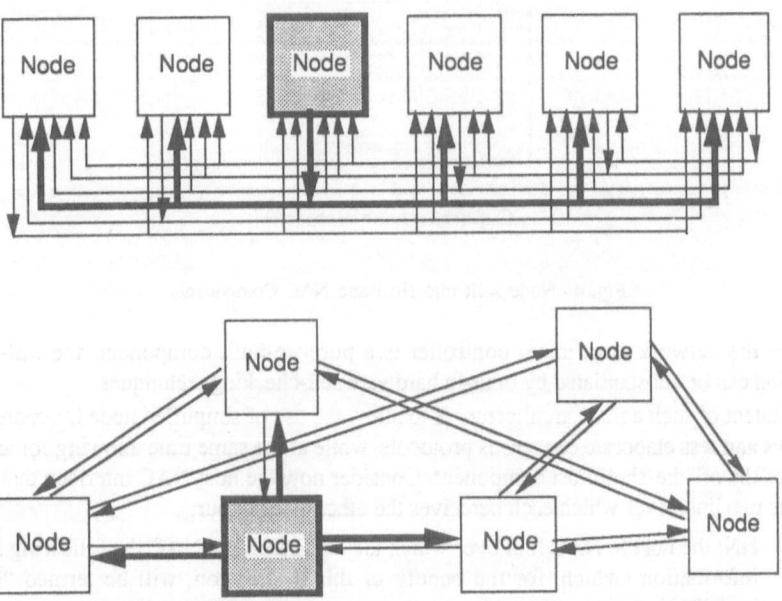

Fig. 3 - Interconnection Topologies suitable for Fail-Uncontrolled Nodes

The main disadvantages of the architectures for fail-uncontrolled nodes shown on figure 3 are that of high cost and lack of extensibility; the addition of a new node implies the addition of a new (set of) transmission channel(s). The alternatives are thus (a) to assume fail-silent nodes or (b) to consider an intermediate solution such as that proposed in the next section.

6.2.3. Network Attachment Controllers

A compromise between the restrictive fail-silent node assumption and the worst-case fail-uncontrolled node assumption can be envisaged whereby each node is considered as consisting of two components (figure 4):

- an off-the-shelf computational component, called a *host*, that may be fail-uncontrolled;

- a purpose-built communication component, called a *network attachment controller* (NAC), that is assumed to be fail-silent.

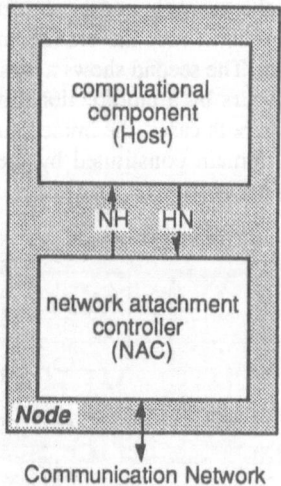

Fig. 4 - Node split into Host and NAC Components

Since the network attachment controller is a purpose-built component, the fail-silence assumption can be substantiated by built-in hardware self-checking techniques.

The intent of such a node architecture is to allow the use of simplified node interconnection topologies and less elaborate consensus protocols, while at the same time allowing for arbitrary failures of the off-the-shelf host component. Consider now the host/NAC interface that can be viewed as two links over which each perceives the other's behaviour:

- HN: the host to NAC link over which the host sends the NAC the following sorts of information (which, for the benefit of this description, will be termed "service items"):

  - messages to be sent over the communication network (resulting from computation on the host);

  - handshakes (requests or acknowledgements) for service items sent over the NH link;

- NH: the NAC to host link over which the NAC sends the host the following sorts of service items:

  - messages received from the communication network (that will affect future computation on the host);

- handshakes (requests or acknowledgements) for service items sent over the HN link.

Whereas the behaviour of fail-silent or fail-uncontrolled nodes was defined in terms of messages sent over the network, the behaviour of *hosts* must be defined in terms of service items on the HN link. For a fail-uncontrolled host, the following possibilities for faulty fail-uncontrolled host behaviour need to be considered:

a) send messages or handshakes that are late (or completely omitted);

b) send messages or handshakes sooner than expected;

c) send messages or handshakes with erroneous content;

d) send unspecified or "impromptu" messages or handshakes.

In the case of handshakes, possibility a) above is of particular importance with regard to controlling the flow of information being multicasted to several, possibly failed, destinations (see §6.5.3) since it difficult to distinguish between the two possible causes of a host handshake being late: (i) the receiving host entity could blocked for a logical reason, which means that the NAC must request the "network" (ultimately, the sending nodes) to stop the flow of information, or (ii) the host could have failed, in which case the sending nodes should be able to continue the flow of multicasted information.

Possibilities b), c) and d) above clearly indicate that, if the NAC is to be able to respect a fail-silent assumption, protection mechanisms must be built into the NAC to shelter it from unexpected or incorrectly-valued HN service items. This militates in favour of the NAC playing a master role in the interactions across the host-NAC interface whereby the host may only transfer information to the NAC at times, and to locations in the NAC memory, dictated by the NAC itself.

The NH link of the host-NAC interface presents fewer problems since the non-delivery or delivery of incorrect NH service items is (will eventually be perceived as) equivalent to the host itself failing. In practice, however, since the NAC needs to be implemented using self-checking techniques to substantiate its fail-silent behaviour with respect to the network, a similar fail-silent behaviour can be assumed for the NH link.

In the remainder of this chapter, the hardware architecture shown in figure 5 will be assumed. Each node is split into host and NAC components; NACs are always assumed to be fail-silent, whereas hosts may be either fail-silent or fail-uncontrolled. The local area network shown in figure 5 may contain redundant communication channels or *media* (cf. figure 2, see also chapter 10). Management of these redundant channels is not considered here — it is assumed that the resulting local area network is (internally) fault-tolerant; in particular, physical network partitioning is not considered.

6.2.4. Stable Storage

In database applications, there is a need for a mechanism that allows a consistent state of the database to be stored while some new tentative computation is carried out. If some logical condition cannot be fulfilled (e.g., due to a bank account being insufficiently funded), such computation may need to be aborted and the previous state of the database restored. The intention is to allow computation to be carried out as a series of atomic steps or *transactions*. By extension, the mechanism (storage device and atomic update procedures) allowing intermediate states to be stored is referred to as *atomic storage*.

This concept has frequently been extended to allow a previous state to be restored if computation cannot complete due to a *fault* condition (as opposed to a logical condition). Intermediate states (or *checkpoints*) of a node's computation are stored in a "safe" place so that, if an error should be detected during subsequent computation, a previous error-free state can be

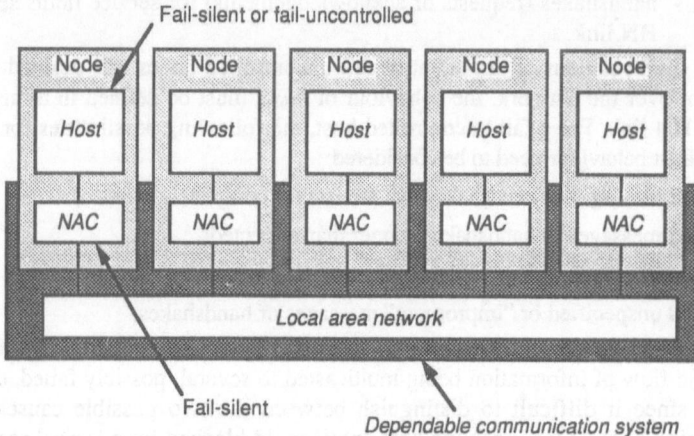

Fig. 5 - System Hardware Architecture

retrieved. As a means for fault-tolerance, the "safe place" in which checkpoints are stored must in itself be fault-tolerant; in particular, since it should be able to survive power outages, such a storage mechanism is referred to as *stable storage*. Stable storage is typically implemented using magnetic disc technology, although semi-conductor *stable memories* have also been implemented [Banâtre et al. 1986, Banâtre et al. 1988].

From the viewpoint of fault-tolerance, it is important to distinguish between these two different motivations for storing the intermediate states of a computation, i.e., a) as a means in transactional applications to allow tentative computation to be aborted should some logical condition not be fulfilled, or b) as a means to restore an error-free state after detection of an error due to a fault. In transactional applications, an atomic storage mechanism is needed even if the underlying system is completely fault-free. System fault-tolerance may or may not be based on the implementation of atomic storage as stable storage.

If the nodes in a distributed system possess stable storage, they can be "repaired" (either manually or automatically) after failure and re-inserted into the network with the assurance that the stored "stable data" is still in a state that is identical to that before node failure. Note however, that such a backward error recovery scheme inevitably leads to a time overhead (which, in the case of manual repair, could be quite long) that can lead to an unacceptable decrease in service availability. Redundancy of data and code is necessary if computation is to proceed while failed nodes are being repaired.

Alternatively, if nodes do not possess stable storage, then the design of the system-level fault tolerance techniques must be based on the assumption that all data stored locally is lost should the node fail. This means that if the node is re-inserted into the network, it must be assumed to have suffered total amnesia; the re-inserted node is thus equivalent to a totally new node. It can only be re-introduced into the system after its local storage has been re-initialized by copying information across the network from other (non-failed) nodes.

It is worthwhile considering the stable storage abstraction in the context of the two extreme assumptions made for hosts in sections 6.2.1 and 6.2.2:

  a) *Fail-silent host with stable storage.* Seen from the communication system, such a
      node either delivers correctly-valued and timely messages or stops sending
      messages until repair is carried out. After repair, information may be retrieved from
      stable storage that is identical to that stored there before failure.

Since the host is fail-silent, an implementation of atomic storage using redundant non-volatile storage media is a reasonable approximation of stable storage.

b) *Fail-uncontrolled host with stable storage.* Seen from the communication system, such a node may deliver incorrectly-valued or untimely messages. However, data written to stable storage is unaffected by host processor failures; therefore, information retrieved from stable storage after node repair is identical to *some* consistent state before failure.

Since the host may fail in an arbitrary fashion, atomicity of updates and non-volatility of storage are insufficient to ensure the stable storage abstraction. Further mechanisms are needed to protect against data corruption due to host processor failure [Banâtre et al. 1988].

In conclusion, although stable storage (together with a transactional model of distributed computation, see section   6.3.2) can be viewed as one possible basis for fault-tolerance in distributed systems, replication of code and data is more appropriate for applications that cannot be structured as transactions or which cannot support the time overheads induced by backward error recovery. However, the stable storage concept does provide an *option* for simplified re-initialization of repaired nodes (see §6.8).

6.3. Models of Distributed Computation

Before presenting the various possible approaches to distributed fault-tolerance by replication in section §6.4, this section introduces some elementary concepts for expressing what is meant by a *distributed* computation.

6.3.1. Software Components

We define a *software component* to be an elementary run-time unit of distributed computation and data encapsulation that, in the absence of replication, resides on a single node. The data encapsulated by a software component is referred to as its *state*. Software components are active logical entities that may communicate with each other by means of messages (only). Even when several software components co-reside on a single node, they do not explicitly share common memory. A distributed application can be viewed as any activity coordinated across multiple software components.

A software component may send messages to other such components thourgh one or more *output ports* and receive messages through one or more *input ports*. For simplicity, a single output port and a single input port will be assumed unless explicitly stated otherwise.

The concept of a run-time *software component* is introduced here to avoid using the very overloaded term "object"; it is used here to reason generally about the run-time view of more specific notions such as single-threaded or multiple-threaded "processes", Eden "Ejects" [Almes et al. 1985], CONIC "modules" [Loques and Kramer 1986], the "active objects" of Emerald [Black et al. 1987], ANSA [ANSA 1989] or Deltase "capsules" (see chapter 7), etc.

Since software components are the basic units into which a computation may be partitioned and allocated over the nodes of the distributed system, it is also convenient to consider them as the basic units by which computation can be replicated to tolerate faults. Replication of software components is discussed in section   6.4.

6.3.2. Transactions

Transactions were first introduced in the field of data-base systems (see [Bernstein et al. 1987] for a full bibliography) to ensure that updates to multiple items of data (or the states of multiple software components) are executed atomically:

- the refusal of an operation on one data item (e.g., a debit from an account is refused if the account has insufficient funds) may imply that related operations on other data items need to be cancelled (or "un-done");
- the potentiality for concurrent multiple access to shared items of data requires that operations on individual data items be scheduled to avoid mutual interference due to interleaving.

The mechanisms that are necessary to enable operations to be undone due to a purely logical reason can also be used to implement backward recovery following a failure or a conflict between operations that was not avoided by the control mechanisms. For this reason, the transaction concept has been extended beyond the database world and is often used as a basis for providing fault-tolerance in distributed systems by means of backward error recovery.

It has in fact been shown that the transaction mechanism is a dual of the *conversation* scheme that was introduced as a structuring mechanism for fault-tolerance in concurrent systems [Mancini and Shrivastava 1989, Randell 1975]. The aim of the conversation scheme is to control the *"domino effect"* that may be caused when, after detection of an error, communicating processes (software components) are rolled back and re-executed from some previously saved state (a checkpoint or recovery point). The domino effect may occur when the state of a process A is restored (rolled back) to some previous state that existed prior to a communication between A and some other process B. If A had sent a message to B before initiating roll-back then, when re-executing, it will send another message to B that, in the general case of non-idempotent messages, will cause B's state to become incorrect[3]. Similarly, if A had received a message from B, then when re-executing, it would require B to re-send the message it had already sent. In either case, B would also have to be rolled back. This could require a further roll-back of A if A and B had interacted after B's last checkpoint yet before that of A.

Conversations provide a means for restricting this domino effect. Once a process has entered a conversation, it is not allowed to communicate with processes not in the same conversation. If each process takes a checkpoint on entering a conversation then the roll-back of any process in the conversation causes the roll-back of only those processes in the same conversation. The mechanisms necessary for controlling the entering and leaving of conversations by processes are analogous to those used by transactions for locking and unlocking data items [Mancini and Shrivastava 1989].

6.4. Replicated Software Components

Replication of data and/or computation on different nodes is the *only* means by which a distributed system may continue to provide non-degraded service in the presence of failed nodes. Even though stable storage within nodes can be used to allow the system to recover (eventually) from node failures and can thus be thought of as a means for providing fault-tolerance, such techniques used alone do not allow distributed system architectures to achieve better dependability than a non-distributed system. In fact, if a computation is spread over multiple nodes without any form of replication, distribution can only lead to a *decrease* in

---

3   It is however possible to envisage the tolerance of repeated input messages by using sequence numbers.

dependability since the computation may only proceed if each and every node involved is operational.

The basic unit of replication considered here is that of a software component (cf. §6.3.1). A *replicated software component* is defined as a software component that possesses a representation on two or more nodes. Each representation will be referred to as a *replica* of the software component (even though the actual representations may in practice be different at a given instant in time). Unless stated otherwise, the term *software component* will refer to the logical entity as a whole (i.e., the *group* of replicas).

There are two issues to replication of software components:

- *inter-replica coordination:* how is the activity of the group of replicas coordinated in order to process errors and give the illusion to other software components that the group is a single (fault-free) software component?

- *group membership management:* how is a software component instantiated as a group of replicas and how is group membership updated as a consequence of failures and repairs?

For clarity, we shall first concentrate on the replica coordination issue: we shall suppose that software components have been instantiated as groups and that management of group membership is restricted to that of replicas *leaving* the group as a consequence of failure. The creation of groups and the management of replicas joining existing groups will be considered in section §6.8.

The degree of replication of software components in the system depends primarily on the degree of criticality of the component but also on how easy (and fast) it is to add new members to an existing group (to replace failed replicas). In general, it is wise to envisage groups of varying size, even though the degree of replication may often be limited to 2 or 3 (or even 1, i.e., no replication, for non-critical components) (see figure 6).

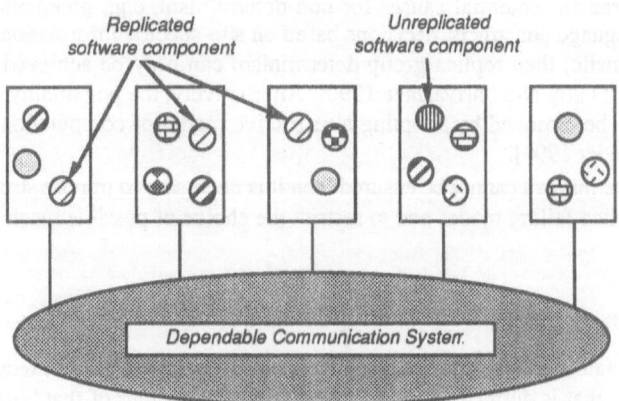

Fig. 6 - Replicated Software Components

Three basic techniques for replicated computation can be identified according to the degree of replica synchronization:

- *active replication* is a technique in which all replicas process all input messages concurrently so that their internal states are closely synchronized — in the absence of faults, outputs can be taken from any replica,

- *passive replication* is a technique in which only one of the replicas (the *primary* replica) processes the input messages and provides output messages — in the absence of faults, the other replicas (the *standby* replicas) do not process input messages and do not produce output messages; their internal states are however regularly updated by means of checkpoints from the primary replica,

- *semi-active replication* can be viewed as a hybrid of both active and passive replication; only one of the replicas (the *leader* replica) processes all input messages and provides output messages — in the absence of faults, the other replicas (the *follower* replicas) do not produce output messages; their internal state is updated either by direct processing of input messages or, where appropriate, by means of notifications or "mini-checkpoints" from the leader replica.

These three different replication techniques will be described in the following sections; the remainder of this section is devoted to concepts that are common to all replica coordination techniques.

6.4.1. Replica Determinism and Replica Group Determinism

A *replica* (of a given software component) is said to be deterministic if, in the absence of faults, any execution of the replica starting from the same initial state and consuming the same ordered set of input messages leads to the same ordered set of output messages.

A *replica group* is deterministic if, in the absence of faults, given the same initial state for each replica and the same set of input messages, each replica in the group produces the same ordered set of output messages. If all replicas in a group consume identical input messages in the same order, then replica determinism is a sufficient condition for replica group determinism.

Replica determinism is difficult to ensure in a truly heterogeneous environment. For each software component, it is necessary to restrict the locations of replicas to a sub-set of nodes that, if non-faulty, guarantee that each replica is deterministic. Even in a homogeneous sub-set of nodes, there remain potential causes for non-determinism, e.g., preemption, use of non-deterministic language constructs, decisions based on site-specific information, etc. If replicas are not deterministic, then replica group determinism can only be achieved by negotiation between replicas [Tully and Shrivastava 1990]. Alternatively, the potentiality for replica non-determinism can be removed by adopting a restrictive model of computation based on *state machines* [Schneider 1990].

If replica determinism cannot be ensured then it is necessary to impose strong assumptions on allowable replica failure modes and to restrict the choice of possible mechanisms for error processing.

6.4.2. Replica Failure Mode Assumptions

If only hardware faults are considered, it can be assumed that the replicas executing on a given host fail in a way that is defined by the assumed failure behaviour of that host. If a fail-silent host fails, then all replicas that were executing on that host will appear to have failed silently or to have "crashed". Conversely, if a fail-uncontrolled host fails, then any or all replicas that were being executed can fail arbitrarily, i.e., send early or late, omit to send some messages, send messages with incorrect content or send extra or "impromptu" messages (see 6.4.2.3 below).

However, it is possible to refine these assumptions and at the same time use a finer failure granularity than that of a complete host [Cristian 1991]. For instance, an incident in a node's local operating system could cause a single replica to crash. Alternatively, if a host becomes overloaded (due to an inappropriately-dimensioned system configuration, i.e., a *configuration fault*), then although the host hardware may be completely fault-free, buffer overflow may cause replicas on that node to fail by omitting to respond; such failures are called *omission failures* [Cristian et al. 1985]. Similarly, replicas may respond too late; these failures are called *late-timing* or *performance* failures. Of course, if a software component contains a residual *design fault*, then all replicas could fail in a quite arbitrary fashion, including both *timing* and *value* failures. Certain replication techniques, although primarily designed to tolerate hardware faults, are capable of tolerating host configuration faults and software component design faults if such faults manifest themselves *independently* on different hosts. Although this may seem reasonable for host configuration faults, this is less evident in the case of software design faults since, by definition, all replicas of an incorrect software component will be identically incorrect. However, it has been observed that some design faults manifest themselves independently in different replicas due to slight differences in the local environment of the replicas at the time of their execution. In [Gray 1986], such faults are referred to as "Heisenbugs" since they "go away when you look at them".

Consideration of the various ways by which components can fail in the time domain and the value domain can lead to the definition of a wide spectrum of failure modes with different severities (e.g., see [Powell 1991]). However, for the particular case of interest here, i.e., software component replicas executing in a distributed message-passing environment, just a few categories of failure modes are sufficient. Replicas that are assumed to suffer only crash failures, only omission failures or only performance failures can be termed respectively *fail-silent, fail-omissive* and *fail-tardy* replicas. Replicas that fail thus can also be collectively termed *fail-restrained* replicas since, by assumption, they only ever send messages that are of correct value; replicas that can fail arbitrarily are termed *fail-uncontrolled* replicas.

It is tempting to consider the possible phenomena that could cause a replica to fail *only* by omitting to send some messages or by sending some messages too late and then to reason about the other possible consequences of those phenomena. For instance, a replica might fail to send a reply because it "lost" a request message. In this case, the lost request message would probably also cause the state of the replica to be erroneous — thus leading ultimately to a failure in the value domain. However, the concept of considering "abstract" failure modes intermediate between sudden silence and totally uncontrolled behaviour is a useful one because the mechanisms necessary to tolerate such restrained forms of failure are not much more complex than those necessary to tolerate total silence; yet, since the corresponding failure mode assumptions are provably less restrictive than total silence, the resulting assumption coverage can only be higher.

Note also that whereas replicas may be assumed to be fail-restrained *or* fail-uncontrolled when executing on fail-silent hosts, all replicas executing on fail-uncontrolled hosts must be assumed to be fail-uncontrolled.

6.4.2.1. Time-Domain Errors. Whether replicas are considered to be fail-restrained or fail-uncontrolled, detection of timing errors is a fundamental part of error processing. Such errors are particularly difficult to process in an *open* distributed environment. One technique that been proposed to simplify processing of timing errors is to keep local clocks approximately synchronized [Lamport and Melliar-Smith 1985, Schneider 1986]. However, this is of no direct use unless the local scheduling of replicas at each node explicitly uses the resulting global time reference to determine the instants at which messages should be sent or received [Lamport 1984]. This cannot be assumed an *open* distributed system. Scheduling techniques on heterogeneous hosts cannot be assumed to be the same, let alone time-dependent. In practice, the only viable basis for dealing with time-domain errors in an open system is the use of time-outs.

Nevertheless, the use of time-outs does not avoid the requirement for upper bounds on replica execution times and communication delays (if such upper bounds did not exist, it is impossible to distinguish between a component that has stopped or is infinitely slow [Fischer et al. 1985]). Since it is particularly difficult to estimate execution times and communication delays, especially in complex dynamically-evolving systems, one is faced with the inevitable problem of dimensioning time-outs sufficiently high so as to achieve an acceptable rate of late-timing failures yet sufficiently low to allow speedy detection.

However, it is important to underline that, although from an error-processing viewpoint, expiration of a time-out can only be attributed to a replica late-timing failure, this does not necessarily mean that the sending node as a whole is faulty and will be irrevocably removed from the system. Properly-designed error-processing protocols will mask such errors but report them to the administration system's fault-treatment facility. The latter can first try to alleviate the incriminated node by moving (some of) its software component replicas to other nodes (load-balancing) and will only passivate the incriminated node if it diagnoses that the number of such reported errors has exceeded a given threshold (that could be dynamically adjusted to account for varying system load).

6.4.2.2. Value-Domain Errors. Value-domain errors only need to be considered when replicas are assumed to be fail-uncontrolled. The only way to detect value errors is to compare equivalent output messages from different replicas. This requires of course *active* replicas that satisfy the replica group determinism condition (cf. §6.4.1). To mask t value errors, then there must be at least $2t+1$ replicas in the replica group. The comparison itself can be carried out either on complete messages or, for performance reasons, on hash-coded representations of messages — called hereafter, *signatures*. Note that in the latter case, it must be assumed that two different messages do not produce the same signatures; there is thus an attendant assumption coverage to take into account when evaluating the achieved dependability.

6.4.2.3. Impromptu Errors. An impromptu error occurs when a replica spontaneously produces an unspecified message. Impromptu errors may be detected in either the value domain or in the time domain. For example, if a time window has been opened in which messages are expected, then any impromptu message that occurs *inside* this time interval will appear to be correct in the time domain. However, the impromptu message will be detected as value-erroneous if compared with the values of messages from other replicas as in §6.4.2.2 above. If an impromptu message is received *outside* any expected-message time window, then the message will be detected as timing-erroneous. Since impromptu errors affect both the time and value domains, it is necessary that there be at least $2t+1$ replicas in the replica group to mask t errors.

6.4.3. Replication Domains

A software component *replication domain* is defined to be the set of nodes on which replicas of that software component are allowed to reside.

Replicas of a given software component can, of course, only be executed on nodes that possess the necessary resources. However, there are often other reasons for restricting a component's replication domain. For example, it may be further restricted to those nodes that not only possess the necessary resources but which guarantee replica determinism (cf. §6.4.1).

Another factor affecting the definition of a component's replication domain could be that the chosen replication technique relies on a particular replica failure mode assumption. If the dependability requirements of the application dictate that the coverage of that assumption be greater than some minimum value, then replicas may have to be confined to nodes with features that support that assumption. For example, if replicas must be fail-silent with a high degree of confidence, then replicas should be restricted to hosts that have been implemented using self-checking hardware.

Finally, equivalence of execution speed is yet another criterion entering into the definition of a replication domain. Even though replicas may have identical value domain behaviours when executed on a given set of nodes, a large dispersion in execution speed would complicate the detection of timing errors. It should also be noted that inter-replicate synchronization would anyway force all replicas to proceed, on average, at the speed of the slowest replica.

6.4.4. Replica Coordination Entities

It is desirable to be able to program a software component without taking into account the fact that it may be instantiated as a group of replicas. The programmer should be able to concentrate on the logical problem or function that the software component is meant to solve or to provide without having to deal with the intricacies of replica coordination.

It is therefore useful to separate this replica coordination functionality and let it be provided by one or more standard system entities. Each software component replica can be thought of as having one or more local "replica coordination entities" (abbreviated hereafter to *rep_entities*) acting on its behalf (figure 7).

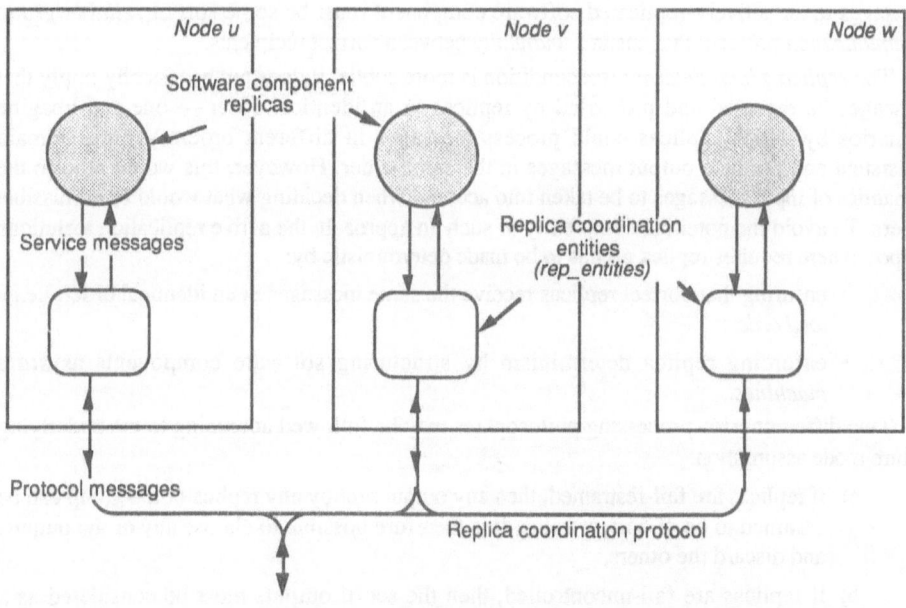

Fig. 7 - Software Component Replicas and Replica Coordination Entities

Rep_entities are assumed to be fail-restrained even if their corresponding replicas are fail-uncontrolled; this separation of function allows a considerable simplification of the distributed error-processing protocols for the fail-uncontrolled case. This means that rep_entities must be executed on fail-silent hardware — either on the host, if the latter is fail-silent, or on the associated NAC (cf. figure 5).

Replicas and their corresponding rep_entities exchange two sorts of *service messages: data messages* corresponding to the user-programmed data exchange between software components, and *handshake messages* for the purposes of flow-control. Messages exchanged between rep_entities are called (replica coordination) *protocol messages,* which may or may not contain embedded data messages.

6.5. Active Replication

Active replication is a technique whereby a software component is installed on multiple nodes such that at all times each replica in the group may, in the absence of faults, provide a service that is equivalent to any of the other replicas in the group. Quasi-instantaneous recovery from detected errors can be achieved if it can be guaranteed that all correct replicas produce the same output messages in the same order over the same output ports — this is referred to as the *output consistency* condition.

Sufficient conditions for output consistency are:

- *input consistency:* the sets of input messages delivered to correct replicas are identical;

- *replica group determinism* (cf. §6.4.1): when starting from identical initial states and processing identical sets of input messages, each replica produces identical output messages in the same order.

The *input consistency* condition implies that the communication protocol used to transmit messages to an actively-replicated software component must be some sort of *reliable group communication* protocol that ensures *unanimity* between correct recipients.

The *replica group determinism* condition is more subtle: it does not necessarily imply that messages be received and processed by replicas in an identical order — one can imagine scenarios by which replicas could process messages in different orders yet still remain consistent and produce output messages in the same order. However, this would require the semantics of input messages to be taken into account when deciding what would be admissible orders. To avoid the potential complexity of such an approach, the active replication technique proposed here requires replica groups to be made deterministic by:

- ensuring that correct replicas receive the same messages in an identical order, i.e., a *total order;*

- enforcing replica determinism by structuring software components as *state machines.*

Two different error-processing philosophies may be followed according to the underlying failure mode assumption:

a) if replicas are fail-restrained, then any output sent by any replica of the group can be assumed to be of correct value; it is therefore possible to choose any of the outputs and discard the others,

b) if replicas are fail-uncontrolled, then the set of outputs must be considered as a whole so that value errors and unexpected outputs may be masked.

In case a), since the output from any fail-restrained replica can only be a correct output, it is possible to relax the output consistency condition and optimize the use of individual replicas by, for example, only sending requests that do not modify the state of the component to just one replica of the group (e.g., the "nearest"). If no response is forthcoming, the request can be re-submitted to another replica. In database terms, such requests are called read-requests. Of course, any inputs to the software component that *do* modify the internal state (i.e., "write" requests) must be delivered to all replicas. Such "read optimization" (see [Bernstein et al. 1987]) allows a decrease in node workload and in message traffic over the network at the expense of imposing a read-write serialization mechanism to ensure consistency of the replicas.

The management of replicas in this way can be carried out in the framework of a transactional approach to distributed computation. However, we prefer not to consider such an optimization, since:

- in the general "software component" paradigm, outputs from a software component cannot necessarily be paired with a corresponding "read-request" input (e.g., a software component need not have *any* inputs but could be programmed to periodically transmit some internally computed value),
- similar mechanisms can be used for managing active replicas with both the fail-restrained and fail-uncontrolled assumptions.

The error processing protocols for fail-restrained and fail-uncontrolled active replicas are presented in the next two sub-sections. In both cases, two "modes" of operation are described. Note however that these separations in explanation are for clarity only and that, in practice, a single *parameterized* protocol can be used to cover both failure mode assumptions and both modes of operation.

6.5.1. Fail-Restrained Active Replicas

As mentioned earlier, the philosophy followed here is to treat the group of active replicas as a logical whole. Any messages sent to the software component must be delivered to all replicas in the same order so that, if the replica determinism condition is fulfilled then, in the absence of faults, each replica will produce the same output messages in the same order. The inter-replica protocol, carried out by the rep_entities, must mask and detect errors that may manifest themselves when the replicas attempt to send or to receive messages.

Since replicas are assumed to be fail-restrained, any message sent by such a replica is, by assumption, of correct value. Thus, from the output message viewpoint, the error-processing activity is reduced to a simple arbitration between the multiple copies of the output messages so that only one copy is delivered to the intended destination(s). This arbitration activity can be implemented in two ways:

a) for each output message, an arbitration protocol between the local rep_entities is executed in order to decide which of them will send the message,

b) all the rep_entities forward every output message; the rep_entities of the corresponding destinations discard all but one of the messages that they receive.

For long messages, the first approach is obviously less demanding in communication activity. Each replica forwards all its output messages to its local rep_entity. For each message, the various rep_entities must mutually decide which of them is to forward the message to its destination(s). The output message selection protocol is built on top of an underlying multicasting service that ensures that messages multicasted by any rep_entity are received by all other rep_entities on non-failed nodes (including the sender) and in the same order. The output message selection protocol may operate in either a competitive or a cyclic mode.

6.5.1.1. Competitive Propagate Mode. In the *competitive mode,* when a rep_entity receives a data message from its replica, it first verifies that the message has not already been sent by another rep_entity. If not, the rep_entity multicasts a *claim* protocol message to the group of rep_entities (including itself). Therefore, if a rep_entity receives its own *claim* message before any others, it may forward the data message; if not, the data message is discarded. As it stands, this simple competitive message selection protocol allows silence, omission errors and late-timing errors to be masked. However, to initiate fault treatment, such errors must not only be masked, they must be detected. This is achieved by a time-out mechanism. If a rep_entity receives a *claim* protocol message corresponding to a data message that it has not yet received from its local replica, a timer is armed. Replica silence, omission errors and late-timing errors are declared if this timer should expire before the local replica has produced the corresponding data message.

The competitive mode gives preference to the fastest replica of the group and can allow other replicas to lag further and further behind. The amount of desynchronization may be implicitly limited by controlling the flow of information to the replicas (when the input queue to the slowest replica is full, flow control will limit the rate of delivery of input messages to the rate at which the slowest replica dequeues input messages). Alternatively, the desynchronization may be explicitly limited if the rep_entities periodically carry out a "rendezvous" during which they wait for all *claim* messages to be received before one of them forwards the corresponding data message. The rendezvous is again time-limited in order to detect and recover from replica silence, omission errors and late-timing errors.

In practice, it has been observed that re-synchronization need not be carried out frequently; the fastest replica automatically slows down because it *must* send a *claim* message whereas the slower replica(s) will not need to send one if the fastest replica's *claim* message has already been received.

6.5.1.2. Round-Robin Propagate Mode. In the *cyclic or round-robin mode*, the rep_entities are configured in a logical ring with an associated token. When a rep_entity receives a data message from its replica that has not already been sent by another rep_entity then, if it possesses the token, the message is forwarded to its destination(s) and the token is transferred to the rep_entity's successor in the logical ring. Rep_entities not possessing the token must store messages until they receive the token; then, all messages already sent (identified by a *last_message* identifier contained in the token) are discarded. Since the round-robin mode treats all replicas "fairly", no further inter-replica synchronization mechanism is necessary.

To ensure token recovery, the interval between receipt of the local data message and receipt of the token is monitored. If time-out occurs, then the rep_entity reverts to the competitive mode by issuing a *claim* message (the rep_entity also reverts to the competitive mode should it receive a *claim* message from a peer rep_entity).

Omission and late-timing errors can be detected by monitoring the time interval between the receipt of a (multicasted) protocol data message by a peer rep_entity and receipt of the local service data message if the former should occur before the latter.

6.5.2. Fail-Uncontrolled Active Replicas

When replicas are no longer assumed to be fail-restrained, the error-processing activity must take account of not only silence, omission errors and late-timing errors but also *value* errors, *early-timing* errors and *impromptu* errors (cf. §6.4.2.3).

To mask t early-timing errors, $2t+1$ replicas are necessary. Even if t replicas of the group send a data message to their local rep_entities at approximately the same time, the latter cannot know immediately whether these messages are the first t messages of a set of $2t+1$ or if they are messages being sent too early or indeed impromptu messages. Each rep_entity must therefore arm a timer and await notification that equivalent messages have been sent within a specified time interval by at least t other replicas.

To process value errors in data messages, the rep_entities must cross-check each data message sent by the local replica with equivalent data messages sent by remote replicas. This cross-checking is referred to here as *message validation*. To mask t value errors, equivalent data messages from $t+1$ replicas must be compared and found to agree before propagating a validated message to its destination(s); since there can be t messages with erroneous values, a total of $2t+1$ messages is necessary (cf. §6.4.2.2) (an alternative approach based on message propagation-*before*-validation, is discussed in annexe G)[4].

The message validation mechanism is built into the output message selection protocol executed by rep_entities. As soon as $t+1$ messages are found to agree then, since there are only supposed to be t errors, it can be safely assumed that the consensus message that is propagated is error-free. It is important to underline that the rep_entity that propagates the validated message must not alter the message in any way — rep_entities must be fail-restrained so that it can be assumed that any message that they do send is a correctly-valued message.

To prevent faults in the remaining t replicas from remaining dormant, it is also necessary to ensure that all replicas are regularly activated and checked either by rotation of the $t+1$ replicas whose messages are compared or by systematic comparison of messages from all $2t+1$ replicas. The latter approach is simpler to implement and can provide acceptable performance if the last t messages are compared after having propagated the consensus value.

As for the fail-restrained replica case, the message-sending error-processing protocol may operate in either a competitive or round-robin mode.

6.5.2.1. Competitive Validate-before-Propagate Mode. In the *competitive mode*, when a rep_entity dequeues a data message from its replica, it first verifies that this message has not already been sent by another rep_entity. If not, the rep_entity multicasts a *claim* protocol message to the group of rep_entities (including itself); the *claim* message includes the signature of the data message received from the local replica. Since the underlying group communication service ensures ordered delivery of protocol messages to all rep_entities, the *claim* messages from the group of rep_entities will be received by all in the same order. Each rep_entity compares the signatures of the sequence of claim messages until $t+1$ signatures are found to be identical; this point in the sequence is termed the *validation point*.

The unique rep_entity that reaches its validation point by means of the signature in its own *claim* message, forwards the locally-received data message to its destination(s) and indicates to its peers that this has been done by means of an *ack* message (the forwarded data message and the *ack* message are sent together in a single atomic operation).

Early-timing and late-timing errors (including omission and silence) are detected and masked by monitoring the time interval between receipt of the local data message and the validation point or vice versa. An early-timing error is detected if the validation point is not reached within a specified interval after receipt of the local data message. A late-timing error is detected if the validation point is first reached by means of messages from the other replicas and the local data message is not received within the specified time interval.

An impromptu error is detected as a value error if the impromptu message is received *within* the time interval (and of course if its signature is different to that of t other signatures) or as a timing error if no timing window has been opened.

As in the case of fail-restrained replicas, the competitive protocol gives preference to the fastest replica of the group so resynchronization is necessary. The resynchronization is again time-limited in order to detect timing errors.

6.5.2.2. Round-Robin Validate-before-Propagate Mode. In the *cyclic or round-robin mode*, the group of rep_entities is configured as a logical ring with an associated token. When a rep_entity receives a local data message that has not already been sent by another rep_entity then, if it possesses the token, the signature of the local data message is sent to all

---

4   Note also that it is possible to ensure *detection* of t value or timing errors with just $t+1$ replicates; a replica group configured in this way effectively constitutes a fail-silent software component.

members of the group and the token is forwarded to the rep_entity's successor in the ring by means of a *turn* protocol message. The token circulates round the ring until it reaches a rep_entity for which the signature of the local data message is identical to that contained in *t* previous *turn* messages. This rep_entity has thus reached its validation point; it forwards the corresponding data message to its destination(s) and informs its peers that this has been done by means of an *ack* message (containing the majority signature). The other replicas reach their validation point after having received the *ack* message and a concording local data message.

To ensure token recovery, the interval between receipt of the local data message and receipt of the token is monitored. If time-out occurs, then the rep_entity reverts to the competitive mode by issuing a *claim* message (the rep_entity also reverts to the competitive mode should it receive a *claim* message from a peer rep_entity). An early timing error is declared if, after reverting to the competitive mode, the validation point is not reached within a further specified interval. Late timing errors are treated in exactly the same way as in the competitive protocol — by monitoring the time interval between occurrence of the validation point and receipt of the local data message if the former should occur before the latter.

6.5.3. Message-Reception Error Processing

Any message sent to a group of active replicas must, in the absence of faults, be delivered to all replicas to ensure that all replicas can produce the same outputs. Therefore, all messages sent to a software component must be multicasted *atomically* to the corresponding rep_entities who must then forward these messages to their local replicas.

However, to ensure end-to-end flow control, replicas may refuse to accept data messages from their rep_entities if their receive buffers are full — replicas thus have to send explicit handshake messages to their local rep_entities to indicate their willingness to accept the incoming data message. Replicas must unanimously accept all data messages to ensure that they remain synchronized. The logic of atomic multicasting with end-to-end flow control would normally dictate that, if any replica cannot accept an incoming data message, then all replicas must refuse the message.

Now, seen from the rep_entities, the flow-control handshake messages are just another sort of service message (cf. figure 7) that a *faulty* replica could omit to send or send too late. We are thus faced with contradictory requirements in the two following situations:

a) if a non-faulty replica cannot accept an incoming data message, then the flow of data messages to all replicas of the group must cease;

b) if a replica fails by omitting or delaying to send handshake messages: data message flow to the non-faulty replicas should continue.

The solution to this contradiction is to limit the time for which situation (a) can persist before declaring that the replica in question is faulty.

Whenever a data message is to be forwarded to the local replica, the rep_entity sets a timer to await the corresponding handshake message and thus limits the time for which a local transient overflow condition may persist. If time-out occurs then the rep_entity requests its peer rep_entities to send back their local status. The way in which these replies are interpreted depends on the replica failure assumption. If replicas are assumed to be fail-restrained, then all handshake messages sent are correct handshake messages. Therefore, if *any* remote rep_entity has received the handshake message, then the message-refusal situation is judged abnormal and the local replica is declared as failed. If replicas are fail-uncontrolled, then the possibility of an erroneously-produced handshake message must be envisaged. Therefore, when a message-refusal condition exists, it must be assumed to be a normal overflow situation until a majority of remote rep_entities have received their local handshake messages.

6.5.4. Error-Reporting

The inter-replica protocol for active replicas masks errors resulting from failures of a certain number of replicas in a group:

- for fail-restrained replicas: t crash, omission or performance errors can be masked by a group of $t+1$ active replicas;
- for fail-uncontrolled replicas: t arbitrary errors (value, timing or impromptu errors) can be masked by a group of $2t+1$ active replicas.

The inter-replica protocol also ensures that any error caused by a replica is detected locally by its corresponding rep_entity. The latter can then follow two different strategies:

- the rep_entity can abort itself immediately after informing the other members of the group (by a *leave* protocol message) and reporting to the fault treatment facility,
- the error can be reported to the fault treatment facility but the rep_entity remains in the group until it is told to remove itself by the fault treatment facility; this allows soft faults to be tolerated in the case where the internal state of the software component is re-initialized after each output message or, equivalently, is non-existent.

6.5.5. Performance Considerations

A major advantage of the active replication technique is that recovery from a detected replica error is quasi-instantanous since the all replicas in the set are maintained in close synchronization. Of course, the price that is paid for this is that the overall processing power must be increased by at least the degree of redundancy.

At first sight, the competitive mode has a performance advantage over the round-robin mode in that it allows message propagation at the earliest possible opportunity (see annexe G). However, this must be weighed against the higher protocol message traffic that is incurred due to the systematic sending of *claim* messages.

6.6. Passive Replication

When replicas are executed only by fail-silent hosts, it is possible to envisage an alternative replication technique that economises host processor utilization by activating redundant replicas only when they are needed to ensure recovery. A passively replicated software component consists of a group of replicas in which one replica, termed the *primary* replica, processes all input messages and provides all output messages. Since, in the absence of errors, only one replica produces output messages, this technique is only suitable for fail-restrained replicas (i.e., replicas that only fail by crashing, by omitting to send some messages or by sending messages late).

The other replicas in the group are *standby* replicas[5]; each consists of a copy of the code of the software component together with a copy of some previous state of the component from which execution can be resumed should the primary replica fail. The internal state of the standby replicas must be regularly updated by the primary replica: this operation is called *checkpointing*. Standby replicas are passive since, in the absence of faults, they carry out no processing other than house-keeping operations following reception of checkpoints. A previously-passive back-up replica attempting recovery is termed a *substitute* replica.

---

5   Whence the alternative name for passive replication — the *primary/standby* technique.

6.6.1. Checkpointing Strategies

Various strategies are possible for the taking of checkpoints. One technique is to implement *transactional checkpoints* whereby interactions between groups of software components are structured as transactions (cf. §6.3.2) and the primary replica of each software component involved in a transaction checkpoints its state to its back-up replica(s) only when changes to this state are committed (i.e., if and when the transaction terminates successfully).

In the absence of a transactional model of computation, recovery can be ensured on a component-by-component basis if checkpointing is organized in such a way as to prevent the domino effect (cf. §6.3.2). Checkpointing must be carried out in a way such that a substitute replica re-executing from the last checkpoint does not need to request re-sending of previously received input messages and avoids sending duplicate output messages.

To avoid having to request re-sending of previously received input messages, either the back-up replicas must maintain a queue of input messages identical to those received by the primary replica since the last checkpoint was taken or, each time the primary replica receives a message, a new checkpoint must be taken. The former approach has the advantage that checkpointing is less frequent and the communication overhead is thus usually lower, especially if the input message queues are created concurrently with the normal inter-component message flow, e.g., by means of a reliable group communication protocol exploiting broadcast channels.

The sending of duplicate output messages can be avoided by means of either systematic or periodic checkpointing.

Systematic checkpointing involves the creation of checkpoints whenever the primary replica communicates some of its internal data to the outside world, i.e., whenever a message is sent. Thus, rollback to the last checkpoint never requires re-sending of an output message.

Periodic checkpointing is a strategy whereby the number of checkpoints is reduced by only taking them say, every n output messages [Borg et al. 1983]. During recovery, any output messages generated by the substitute replica are checked against a log of previously sent messages and only sent over the network if no equivalent message is found.

Correct recovery using periodic checkpointing requires that replicas be deterministic (cf. §6.4.1) and that messages be received by all replicas in the same order (as for the active replica strategies) so that the substitute replica produces the same messages as those that were produced by the primary replica before its failure. In the case of transactional and systematic checkpointing, this requirement for replica determinism and identical order of input messages is unnecessary since *any* execution based on any order of input messages that respects causality is a valid execution.

Although systematic checkpointing entails more overhead than either periodic or transactional checkpointing, its capacity to accommodate non-deterministic processing and its suitability for the implementation of independently fault-tolerant software components (unlike transactional checkpointing) are very important advantages. It is thus this technique that has been implemented in Delta-4.

In the systematic checkpointing technique, it is important that the transfer of checkpoints to the standby replicas and the sending of data messages by the primary replica to their destination(s) be carried out as a single atomic operation to avoid the domino effect. This could be done by using an atomic multicast protocol to send a combined data and checkpoint message to both the standby replicas and the designated data-message destination(s). However, since checkpoints may be quite large, this would put an unnecessary load on the latter who would have to unpack the checkpoint only to discard it. An alternative approach is illustrated on figure 8 which shows a passively replicated group of three replicas: the primary replica and two back-up replicas.

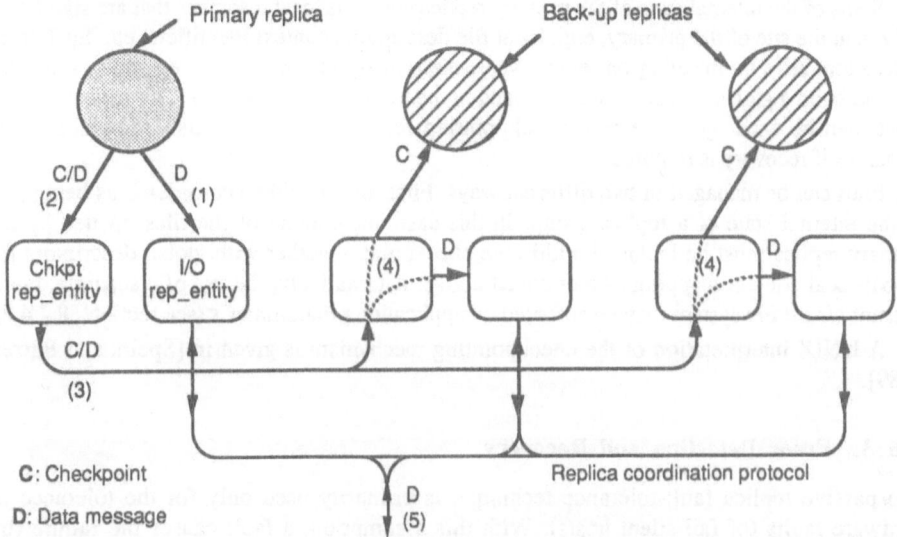

Fig. 8 - Systematic Checkpointing Technique

In this scheme, there are two rep_entities for each replica: a checkpointing rep_entity and an I/O rep_entity[6]. Whenever the primary replica forwards a data message to its I/O rep_entity (1), it also forwards the message to the local checkpoint rep_entity to indicate that a checkpoint must be taken (2). The data message and the checkpoint are atomically multicasted together (3) to the checkpoint rep_entities of the standby replicas. The latter forward the data message to their local I/O rep_entities and update the internal state of the standby replicas (4). All I/O rep_entities are thus in possession of the data message produced by the primary replica. They can thus carry out a message-sending arbitration protocol (5) in exactly the same way as in an active replication technique for fail-restrained replicas (cf. §6.5.1). If the competitive propagate protocol is used in a "rendezvous" mode, then it is ensured that all I/O rep_entities will be synchronized before the data message is sent to its destinations. Since, in the absence of faults, the primary I/O rep_entity can enter the arbitration competition before or while the checkpointing is carried out, its *claim* message will be sent long before the other I/O rep_entities are ready to compete. Consequently, the former will win the competition outright and no other *claim* messages will be sent. If the primary I/O rep_entity should fail, the time-limits on the rendezvous will ensure that the data message is sent by one of the other I/O rep_entities.

6.6.2. Taking Checkpoints

A checkpoint consists of a snap-shot of the "internal state" of the primary replica. Access to this internal state is inevitably specific to host type, local execution support system and the run-time representation of replicas. It must consist of the data space of the replica together with all system information specific to the replica: processor registers, stack pointer, status information, etc. Furthermore, if input messages are multicasted to all I/O rep_entities, checkpoints muwt include information as to which input messages have been processed by the primary replica since the last checkpoint was taken and should thus be discarded from the back-up input queues.

---

6   For the purposes of the explanation, it can be imagined that replicates and rep_entities are separate processes.

Some of the internal state of the primary replica concerns local resources that are significant only in at the site of the primary, e.g., local file descriptors, context identifiers, etc. Such local descriptors have no meaning on remote sites; the primary checkpointing rep_entity must use them to build a global context that is included in the transferred checkpoint. This global context must then be used by the remote checkpointing rep_entities to establish equivalent local resources if recovery is required.

Files can be managed in two different ways. First, files could be considered as being part of the internal state of a replica group. In this case, the content of the files opened by the primary replica must be included within the checkpoint, together with global descriptors for global-local context mapping as mentioned above. Alternatively, the use of a separate, fault-tolerant *global* file system may be enforced on application programmer's (see section 8.2.4.2).

A UNIX interpretation of the checkpointing mechanism is given in [Speirs and Barrett 1989].

6.6.3. Error Detection and Recovery

The passive replica fault-tolerance technique is primarily used only for the tolerance of hardware faults (of fail-silent hosts). With this assumption, a fault causes the failure (by silence) of *all* entities executing on the faulty host. Therefore, error detection can be reduced to the detection of silence of any entity executing on a given host. A set of system entities can implement a node *group membership protocol* based on the exchange of "heartbeat" or "I'm alive" messages. Alternatively, an equivalent node group membership service can be offered as a facility of the communication system (see section 6.9.2). With such a node group membership service, each checkpoint rep_entity of a passive replica group can be informed of the failure of any of the hosts supporting a replica of the group. If the host supporting the primary replica fails then a substitute replica must be selected to carry out recovery from the last checkpoint. This selection may be carried out by a dynamic "election" (for instance, by a competitive protocol such as that used for active replica groups, cf. §6.5.1) or be based on a pre-established ordering between the back-ups. In the latter case, the pre-established ordering must be updated whenever any of the hosts supporting a member of the group fails.

A reduction in failure granularity from that of a complete host down to individual replicas can only be achieved if the failure of an isolated primary replica can be detected. The very principle of passive replication precludes the mutual observation techniques used for replica-level error-detection in active replica groups. Detection of primary replica crash, omission or performance failures could be achieved by several techniques:

a) by monitoring the time interval between a service request and the corresponding reply (a *client-server* model of computation is necessary for this to be possible) — the monitoring can be carried out either by (a representative of) the client issuing the request or by the I/O rep_entity of the primary server who forwards the client's request to the primary server replica[7];

b) by requiring an "I'm alive" response to a periodic local interrogation from the primary's checkpoint rep_entity (this enables the detection of crashes only);

c) by relying on facilities built into the local operating system (e.g., process termination signals, interrogation of the table of active processes, etc.) (this enables the detection of crashes only).

---

7   Note that by the very principle of passive replication, the time needed for carrying out roll-back recovery implies that a performance failure of the primary server replica can only be tolerated if such a failure is defined with respect to a replica response delay set to less than half the maximum response delay (or deadline) permissible for the server as a whole.

The recovery action itself is initiated by the checkpoint rep_entity of the selected substitute back-up replica. The required replica data and stack areas are allocated and initialized from the information contained in the checkpoint. Similarly, the processor registers and stack pointer are updated from the stored values and the substitute replica starts execution at that point in the software component code at which the checkpoint was taken.

6.6.4. Performance Considerations

Passive replication has a few *potential* performance advantages over active replication. First, since hosts supporting passive replicas do not carry out redundant computation (other than that necessary for house-keeping by rep_entities), the computation load is lower than the active replication case; replicas that are active on a host will not be severely penalized by the fact that the host is also supporting passive replicas. Furthermore, when systematic or transactional checkpointing is used — thus permitting input messages to be multicasted to all replicas without ensuring identical orders of reception — the primary rep_entity can submit an input message to the primary replica as soon as it is received, i.e., without having to wait until it is sure that the other rep_entities have received the same message. In the absence of faults, the speed of response to an input message can thus be faster than in the case of active replication.

However, it is important to bear in mind that these potential performance advantages must be weighed against two separate time overheads:

- a *permanent* communication and processing overhead to provide back-ups with checkpoints for backward error recovery;
- a *temporary* processing overhead due to rollback and re-execution from a previous checkpoint, when a fault occurs.

The permanent checkpointing overhead could be prohibitively high for software components with large internal states. However, several optimizations are possible to at least decrease the communication contribution to this overhead (to be weighed against an increase in the processing contribution):

a) As mentioned earlier, the frequency of checkpointing can be decreased if multicasting is used to ensure that all input messages are delivered to both the primary and all back-up I/O rep_entities.

b) Data compression algorithms can be applied to checkpoints before transmission to avoid sending "unused" parts of a replica's data space. Similarly, unused areas of stacks and dynamically-allocated data spaces can be excluded from the checkpoint. If a replica consists of multiple processes, then only the data spaces of those processes that have been scheduled since the last checkpoint need to be included.

c) With appropriate support from system hardware and/or operating system, it would be possible to identify and include in the checkpoint only those memory "pages" that have been modified since the last checkpoint. Alternatively, the primary checkpoint rep_entity could keep a copy of the last checkpoint and, after taking a new checkpoint, only transmit the changes that are identified by comparison with the copy.

Despite these possible optimizations, the overheads due to checkpointing and rollback will usually outweigh the potential performance advantages outlined at the beginning of this section. Even in the fault-free case, it is to be expected that passive replication will provide a lower performance than that of active replication.

6.7. Semi-Active Replication

The previous section has pointed out that active replication has significant performance advantages over passive replication since (a) the communication overheads due to checkpointing are avoided and (b) the time to ensure recovery is lower since redundant computations are executed in parallel. However, passive replication (using transactional or systematic checkpointing) has the important advantage of not requiring replicas to be deterministic. The technique discussed in this section attempts to make the best of both worlds in order to provide speedy recovery despite potential non-determinism [Barrett et al. 1990].

As its name implies, the *semi-active replication* technique is in effect a hybrid between active and passive replication. One of the replicas of the group is termed "leader" and the others "followers"[8]. In the absence of errors, only the leader replica produces output messages (like the primary replica of a passively-replicated group). Since, in the absence of errors, only one replica produces output messages, this technique is only suitable for fail-restrained replicas[9].

On the contrary to the passive replication technique, the other replicas (the followers) are not completely inactive (whence the term "semi-active"); they receive the same inputs as the leader and autonomously execute all *deterministic* computation and update their local state accordingly. The leader is responsible for taking all *non-deterministic* decisions and informing the followers of these decisions by means of *notification* messages or "mini-checkpoints" (figure 9).

As will be seen later, the semi-active replication technique does not require messages to be delivered to all replicas in identical orders.

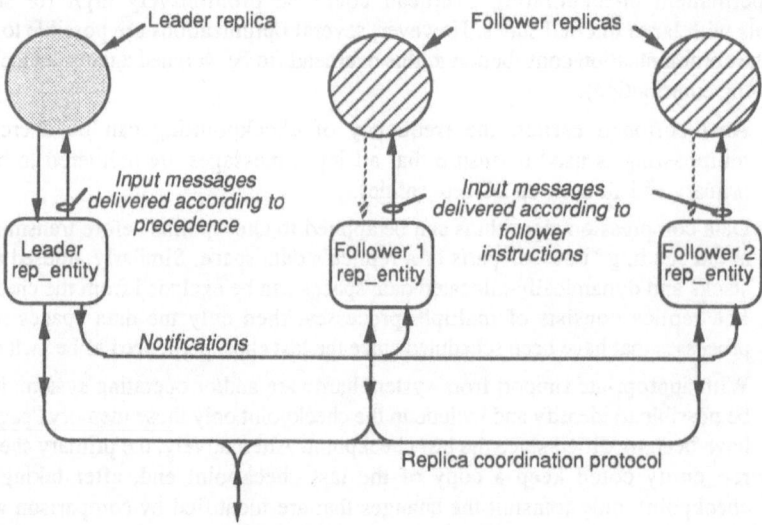

Fig. 9 - Leader-Follower Technique

The non-determinism that the leader-follower technique aims to resolve can stem from:

---

[8]   Whence the alternative name for passive replication — the *leader/follower* technique.
[9]   A combination of active and semi-active replication is considered in section §7.6 that, in certain applications, allows fail-uncontrolled replicas to be accommodated.

a) purposely-introduced preemption in order to improve performance;

b) a requirement to use off-the-shelf software that was not specifically designed with active replication in mind and so cannot be assumed to be deterministic.

In both cases, the rep_entities associated with the group of replicas must carry out a protocol that forces follower replicas to take the same decisions as the leader replica.

6.7.1. Non-Determinism due to Purposely-Introduced Pre-emption

Preemption allows computations of higher priority or *precedence* (see chapter 5) to displace lower-precedence computation. Preemption may be introduced at two levels:

- in the handling of message queues: messages of high precedence may be allowed to overtake messages of lower precedence;

- in the interruption of processes: a high-precedence event within a software component may be allowed to interrupt a lower-precedence process.

6.7.1.1. Message Preemption. Consider a sequence of two input messages multicasted to the rep_entities of the leader replica and the follower replicas. Suppose that the first message to be delivered is of lower precedence (message "*LO*") than the second one (message "*HI*"). To ensure replica group determinacy, the requirement is that all replicas should process the messages in the same order (cf. §6.5). However, in the absence of inter-replica coordination, some replicas could already have started processing message *LO* before message *HI* arrives and would therefore process *HI* after *LO*. Those replicas that have not already started to process message *LO* would find message *HI* at the head of their input queues and would thus end up by processing *HI* before *LO*. The replica group would therefore not process messages in the same order and replica group determinacy would not be ensured.

The leader-follower solution to this situation is to allow only the leader replica to process messages as and when they arrive. Each time that the leader accepts a message *M* from its rep_entity, the latter informs the follower rep_entities by means of a *notification* of the form (cf. figure 9):

• "Present message *M* to your local replica"

Since the leader rep_entity enforces the order of processing of input messages onto the follower rep_entities, messages sent to a semi-actively replicated component need not be multicasted with assurance of identical order — the order adopted by the leader will be enforced on the followers.

An alternative solution would be for messages to the group to be sent only to the leader rep_entity and for the latter to forward these to the follower rep_entities as and when the leader replica consumes them. However, since the notification messages will generally be much shorter than input messages, the performance advantages of concurrent multicasting of input messages would be lost.

6.7.1.2. Process Preemption. Consider now the case of process interruption. It may be required that a low-precedence process be interrupted by the arrival of a high precedence message or signal. Clearly, such preemption must be synchronized across all replicas if replica group determinacy is to be ensured (see, for example [Frison and Wensley 1982, Sheridan 1978]). The leader-follower technique enables such synchronization to be achieved by forcing the follower computations to be preempted at the same point as the leader's computation.

The technique makes use of the concept of a *preemption point* that is a predefined point in the computation of a software component at which it may be preempted. Each time the leader reaches a preemption point, the leader rep_entity increments a counter. When a message *M* arrives at the leader rep_entity, a check is made to determine whether M requires the leader to be preempted. If so, the preemption point P at which this will take place is selected (the current counter value plus 1 represents the next preemption point). The leader rep_entity informs the follower rep_entities by means of a *notification* of the form:

- "Present message M to your local replica at preemption point P"

Since the preemption point code inserted in the software component must be executed more often than the maximum allowable preemption delay, it is essential that the normal, non-preempting path through this code be efficient.

6.7.2. Non-Determinism in Off-the-Shelf Software

The use of off-the-shelf application software is often a commercial necessity. However, it is not always easy to provide transparent fault-tolerance in the underlying hardware. The only totally transparent way of doing so is to build a tightly-synchronized fault-tolerant machine constructed to provide an interface to the application that is identical to that of the (non fault-tolerant) machine for which the application software was written. This approach to fault-tolerance is diametrically opposite to ours and as such, several advantages of our approach would be lost (cf. section §6.1).

The problem with off-the-shelf software in our approach is that it will probably not have been designed with fault-tolerance in mind — in particular, such software may use non-deterministic language constructs or host-specific information that would violate replica determinism. Two cases need to be considered, the case of "white-box" off-the-shelf software whose source code is easily available and "black-box" software whose internal structure is unknown.

6.7.2.1. White-Box Software. A white-box application is one whose source code is easily available and can thus be inspected to implement mechanisms to intercept non-deterministic decisions. An example would be an Ada application using non-deterministic constructs such as *Interrupt*, *Delay* or the *Clock* function[10].

To implement the required system call interception mechanisms, the rep_entities must be placed "between" each replica and its local executive. When the leader rep_entity intercepts a system call C, it informs the follower rep_entities (who will intercept the same call) of the corresponding system reply R by means of a notification of the form:

- "When you intercept call C, substitute the reply R"

A similar mechanism could be used if the leader reads some non-replicated peripheral and also to ensure that preemption occurs at the same point in all replicas.

The number and frequency of such leader-follower notifications therefore depend on the White Box. Exceptionally, it may be unacceptably high and the White Box could then only be supported after some source-code modification to reduce the frequency of non-deterministic actions.

6.7.2.2. Black-Box Software. A black-box application is one whose internal structure is totally unknown, although a description of its external interfaces is normally available. Black boxes can be used if the rep_entities act as "front ends" to interface each replica to the rest of the system.

A typical example is that of a commercial database system such as Oracle.

---

[10] Note that we do not assume that host clocks are necessarily synchronized to some common reference.

Passive replication of an Oracle database would be unsuitable because of the difficulty of checkpointing its multiple processes, shared memory and disc data.

Active replication would be more suitable but since we do not know how Oracle replicas would process concurrent inputs, the risk of non-determinism eliminates this choice. Even if inputs are presented to Oracle replicas in the same order, we cannot be sure that they will be processed in the same order due to scheduling decisions carried out within the black box. This is particularly important for concurrent *lock* requests. One solution would be for rep_entities to present inputs to replicas only after having received the reply for the previous input. However, this sequentialization would lead to rather poor performance.

Semi-active replication allows more concurrency but the leader and follower rep_entities must be more complex. The leader rep_entity needs to be able to interpret incoming requests and local responses and instruct the follower rep_entities according to their semantics, e.g.:

- "Pass *lock(item)* to your local replica since leader has granted this lock"
- "Discard *lock(item)* since leader says item is already locked"
- "Discard *read(item)* since leader has already replied"
- "Takeover the leader role since my replica has crashed"

The important point to grasp is that such protocols cannot be generic but specific to a particular black box. Another black box would require different programming of the rep_entities and a different set of *notifications*.

6.7.3. Error-Detection and Recovery

The semi-active replication technique, like passive replication, is primarily intended for the tolerance of hardware faults (of fail-silent hosts) — error detection can thus be reduced to the detection of silence of any entity executing on a given host and the requirements for a node group membership facility are the same as those already discussed in section §6.6.3 for passive replication.

A reduction in failure granularity from that of a complete host down to individual replicas can also be considered in the same way as for the passive replication technique. Alternatively, since follower replicas are in fact active, the set of replicas can be managed so as to detect excessive desynchronization between the leader and the follower replica(s).

The recovery action is simpler than in the passive replication case since — by its very principle — the semi-active replication technique ensures that the internal state of follower replicas is *almost* consistent with that of the leader replica. When the leader replica (or it's host) is detected as having failed, a follower replica is selected to take on the role of leader and brings itself up to date by processing the messages present in its input queue. The selection of a new leader may be carried out either by a dynamic election or be based on a pre-established ordering between follower replicas.

6.7.4. Performance Considerations

The whole purpose of the semi-active replication technique is to be able to reap the performance advantages of the active replication technique and the ability to accommodate potentially non-deterministic processing like in the passive replication technique.

In the presence of faults, the semi-active replication technique will ensure a recovery delay that is limited to the maximum allowable skew between leader and follower replicas.

In the absence of faults, the semi-active replication technique may provide a better performance than either the active or passive replication techniques. First, the relaxation of the constraint on input message order (identical order between replicas is not required) means that, like in the passive replication case, the leader rep_entity can submit an input message to the leader replica as soon as it is received, i.e., without having to wait until it is sure that the other rep_entities have received the same message. Second, the overheads due to the transmission of notification messages can be expected to be much smaller than those due to checkpoints in the passive replication technique. In fact, since it is expected that notification messages will be much shorter than normal input messages then, due to the aforementioned advantage of immediate processing of input messages by the leader replica, the semi-active replication technique should be of better fault-free performance than the active replication technique.

6.8. Group Management and Fault Treatment

The installation and the management of groups are the responsibility of system administration (see sections 8.2 and 9.5). There are three sorts of groups to be managed:

- the group of fault-free *nodes* or *stations* in the system,
- groups of software component *replicas*,
- groups of *software components* (each of which may or may not be replicated).

Membership of the *node group* can be managed as a function of an underlying multicast protocol (cf. §6.9.2). When a message is multicasted, explicit acknowledgements from the designated destinations enable the presence of nodes to be established. Multiple retransmissions can be considered to resolve the ambiguity between lack of acknowledgement due to node failure or due to a transmission error (see chapter 10). Nodes may leave the node group either in an orderly fashion (by issuing an explicit disconnect command) or suddenly — due to local error detection by fail-silent hardware. Nodes must enter the node group by means of an explicit join procedure that enables their presence to be detected consistently by all other nodes in the group.

Membership of *replica groups* must first be established when a group of replicas in brought into being for the first time. The number of replicas of a particular component is determined by a replication policy that takes into account whether hosts are assumed to be fail-silent or fail-uncontrolled and the cardinality of the specified replication domain of that component. Memberships of replica groups are managed by "replication domain managers" that, like any other software component, may also be replicated. The recursion stops with a "replication superdomain manager" whose member replicas are installed at system (re-) boot time (see sections 8.2 and 9.5).

Logically-distinct *software components* (i.e., *not* replicas) may also be gathered together for some cooperative interaction (e.g., a group of servers providing a "similar" service or a group of transaction managers). Groups such as these provide useful communication abstractions that can simplify distributed application programs. The multipoint associations of the Delta-4 MCS communication system provide such a group abstraction (see section §8.1.3.3). Note that the notions of replica groups and software component groups are orthogonal — each member of a software component group may or may not be replicated.

The replica coordination techniques discussed in sections 6.5 to 6.7 were concerned with the processing of errors so as to hide replica failures from the rest of the system. From the group management viewpoint, the detection of errors may or may not result in the immediate withdrawal of a replica or a node from its corresponding group. Locally-detected errors (i.e., by a fail-silent host or by a NAC) lead to the immediate removal of the node from the set of working nodes (cf. §6.6.3). Similarly, the consequence of a *replica* failure is (usually) the removal of a replica from its group (cf. §6.5.4). Note that failures of multiple replicas (of different software components) residing on the same host would suggest that the fault lies in the host hardware or in its configuration. In this case, a system-level diagnosis function should conclude by a declaration of host failure.

The set of functions necessary for system-level diagnosis and coordination of the actions necessary to remove and re-create group members is referred to here as *fault treatment*. More specifically, fault treatment consists of:

- fault diagnosis,
- fault passivation,
- system reconfiguration, and
- system maintenance.

Fault diagnosis is necessary to (a) localize the fault (at host level or replica level) and (b) to decide whether the fault is solid or soft (cf. chapter 4). If fault diagnosis should conclude that a solid fault has occurred, then fault passivation must be carried out.

Fault passivation is necessary if it is judged that the faulty entity could cause further errors. Fault passivation is carried out automatically and autonomously in the case of hardware-detected errors that result in silence of a host or a NAC. However, in the case of replica failures whose cause is later diagnosed to be a solid host fault, an explicit fault passivation action must be carried out.

System reconfiguration can be envisaged if there are sufficient redundant resources. It entails re-allocation and re-initialization of the software component replicas that have failed in order to restore the level of redundancy required for the error-processing protocols to function correctly despite further faults[11]. If re-allocation is not possible, then some software components may either have to be abandoned in favour of more critical ones. Alternatively, fault-tolerant operation is degraded to fail-safe operation to ensure safety and/or integrity of the distributed application(s). In the absence of sufficient resources, system reconfiguration will have to be deferred until a node recovers (following maintenance).

Re-allocation of software component replicas is achieved by means of a *cloning* operation that creates a new replica on a specified node. Three sub-operations can be identified:

a) creation of a *template* of the software component at the new location; this can sometimes be done in advance of an actual cloning request in accordance with some application-specific contingency plans,

b) creation of a copy of the component's persistent data or "state" at the new location (this is equivalent in effect to the checkpointing operation necessary for passive replication, cf. §6.6),

c) activation of the new replica whilst ensuring group-consistency; this involves the automatic management of the dynamic, configuration-dependent associations between replicated components.

Note that in the case of nodes possessing stable storage (cf. section 6.2.4), node-recovery can allow sub-operation a) above to be carried out completely locally (from a locally-stored template). Similarly, sub-operation b) above may be replaced by an operation recovering some previously-stored consistent state of the replica if the distributed computation is carried out according to a transactional model (cf. section 6.3.2).

---

[11] Note that in the event of a fault diagnosed as a *soft fault*, the re-allocation and re-initialization of replicas can be carried out on the same node as where they resided before failure.

6.9. Communications Support

The previous sections dealt with techniques based on "macroscopic" replication of software components and software-oriented error processing and fault treatment, in a distributed environment. In Delta-4, these techniques rely on basic services such as inter-replica coordination and group membership management, as explained in section 6.4. In the sections that followed, it became apparent that providing these services places some demands on the communications support system, depending on the particular replication technique.

The implementation of distributed fault-tolerance in Delta-4 benefits from the availability of high-quality communication services, such as the ones materialized by protocols designated as *reliable broadcast* or *multicast* [Birman and Joseph 1987, Chang and Maxemchuck 1984, Cristian et al. 1985]. Furthermore, since openness and versatility are desired, the reliable communication service is designed to operate over widely-used local area networks.

This section surveys the properties of a communications system that are desirable to support fault-tolerance based on groups of replicas of components residing in different nodes of the system. The group communication service will be discussed in more detail in chapter 10.

6.9.1. Support for Replication

The replication techniques presented earlier in this chapter have different inter-replica interaction requirements. These requirements are satisfied by appropriate properties of the communication service.

Active replication has a determinism requirement that obliges messages to replicas to be delivered to all of them and in the same order: this is called unanimity and total order (see section 10.1). A service providing these properties is called *atomic*.

Passive and semi-active replication rely on a privileged participant, which is the representative of the replica set. The presence of such a representative allows the unanimity and total order requirements to be relaxed. This can be achieved by using auxiliary protocols at a higher level that take advantage of semantic knowledge. However, in the techniques exploited in Delta-4, at least unanimity may be advantageously preserved, given that: follower replicas have to execute the same commands as the leader replica, checkpointing to standby replicas affects all equally, and changes in the replica set (re-insertion, takeover, etc.) should be perceived consistently. This is especially important when there are more than two replicas in a set. A service providing unanimity alone (and at most ordering messages from individual senders in the order sent, i.e., FIFO) is called *reliable*.

Protocols that order messages according to a cause-effect relationship are called *causal* protocols. For the sake of generality, the communications service should provide causal order, since it is the genuine ordering of events in a distributed system (see section 10.1). However, the cost of causally ordering messages may be avoided in certain settings (see section 9.5.1 for a detailed discussion). In short, this happens when it is shown that there is not a causal relation between senders in different nodes, either because of a restricted concurrency of the computation model, or because the semantics of the application is known not to require such an order. Simpler, non-ordered or FIFO protocols can then be used.

6.9.2. Support for Groups

A modified form of broadcast, where messages only arrive at a subset of the possible system destinations, is called multicast. It is the basis for *group communication*, and is an efficient way of disseminating information.

Multicasting efficiency can be easily increased, if *logical addressing* is used. This allows delivery of messages addressed to "whom-it-may-concern", i.e., with transparency (to the sender) of the number and location of recipients. In Delta-4, system-level logical designation is mapped into communication-level logical addressing.

The group paradigm in Delta-4 allows independent sets of replicas to be supported simultaneously, whether they use the same or different replication techniques. Such sets of replicas could be, for example, several independent fault-tolerant applications in the same system, or several groups working in parallel for the same fault-tolerant application, such as groups of replicated clients, accessing the same group of replicated servers. Several group communication services are provided, to adequately support the different techniques.

Inside a group, it is important for participants to observe the changes in its composition in a manner consistent with their semantics. Group composition must be known, for example, to check whether the actual members gather the necessary functionalities to execute a distributed application, to activate recovery procedures, to reestablish the level of redundancy, etc. In consequence, the most general service of group management, is to provide a *consistent group view* to participants, i.e., each change in group composition is indicated, in a total order, to all participants in that group. This may be used to facilitate the implementation of high-level group management procedures, like group membership, group replication and group cooperation management.

As a rule, any replication technique may take advantage of logical group communication facilities, enhanced with group housekeeping, like the consistent group view property. These simplify implementation of error detection and recovery and error masking protocols. This observation will be confirmed in chapters 8 and 9.

The group communication service of Delta-4, called xAMp, is discussed with detail in chapter 10.

6.10.   Conclusion

In this chapter, after having discussed the failure assumptions that can be made for the different elements of a distributed system, we have outlined a simple hardware architecture — based on fail-silent network attachment controllers — that can accommodate the worst-case failures of fail-uncontrolled computers. Three basic software component replication strategies have been discussed — table 1 summarises the relative merits of each.

Table 1 - Comparison of Replication Techniques

Replication technique	Recovery overhead	Non-determinism	Accommodates fail-uncontrolled behaviour
Active	*Lowest*	*Forbidden*	*Yes*
Passive	*Highest*	*Allowed*	*No*
Semi-active	*Low*	*Resolved*	*No*[12]

In the Delta-4 Open System Architecture (OSA, see chapter 8), the active replication technique is implemented within the MCS communication system. The inter-replica protocol is

---

[12] An extension of the semi-active replication technique to accommodate fail-uncontrolled behaviour is presently being investigated (see section 7.6).

situated within the MCS session-layer and the replication coordination entities of section 6.5 are in effect protocol entities executed on the fail-silent NAC hardware (see section 8.1.4.3). Since it is implemented within the communication system of OSA, active replication is therefore possible in OSA independently of the host computational support environment. The passive replication technique is also available in OSA when the Delta-4 Application Support Environment is used (see chapter 7). The checkpoint rep_entities (cf. figure 8, section 6.6) are implemented as part of the object envelopes of replicated Deltase capsules and use is again made of the MCS session-layer inter-replica protocol whose protocol entities now play the role of the I/O rep_entities of figure 8. Semi-active replication is used in OSA in order to implement a dependable database system (see section 7.6); the leader/follower rep_entity functionality (cf. figure 9, section 6.7) is split between a Deltase *transformer* and the MCS inter-replica protocol.

The Delta-4 Extra-Performance Architecture (XPA, see chapter 9) is intended to accommodate only fail-silent hosts. The semi-active replication technique was pioneered in this architecture since, as explained in section 6.7, two of the primary objectives of semi-active replication are to allow high performance and to resolve non-determinism due to preemption. The leader/follower rep_entities are implemented as part of the XPA group managers (see section 9.6.1.3).

The present implementations of passive and semi-active replication assume a failure granularity equal to that of a complete host — reduction of the failure granularity down to the level of individual replicas (cf. section 6.4.2) is under consideration.

Chapter 7

Delta-4 Application Support Environment (Deltase)

The Delta-4 system architecture is intended for open, dependable, distributed systems. The purpose of Deltase, the *Delta-4 Application Support Environment*, is to provide a *homogeneous computing environment* for the support of distributed applications in Delta-4 systems.

This chapter covers the concepts of Deltase, and its generic functions and features. The actual functions and features in any particular implementation will be described in the literature related to that implementation. The prototype implementations within the Delta-4 project are described in the *Delta-4 Implementation Guide* [Delta-4 1990].

7.1. Purpose and Background

The decomposition of an application into a number of *language-level* program modules, which interact with one another by means of procedure calls, is a widely-used methodology for applications that are implemented as a single program and executed on a single machine. The aim of Deltase is to support this application structure in *distributed* systems; that is, to extend, to distributed fault-tolerant systems, an application structure that is widely-used within applications for execution on a single computer.

Deltase provides a uniform *virtual execution environment* for the execution of, and interactions between, *language-level* program modules (see figure 1). This virtual execution environment can be mapped onto a distributed system (or a part thereof), but conceals both the physical configuration of the distributed system and the *mapping* of the program modules onto the hosts.

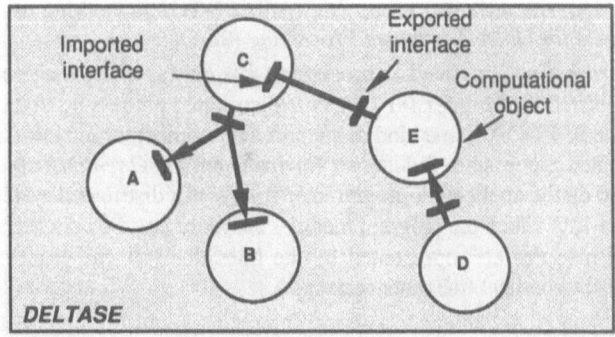

Fig. 1 - Programmer's View of Object-Based System

Deltase is *generic*, and many different realizations are possible. Each host machine that supports this virtual execution environment requires an implementation of Deltase, which would normally interface to the *Local Executive* (LEX), and support one of the languages already available for that host. The Delta-4 *Open System Architecture* (OSA) allows heterogeneous hosts within a single distributed system, with inter-working between the implementations of Deltase for the different hosts, LEXes and language systems.

Deltase is intended to allow migration to future technology, by permitting implementations for new hosts, LEXes, language systems, etc., as these become available.

While the Delta-4 system architecture is aimed primarily at large (distributed) systems, Deltase is intended to cover a wide range of systems, from a single machine (without fault-tolerance) up to large-scale Delta-4 systems. While this chapter is concerned mainly with the use of Deltase in distributed fault-tolerant systems, applications written for use with Deltase may equally be executed within a single machine with a suitable implementation of Deltase. (Since fault-tolerance and distribution are made *transparent* to the application programmer — (see §7.2), such applications can equally-well be executed on systems that do not provide such features.)

Both the Delta-4 *Open System Architecture* (OSA) and the Delta-4 *Extra Performance Architecture* (XPA) include Deltase; Deltase is mandatory for XPA, but is optional for OSA. The requirements for the support of fault-tolerance (in both OSA and XPA) and real-time behaviour (in XPA) impose particular constraints on Deltase.

7.1.1. Open Distributed Processing (ODP)

Deltase is based upon the concepts of the emerging work on *Open Distributed Processing* (ODP), and its related standardization; see [ECMA TR/49]. ODP is concerned with defining both a *generic architecture* for distributed processing systems, and the standards to support this generic architecture.

The ODP standardization work in the *International Standards Organization* (ISO) is concerned with the definition of a *Reference Model for ODP* (RM-ODP); this work is undertaken by ISO/IEC JTC1/SC21 WG7. This reference model is to ODP, what the *Open Systems Interconnection (OSI) Reference Model* was to OSI. The RM-ODP is concerned with architecture that is intended to be applicable to most kinds of (distributed) application.

The ODP standardization work is supported by the *European Computer Manufacturers Association* (ECMA); the technical group ECMA/TC32-TG2 is working on standards for a *Support Environment for Open Distributed Processing* (SE-ODP).

Distributed processing is defined as processing that may span separate computer address spaces, and *open distributed processing* concerns distributed processing that conforms to the requirements of the RM-ODP. Some kind of support environment, or infrastructure, is required to support distributed processing. A *Support Environment for Open Distributed Processing* (SE-ODP) is based on the application programmer's view of a distributed system. This view is illustrated in figure 1, in which the program modules are represented by circles, and interactions between program modules are represented by the lines that link the circles, within an all-pervading support environment (the outer rectangle).

An SE-ODP is an *environment* to support interactions and bindings between program modules. Each program module represents a separate address space, and these address spaces may be located in a number of physically separate machines; a number of address spaces may be co-located within a single machine. An SE-ODP facilitates the construction, operation and maintenance of distributed applications. Standardization of the SE-ODP is necessary where there is a requirement for heterogeneity and multi-vendor procurement.

An SE-ODP may be viewed as a means to provide interaction between distributed program modules, and nothing more. If provided in a sufficiently general way, this *one* function (interaction between distributed program modules) is a sufficient basis for organising the provision of arbitrary function. Therefore, the SE-ODP is an *"enabler"* for distributed processing in general, and need itself have no other function.

The role of an SE-ODP is thus different from that of a conventional operating system, which manages resources and offers rich functionality. Under an operating system, the application interface is that between an application and the operating system. For an SE-ODP, the application interfaces of concern are those *between program modules*.

The current version of Deltase is a *prototype SE-ODP*, providing Deltase with a firm conceptual basis. The long-term objective is the convergence of Deltase with the relevant SE-ODP standards. Conformance to ODP standards will provide an assurance to potential users that they will not become locked into a proprietary system architecture.

7.1.2. Viewpoints

The concept of viewpoints is one that has already emerged from the ISO RM-ODP work. Viewpoints provide a framework for the discussion of distributed processing systems, based on the recognition that a given distributed system can be considered from a number of different points of view. Each viewpoint corresponds approximately to the view that some group of people within an enterprise (or user organization) will have of the distributed processing system:

- The **enterprise viewpoint** is essentially that taken by those responsible for the enterprise (or user organization) as a whole, and is therefore concerned with models of the overall objectives of the enterprise, and the part played by the distributed processing system in achieving those objectives. This viewpoint is concerned with what the system does for the organization, and how the system interacts with the rest of the organization.

- The **information viewpoint** is typically the view of a system designer or systems analyst, and is concerned with models of information structures and information flow. From the information viewpoint, the system specification is of a functional, rather than algorithmic, nature.

- The **computational viewpoint** is that of the application programmer, and models the application software as a set of objects that interact with one another in a defined manner. These objects would typically represent language-level program modules. From the computational viewpoint, these program modules are (by default) independent of the particular computing environment on which they will be executed.

- The **engineering viewpoint** is typically that of a system programmer, and is concerned with how to *engineer* support for the distributed application described by the computational model. The engineering viewpoint is concerned with achieving the required quality attributes for the distributed processing system; in particular, dependability (in its various forms), performance, and real-time responses. The engineering viewpoint is also concerned with the provision of abstract mechanisms to enable the application program modules to be independent of the computing environment.

- The **technology viewpoint** is that of the system builder, and models the computing environment in terms of its basic commodities (computing hardware, operating system software, communication systems, etc.).

Each of these viewpoints gives a different view of a given distributed processing system; these views are not layers, but are views that focus on different concerns.

Deltase, and SE-ODPs generally, are primarily concerned with the *computational and engineering viewpoints*. These are described in the next two sections.

7.2. Computational Model

This section concerns the computational viewpoint, and describes the computational model supported by Deltase. This model defines the rules for the behaviour of computational objects that can be supported by Deltase. Only those computational objects (and by extension, distributed applications) that conform to this model can achieve the full benefits offered by Deltase: transparent distribution, transparent fault-tolerance, portability, independence from LEX, etc.

The model is based on the support of *language-level computational objects*. Since Deltase is a prototype SE-ODP, the computational model primarily concerns the *interaction* between computational objects.

In the *black-box* view, a distributed application (figure 1) is viewed from the outside; the computational objects are seen as *black boxes*, which interact via defined interfaces. This is the view of, say, an application designer.

The *white-box* view is of *one* computational object, from inside that object; other parts of the system are seen only by the *interfaces* used for interaction. This is the programmer's view, and concerns the programming of a computational object in such a way as to conform to the Deltase computational model.

Many of the concepts on which this computational model is based are inherited from ODP and the *Advanced Network System Architecture* (ANSA, see [ANSA 1987, ANSA 1989]). However, there are a number of areas where the Delta-4 project has contributed, particularly in recognising the importance of language mappings.

7.2.1. Distribution Transparencies

Deltase offers a virtual environment that conceals, from the computational objects, the *consequences* of distribution and of the support of distributed fault-tolerance. Those consequences of distribution that are hidden from the application program are referred to as the *distribution transparencies*, as follows:

- **Access transparency** allows interactions between computational objects to be programmed independently of the actual low-level mechanisms used to implement the interaction. This means that the programmer is not concerned with the characteristics of the LEX.

- **Location transparency** ensures that bindings between objects are independent of the communication route that connects them. This means that the programming of interactions between objects is independent of the relative locations of the interacting objects.

- **Migration transparency** allows the re-location of an object in a distributed system without making that change of location (migration) apparent to the interacting objects.

- **Replication transparency** allows replicas of an object to be used, without the existence of the replicas being apparent to either the replicated object or to any other object that interacts with the replicated object. Within the Delta-4 system architecture, replication is essential for the provision of fault-tolerance. Replication may also be used in Delta-4 and other architectures to increase performance.

- **Fault transparency** allows inter-object interactions to be programmed without regard to the possible existence of certain classes of fault within the distributed system.

- **Concurrency transparency** allows the concurrent use of a resource, without permitting inconsistent use of that resource.

- **Language transparency** allows the programming of inter-object interactions to be independent of the language system used, such that the programming of one object is independent of the language system used for the generation of the other object.

- **Machine transparency** allows the programming of inter-object interactions such that the programming of an object is independent of the particular type of host machine on which any other object may be executed.

These distribution transparencies hide differences that might arise from differences in processor hardware, in operating system software, in programming languages or in software development systems used for constructing or executing the application objects. In short, the distribution transparencies hide the consequences of *distribution* from the objects.

From the computational viewpoint, the distributed application is independent of the mapping of the computational objects onto the underlying computing environment; similarly, the replication of objects is transparent, so that this description makes no further reference to replicated objects.

7.2.2. Computational Objects

From the computational viewpoint, a *distributed application* consists of a number of (language-level) program modules, which can interact with one another. This is illustrated in figure 1.

The important feature of this computational model is that the program modules are *computational objects*; each computational object encapsulates some *state* (data) and the *operations* on that state.

From the computational viewpoint, these operations provide the only means whereby other computational objects may read, or cause changes to, the state encapsulated by a computational object. Direct manipulation, by one computational object, of the state of another computational object is not permitted, and, in the general case where two interacting computational objects are executing on distinct host machines, will not be possible.

The following text, adapted from [Golberg and Robson 1981], provides a clear description of the principles of object interactions:

> "The set of *operations* to which an object can respond is called its interface with the rest of the system. The only way to interact with an object is through its interface. A crucial property of an object is that its private memory can be manipulated only by that object's own activity. A crucial property of *operations* is that they are the only way to invoke an object's activity. These properties ensure that the implementation of one object cannot depend on the internal details of other objects, only on the *operations* to which they respond."

(This text has been adapted slightly to bring its terminology into line with that used in this book. In particular, the word "message" in the original has been changed to *operation*; from the computational viewpoint, interactions take place by the invocation of operations and the return of results. From the engineering viewpoint, such operations will be seen as the passing and return of messages.)

The implementation of the object's encapsulated state, and of the operations supported, is not visible to other computational objects.

7.2.3. Service Interfaces

From the computational viewpoint, object interactions are based on the offering and invocation of services, via clearly-defined interfaces.

The set of operations supported by a computational object (and made available for invocation by other computational objects) constitutes the interface to that object. In many cases, these operations fall into a number of distinct sets, and it is convenient to treat such a set of operations as a distinct interface, referred to as a *service interface*.

The distinct service interfaces supported by a computational object give different "views" onto that object. For example, in the case of an object offering a file service, there might be a general interface providing the common file-handling operations, and a specialized interface for management of the file service (available to only a limited group of users). Different service interfaces may be made available to different groups of users.

A *service interface specification* comprises the set of specifications of the operations within that service interface; each operation is defined in terms of:

- what parameters (including the type of each parameter) must be supplied when invoking the operation;
- what results will be returned (again, including the type of each parameter in the result);
- what *visible* effect this operation will have on the rest of the system;
- what effect (if any) this operation will have on the results of subsequent operations on that object;
- any constraints on the ordering of operations (e.g., a file cannot be accessed until it has been opened, etc.).

The specification of a service interface should exist independently of any service user(s) and service provider(s), since there should be many possible implementations of a computational object to support that interface. Within a system each service interface specification is assigned a *name* (the service interface type name) that is unique within the chosen naming context.

A *service provider* is a computational object that *provides* the services defined by a particular service interface, as defined by the service interface definition.

A *service user* is an object that *uses* (some or all of) the services defined by a particular service interface.

A service interface is *asymmetric*; the service user and service provider have complementary relationships to the service interface; the *service user* invokes operations on the *service provider*.

Figure 2 shows the elements involved in the interactions between two computational objects A and B. Object A is the *service user*; object B is the *service provider*. This interaction is constrained by a particular service interface; the only interactions supported are the operations and responses defined by that interface. A *D-association* is the logical path via which the computational objects interact; (see §7.2.5).

A given computational object may, of course, be both a user of some services and a provider of other services; the terms "service user" and "service provider" refer only to their relationship to the particular service interface, and not to the computational object as a whole.

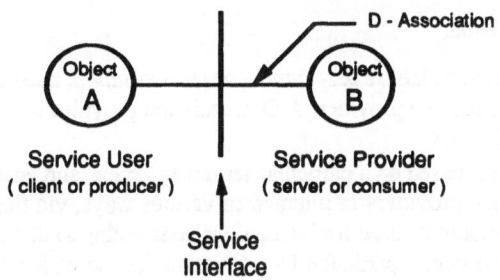

Fig. 2 - Simple Object Interaction

7.2.4. Remote Procedure Call (RPC)

The invocation of an operation in another computational object, with the passing of parameters and the return of results, is a direct extension, to distributed systems, of the widely-used procedure call mechanism provided by languages such as "C", Ada, etc.

The *Remote Procedure Call* (RPC) mechanism provides a means of programming the interactions between objects using existing language constructs, with each operation in a service interface treated as a separate procedure; the operation is invoked by a call on that (remote) procedure. This concept may be applied to both the service user (which invokes the operation by means of the procedure-call mechanism) and the service provider (in which the operation is written as a procedure).

The remote procedure call approach means that the same language construct, the procedure call, can be used both for the invocation of a service that is provided internally by a local procedure, and for the invocation of a service provided externally by another object.

At the programming-language level, the invocation of a remote operation is expressed by means of a procedure call with the parameters as language-level typed entities. The important point to note is that inter-object interactions can be programmed directly in the high-level language, with all the language-based type-checking mechanisms at both the calling and the called objects. The use of RPC provides an assurance that both the service user and the service provider conform to the same interface (the service interface specification); this is an important feature for distributed applications, and one that is not provided where the objects generate their own messages. Remote procedure calls provide a way of abstracting away from messages, and using language-level facilities for programming the interactions between computational objects.

By definition, a *local* procedure call (when the call is part of the same program as the procedure called) causes immediate entry to the called procedure. In order to keep the semantics of a *remote* procedure call as close as possible to the semantics of a local procedure call, a remote procedure call (where the called procedure is not part of the same program as the procedure call) also causes the remote procedure to be executed (and any results returned), before execution of the calling program is permitted to proceed beyond that remote procedure call.

Additional information on remote procedure calls can be found in [Birrell and Nelson 1984] (an important early paper on the implementation of RPCs), and the ECMA RPC standard [ECMA 127].

7.2.5. D-Associations

For two objects to interact, a *Deltase association*, or *D-association*, must be established between the service user and the service provider. A D-association provides the logical path via which computational objects interact.

Each D-association is based on a particular service interface, and enables a logical group of service users and service providers to interact, in various ways, via that service interface. A given D-association can only be used for interactions conforming to that service interface. Each computational object that joins a particular D-association does so with a defined role, either as a service user or as a service provider.

There may be two or more D-associations based on a given service interface type, each of these D-associations having its own set of service user and service providers.

Two computational objects can interact (via a particular service interface) only when both are members of the same D-association and have complementary roles (one as service user, the other as service provider). Finding the required D-association is achieved by the "trading" of service interfaces (see §7.3.3).

One computational object may be a member of several D-associations, with roles according to whether the computational object is a user of those services or a provider of those services. The white-box view of a D-association is a channel through which service requests are sent (in the case of a service user) or received (in the case of a service provider).

In order to support a variety of interaction styles, and, in particular, producer-consumer interactions (see §7.2.6), a D-association may include *multiple service providers* as well as multiple service users. In the *black-box* view, a D-association is modelled as a *multi-point virtual circuit*. This is illustrated in figure 3.

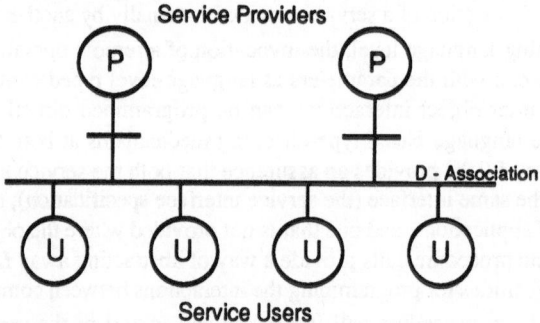

Fig. 3 - D-Associations

For a replicated object, all replicas will be members of the same set of D-associations, although this replication is not visible from the computational viewpoint.

A distinction is made between *creating* a D-association and *providing a service* via that D-association. A D-association is normally created by a service provider, but may, instead, be created by an *agent* (which then makes no further use of that D-association). Once a D-association is established, users and providers of that service may join it.

When a service user issues a service request, that request is sent to all service providers on the selected D-association. Multiple service providers are permitted only on those D-associations where none of the operations results in a reply to the service user, since a service user cannot be expected to handle multiple replies. This type of D-association is intended for the handling of producer-consumer operations, allowing a single service request to be multicast to a number of consumers.

Interface trading is the process of finding a suitable D-association, based on a particular service interface.

For a service user, interface trading is concerned with finding a D-association based on the required service interface, and providing access to a suitable service provider. This is done by searching a catalogue of D-associations (which is effectively a catalogue of services offered); the D-associations are identified by the service interface type name.

This approach to the establishment of D-associations is suitable for use in open systems, and does not rely on prior knowledge of the identity of either the D-association or the service provider.

From the computational viewpoint, interface trading may be invoked explicitly by the application programmer to create, join or leave particular D-associations. By default, interface trading will be carried out transparently. The *trading system* is discussed in section §7.3.3 as part of the engineering model.

7.2.6. Interaction Styles

There are two distinct styles of interaction (illustrated in figure 4), as follows:

1) **Client-server** interaction operates on the principle of a *handshake*; the service user invokes an operation on the service provider; on completion of the specified operation the result is returned to the service user. This is the interaction style of the remote procedure call, as described above. A "null" reply may be used to indicate completion of the invoked operation without passing any results.

2) **Producer-consumer** interaction is a *fire-and-forget* interaction; the producer (service user) invokes an operation on the consumer (service provider), but there is *no feedback* from the consumer to the producer. This style of interaction allows the call from the producer to be multicast to a number of consumers.

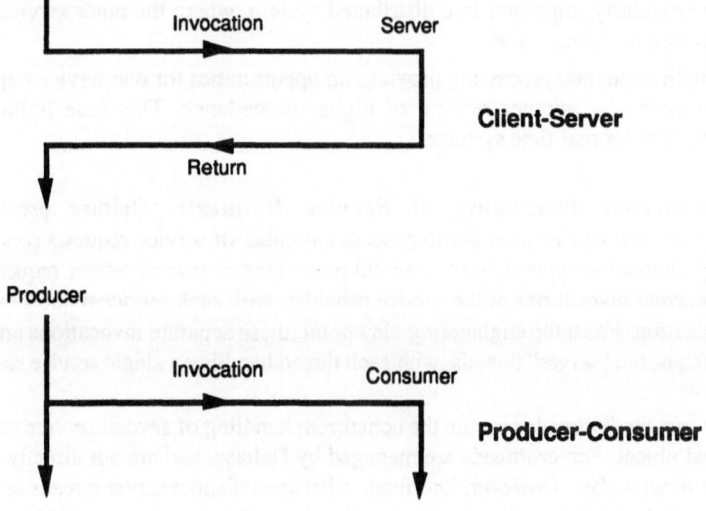

Fig. 4 - Client-Server and Producer-Consumer Interactions

The client-server operation is typically associated with a one-to-one interaction, as typified by a remote procedure call. A single D-association may be used independently by a number of service users (that are members of that D-association) to invoke operations on a service provider; this is the typical *server* situation, as in the case of a file-server.

By exploiting the "null" reply option, client-server operation may be extended to a single invocation of a number of servers; the "reply" condition would normally be that a null reply be received from all servers.

The producer-consumer style allows a number of different modes of operation. In particular, when used over a D-association that permits a number of consumers (service providers), an operation may be multicast to a number of visibly-distinct destinations; that is, the destinations are distinct from the computational viewpoint (and not just a single set of replicas).

One example of the use of the producer-consumer style is the dissemination of "alarm" messages, in an industrial plant, to a number of distinct alarm processing components. Another example is the "commit" operation of a distributed transaction.

These interaction styles are properties of the *operations* concerned. A given service interface may include both client-server operations and producer-consumer operations. Only when the service interface fulfils certain conditions may the D-association support more than one service provider. These conditions are now described.

7.2.7. Concurrency, Threads and Parallelism

A service provider will typically handle service requests from a number of service users, and may thus be faced with the requirement to handle a queue of service requests.

The simple solution is to handle one service request at a time, and for some service providers this may well be the best solution. However, for systems where overall performance is important, such a solution ignores the following:

1) For many service providers, the servicing of a request is not simply a matter of computation, but involves either invocation of other services, or access to a local device — such as a disc. In these cases there will be time during which the service provider is suspended; such time could be utilized in servicing other requests. This is particularly important in a distributed system, where the other services may be executed by another host.

2) Simple sequential processing provides no opportunities for one service request to be preempted by another request of higher precedence. This case is particularly important for real-time systems.

7.2.7.1. Concurrent Processing of Service Requests.
Deltase provides the mechanisms to allow a service provider to process a number of service requests *concurrently*. From the computational viewpoint, the concurrent processing of several service requests is seen as several concurrent invocations of the service provider, with each service request handled by a separate invocation. From the engineering viewpoint, these separate invocations are handled by separate threads, the "server" threads, with each thread handling a single service request (see §7.3.5).

Server (or *spawned*) threads provide the concurrent handling of several service requests by a computational object. Server threads are managed by Deltase, and are not directly visible to the application programmer. However, interference between distinct server threads could occur through the use of shared data within the computational object.

The application programmer must be aware of the possibility of multiple concurrent invocations when programming a computational object. Persistent shared data may be accessed on behalf of any service request, and may, therefore, be subject to conflicting requests from concurrent invocations of the computational object. It is essential that such shared data be protected from the inconsistencies that could arise. However, this protection is the responsibility of the application programmer, using existing techniques (semaphores, monitors, etc.).

7.2.7.2. Concurrent Processing within a Service Request. *Forked* threads provide a facility for distinct threads within a computational object. These threads are directly visible to the application programmer, since they are explicitly created and deleted by program commands.

A remote procedure call causes the calling thread to be blocked; other forked threads within the same computational object may execute if there is processing to be done.

Each forked thread is associated with a particular server thread, and thus with a particular invocation of the computational object. At the start of the execution of a service request, there are no forked threads; any that are required must be created. A forked thread may be deleted when its activity is complete. When execution of that service request is complete (that is, the server thread becomes free), all remaining forked threads (associated with that server thread) are destroyed.

The commands to control forked threads are defined by a service interface, and invoked by procedure calls; to the computational object, the handling of forked threads appears the same as other services. The commands provided include those for creating, deleting and activating forked threads within that server thread. The scheduling of threads within a computational object is performed by Deltase.

As described above, the control of access to shared data within the computational object is the responsibility of the application programmer.

7.2.7.3. Parallelism. Within a distributed system, true parallelism may occur when services requested by one computational object are executed on different hosts. The achievement of this is outside the control of the application programmer, and will depend on how the system is configured; the location of the service provider is deliberately hidden through location transparency. However, the application programmer is able to create opportunities for parallelism; if service requests are invoked sequentially (with one completing before the next one is invoked), then there are no opportunities for parallelism.

The programmer has two alternative strategies that may be used to offer "true" parallelism, where services invoked are executed on other host machines, so that such services may be executed in parallel with each other and with the capsule invoking these requests.

Forked threads may be used in association with remote procedure calls, to achieve several concurrent service requests from a given service user. When an RPC is used within one thread, that thread will be blocked (suspended) until the result is returned, but it is only that thread that is blocked. Other threads that are not blocked may execute, and, in the course of their execution, make an RPC, causing that thread to be blocked. The result is that, at a given time, there may be a number of outstanding service requests originating from a single server thread.

Delta-4 has explored an alternative mechanism for use with languages that do not support multiple threads; this is the *Remote Service Request* (RSR) mechanism. RSR is a procedure-call-like method of invoking a remote service, without blocking the calling thread; this allows execution of the calling thread to continue whilst the remote service is fulfilled. A remote service request carries an implicit pledge to synchronize with the service completion and accept any results, by means of a *Remote Service Wait* (RSW) call. The means of guaranteeing that this pledge is kept, and that its context is appropriate, are language-dependent issues. RSR does not require changes to the language compiler, and carries with it the benefits of expressing the parameters for the remote service as language-level entities, in the same way as for RPC.

The RSR construct provides a mechanism for the parallel invocation of a number of remote procedure calls, without the need for explicit creation of separate (forked) threads and without the need for language extensions. The construct is not elegant to use, relying on the application programmer to invoke RSW within a valid context; as defined, the RSR mechanism is unstructured.

To provide the facility for multiple remote procedure calls in an elegant manner (and therefore less prone to error), a "COBEGIN" construct has been defined as an extension to "C", to provide a structure for programming multiple remote procedure calls without the need for creation and deletion of forked threads.

The COBEGIN construct allows a set of remote procedures calls to be specified; these calls are to be invoked without waiting for the replies, so that the calls may be executed in parallel. The calling thread continues until a specified point is reached, and then waits until the results have been returned from all the set of remote procedure calls.

A pre-processor is used to convert each COBEGIN construct into a number of as RSR and RSW operations, so that no changes to the "C" compiler are required.

RPC, RSR/RSW and COBEGIN are alternative methods of programming the invocation of interactions at the language level within a service user; there is no means whereby a service provider can distinguish between a service request invoked via RPC and one invoked via RSR or COBEGIN.

7.2.7.4. Use of D-associations. For a given service interface, a single D-association is sufficient to meet the requirements of a multi-threaded service user; a D-association can support several concurrent service requests, serviced by concurrent operations in the service provider. These service requests arise not only from concurrent operations in one service user, but also from several different service users.

7.2.8. Interface Definition Language

Where the service user and service provider both use the same programming language, the service interface may conveniently be defined as a set of procedure definitions in that language.

Deltase currently uses a *"C" Interface Definition Language* (CIDL) [Delta-4 1990] for defining the interfaces between objects written in "C". Similarly, Ada package definitions provide an interface definition for objects written as Ada packages. However, to support interactions between an object written in "C" and one written in Ada, it is necessary to produce two equivalent interface definitions, one in "C" and one in Ada.

To support interactions between objects written in different languages, a service interface specification should be in a form that can be readily converted into equivalent language-dependent forms, so that the correct use of that interface can be assured for the different languages concerned. The aim is that, eventually, the specifications of service interfaces should be written in a standardized *Interface Definition Language* (IDL). An IDL is a language used for defining service interfaces in such a way that the definitions are not dependent on any particular programming language.

Software tools will be required to mechanize the transformation of a service interface specification (in IDL) into its *representation* in the actual programming language of implementation. This representation is a description of a set of operations, together with their parameters, expressed as the appropriate construct in each individual programming language.

Only by such software tools can the necessary levels of assurance be given that the interface definitions in different languages are equivalent.

Once such tools are available, the benefits of any interface checking (such as type-checking) carried out by the relevant compilers will be inherited into the distributed environment, even where service user and service provider are implemented in different programming languages. At present, this highly desirable check on the correctness of distributed interactions can be applied in full only where service user and service provider are written in the same language; this is a considerable advance on traditional (message-based) methods.

Configuration control also becomes necessary where the use of service interface definitions becomes widespread, with the aim of ensuring that both the service user and the service provider conform to the same *version* of the interface definition.

7.2.9. Remote Procedure Specifications

Because these interfaces are for use in distributed systems, there are certain restrictions on these interface definitions resulting from the separation of the interacting objects. In particular:

1) The interface must be defined in such a way that all the information required by the service provider is included in the parameters; with a remote procedure call, the called procedure (in the service provider) cannot access the context of the calling procedure (in the service user).

2) The information passed as parameters must be meaningful in the context of the called procedure; pointers, and other local information may not be meaningful in another context. This applies equally to parameters returned as results.

3) The specification of each operation must be compatible with the syntax of procedure calls in a wide range of programming languages. Thus, few languages permit variable numbers of parameters, or different types for a given parameter. The full benefit of the remote procedure call paradigm can only be derived where the procedure can be invoked within the scope of the available language facilities.

This model assumes that the response from the remote procedure will always reach the computational object from which the remote procedure was invoked; that is, there is an assumption that, in a Delta-4 system, a remote procedure call can be treated (by the applications programmer) exactly like a local (i.e., internal) procedure call. This simple model makes no provision for handling the situation where the return from the remote procedure is lost, due to failure of the machine on which the remote procedure is executed, or for any other reason. In the event of loss of the response, the thread from which the remote procedure was invoked would be suspended indefinitely; this would have a knock-on effect, where the thread is providing a service for another object.

With a local procedure call, both the calling program and the called procedure are executed by the same machine, without requiring the use of a communication system. If the called procedure is lost through machine failure, then so is the calling program. Thus, for a local procedure call, there is no point in making provision for loss of the return from a called procedure.

A non-fault-tolerant distributed system represents the other end of the spectrum of possible ways of handling loss of response from a (remote) procedure call. To prevent the system from locking up, on failure of a machine or partitioning of the network, the applications must be able to recover from the failure of a remote procedure call.

For applications to be used on Delta-4-based systems, there is a range of possibilites between these two extremes; for instance, if the remote service is replicated, the designer may assume that it will exhibit sufficient availability without the need for special precautions. There is a compromise here between providing transparency to distribution, and yet being able to recover from failures that arise because of distribution.

Where the degree of transparent availability achieved through Delta-4 fault-tolerance mechanisms is inadequate, then the recovery mechanisms provided by Deltase may be used to handle the consequences in an application-specific manner; fault-tolerance is no longer fully transparent for applications that do so.

The recovery mechanisms are based on the use of *event handlers*, as described in §9.7.4. Thus an event is raised if no response from the remote procedure has arrived at the calling program within a specified time.

Event handlers provide a powerful mechanism for handling exceptional conditions; in the Delta-4 Extra Performance Architecture, where timeliness is of concern, the event handling mechanism is used to invoke any necessary recovery action when the required time constraints are not met.

The event handling mechanism provides the following options which are somewhat analogous to the facilities provided by the Ada programming language:

1) application-specific event handlers (visible to the applications programmer) may be provided (as part of the application) for handling specific events;

2) default event handlers may be included as part an application for handling specific events;

3) if there is no event handler for a specific event, then the event is passed to the client invoking the service on which the event was raised.

For further information on event handlers, the reader is referred to §9.7.4.

7.2.10. Application Programmer's View

The computational viewpoint is the application programmer's view; this is the view of the system as it appears from the source code of a program module written in a high-level language.

In order to achieve the properties offered to computational objects, *all* external interactions must follow the rules laid down for interactions between computational objects. For a given computational object, all the services used must be accessed via service interfaces, as described. This precludes the use of "standard" libraries that make direct calls on the local operating system.

The requirements for portability of a computational object between implementations of Deltase for different environments are:

1) the use of a standard programming language, and the program development systems to support such a language;

2) availability of the services used by that object in any system to which the object is ported;

3) all external interactions to take place via service interfaces, as described above; such interactions may be programmed using only language-level entities with RPC or RSR.

Departure from these rules (for example, for performance reasons) may reduce the portability of that computational object and its transparency to distribution.

7.2.11. Interactions with the non-Deltase World

To provide commercially-viable solutions for distributed computing systems in the real world, then applications based on Deltase have to be able to interact with existing software and with input-output device drivers. While the details are the concern of the engineering viewpoint, some of these issues will be directly visible to the application programmer from the computational viewpoint.

7.2.11.1. Transformers. The term *transformer* is used to refer to an entity that interacts both with the Deltase world of computational objects and explicitly with some part of the non-Deltase world. In particular, a *transformer object* is a *"half-object"*, that has the properties of a computational object from one view (and can therefore be modelled as such in the computational viewpoint), but also interacts with the non-Deltase world. To the Deltase world a transformer object appears as a computational object, and interacts with other computational objects via the Deltase mechanisms. From the other side, the transformer interacts explicitly with some other computational world.

An important application of transformer objects is to provide an interface from the Deltase world to an existing (non-Deltase) software package. To make services provided by a non-Deltase application (such as a database, see §7.6) available in the Deltase world, a transformer object appears to the Deltase world as a *service provider*, and to the non-Deltase world as a *service user*. This provides a method by which computational objects can gain access to the services of existing software packages. Typically, the software accessed in this way will be proprietary and commercially important; two examples are:

- software offering standard OSI services such as FTAM;

- database software supporting standard interfacing methodologies such as SQL.

Equally, a transformer object may be used to make the services provided in the Deltase world available to the non-Deltase world. In this case, the transformer appears to the Deltase world as a *service user*, and to the non-Deltase world as a *service provider*. There is the possibility of providing a service to an existing service specification; the server would have the benefits of Deltase expressive power and representational transparency, together with Delta-4 dependability.

Another example of the use of transformer objects is in the interfacing to input-output devices or device driver software; that is, in interfacing to the real world. This is discussed in chapter 12.

The transformer concept also provides a method of inter-working between the Deltase world of computational objects and "other" computational worlds, such as those of X-windows or MMS (Manufacturing Messaging Service).

An application programmer developing a transformer object will necessarily see interactions both with the Deltase world (via service interfaces), and with the non-Deltase world. However, to emphasise the point made earlier, Deltase can guarantee the properties and transparencies, described for the computational model, only where the computational objects are "pure", i.e., *all* their external interactions conform to Deltase. The implications of transformer objects on fault-tolerance are discussed in chapter 7.

7.2.11.2. Interactions with the Local Environment. The term "local environment" here refers to the local operating system.

To give objects a consistent view of externally-provided services, those services provided by the local environment (and invoked explicitly by the object) should be invoked in the same way as services provided by other objects; that is, the interactions between an object and the local environment should be modelled as interactions between objects, via a defined service interface. An object should not be able to distinguish between a service provided by another object, and a service provided by the local environment.

Thus, the (visible) parts of the local environment are accessed via defined service interfaces, through which the computational object accesses the services provided locally.

Wherever possible, these interfaces should be generic, i.e., not dependent on any particular local environment. One of the aims of Deltase is to make possible the creation of computational objects that are independent of local environment. An important contribution to this objective is by ensuring that an object's explicit interactions with its local environment may be programmed in the same way as interactions with other objects.

7.3. Engineering Model

7.3.1. Introduction to the Engineering Model

The engineering viewpoint is the *system programmer's view* of the distributed system, and is concerned with the *engineering* necessary to support the *virtual machine* (computational viewpoint) on the available computing environment (technology viewpoint). This support must be provided in such a way that the required *distribution transparencies* and *quality attributes* (performance, dependability, etc.) are achieved.

While many of the concepts described in this section are inherited from ODP and ANSA, Delta-4 has a distinctive contribution to make in the application of these concepts for the support of dependability and real-time in distributed systems.

From the engineering viewpoint, Deltase is seen to support the computational model by a combination of:

1) **generation support**: in the transformation of the language-level computational objects into *capsules*; capsules are executable Deltase-based software components — the capsule generation software is referred to collectively as *Deltase/GEN*;

2) **run-time support**: the Deltase run-time support subsystem, which includes additional capsules that are not visible to the application programmer — the Deltase run-time support software is referred to collectively as *Deltase/XEQ*.

Within Delta-4, the term software component is used generally to refer to the executable units, or modules, of a distributed application. The term *capsule* is used to refer to a software component that is the representation, in the engineering viewpoint, of one (or more) computational object(s), all of which interact *only* via Deltase. The term "capsule" is taken from the ECMA ODP work [ECMA TR/49]. The term *transformer capsule* is used for a capsule that contains one (or more) transformer objects, so that there are some potential interactions that are not under the control of Deltase.

A capsule is seen as an engineering object. In Delta-4, the ODP concept (representing an *address space*) is mapped onto a software component. From the engineering viewpoint, a system is modelled as a number of interacting capsules.

In the technology viewpoint, a capsule is mapped onto an executable unit supported by the local operating system, such as a UNIX process.

From the engineering viewpoint, a capsule is seen to contain one or more computational objects that are surrounded by additional software to perform the transformations and mappings that take place at run-time. This additional software is referred to as the *envelope*. The process of generating the envelope can be automated by suitable software tools, thus preserving the view that Deltase supports language-level computational objects. Each capsule contains one envelope, which represents all code added to the computational objects to produce the capsule; such code includes libraries and automatically-generated code, for all computational objects (and transformer objects, if any), within that capsule.

There is an analogy between these mechanisms and the compile-time and run-time support subsystems of a high-level programming language. Taken together, these subsystems provide the programmer with the illusion of preparing code to run on a machine which "understands" the high-level language program, whereas much of the "understanding" is resolved prior to any execution being possible, during compilation to executable code.

Figure 5 shows two phases, or "Epochs", in the generation and installation of a capsule.

1) The **capsule generation phase**, in which computational objects are transformed into capsules by Deltase/GEN, and the envelope is generated from a combination of selected libraries and automatically-generated code. A capsule is generated for use in a particular environment.

2) The **capsule installation phase**, in which the capsule is loaded onto the host (or hosts) on which the capsule is to be executed. The initialization code is executed, during which time all the default trading takes place, to establish D-associations for services provided and to join D-associations for services used.

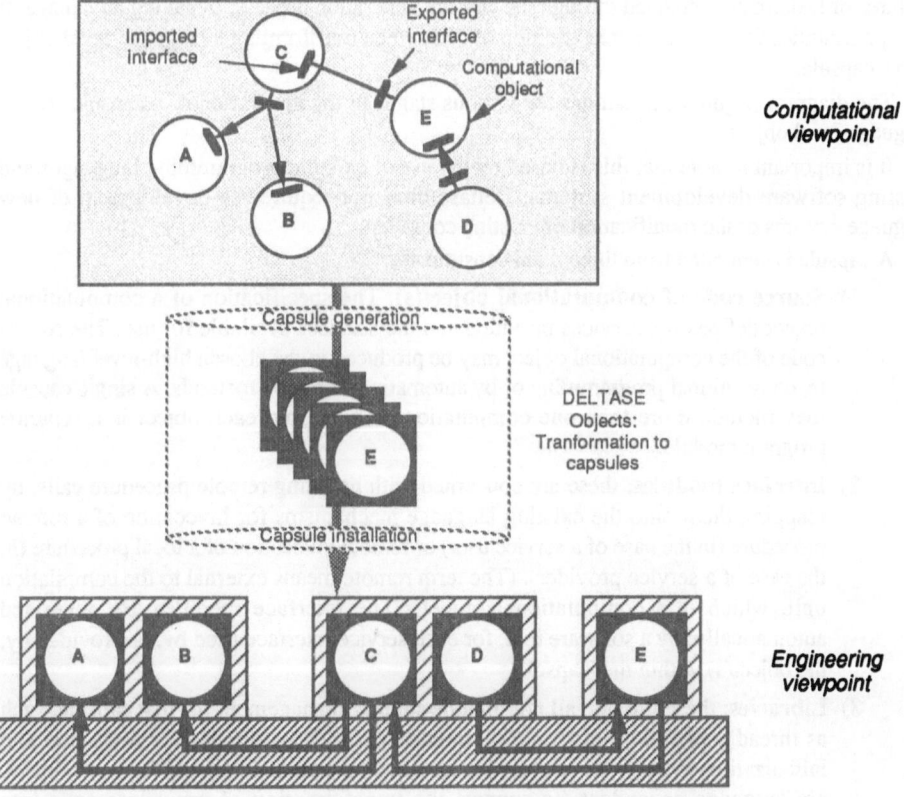

Fig. 5 - Relationship of Computational and Engineering Viewpoints

Once this installation phase is complete, the capsule is available for execution as part of the run-time system.

The *Deltase run-time support subsystem, Deltase/XEQ,* provides a number of services that are *always present,* and is itself a distributed application. In accordance with the best engineering practice, it is therefore based on defined Deltase principles, namely those supported by Deltase; Deltase makes recursive use of itself.

From the computational viewpoint, Deltase/XEQ consists of a number of computational objects, which may then be transformed into capsules, with all the benefits (distribution transparencies, portability, etc.) that Deltase seeks to confer on distributed applications.

Most of the capsules in Deltase/XEQ are not normally visible to the application programmer; in particular, the capsules that provide the trading activities. Interactions with Deltase/XEQ are visible in the engineering viewpoint, and are invoked from the code that constitutes the envelope. (From the computational viewpoint, these are two separate applications that co-exist but are not visible to one another.)

7.3.2. Generation of Capsules

7.3.2.1. Introduction. Deltase provides support for *language-level objects,* and the generation of the capsules for a particular system is an essential part of Deltase. Many of the features of Deltase are achieved through the capsule generation process, by taking advantage of the opportunities for the transparent addition of code when transforming a language-level object into a capsule.

The diagram (figure 6) illustrates the various stages in the generation of a capsule from a language-level object.

It is important to note that this is based on the use of existing programming languages and existing software development systems; Deltase does not require the development of new language systems or the modification of existing compilers.

A capsule is generated from three main constituents:

1) **Source code of computational object(s):** The specification of a computational object defines the services provided and the services available for use. The source code of the computational object may be produced in the chosen high-level language by conventional programming or by automatic generation methods. A single capsule may include more than one computational object, but each object is a separate program module.

2) **Interface modules:** these are concerned with handling remote procedure calls, by mapping them onto the existing language mechanisms for invocation of a remote procedure (in the case of a service user) or remote invocation of a local procedure (in the case of a service provider). (The term remote means external to the compilation unit, which is a computational object.) The interface modules are generated automatically, by a software tool, for each service interface used by, or provided by, the object(s) within that capsule.

3) **Libraries:** these provide all non-unique support management for that capsule, such as thread scheduling, dispatching, interfacing to the communication system, and initialization code. Some of these libraries are dependent on the LEX; other libraries are language-dependent (to support the transformation of procedure calls into messages, and vice-versa). Selection of a capsule's fault-tolerance model will result in the appropriate library being used for those parts of the envelope that are dependent on the fault-tolerance model. These libraries are not visible to the

computational object itself, but form an essential part of the infrastructure to support
the computational object(s) within a capsule.

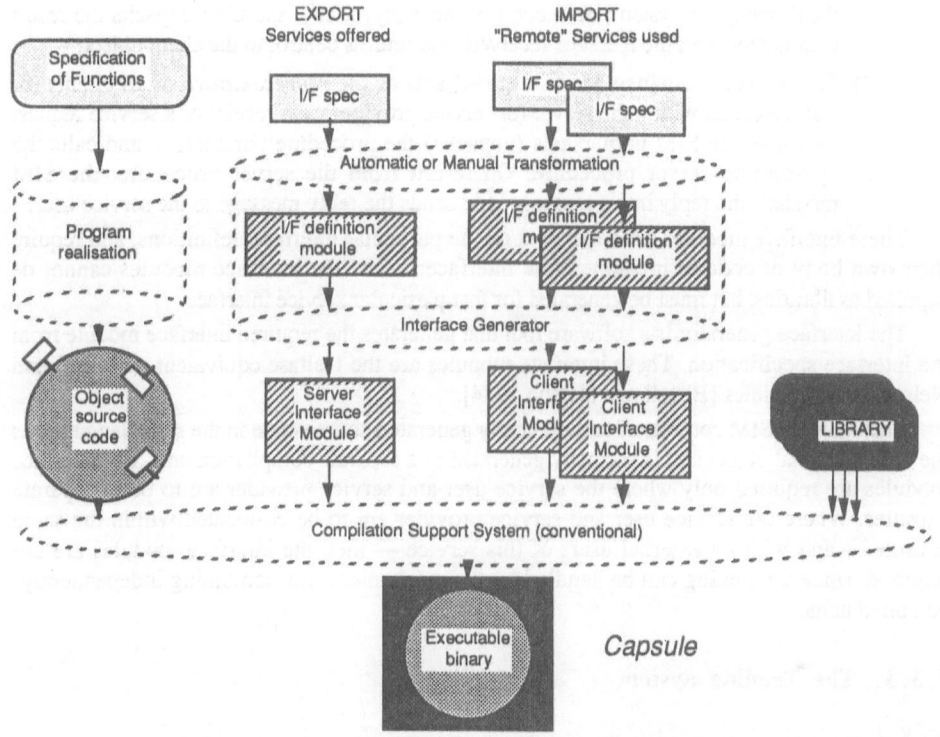

Fig. 6 - Capsule Generation by Deltase/GEN

The interface modules and the libraries, together, constitute the *envelope* and implement the
Deltase abstractions that the programmer is encouraged to use.

The output produced by Deltase/GEN is a file containing a binary representation of the
capsule for execution in a particular local execution environment. This file provides a template,
that may be used to instantiate instances of that capsule wherever that run-time environment
exists.

Because Deltase supports language-level objects, which provide opportunities for the
addition of environment-dependent code during the generation process, portability of capsules
(executable binary) is not a requirement for Deltase objects.

However portability could, in the future be based on an "architecture-neutral" intermediate
code version of the object; this would allow the addition of the environment-dependent code
(the envelope), while *hiding* the original source code and therefore protecting the intellectual
property rights of the originator. This approach implies a subdivision of the generation phase
into two (or more) separate phases.

7.3.2.2. Interface Modules. The language-specific definitions of the interfaces are used
both by the programmer who is constructing the specified object functionality in the chosen
language, and by the major Deltase transformation component, the *interface generator*.

There are two types of interface module:

1) The **Client Interface Module** (CIM) is the server's (or service provider's) *representative* within the client (or service user). The CIM marshals (assembles) the procedure call parameters into a message, sends the message to the server via the local operating system; on receipt of the reply, if any, the CIM unpacks the result parameters from the message received, and returns control to the client object.

2) The **Server Interface Module** (SIM) acts as the *representative* of all clients (or service users) within the server (or service provider). On receipt of a service request message, the SIM unmarshals (unpacks) the procedure parameters, and calls the appropriate server procedure. On return from the server procedure, the SIM marshals the reply into a message, and sends the reply message to the service user.

These interface modules are dependent on the particular interface definitions, and require their own body of code to implement the interface; hence the interface modules cannot be supplied as libraries, but must be generated for that particular service interface.

The interface generator is a software tool that generates the required interface module from the interface specification. These interface modules are the Deltase equivalent of Birrell and Nelson's *stub* modules [Birrell and Nelson 1984].

Each CIM or SIM consists of automatically generated source code in the same language as the computational object and is normally generated as a separate compilation unit. The interface modules are required only where the service user and service provider are to be in separate capsules. Where the service user and service provider are to be co-located within the same capsule — and with no external users of this service — then the interface modules are not required, since the linking can be handled by the mechanisms for combining independently-compiled units.

7.3.3. The Trading System

7.3.3.1. Introduction. As described earlier, a service user must establish a binding to a D-association before those services can be invoked. The purpose of the trading system is to enable such bindings to be established (and deleted) *dynamically*, based on matching the service interface type required with the service interface type of the available D-associations. The trading system enables a service user to establish a logical path to a service provider for the purpose of invoking services and receiving replies, without either the service user or the service provider having to know the location or the identity of each other.

The trading system is based on:

1) a *trading catalogue* which holds information on the available D-associations;

2) a number of *traders* that offer the trading service and are responsible for execution of the operations on the trading system.

Logically, the trading catalogue is a distributed catalogue, but it may be partitioned into a number of separate named catalogues, or *trading scopes*. There is a separate record in the trading catalogue for each available D-association; this record holds the name of the particular service interface type on which that D-association is based. When a new D-association is created, its record is created in the specified partition of trading catalogue. The use of distinct trading scopes allows the service interface type name to be context-relative. The trading catalogue is analogous to a "Yellow Pages" telephone directory, which gives a list of service *offers*.

The two principal operations on the trading system are the export and import operations as follows (figure 7):

1) An *export* operation is used by a *service provider* to create a D-association and to act as a provider of that service. The export operation specifies both the name of the service interface type (on which the D-association is to be based) and the name of the catalogue (or trading scope) where a reference to the D-association is to be lodged (for use by subsequent import commands).

2) An *import* operation is used by a *service user* to join a D-association for access to the services provided — as defined by the service interface type. An import operation causes the specified named catalogue(s) to be searched for a D-association supporting the specified service interface type. Having selected a particular D-association providing the required service, a *binding* is established between the service user and the selected D-association.

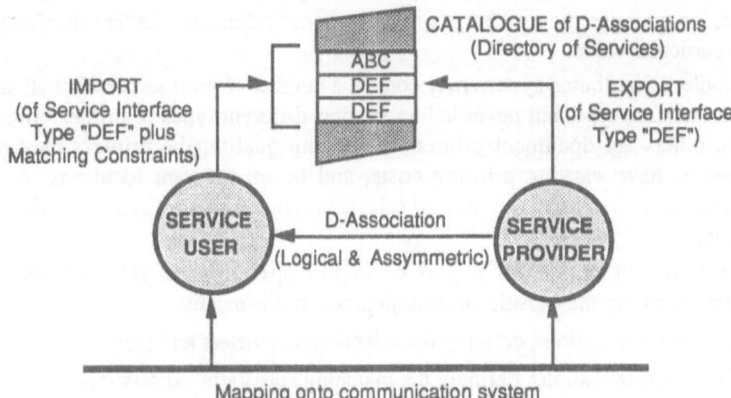

Fig. 7 - Principles of Trading

Most import and export operations are carried out, transparently to the programmer, as part of the capsule initialization code (which is created by Deltase/GEN and is executed as soon as an instance of a capsule becomes "live"). This initialization software creates (by means of an export operation) a D-association for each service *provided* by the capsule, and establishes a binding to (i.e., joins — by means of an import operation — a D-association for) each service *used* by the capsule. The capsule remains a member of all of these D-associations for the whole of its life. These default operations would not be visible to the application programmer.

Each capsule is given a preset binding to its local trader, so that a capsule can always perform export and import operations, as essential prerequisites to interactions with other capsules.

However, for certain specialized computational objects, it may be necessary for the joining and leaving of D-associations to be under direct program control, at run-time. In these cases, *import* and *withdraw import* operations will be written into the code of the computational object.

When a capsule joins a D-association, the communication paths (and associated resources) are set up using information from the catalogue; the D-association is mapped onto the communication system or the facilities of the local execution environment.

7.3.3.2. Constraints Matching. The aim of interface trading is to select a suitable D-association for a particular import request. Where there is (within the specified trading scope) more than one D-association matching, the selection can be refined by means of *matching constraints*. In the absence of any matching constraints, the first D-association supporting the specified service interface type will be selected. Matching constraints may also be used where specific constraints apply to any D-association selected, so that even if there is only one D-association matching the service interface type, it will not be selected unless the matching constraints also allow selection.

Associated with each service interface type are a number of selection parameters; these parameters are associated with the service interface, but do not form part of the service interface as visible to the application programmer. When a D-association is created, values are assigned to these selection parameters.

An import operation includes a *matching constraint expression*, expressing relationships between the selection parameters. These expressions are evaluated for each compatible D-association to find the "preferred" D-association. It may happen that *none* of the D-associations is suitable, if, for example, the matching constraints specify that a particular selection parameter must have a particular value.

For example, a distributed system may contain a number of print servers that all support the same service interface type, but nevertheless support different types of printer. Some may be line printers, others A4 document printers of varying quality; the printers may operate at different speeds, have varying printing costs, and be in different locations. A matching constraints expression may then be used to select, say, the nearest printer capable of printing A4 documents.

Where the default option for export or import operations is selected, the matching constraints are set during the capsule generation process; this means:

- for export operations, defining the selection parameters and their values;
- for import operations, defining the matching constraint expressions.

Where these operations are programmed, for dynamic invocation of these operations, then the corresponding information must be included in the computational object concerned.

7.3.3.3. Access Control. The trading system handles the creation and joining of all D-associations, and is thus able to control both the service offered and the access to those services. A particular service user can access a given service only if that service user is permitted to join the D-association through which the service is accessed. Similarly, a service provider can offer a service only if the export operation is accepted.

A trading system may control not only the creation of D-associations, but also which capsules may join which associations.

7.3.3.4. Negotiation of Transfer Syntax. Interactions between heterogeneous environments require the use of an architecture-neutral transfer syntax, such as ASN.1, for the encoding of the RPC parameters and results.

The code to perform the encoding and decoding can be generated automatically, as part of the envelope, and forms an essential part of the support for the transparency to heterogeneous machines and heterogeneous language systems. However, such encoding and decoding impose an overhead on interactions, and an objective is to avoid such conversions where the interacting capsules use an identical native transfer syntax. The native transfer syntax depends not only on the local execution environment, but also on the language systems, since different compilers may represent a given data-type in different ways even where the processors are identical.

The trading system has an opportunity to provide information for determining whether use of an intermediate transfer syntax is necessary or not. On an import operation, the service user could be informed whether the service provider uses an identical transfer syntax or not. In the latter case, conversion to an intermediate transfer syntax will be necessary, for both the invocation request and its reply. This approach would also require that the client and server interface modules (CIM and SIM) could handle both their native transfer syntax and the architecture-neutral transfer syntax.

7.3.3.5. Epochs of Trading. Trading can occur at several different times, or epochs in the life of a capsule. The only requirement is that the binding to the D-association be established before the first invocation of an operation of the interface. This can be:

1) during *construction* — when capsules are generated;

2) during *configuration* — when an instance of a capsule is installed on a host for execution;

3) during *initialization* — when capsules are first entered, D-associations are established by execution of the initialization code that was automatically generated;

4) during *execution* — when capsules are active; the program itself invokes operations on the trading system.

7.3.3.6. Joining a D-Association. There are a number of circumstances in which a capsule might require to join an existing D-association other than by an import operation:

1) where there are to be a number of distinct *service providers* on that D-association, in which each receiver is to act as a service provider; for example, in the handling of system alarms;

2) where a capsule has been created by cloning, and therefore needs to join all of the D-associations of which its other replicas are members;

3) where a capsule has been given a reference to a D-association, as a parameter, to allow that capsule to join the specified D-association for access to the service. (This is like passing a file reference, to enable a print server to print the file without the invoker having to pass the whole file.)

In each of these cases the particular D-association to be joined would be identified by reference, rather than found by the normal operation of the import command.

Additional operations on the trading system are required for the following:

1) to enable an agent to *create* a D-association (without joining it as either service user or service provider);

2) to enable a service provider to *join* an existing D-association (for use when there is more than one service provider on that D-association);

3) to enable a service user to *leave* a D-association;

4) to enable a service provider to *cancel* its offer of service, closing the D-association if there is no other service provider on that D-association.

7.3.3.7. Trading System as a Distributed Application. The Deltase trading system is a distributed application, and may be implemented as a Deltase-based application, which is not normally visible to the application programmer. Where the trading system is accessed directly from an applications program module, the operations on the trading system are invoked via service interfaces, in the same way as other interactions.

Each capsule must have a preset local binding to the D-association for operations on the trading system, so that a capsule can always interact with the trading system.

There are many possible implementations of the trading system to provide the required functionality. As a distributed application, a trading system can take advantage of the implementation choices offered by Deltase; such choices include the grouping of computational objects into capsules, the fault-tolerance model used by each capsule, and the distribution of the capsules among the available hosts. One constraint is that there is normally a component of the trading system on each host, both to provide a purely *local* trading service, and to provide the interface to the rest of the trading system — the *global* trading system.

From the computational viewpoint, access to a service is the same, whether the service be local or global; this amounts to a selection of the trading catalogue. However, the choice can affect both the *performance* (access to a globally-provided service will normally take longer than a locally-provided service), and *fault-tolerance*. The choice can be made by the system designer, and hidden from the application programmer.

7.3.3.8. Local and Global Trading. The term "local" here means "within the same host machine". When the local portion of the trading catalogue is selected, the service will be provided locally, *and will not be replicated.*

The diagram (figure 8) shows a service user (client) and service provider (server) performing *local* trading, by reference to the local portion of the trading catalogue. This is of course only possible when the two capsules reside on the same host. In this case, a D-association is mapped onto the Inter-Process Communication (IPC) facilities of the local operating system.

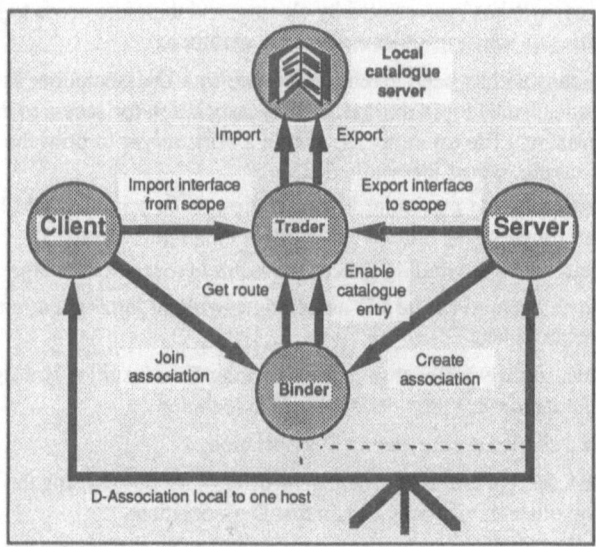

Fig. 8 - Local Trading

Figure 9 shows a service user and service provider performing *global* trading and binding, by reference to a global portion of the trading catalogue. The service user and service provider may be replicated, as required, and may reside on either the same host or on different hosts. Global trading is the normal case, and the D-association is mapped onto the underlying communication system, and inherits its properties.

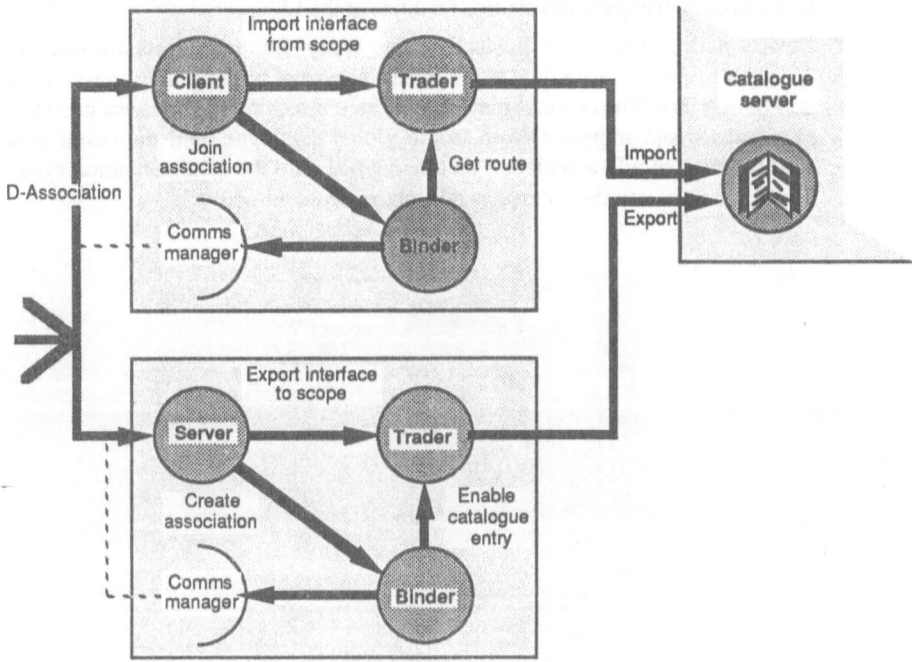

Fig. 9 - Global Trading

7.3.4. Implementation of Remote Procedure Calls

7.3.4.1. Configurations. This section follows the course of a remote procedure call, to show how all the various actions, which have been introduced separately, fit together. The word "remote" refers to a procedure outside the language-level computational object that made the procedure call.

Figure 10 shows the same two computational objects transformed into capsules in three different ways, with different implementations of the remote procedure call; from the computational viewpoint, these three cases are all indistinguishable.

The three cases are as follows:

1) **Two objects in one capsule**: In this case, the service user and the service provider objects have been combined into a single capsule, and generated on the assumption that the use of this particular service will not be offered to other capsules. This case therefore simplifies to that of combining two source modules into a single compiled entity. The D-association is replaced by a conventional inter-module binding, established by the compiler and linker. This diagram illustrates why it is not always appropriate to map computational objects to capsules on a one-to-one basis.

2) **Two capsules in the same local execution environment**: In this case, each computational object has been generated into a separate capsule, on the assumption that both capsules will be installed on the same machine. The default operations on the trading system ensure that their service interfaces will be exported to, and

imported from, the local catalogue. The result is the establishment of a purely-local D-association, mapped onto the IPC facilities of the LEX.

3) **Two capsules in separate hosts**: The two computational objects are generated as separate capsules, according to the particular type of host machine on which each is to be executed. The capsules may be replicated if required. The service interface is exported to, and imported from, some global catalogue, and the result is the establishment of a D-association that is mapped onto the communication system. Interactions between these capsules take place across the network.

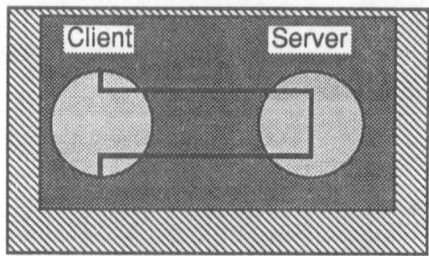

Case 1: Two objects within one capsule

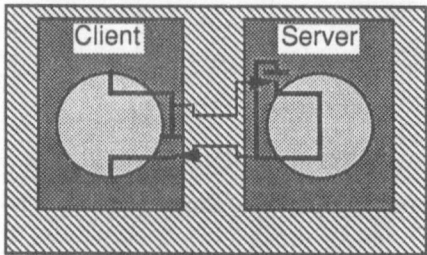

Case 2: Two capsules in the same local execution environment

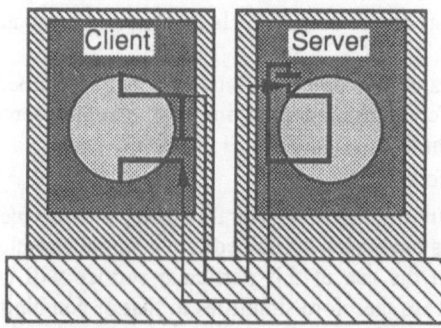

Case 3: Two capsules in separate hosts

Fig. 10 - Three Versions of Inter-Object Procedure Call

7.3.4.2. Execution of an RPC from the Engineering Viewpoint. This section follows through the execution of an RPC, as seen from the engineering viewpoint, for the case where the service user and service provider are in separate capsules (figure 11).

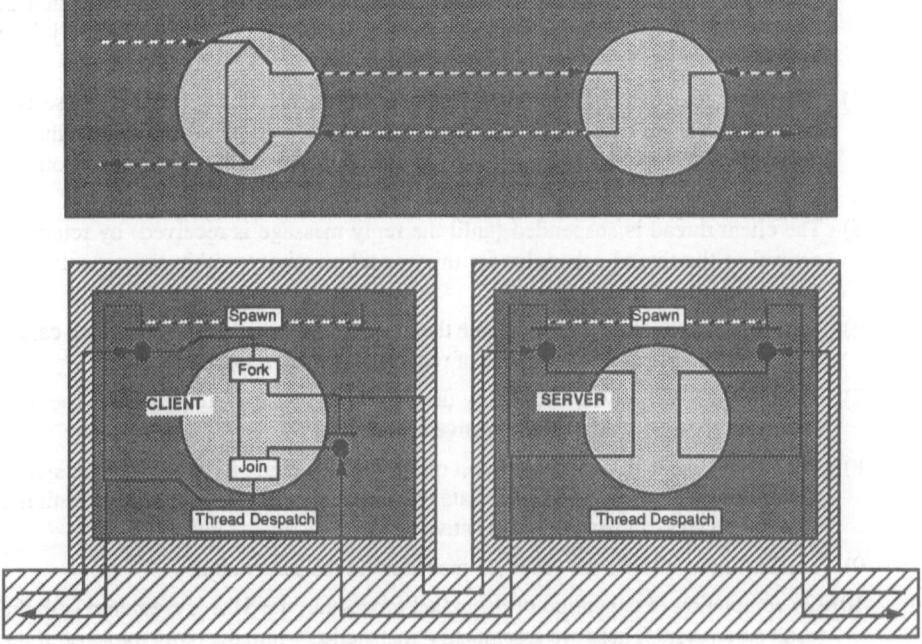

Fig. 11 - RPC in Engineering Viewpoint

At the language level, invocation of a remote procedure is the same as invocation of an external procedure. The procedure interface must have been defined in such a way that all the necessary information is passed in the parameters (since the context of the caller will not be accessible to the remote procedure) and that only parameters that are meaningful in the remote environment are specified. The remote procedure is defined as an external procedure, and therefore in a separate compilation unit. A *Client Interface Module* (CIM) is the local representative of the remote procedure, and forms the separate compilation unit required by the language system. A CIM is generated automatically from the service interface specification, compiled separately, and linked with the other modules of the capsule.

The capsule containing the service provider (server) is generated in a similar way, with the *Server Interface Module* (SIM) being the local representative of all remote service users.

The establishment of a logical interaction path (via a D-association) is described in section §7.3.3; typically, this would be done by the capsule initialization software of the service provider and service user.

At run-time, the sequence of events in the execution of a remote procedure call is as follows:

1) Invocation of the remote procedure results in a call on the corresponding procedure in the CIM.

2) On procedure invocation, the CIM assembles (marshals) the procedure call parameters (including the identity of the particular procedure) into a message.

3) In the general case, the client and server may have been generated using different language systems, and may be executed on hosts of different type. The procedure call parameters must therefore be converted by the CIM into a *transfer syntax*, such as ASN.1, which will be understood by the SIM. Where both the client and server use the *same* parameter formats, then the conversion is not necessary, with

consequent saving of time. The trading system should provide an indication of whether the use of a transfer syntax is necessary, at the time that the client is joined to the D-association.

4) The message is passed, via the LEX and the communication system, to the server, depending on the destination given by the trading system when joining the D-association. The CIM is dependent on the local operating system and on the interface to the communication system.

5) The client thread is suspended (until the reply message is received) by returning control to the thread scheduler for this capsule; activity within the capsule may proceed in other threads.

6) The server may not be able to handle the new request immediately, in which case it will be added to the server's queue of requests awaiting service.

7) The message is received initially by the *Server Interface Module* (SIM). The SIM acts as the local-to-server representative of the client.

8) The SIM unpacks the service-request message, converting from the transfer syntax to native syntax where an intermediate transfer syntax has been used; the result is a set of procedure call parameters in native form.

9) The SIM now invokes the specified procedure in the server object.

10) On completion of execution of the called procedure, the SIM is reentered with the result parameters. These are assembled ("marshalled") into the reply message; if the request message used the native transfer syntax, then so does the reply, otherwise the message is converted into the intermediate transfer syntax.

11) The reply message is then returned to its originator, using the communication system if the original was sent via the communication system and the local operating system.

12) The reply message may have to wait before it can be received by the service user. When the reply is received, the thread that was suspended (when the call was made) may resume execution from the point of suspension in the CIM.

13) The CIM unpacks the result parameters from the reply message, converting them to native syntax if an intermediate transfer syntax was used.

14) The results are returned to the computational object, and execution of the application code resumes at the point in the calling thread immediately after the procedure call, as if returning from a local procedure.

7.3.4.3. Implementation of Transparencies. The interface modules (CIM and SIM), along with the Deltase libraries that they use, play a major part in achieving the distribution transparencies introduced above:

- the underlying operating system primitives are hidden, giving *access transparency*;
- the bindings established via the trading system are provided in such a way as to give *location transparency*;
- by producing the client and server interface modules in the language of implementation of the respective client and server, and by use of an intermediate transfer syntax, *transparency of heterogeneous languages and machines* can be achieved;

- by making use of the services of the *Dependable Communication System* (DCS), and enclosing the local activity necessary for fault-tolerance, *replication transparency and fault transparency* may be provided.

7.3.5. The Handling of Threads

Concurrency of execution within a capsule is provided by *multiple threads*; each thread is a distinct "locus of control" within a given computational object. As described above, Deltase supports two types of thread (*server* threads and *forked* threads). The management of both types of threads is provided by Deltase as part of the envelope, created by Deltase/GEN.

Server threads are long-term threads, treated by Deltase as separate resources. Each incoming service request is assigned to a server thread that is currently free; each server thread can handle only one service request at a time. When execution of that service request is complete, the thread is suspended until it is required for execution of another service request. Server threads enable a capsule to handle several service requests concurrently.

Deltase treats all server threads as equivalent, so that an incoming service request may be assigned to any free server thread within that capsule. Such threads must not preserve any knowledge of previous assignments; any history that must be preserved is represented in the persistent data that is shared by all threads within the capsule.

When a capsule is initialized, several server threads are "spawned" to become dormant Deltase resources, for animation by a thread dispatcher in fulfilment of service requests. Should all the server threads become simultaneously animated, then any further service request must be retained (on a queue) until a server thread become available.

Other strategies could be adopted; a slower alternative would be to create new server threads on demand, deleting each on completion of its service.

Server threads provide concurrent invocation of independent services within one object; there is no true parallelism between these threads since, within a single processor, only one of these threads can be executing at a given time. However, the use of distinct server threads allows the management of concurrency to be provided by Deltase, freeing the application programmer of this responsibility, and allowing competition for access to a resource to be controlled locally.

7.4. Deltase Support for Distributed Fault-Tolerance

7.4.1. Introduction

Deltase combines ODP concepts on distributed systems with the Delta-4 approach to fault-tolerance. The support of fault-tolerance concerns the computational, engineering, and technology viewpoints.

- The **computational viewpoint**: The Deltase *computational model* is based on the requirement that any computational object conforming to this model can be made fault-tolerant, transparently, under Deltase. The Deltase computational model imposes constraints on the behaviour of a computational object, in terms of its external interactions and internal scheduling.
- The **engineering viewpoint**: It is from the *engineering viewpoint* that the mechanisms for the support of fault-tolerance are modelled. These mechanisms are provided by a combination of *generation support (Deltase/GEN)* and *run-time support (Deltase/XEQ)*.

- The **technology viewpoint**: It is from this viewpoint that the special subsystems for the support of fault-tolerance are modelled, such as the *Dependable Communication System* (DCS) which provides the essential distributed consistency (see §7.5.3), and the *fail-silent processors* that are required for the support of some of the models of fault-tolerance.

7.4.2. Replicated Capsules

The general Delta-4 approach to fault-tolerance is based on the support of *replicated software components*, (see §7.5). The *Deltase* approach to fault-tolerance is based on the support of replicated capsules, such that the support of fault-tolerance is transparent to the *computational object(s)* enclosed. A replicated capsule ensures survival of the state contained therein despite the loss of one of the nodes on which that capsule is represented.

A replicated capsule consists of a group of replicas, whose activity must be coordinated in such a way as to give the appearance of a single capsule. For the support for fault-tolerance to be transparent to a computational object, both the existence of the group of replicas and the activities to coordinate the group must be independent of the computational objects.

Delta-4 supports a number of different models of fault-tolerance (see chapter 6) for use under different circumstances, and the envelope is generated according to the particular fault-tolerance model to be used for that capsule. The activities of a group of replicas are coordinated in different ways for the different models of fault-tolerance.

Fault-tolerance requires that the group of replicas be able to survive the failure of a host (on which one of these replicas is executed) with consequent loss of that replica. For a given replicated capsule to survive over a "long" period, it is essential that replicas lost by host failure be replaced (see chapter 6). The process of replacing a lost replica is referred to as *cloning* (i.e., creating a clone of the remaining replica(s)), and involves the transparent creation of a new replica that is identical to the surviving replicas. The new replica is created on a host where there was no replica of that capsule prior to the start of the cloning process.

The set of possible hosts on which a given capsule can be installed is referred to as the *capsule replication domain*, (see section 6.4). All hosts within a given capsule replication domain are of the same type (and therefore support identical instruction sets), and with equivalent local execution environments.

D-associations are defined in such a way that, if a replicated capsule is a member of a D-association, then all replicas of that capsule are members of that D-association.

7.4.3. Determinism

The execution of a section of code is said to be *deterministic* if, for a given initial state and input message, execution of that section of code always results in the same final state and output message.

Replicas of a capsule do not automatically behave deterministically, even though all replicas of a given capsule:

- have identical code;
- are executed on hosts with identical instruction sets;
- start from the same initial state.

Figures 12 and 13 show two replicas with identical initial states, and an asynchronous event that is allowed to influence the sequence of processing. The asynchronous event might be any number of the following: some sort of time-slicing within the capsule, the end of a time-out, or an indication of message arrival. The replicas are loosely synchronized, and the asynchronous event occurs at different times with respect to the execution of that code:

- *after* message output in the replica on the *left* (in figure 12);
- *before* message output in the replica on the *right* (in figure 12).

In figure 12, the event is allowed to have immediate effect, and affects the two replicas in different ways resulting in different final states. However, in figure 13, the asynchronous event is trapped, and only allowed to take effect *after* the first message output; the result is that both replicas reach the same final state.

To ensure that behaviour of replicas is deterministic, asynchronous events have to be controlled; asynchronous events, which cause a transfer of control within the capsule at an arbitrary point, are not deterministic and cannot be permitted.

A *non-deterministic decision* is one for which re-execution would not necessarily lead to the same decision. With a capsule, all non-deterministic decisions are taken within the envelope, the execution of the code within the computational object being deterministic.

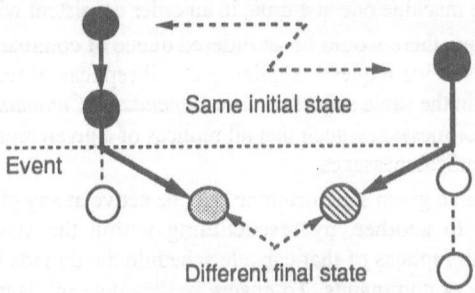

Figure 12: Non-deterministic behaviour

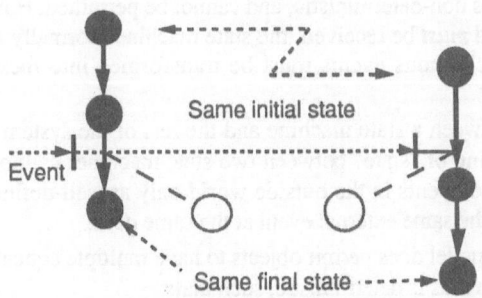

Fig. 13 - Deterministic Behaviour

7.4.4. Replica Execution

The term *replica group determinism* is used to refer to the deterministic behaviour of the replicas of a replicated capsule. The code to provide overall control of the capsule is dependent on the particular fault-tolerance model to be used, and is generated as part of the envelope during capsule-generation.

7.4.4.1. Active Replica Model. The active replica model (whether on fail-uncontrolled or fail-silent hosts) uses the *state machine* model to achieve long-term replica group determinism, without negotiation between the replicas.

Schneider [Schneider 1990] defines a state machine as follows:

"A *state machine* consists of *state variables*, which encode its state, and *commands* which transform its state. Each command is implemented by a deterministic program; execution of the command is atomic with respect to other commands, and modifies the state variables and/or produces some output. A client of the state machine makes a request to execute a command. The request names a state machine, names the command to be performed, and contains any information needed by the command. Outputs from request processing can be to an actuator (e.g., in a process control system), to some other peripheral device (e.g., a disc or terminal), or to clients awaiting responses from prior requests. (...) Requests are processed by a state machine one at a time, in an order consistent with causality."

For each state machine, there would be an ordered queue of commands to be serviced; the term command includes service requests, replies, etc. All replicas of that state machine must process these commands in the same sequence. The *Dependable Communication System* (DCS) ensures ordered delivery of messages, such that all replicas of a given replicated capsule receive identical sequences of identical messages.

Only one thread within a given state machine may be active at any given instant. Changing from one active thread to another, by rescheduling within the state machine, must be deterministic, such that all replicas of that capsule schedule the threads in the same sequence, given the same sequence of commands. To ensure replica determinism, thread rescheduling may only occur at identifiable points in the execution of the active thread, such as sending or receiving a request. The new thread is determined by the next "command" in the sequence, with all replicas receiving the same sequence of commands (messages).

"Preemption", causing a change from one active thread to another at points other than these clearly-defined points, is non-deterministic, and cannot be permitted. Having arrived at the new state, the next command *must* be received; the state machine (normally the whole capsule) can do nothing else. Asynchronous events must be transformed into messages that are sent as commands via the DCS.

All interactions between a state machine and the rest of the system must take place using commands. Direct sharing of "state" between two state machines cannot be permitted. A state machine may respond to events in the outside world only at well-defined points, at which all replicas can respond to the same external event at the same point.

The state machine model does permit objects to have multiple concurrent threads, *provided* that the scheduling of threads is deterministic, such that:

- all replicas schedule equivalent threads in the same sequence,
- the transfer of control from one thread to another occurs at the same point in all replicas.

Under Deltase, each capsule is treated as a separate state machine, the state variables comprise the total state of that capsule, and the commands include both the service requests (for services offered by that capsule) and the replies (from services used by that capsule).

7.4.4.2. Leader-Follower Model. This model requires fail-silent hosts, since correctness depends upon the behaviour of a single replica (but see section 7.6); this is the only fault-tolerance model used in the XPA variant of the architecture (see chapter 9). One of the replicas operates as the leader, and the other replicas operate as followers. It is the leader that determines the course of execution within the capsule; all non-deterministic decisions are taken by the leader. The leader informs the followers of the course of execution to follow, by the use of short messages (which are not visible to the computational objects).

In its simplest form, the leader-follower model may be considered as a variant on the state machine model, the difference being that under this variant, the leader can *select* which message to process next, rather than be obliged to accept the next message from the ordered queue. Such a decision is not deterministic, but since the host is fail-silent and the decision is passed to the followers, all replicas will process the same message next.

In a more sophisticated form, the use of *preemption points* enables the queue of incoming messages to be scanned more frequently than would otherwise be possible. Without preemption points, the message queue can be scanned only when the envelope is entered during the normal course of execution of a computational object. Preemption points allow the processing of one service request to be *preempted* by a service request of higher precedence; at each preemption point, the envelope is entered to inspect the message queue and to permit rescheduling. Preemption points may be set by a software tool that plants a suitable statement in the source code of the computational object; hence preemption points occur at identical points in all replicas of a given capsule, and preemption can be assured to occur at the same point in all replicas.

The support of preemption points is an important part of the support, by Deltase, for the combined requirements of fault-tolerance and real-time behaviour.

With the leader-follower model, all *non-deterministic* decisions are taken by the code within the envelope of the leader. The code within the computational object must be executed deterministically; once the leader has decided on a particular path, then all replicas must follow that path until the next change of path, as determined by the leader.

7.4.4.3. Passive Replica Model. This model also requires fail-silent hosts. Since only one replica is active at any given time, it is this replica alone that determines the execution path; the state in the standby replicas is updated by means of checkpoints, (see chapter 6). Code within the envelope generates (and sends) checkpoint messages when the replica is in active mode, and receives (and processes) checkpoints when in standby mode.

7.5. Engineering Support for Fault-Tolerance

This section covers those aspects of the engineering model where support for fault-tolerance is visible.

7.5.1. Transformer Capsules

A *transformer capsule* is a software component generated (by Deltase/XEQ) from one or more elementary transformers. (A transformer capsule may also include one or more computational objects.) A transformer capsule will thus have at least some *programmed* interactions with the non-Deltase world.

Because of such interactions with the outside world, the deterministic behaviour of replicas (of a transformer capsule) cannot be guaranteed; the result of an external interaction may, or may not, be identical for all replicas of the transformer capsule.

Where the transformer capsule is merely forming a bridge between two replicated worlds (for example, between Deltase capsules and replicated MMS software components), there is replicated behaviour on both sides of the transformer capsule.

However, where the transformer capsule is providing an interface to local services, then the transformer is also having to reconcile:

- the Deltase world: portable and generic software, with transparent support of replicas, and

- non-replicated local environments: existing applications, local operating systems, device drivers, etc.

A (replicated) capsule, offering a global service, may have to use a locally-provided service based on a local device. One example is that of a global file server that ultimately has to use a number of local file servers for persistent storage (on disc). Another example is that of making available, to users in the Deltase world, the services of proprietary database software (provided as a local server that accesses the disc directly; see §7.6). None of the Delta-4 support mechanisms can assure that these local servers are equivalent, in the replica sense; each of the local servers may have a different set of (local) clients.

A *transformer capsule* may be used to provide a Deltase "front end" to each of the local servers, each transformer capsule being co-located with its associated local server. These transformer capsules provide an interface to the Deltase world, but they cannot be assumed to exhibit identical behaviour.

These transformer capsules are not replicas, but *rivals* (see chapter 12); each presents a *distinct* service interface (although these interfaces will all be of the same type). A (replicated) application module, interfacing to *all* the given set of rival transformer capsules, can derive an agreed behaviour from the observed behaviour of the rivals; this agreed behaviour can then be made available to other computational objects via a service interface (see figure 14).

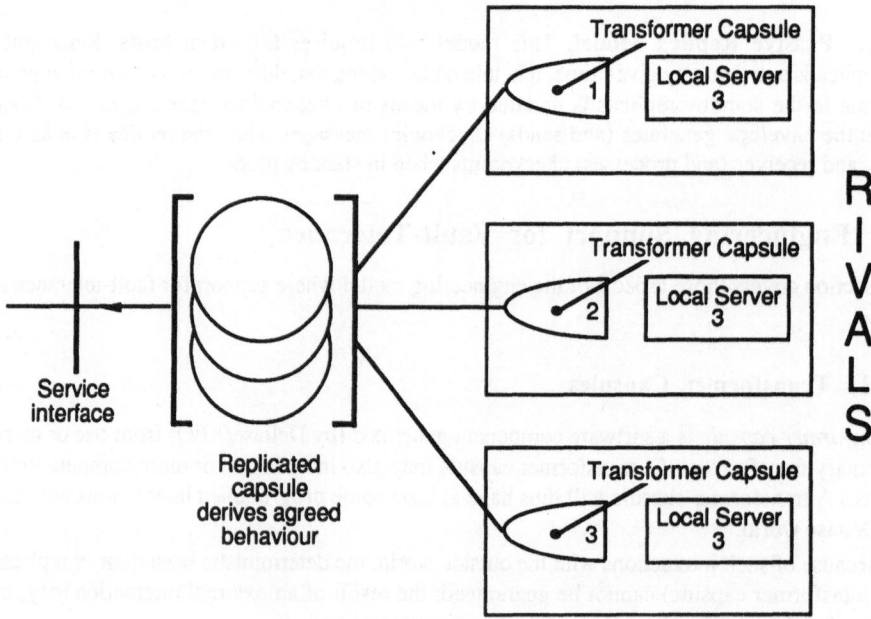

Fig. 14 - Rivals

In practice, various simplifying assumptions may be made; a common assumption is that the transformer capsules can be treated as replicas (which would normally be true for "most" of the time).

The use of transformer capsules as rivals, with an application module to derive an agreed behaviour, leaves the two worlds (Deltase and the local environment) unaltered.

7.5.2. Capsule Installation and Cloning

This section concerns capsule installation and capsule cloning. These topics are closely related to the system administration software (see section 8.2 for OSA system administration, and section §9.5 for XPA system administration).

Deltase/GEN produces a "file"; this file acts as a template, from which an instance of that capsule can be established on a host machine where the capsule is to be executed.

Installation is the process of establishing a clean instance of a capsule on a selected host; that instance will start executing from its initial state. If the capsule is replicated, then the replicas start together, from their initial execution point, in loose synchronism.

Cloning is the process of establishing a new *clone* of an existing partially-executed instance of that capsule. Following the loss of a host, typically because of a fault on that host, all the capsules executing on that host are lost. The level of fault-tolerance of the services provided by those capsules is then reduced. In order to restore the level of fault-tolerance, it is essential to create new replicas to replace those that were lost.

A new replica can only be installed on a host of the particular type for which the capsule was generated, and with the particular type of LEX. The new replica must be set up to have the *same state* as the existing replica(s) of that capsule, so that a new replica can play its full part (depending on the particular fault-tolerance model) as soon as possible. This will then minimize the "window of risk" during which the number of replicas is below the level specified. Any break in service (while cloning takes place) must also be kept within the timing constraints acceptable for that service.

The generic components responsible for the installation, cloning and termination of capsules are as follows:

1) **Replication Domain Manager (RDM)**. The RDM is responsible for the overall control of the installation, cloning and termination of capsules across all hosts in a Delta-4 system. Events, such as station failure, which may require the cloning of one or more software components, are passed to the RDM; it is the RDM that implements the cloning policy (see section 8.2) and determines on which host a new clone is to be created. The RDM is, itself, a software component, and can be replicated for fault-tolerance.

2) **Local factory**. There is a local factory on each host in the system, to carry out the instructions of the RDM on that particular host. A local factory is an example of a transformer capsule, appearing to the RDM as a capsule, but interacting directly with the LEX to perform the cloning (see figure 15).

3) **Object Manager Entity (OME)**. The OME is part of the envelope, to carry out those operations that can only be done from within the capsule. The OME consists of a set of library procedures, which are included in the envelope by Deltase/GEN.

The creation of a new replica of a capsule involves combining information from three sources:

1) **Initial state** (including code and related data) is taken from the file created by Deltase/GEN.

2) **Current state** information is taken from the existing replica(s). The means by which the current state information is supplied varies according to the fault-tolerance model used.

3) **Local state** information is data relating to the allocation, by the LEX, of resources in the local environment, such as buffer addresses or file identifiers. Such information is not visible to the application programmer, but is necessary for the library procedures that interface to the LEX. This information is only valid locally, and must be handled locally, by the OME.

The operational mode of the new replica depends on the particular fault-tolerance model used for that capsule.

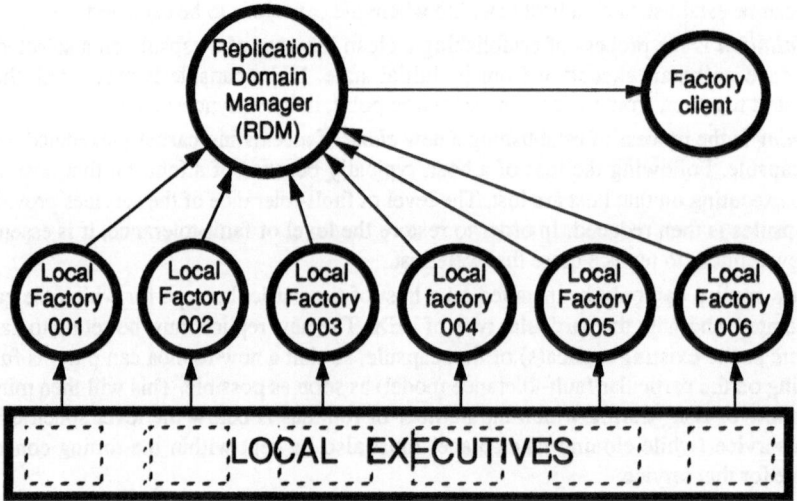

Fig. 15 - RDM and Local Factories

7.5.3. Dependable Communication System

Within a distributed system, a communication system is used to interconnect the distinct hosts that execute the applications software. The Delta-4 approach to fault-tolerance is based on the use of a communication system that provides explicit support for fault-tolerance; such a communication system is called a *Dependable Communication System* (DCS). A general objective of a DCS is to offer *distributed consistency*, to ensure consistent behaviour in distinct nodes.

To support fault-tolerance, Deltase must be used with a suitable DCS. In certain implementations it may be possible for some of the properties necessary for the support of fault-tolerance to be provided by suitable additions to the Deltase libraries, although there would normally be a performance penalty for doing so. Deltase hides the communication system from the application programmer, and neither the communication system nor its use is visible from the computational viewpoint.

Two dependable communication systems are being developed within the Delta-4 project:

1) the *Multipoint Communication System* (MCS, see section §8.1), used in the *Delta-4 Open System Architecture* (OSA), is an OSI-based communication system, for use in open dependable distributed systems;

2) the *Collapsed Layer Communication System* (CLCS, see section §9.6.4), used in *Delta-4 Extra Performance Architecture* (XPA), provides higher performance and support for real-time behaviour in systems having the necessary homogeneity.

7.6. Dependable Databases

Transformers can be used to include *commercially-available database software* within the Delta-4 system architecture, in such a way that the databases can be made *fault-tolerant*, by using replicated copies of the database. The database software is used without modification.

Clients use the standard database call interface for the particular database software, but this interface is treated as a Deltase service interface. Invocation of one of these procedures results in a service request to the database transformer. In principle, different clients could use different interfaces to the database. A number of clients may enjoy concurrent access to the replicated database.

A *transformer* is used to represent the database as a Deltase object, enabling other Deltase objects to use the services provided by the database software. The principle is illustrated in figure 16. Although the principles are generic, the details of this transformer are dependent on the particular database software.

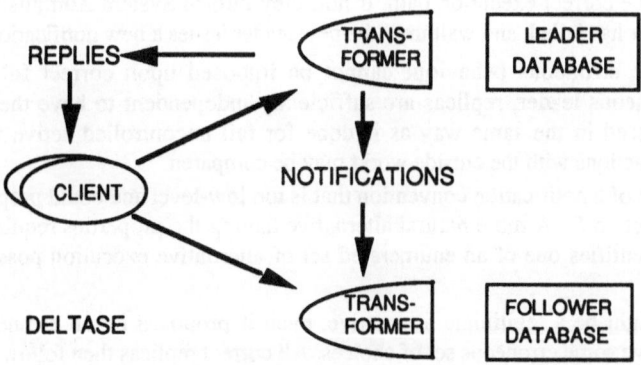

Fig. 16 - The Dependable Database

The database transformer is not a true state machine, and replicas of the transformer differ in internal operation, though not in externally visible effect. In the terminology of [Bond et al. 1987], they do not follow *identical* execution paths, but they do follow *equivalent* paths leading to identical final states.

To ensure replica group consistency, all changes to the database must be executed in the same order in all replicas of the database. This is achieved by using variants of the *leader-follower model* of fault-tolerance (see chapter 6). However, different variants of the model are required to support the database server on fail-silent and fail-uncontrolled hosts.

The variant for *fail-silent* hosts is considered first. The database transformer in one host acts as the leader, and all other instances of the database transformer (in separate machines) act as followers. The leader transformer performs updates in an opportunistic order, and then notifies the followers of the order in which the updates are to be performed. Since the leader host is fail-silent, these notifications are either correct or not produced. If they are not produced, the (first) follower takes over the role of leader.

Locks (including "shared lock" commands preceding read requests to the database) have to be treated in the same way as updates. Pure read requests, which do not change the state of the database, need normally be executed only by the fail-silent leader; only if the leader fails need the new leader execute the read. However, some read requests are implicitly also update requests and must therefore be performed by all replicas; for example, they may increment a pointer to the data being processed.

The leader also replies to all client requests; when it replies (using an atomic or reliable multicast message — see section 9.6.4), the notification to the follower(s) is included in the same multicast. The effect is that the client receives a reply from the leader if, and only if, the follower(s) receive a notification. So, if the leader fails, the (first) follower can avoid duplicating replies; hence the client receives exactly one reply in all circumstances.

Dependable databases can also be supported on *fail-uncontrolled* hosts. The leader-follower model can still be used provided it is combined with the active replication model, as follows:

1) The conventions of notification can be arranged such that an erroneous leader cannot mislead a correct follower. In the leader-follower model, replicas achieve deterministic execution by using the notification as a basis to choose identically from a set of alternative execution paths, any one of which is equally correct.

2) The notification must be expressed so that all correct replicas can easily determine whether it specifies one of these correct paths. If it is correct, there is no problem, but even if it is erroneous, all correct followers may be able to use it to identify a unique correct execution path. If not, they inform System Administration that the leader has failed, and wait until the new leader issues a new notification.

3) Since erroneous behaviour cannot be imposed upon correct followers by an erroneous leader, replicas are sufficiently independent to have their correctness assessed in the same way as is done for fail-uncontrolled active replicas; their interactions with the outside world may be compared.

An example of a notification convention that is too low-level and could propagate error is: "Jump to instruction I". A more natural alternative having the properties required is "choose K", where K identifies one of an enumerated set of alternative execution possibilities. Two cases then arise:

- K might be a legitimate alternative, even if proposed by an erroneous leader to resolve some erroneous set of choices. All correct replicas then follow the path K.

- If K is not one of the legitimate alternatives, each follower independently derives an opinion on the correctness of the leader, and expresses this opinion to permit it to be validated. In this case each follower can continue in the short term by deterministically resolving an acceptable alternative by, for instance, evaluating K *modulo H*, where H is, in the opinion of the follower concerned, the number of legitimate alternatives available.

In both these cases, all correct replicas will take the same path and express the same opinion in all interactions with the outside world. These interactions are therefore validated, and any incorrect outputs are detected and masked, assuming the correct replicas are a majority.

This strategy does not automatically bound the latency of errors, since, as with other applications, many normal interactions with a database do not reveal much of the internal state. Thus, for example, a fail-uncontrolled host can return a perfectly correct acknowledgement of a database update request, having miswritten the update to disc. There are two ways to reduce the error latency:

1) The error latency can be bounded by periodically reading the whole database, so that every piece of data is validated.

2) The reply to the update can contain extra replica state information, which is validated but not returned to the client. This reduces the probability of latent errors.

Another problem with databases and some other applications is that users may be permitted to read data without first locking it. This gives the user some performance benefit and assured freedom from deadlock, but the data read is not dependable, since it may be inconsistent or incorrect at the time of use. No model can confer full dependability on such operations, unless it effectively locks the data before the first read, which interferes with the user's semantics.

Database users may also be permitted to read multiple rows of data without specifying the order in which they are to be presented. The database replicas can return differently ordered replies to such a request. The validation routine that compares such replies must recognize them as effectively identical so long as they contain the same rows of data, regardless of order.

The validation of notifications and replies has a performance cost that only arises for databases on fail-uncontrolled hosts. However, the performance cost can be reduced if the leader sends the notification before it starts processing the update. The validation and processing of the notification in the followers then normally overlaps with the processing of the update in the leader, so that replicated replies to the client are available for comparison as early as possible.

Chapter 8

Open System Architecture (OSA)

The aim of Delta-4 *Open Systems Architecture* (OSA) is to offer the whole range of the Delta-4 distribution and dependability techniques in an environment in which heterogeneity must be accommodated. OSA is open to heterogeneous hardware and local executives. In particular, it is able to accommodate off-the-shelf host computers that have not been specialized for fault-tolerance and have *a priori* no built-in mechanisms for that purpose. The Delta-4 OSA infrastructure therefore provides error-processing techniques that do not require restrictive assumptions concerning the failure modes of individual nodes (OSA can accommodate *fail-uncontrolled* host computers). OSA implements all Delta-4 replication techniques (active, passive and semi-active replication).

The basic components of the Delta-4 Open Systems Architecture are:

- A communication system, the Multipoint Communication System (MCS), which was designed according to the ISO/OSI philosophy. MCS includes at any level of this model either new communication protocols, or extensions to existing ISO protocols, in order to achieve the communication requirements implied by the Delta-4 dependability models. MCS provides support for error-processing and basic reconfiguration mechanisms for fault-treatment.

- An administration system that provides a complementary set of mechanisms to handle the system's complexity from the administrative point of view, to support planning and system integration, to assist in daily operations and to help with fault treatment and maintenance. Complying with ISO management framework, it covers the areas of fault-, performance-, configuration- and security management, including all system components and the application itself.

In OSA, applications can be built either directly above these two basic components or preferably above an *Application Support Environment* that will provide the application programmer with a higher level of abstraction and with powerful tools for building his application. Deltase (see chapter 7) is of course a good candidate as such an environment and is available in OSA. Other computational models can be used; in particular, OSI application layer protocols such as MMS (Manufacturing Message Specification) are also available

8.1. Multipoint Communication System (MCS)

8.1.1. Introduction

This section provides a presentation of the Delta-4 *Multipoint Communication System* (MCS). The communication requirements specific to Delta-4 systems, both for fault-tolerance and distribution aspects, are expressed. A multipoint communication model, taking no account of consideration of fault-tolerance, is then defined. An implementation of this multipoint communication model, the Delta-4 MCS, is finally presented, with the addition of fault-tolerance aspects. The architecture of the MCS is described, and an overview of naming and mapping within the MCS is provided.

8.1.2. Delta-4 Communication Requirements

8.1.2.1. Requirements for Distribution. In the Delta-4 architecture, a distributed application is seen as a set of run-time components communicating amongst themselves by means of messages (and only by these means). Even without considering the problem of replication, it may be uncomfortable (although not impossible) to build such distributed applications above point-to-point communication using physical designation, as it is offered by OSI-based systems.

It is often the case within distributed applications that a function is performed by interactions among more than two application entities; these interactions involve communications among the application entities. Current communication standards only allow point-to-point communication, and such multi-peer interactions must be mapped onto a set of peer-to-peer interactions. Furthermore, if some of these interactions must be linked together to fulfil some global requirements (global order, atomicity, consistency,...), protocols must be added to restore this global view.

It is possible within a distributed application to group entities according to the function they execute or to the service they provide together. It would therefore be natural to map such application groups onto similar communication instances, such as associations allowing communication between more than two entities.

Here are some examples of communication between more than two entities within distributed applications:

- In client-server models (Deltase, the Delta-4 Application Support Environment, uses such a model, see chapter 7), a service can be used by many clients. Even if the client-server interactions are point-to-point interactions for every pair (client, server), it is useful to group all interactions related to the same service within a single communication instance associated to the service. This allows communication resources to be economised and decreases the overheads due to communication resource allocation and release every time a new client wants to use the service (this is required only on the client side). It allows the server to have a global perception of all communications related to the same service, and according to the quality of the communication service (order, atomicity,...), to have more facilities in the management of global events affecting the service (e.g., locks, stamps,...). In addition, with such a use of multi-peer communication, the dynamic aspects of communication group membership are important: clients must have the opportunity to join dynamically the communication instance associated with the service when they need to use the service, and to leave it when they do not need it any more.

- Distributed transactions and distributed database management are examples of applications where some operations need to be coordinated in a global way among all entities. The possibility of multicasting some information during certain phases in an atomic way (e.g., commit phase) is of major help, and is rather expensive to implement by means of point-to-point interactions.

- In many applications, data produced at one node can be used by several components residing on different nodes. Alarms and incidents that affect global distributed processing can be forwarded to several nodes for different purposes (archiving, processing, reporting, ...). In this case, the same information must be transmitted towards multiple receivers and even in an OSI-based system that uses a physical medium allowing data multicasting, high level protocols are not able to take into account such requirements.

- Load balancing within distributed systems often requires the migration of software components from nodes to others and then messages to be delivered to a dynamic, time-varying set of receivers with no a-priori knowledge of their physical location.

All these considerations have led the Delta-4 project to work intensively in the field of multipoint communications, and to define some new communication models, services and protocols, which are described in this chapter and in chapter 10.

8.1.2.2. Requirements for Fault-Tolerance. The implementation of host fault-tolerance techniques based on software component replication leads to important requirements on the communication system. The communication system must provide services that are amenable to communication between replicated software components. The Delta-4 objective of providing transparent fault-tolerance management implies that the programmer of these components must be able to ignore the redundancy involved. This means that the visibility that is given to the application programmer when he programs a communication interaction — a source entity forwarding a message to a destination entity — is that of one single message being sent by the source entity and one single message being received by the destination entity (whatever the degree of replication of the interacting entities) (figure 1).

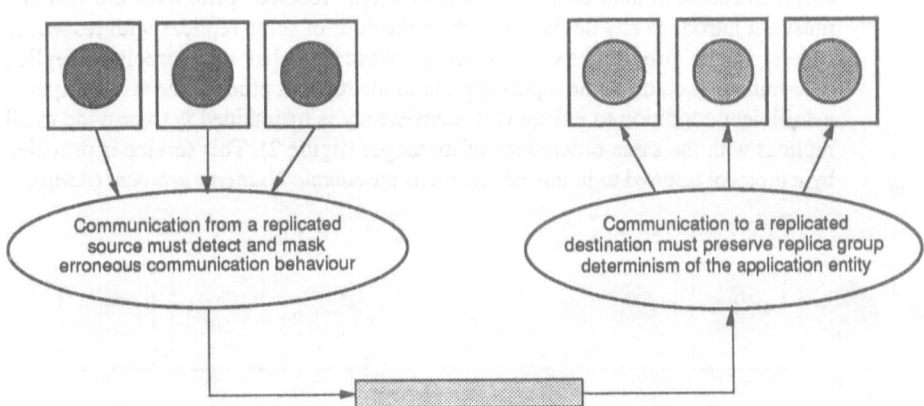

Fig. 1 - Communication between Replicated Application Entities

To preserve this mutual ignorance of replication by the code of interacting replicated entities, the communication system has therefore to take into account the replicated nature of these entities, and perform some actions that are made necessary by this replication:

- The communication system has to manage the redundancy of information sent by replicated sources. Every time the application programmer invokes a "send" primitive in a program, the replicated execution of this program results in the delivery of this message to the communication system on as many stations as there are replicas. According to the failure assumptions that are made on the concerned host computers, and to the error detection strategy that was selected for the concerned application entity, the communication entities located on these stations then have to reach an agreement to determine the message that will be forwarded to the destination. This agreement may include some mechanisms to detect erroneous communication behaviour resulting from faults in the host environment. Such detection should be associated with error reporting to administration entities, and as far as possible must achieve masking of errors to application entities. This agreement is a distributed agreement, and thus it involves some additional protocol as compared to standard communication protocols; this additional protocol is referred in this document as *Inter Replica protocol* (IRp) (figure 2).

To reach agreement, the IRp entities must exchange some protocol data units; a distributed agreement is considerably easier if these exchanges can rely on underlying transmission protocols that provide atomicity and order. For example, a distributed vote among three values that are sent onto the network by three different IRp entities is quickly performed if these three entities are assured to receive all three values in the same order. In the same way, when the agreement is reached, one IRp entity has to propagate the value to the destination (eventually replicated), together with an indication to the peer IRp entities that this propagation has taken place. This set of actions (propagate the value to the destination, inform the peer entities) must be atomic, and it is highly beneficial if it can be done by using a single atomic multicast primitive.

- The communication system has also to preserve the consistency of all copies of a replicated entity when it forwards some messages to this entity. As the application programmer must not be aware of replication, the provision of messages by the communication system to the component when "receive" primitives are invoked must not introduce any deviation in the behaviour of some replicas with respect to others. If the program excludes some constructs that may preclude replica determinism (i.e., the same inputs applied to the replicas produce the same outputs), a sufficient condition to ensure that consistency is maintained is to provide to all replicas with the same ordered set of messages (figure 2). This service is provided by a protocol referred to in this document as the *Atomic Multicast protocol* (AMp).

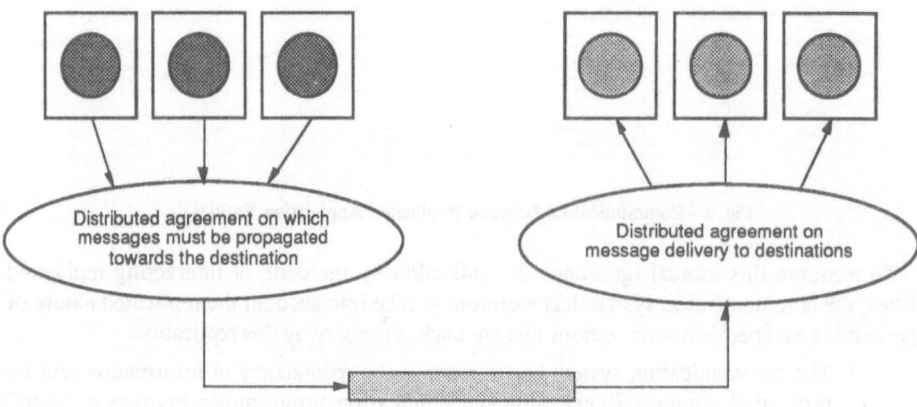

Fig. 2 - Distributed Agreement Protocols.

It therefore appears that the communication requirements of interacting replicated components can be fulfilled by an *inter replica protocol*, which relies on the services provided by an underlying *atomic multicast protocol*.

8.1.3. Multipoint Communication Model

8.1.3.1. Introduction and Definitions. The aim of this section is to define a multipoint communication model that covers the requirements expressed above for multi-peer communication between different application entities within distributed applications. It refers to existing standards, in particular the Open Systems Interconnection Reference Model [ISO 7498-1] and related standards, and to the work that was carried out in ISO/TC97/SC21 about multi-peer data transmission before this item of study was suspended (this work had produced a working draft addendum to ISO 7498-1 on Multi-Peer Data Transmission [ISO N2031]).

Some terminology from the latter document that was found convenient to use in this chapter and following ones is repeated here:

instance of communication	a single instance of a connectionless transmission or a single instance of a connection
multi-peer data transmission	transmission of a data unit to one or more destinations
multi-peer group	a group of peer entities which are mutually willing and able to be senders or receivers of multi-peer data transmissions with other members of the group
defined group	a multi-peer group whose total membership is known to members of the group
partially defined group	a multi-peer group whose total membership is known to some of the members of the group
undefined group	a multi-peer group whose total membership is not known to any members of the group
static group	a multi-peer group whose membership can only be altered outside the operations of an instance of communication between the entities of the multi-peer group
dynamic group	a multi-peer group whose membership may be controlled and altered by some or all members of the group
invoked group	the sub-group of a multi-peer group with which communication is attempted in an instance of communication
active group	the sub-group of a multi-peer group which is participating in an instance of communication
sub-group	some or all of the members of a group
central entity	an entity that is able within an instance of communication to transmit data to all other members of the active group. An entity that is not a central entity can only send data to the central entity.
centralized communication	a type of communication where, within an instance of communication, there is only one central entity and it cannot change.

roving centralized communication	a type of communication where, within an instance of communication, there can be many central entities but there can only be one central entity at any one time.
multi-centred communication	a type of communication where, within an instance of communication, there is more than one central entity.
decentralized communication	a type of communication where, within an instance of communication, all entities of the invoked group are central entities.

8.1.3.2. Multipoint Communication Model. The objective is to allow groups of communications entities to communicate amongst themselves within a multi-peer communication instance. An instance of communication is defined as either a single instance of a connectionless transmission or a single instance of a communication.

Connectionless communication, as defined in [ISO 7498-1], allows the transmission of copies of a data unit to several (more than one) destination addresses. However in that case, every data unit is transferred independently from others by the layer (or layers) that provides the connectionless service. All information that is necessary to transmit a data unit (destination address, quality of service, options,...) is presented to the connectionless service provider together with the data unit to be transmitted, in a single access to the service. The service provider does not have to establish a relationship between this access and others. This means that some important characteristics that may be required by applications (e.g., order) are not provided. This chapter does not therefore consider the connectionless case and concentrates on the definition of a multipoint communication model that would provide a service similar to connection-oriented communication, as defined in the OSI Reference Model [ISO 7498-1].

According to [ISO 7498-1], a connection is an association that is established for transferring data between two or more peer entities. This association is established between the peer entities themselves and between each entity and the layer that provides the connection-oriented service. The use of the connection-oriented service by peer entities has three distinct phases:

1) connection establishment;
2) data transfer;
3) connection release.

An OSI connection has the following fundamental characteristics:

a) it implies the establishment and maintenance of a three or more party agreement concerning the data transfer between the concerned peer entities and the service provider;

b) it allows the negotiation, between all concerned parties, of the parameters and options that will be used during data transmission;

c) it provides the connection identifier which is the means to avoid, during transmission, the overhead due to address transmission;

d) it provides a context in which all successive data units transferred between peer entities are logically linked, and it allows sequence preservation and flow control for these transmissions.

These definitions, if strictly applied to multi-peer groups, would not give the sufficient flexibility to meet the requirements of some distributed applications:

- Respecting phases 1), 2) and 3) only allows communication to take place among members of static groups. The membership of a communication instance cannot be changed during the lifetime of the connection. All members that want to participate to the instance of communication must agree on the connection establishment during the same phase, and all members must release the connection together.

- Characteristics a) and b) imply that the negotiation to establish the communication instance (the connection) takes place among all the members of the group. According to the number of group members, this can be the cause of a considerable overhead.

To allow a more dynamic scheme in the case of multi-peer connection-oriented communication, it seems that these general definitions should be slightly extended; these extensions are proposed here.

8.1.3.3. The Multi-Peer Connection (or Association)

8.1.3.3.1. Main Features. A multi-peer connection is an association that is established for transferring data between two or more peer entities. This is exactly the OSI definition; but whereas in OSI a connection only exists if the peer entities have decided to establish it and have negotiated to do so, and ceases to exist as soon as one of these entities releases it (thus the three phases: establishment — data transfer — release), it is possible to define more phases in the lifetime of a multi-peer connection.

The more general case to be considered is the one of dynamic multi-peer connections. In such communication instances, members of the communication group are allowed to join and leave the connection dynamically during the life of the connection. The consequence is that a multi-peer connection exists as soon as one entity has decided to join it. In other terms, the connection establishment primitive, which in the OSI case means "I want to communicate with entity X", simply means in this multi-peer context "I want to join (to participate in) communication instance I".

This dynamic aspect avoids the necessity for a new negotiation to take place between all active members of the communication group every time a new member wants to join the group. Apart the fact that such a negotiation would be a non negligible performance penalty, it appears impossible to change communication parameters and options of an instance of communication during its life. However, there is still the need that all peer entities participating in the same communication instance use the same communication parameters and options. Several possibilities exist to ensure this:

1) All possible members of a multi-peer communication instance have means to a-priori agree on these parameters and options; this implies that these parameters and options are defined before the actual instantiation of the connection. In that case, the negotiation when joining the multi-peer connection can be reduced to a check that the new member wants to use the right parameters and options for the connection it wants to join. This check can be carried out either by the service provider, or by a particular entity that coordinates the multi-peer connection. The first case assumes that the provider has some means of knowing these parameters and options at the very beginning of the communication instance (when the first member joins it). The second case implies that the first member to join the connection must be the coordinator, that the provider knows that there is a coordinator and knows its identity, and then again has some means to get this knowledge at the very beginning of the connection life.

2) The new members get the knowledge of communication parameters and options when they join the connection. This knowledge can be obtained from the provider or

from a coordinating entity. It still implies, as in 1), that the provider has some means of knowing some options at the very beginning of the connection life. However this case does not seem very realistic, because it would make it rather difficult to program such entities (which get all communication parameters and options only when they want to start communicating).

In any case, it appears that dynamic multi-peer connections must have some existence before the actual instance of communication; the main operations that can occur during the lifetime of a multi-peer connection are then:

1) create a multi-peer connection;
2) open a multi-peer connection;
3) join a multi-peer connection;
4) multi-peer data transfer;
5) leave a multi-peer connection;
6) close a multi-peer connection;
7) delete a multi-peer connection.

8.1.3.3.2. Create a Multi-Peer Connection. This operation consists of creating an administrative view of a multi-peer connection before the actual use of the corresponding communication instance. This administrative view includes all parameters and options that will be used when communicating with this multi-peer connection. Some of these parameters can be used for membership control (access rights, restricted membership,...). A minimum membership can also be defined, that is to say a list of members that must join the connection before any data transfer service is available on the connection.

No assumption is made on how this administrative view is represented and where it is stored. It is simply assumed that application entities have means to access this information, and that the service provider is also able to get them when needed.

8.1.3.3.3. Open a Multi-Peer Connection. It is conceivable that several peer entities located in the same end system use the same multi-peer connection (an end system is a node of the distributed system where application entities execute and use the communication system). In that case, the service provider must not allocate several times the resources it needs to manage the multi-peer connection (connection context). Opening a multi-peer connection therefore consists in allocating the resources that are necessary to instantiate the connection in an end system. The open operation also consists of creating the necessary underlying means to allow communication between the end system and all other end systems where the connection is already open. These means need to be created only once in the lifetime of the connection in a particular end system, and this operation has not to be repeated every time a new entity joins the connection on this end system.

The open operation can be implicitly invoked when the first entity wants to join the connection on the end system, or can be explicitly invoked before any entity joins it.

During the opening of the connection, the service provider gets the parameters and options of the connection from the administrative view that was created by the create operation.

8.1.3.3.4. Join a Multi-Peer Connection. A peer entity can join a multi-peer connection at any time in its lifetime. The entity provides to the service provider the communication parameters and options, and the provider checks that these parameters and options correspond to those of the connection, or transmit them to some coordinator that will perform the checking.

If the parameters and options proposed by the entity do not fit with those of the connection, the join operation is rejected by the service provider.

If a minimum membership is necessary for the data transfer services to be available on the connection, the join confirmation from the provider to the joining entity can be delayed until all required members have joined the connection.

According to whether the multi-peer connection is a defined group, the join operation may be notified to the other active members of the group. If the group changes are not notified to other members, and if the connection is already open on the end system, the join operation can be a local operation that consists only in checking parameters and options and allocating some additional resources for the entity that joins the connection.

8.1.3.3.5. Multi-Peer Data Transfer. As soon as one entity has successfully joined a multi-peer connection, multi-peer data transfer services are available to this entity to communicate with other entities.

Several modes of multi-peer data transfer can be defined:

- *Broadcast transfer:* a data unit transferred on a multi-peer connection using this mode is delivered to all entities that have joined the connection. The sender of the data unit can even be included in the broadcast group.

- *Multicast transfer:* a data unit transferred on a multi-peer connection using this mode is delivered to all entities that are explicitly specified by the sending entity. This specification can take the form of a list of addresses of entities, or of the address of a sub-group of the active communicating group.

Data transfer can be decentralized, centralized, roving centralized or multi-centred. If the communication is roving centralized, some service primitives must be associated with the data transfer services to allow the communication centre to change.

The quality of service of multi-peer data transfer is a parameter of the multi-peer connection. Important properties can be defined in this parameter, such as order properties, priority criteria, etc.

8.1.3.3.6. Leave a Multi-Peer Connection. An entity can decide to leave a multi-peer connection at any moment of its lifetime. This departure can be notified to other members of the connection, according to connection options. If a minimum membership is required for the connection, the provider can decide to force a leaving of all other members and a closing of the connection on all end systems when this minimum membership is no longer available.

Closing the connection on an end system can be performed automatically by the provider if all entities have left it on this end system. Conversely, an option of the connection can be to keep it open, to avoid further re-opening when entities re-join it later on.

Two types of connection leave can be defined:

- *Normal leave:* this is requested by the members and must be done while preserving some properties of on-going data transfer.

- *Abnormal leave* (or abort): this can be requested by the members or provoked by the provider on some abnormal events, and which can be done with loss of some data transfer properties.

8.1.3.3.7. Close a Multi-Peer Connection. This operation deletes the instance of a multi-peer connection on an end system. According to options, if some entities had joined the connection on this end system, they can be forced to leave the connection, or the close operation can be rejected. The close operation also relinquishes the underlying means that were used to allow communication with other end systems where the connection is open.

8.1.3.3.8. Delete a Multi-Peer Connection. This operation deletes the administrative view of a multi-peer connection created by the create operation. This deletion is only possible when no communication instance of the multi-peer connection exists in the distributed system.

8.1.4. The Delta-4 Communication System

8.1.4.1. Communication models. The Delta-4 Multipoint Communication System (MCS) is a layered communication system designed to meet the Delta-4 communication requirements as expressed in section 8.1.2, for distributed applications running on host computers interconnected by Local Area Networks.

The Delta-4 MCS proposes an implementation of the multipoint communication model that is introduced above, by defining multipoint communication protocols at the appropriate levels of the OSI reference model.

The Delta-4 MCS meets the Delta-4 requirements for fault-tolerance by including at the bottom of the session layer an Inter-Replica protocol (IRp). This protocol is able to handle communication from and to replicated application entities and hides from them the management of this replication. The application entities communicate between themselves either by using the multipoint communication model, or by using the ISO bi-point communication model (figure 3).

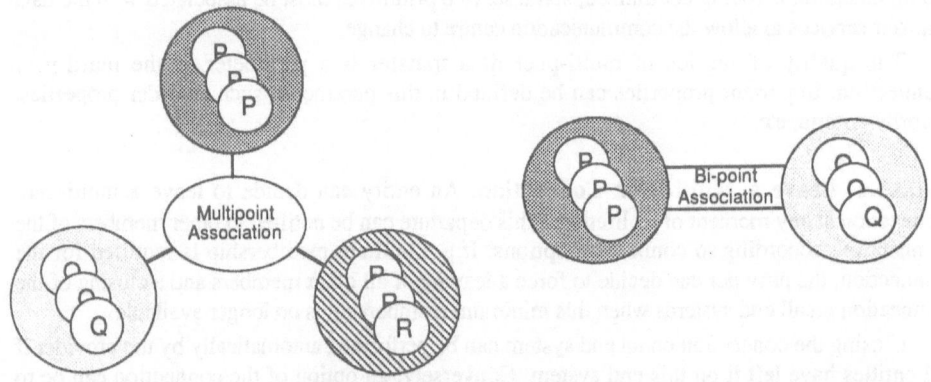

Fig. 3 - MCS Communication Models

Above services provided by IRp, high layers implement either the multipoint communication model or the more classical OSI bi-point communication model (thus allowing applications designed with this last model to take advantage of replication facilities offered by IRp).

The inter replica protocol is built upon multipoint communication services that are provided by multipoint communication protocols in layers 2 to 4 of the Reference Model. When no replication has to be managed, the IRp acts as an inactive protocol, and allows both high layers multipoint and bi-point model to use multipoint services of low layers.

The low layers multipoint protocols use multicast facilities provided by standard Local Area Networks, such as Token Ring [ISO 8802-5], Token Bus [ISO 8802-4] and FDDI [ISO 9314].

There are two functional parts to MCS: the Multipoint Communication Protocol stack (MCP) and the Network Administration System, or Multipoint Communication Management (MCM) which manages these protocols. The latter is dealt with as part of the more general notion of system administration that is covered in section 8.2. This section presents the MCP stack, the services it offers and its architecture.

8.1.4.2. MCS Multipoint Communication Model

8.1.4.2.1. Definitions. In the following, an end system within a Delta-4 distributed system is termed a *station*.

An *application process* is an element within a real open system that takes part in the execution of one or more distributed information processing tasks. An *application entity* is that part of an application process that deals with the communication system. An application entity represents a particular communication activity of the application process.

The application entities have access to the MCS through *M-SAPs* (MCS Service Access Points): the M-SAP is an individual access point to the higher-level services offered by the MCS.

At the high level (ACSE[1]/presentation) a multi-peer connection is termed a *multipoint association*. The application entities are linked by *associations* and communicate between them by using the data transfer services available on these associations. When an application entity joins an association through an M-SAP, MCS creates an *endpoint* to that association, which is fully identified by the couple (M-SAP name, association name) (figure 4).

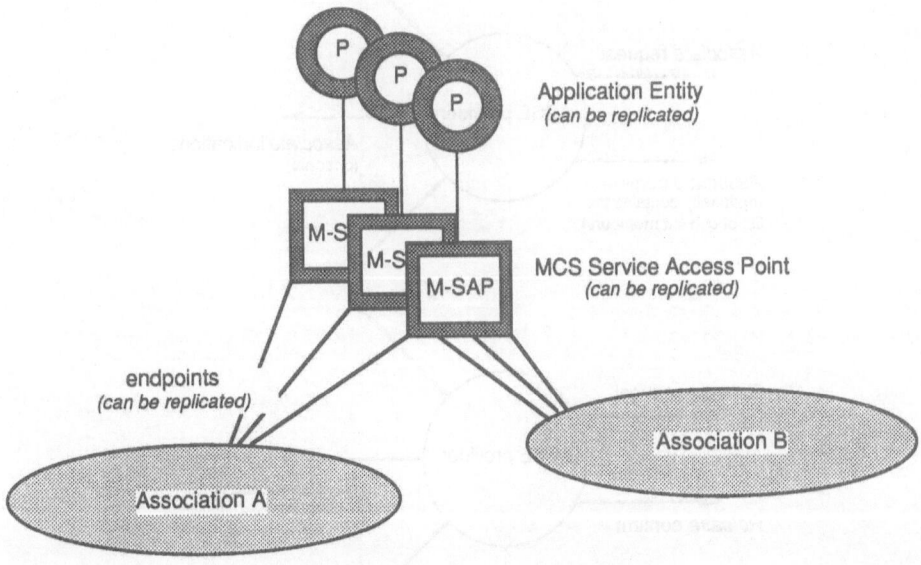

Fig. 4 - Main MCS Communication Objects

---

1   Assocation Control Service Element

8.1.4.2.2. Create and Delete an MCS Multipoint Association. A multipoint association represents a group of communicating application entities. The creation of such an association is an administrative operation that can be requested by a single application entity, or by a human operator. On creation, the characteristics of the association (name, available services (sequencer, token,...)) are defined and stored in a global administrative data base, the global Management Information Base (MIB). The name of the association can be obtained from a global name server, because association names must be unambiguously allocated in a single naming space within a Delta-4 system.

MCS multipoint associations have a dynamic membership, which is not controlled by any coordinating entity. There is no minimum membership for MCS associations. According to an association option, the changes to the association membership are notified to the association members.

8.1.4.2.3. Join and Leave, Open and Close an MCS Multipoint Association. Application entities can unilaterally join or leave MCS multipoint associations. When joining an association, the application entity must provide the parameters and options related to this association, which are checked by the ACSE provider (figure 5). If the association has not yet been opened on the station where the application entity wants to join it, the ACSE provider automatically opens the corresponding association and session connection. To do so, it obtains the administrative view of the association using support from the System Management Application Process (SMAP).

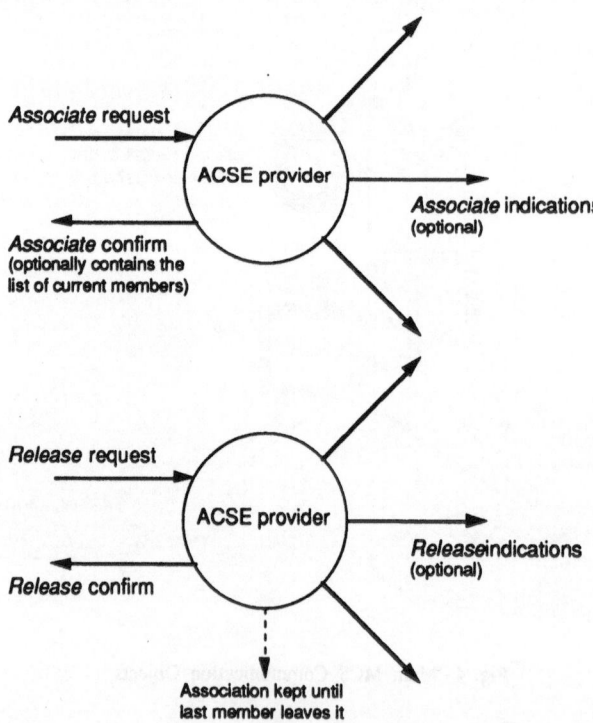

Fig. 5 - MCS Association Joining and Leaving

When all application entities have left an MCS multipoint association on a particular station, the association is automatically closed by the ACSE provider.

If the association has the defined membership option, join and leave operations are notified to all active members of the multipoint association. New members joining such an association also get the knowledge of the current active membership.

8.1.4.2.4. Data Transfer on Multipoint MCS Associations. There are three different modes for transferring information over an MCS multipoint association:

- *Inclusive multicast:* the information is sent to all members of the association including the sender.
- *Exclusive multicast:* the information is sent to all members of the association except the sender.
- *Selective multicast:* the information is sent to a sub-set of members of the association designated either by their logical names or by a common sub-group identifier.

Ordering of message delivery over a multipoint association is such that two application entities receiving two messages from an association receive them in the same order.

All entities connected to a multipoint association may both send and receive messages. A multipoint association is characterized by the type of dialogue that manages the interactions between the members of the association:

- *Decentralized dialogue:* all the entities are simultaneously able to initiate message transmissions.
- *Roving centralized dialogue:* only one entity at a time is able to send data (i.e., the one that possesses the data token of the association). Primitives are provided to allow the association token to circulate among association members.

Multipoint associations also provide a sequencer service that allows the members of the association to get consecutive positive ticket values. This enables the application to stamp and thus to order detected events that modify the state of the distributed application. Up to two sequencers can be used on an MCS multipoint association.

8.1.4.3. Concepts for Fault-Tolerance

8.1.4.3.1. Error Detection and Masking. The Delta-4 communication system meets the requirements created by the replication of application entities on different host computers of a distributed system. It is able to handle communication between replicated application entities in a way that is transparent to the programmer of these entities. This transparency is sometimes termed *invisible multicasting*.

The inter replica protocol (IRp), which acts as a bottom sub-layer of the MCS session layer, handles this type of communication. It includes error detection and masking mechanisms for replicated communication issued from application entities actively replicated on fail-uncontrolled hosts. These mechanisms also allow error masking for replicated application entities (either active replication, or passive, or semi-active, see chapter 6) running on fail-silent host computers; in this case, the error detection mechanisms are located in the host fail-silent environment.

The inter replica protocol relies on the services provided by the low layers of the MCS, essentially the atomic multicast service, and the notification of station failures.

8.1.4.3.2. Impact on Communication Objects. A replicated application entity is one that has been replicated on different stations for the purposes of fault-tolerance. When such a replicated application entity uses MCS services, it does so through an M-SAP (MCS Service Access Point). Generally, Service Access Points have only a local significance, in the station where the service is accessed. However, in the case of replicated entities, the service is accessed by the same application entity in several stations in the system. This is the reason that leads us to consider in MCS that M-SAPs used by replicated entities are themselves also replicated. In the same way, all endpoints that would be created by a replicated application entity through a replicated M-SAP are also considered as replicated endpoints (figure 4).

8.1.4.3.3. Services for Fault-Tolerance. As M-SAPs can have a global significance, they must be known and defined in a global way. M-SAPs are defined by administrative primitives before being used, and this administrative view of M-SAPs is stored in a "Global Management Information Base" or "Global-MIB". M-SAPs have some attributes that are used by IRp to coordinate the replicas of the M-SAP (and then to control the replicated behaviour of the corresponding AE). Some of these attributes are:

- the degree of replication of the M-SAP;
- the replication domain of the M-SAP (the list of stations where a replica of the M-SAP exists);
- the allowed desynchronization delay of data transfer request issued through this M-SAP;
- the type of error detection mechanism to be used;
- ...

The IRp provides to administration entities some means to manage fault treatment and reconfiguration:

- error reporting (voting disagreement between replicas, early or late timing errors, abnormal overflow of one replica,...);
- primitives allowing dynamic reconfiguration of M-SAPs (dynamic creation or deletion of M-SAP replicas), transparently to the replicated AE.

As explained previously, the IRp fully relies on the atomic multicast services that are offered by low layers, and which are built on the basic atomic multicast protocol (AMp). In particular, the AMp is able to detect and signal, both to upper layers (and thus, IRp) and to administrative entities (SMAP), the failure of stations in the Delta-4 distributed system. This property is intensively used by error recovery and fault-treatment mechanisms.

In addition, IRp uses the AMp facilities to provide to replicated AEs a global order for all data units received from several associations through the same replicated M-SAP. This property, termed "deterministic receiving", is an attribute of the M-SAP and can maintain determinism of a replicated AE communicating through the same M-SAP on several MCS associations.

8.1.4.4. Structure of the Multipoint Communication Protocol Stack. The MCP stack is a communication stack that follows the layering principle of the OSI Reference Model. No standardized multipoint protocols yet exist. Consequently, Delta-4 proposes a hierarchy of specific protocols for providing the multipoint communication services. Despite the specific nature of these protocols, there is a total compatibility with ISO standards for the physical layer and medium access control protocols, thus allowing the coexistence of the MCS and ISO protocols on the same LANs. High layer ISO standards can also work on top of IRp (with a logical addressing scheme). This allows applications using such standards to be ported on to the MCS stack and to benefit from the fault-tolerance techniques provided by Delta-4.

Figure 6 shows the layering of the Delta-4 MCS; those protocols that were designed and developed within the Delta-4 project are put in bold characters, those protocols that follow existing standards are put in italic characters.

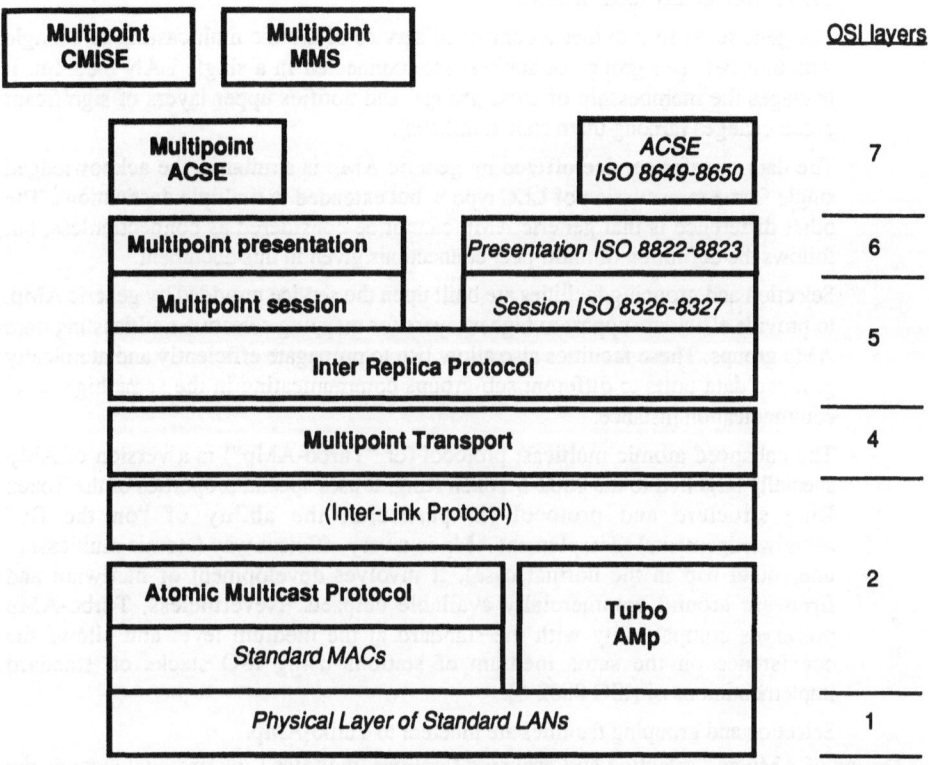

Fig. 6 - MCS Layers

The services and protocols for every layer are described in the MCS section of the Delta-4 Implementation Guide [Delta-4 1991]. The services and protocols are not described in detail when they conform to existing standards. The reader can refer to these standards. The communication profiles that are selected for these protocols are inherited from the ESPRIT project 2617 (CNMA), and are described in detail in the CNMA Implementation Guides.

The protocol and the service provided by the bi-point session conform to ISO standards 8326 and 8327 (FU kernel and FU duplex).

As an exercise to port application entities using ISO communication, and to let these AEs benefit from the Delta-4 fault-tolerance techniques, the use of the application level Manufacturing Message Specification [ISO 9506] by replicated Application Entities was prototyped during the Delta-4 project. A multipoint MMS, using MCS multipoint facilities, has also been specified and implemented [MP MMS].

The MCP stack uses standard communication mediums and the physical layer conforms fully with these standards. The atomic multicast protocol, which is one of the main Delta-4 specific protocols, constitutes the basis for the management of multipoint communication and communication between replicated entities. Two versions of this protocol have been specified within the Delta-4 project:

- The "generic" atomic multicast protocol is a software implementation of AMp. It is medium-independent, and is built upon standard implementations of MAC protocols for Token Ring [ISO 8802-5], Token Bus [ISO 8802-4] and FDDI [ISO 9314] Local Area Networks. It allows the use of commercially available chipsets for these LANs without any modification.

  The generic AMp provides a confirmed service of atomic multicasting of a single data unit between groups of stations interconnected in a single LAN medium. It manages the membership of these groups, and notifies upper layers of significant group changes (among them station failures).

  The data transfer service offered by generic AMp is similar to the acknowledged single frame transmission of LLC type 3, but extended to multiple destinations. The other difference is that generic AMp cannot be considered as connectionless, but follows the definition of multi-peer connections given in this document.

  Selection and grouping facilities are built upon the service provided by generic AMp, to provide efficient support to higher layers for mapping selective multicasting onto AMp groups. These facilities also allow IRp to propagate efficiently and atomically grouped data units to different sub-groups communicating in the same high-level communication instance.

- The enhanced atomic multicast protocol (or "Turbo-AMp") is a version of AMp specially targetted to the 8802-5 Token Ring. It uses special properties of the Token Ring structure and protocol (in particular, the ability of "on the fly" acknowledgements) to implement AMp in a very efficient way (atomic multicast in one round trip in the normal case). It involves development of hardware and firmware around commercially available chipsets. Nevertheless, Turbo-AMp preserves compatibility with the standard at the medium level and allows the coexistence on the same medium of stations using ISO stacks on standard implementations of [ISO 8802-5].

  Selection and grouping facilities are inherent to Turbo-AMp.

On top of AMp and selection and grouping facilities, the Inter-Link protocol extends the service to interconnections of LANs. The transport protocol is a light connection-oriented protocol that adds to underlying services functionalities for segmenting and reassembling, and data flow control.

8.1.4.5. Addressing and Mapping of Communication Objects. The MCP stack uses at any level a logical addressing: communication objects are identified by a logical address that is allocated in a space of 2 power 20 possible addresses.

The communication objects that are visible to the user have a logical address in that space:

- An M-SAP is identified by a unique logical address; if this M-SAP is replicated, all the copies of this M-SAP have the same address (even though they are located on several distinct physical stations).

- An association is identified by a unique logical name; when an association is opened on several stations, the identification of this association is the same in any one of these stations.

- An endpoint is fully identified by the couple (M-SAP logical address, association logical name) constituted by the logical name of the association to which this endpoint is related and the logical address of the M-SAP through which it was created.

The atomic multicast protocol manages logical gates: a gate defines a group of stations within a single LAN segment.

The transport layer manages multipoint transport connections: there is at most one endpoint of a transport connection on one station. There is a one to one mapping of transport connection endpoints onto logical gates. Every time a transport connection endpoint is created on a station, the corresponding AMp logical gate is also opened.

There is a one to one mapping of associations onto transport connections. When an association is opened on a station, the corresponding transport endpoint is created on the station, and therefore the logical AMp gate is opened.

Because of this one to one mapping of associations to transport connections, and to AMp gates, all these communication objects have the same name.

When an M-SAP is replicated, IRp entities that manage this M-SAP need sometimes to communicate among themselves, out of the control of AEs. A replicated M-SAP is therefore mapped onto a transport connection, and then onto an AMp logical gate, which has the same name as the M-SAP address.

It therefore appears that association names and M-SAP addresses are allocated in a unique naming space.

The selection and grouping facilities provided above AMp are used to give to the low level the visibility of high level endpoints and sub-groups, to allow a more efficient implementation of selective multicasting (figure 7). This means that joins and leaves within MCS multipoint associations are made known to transport and to underlying facilities, which offer services to maintain consistent views of sub-groups of atomic multicast groups.

8.2. System Administration

The Delta-4 multipoint communication protocol system, described in section 8.1, which provides the basis for dependability, and the Deltase support of its high-level transparent usage by programmers, described in chapter 7, are supplemented by a third concept, the Delta-4 System Administration. The properties of the envisaged application areas and the fault tolerance approach of Delta-4 require powerful mechanisms to handle system complexity from the administrative point of view, to support planning and system integration, to assist in daily operations and to help with fault-treatment and maintenance. The set of such facilities is referred to here as system administration.

This section outlines the main functional requirements of system administration, the applied structuring concepts and the derived architecture. Within OSA, the Delta-4 approach to system administration follows the ISO/OSI related work on systems management.

8.2.1. Functional Requirements

The management of distributed systems is currently under intense discussion in the academic field as well as in the immense ISO standardization work, international multi-vendor initiatives (MAP, CNMA, OSI/Network Management Forum) and network management product developments (IBM's Netview, HP's OpenView, DEC's EMA — Enterprise Management Architecture).

The functional view of Delta-4 system administration goes beyond the scope of these activities as a consequence of the sophisticated fault tolerance approach. This consequence also implies that administration is not limited to the management of communication resources, but encompasses all kinds of data processing resources.

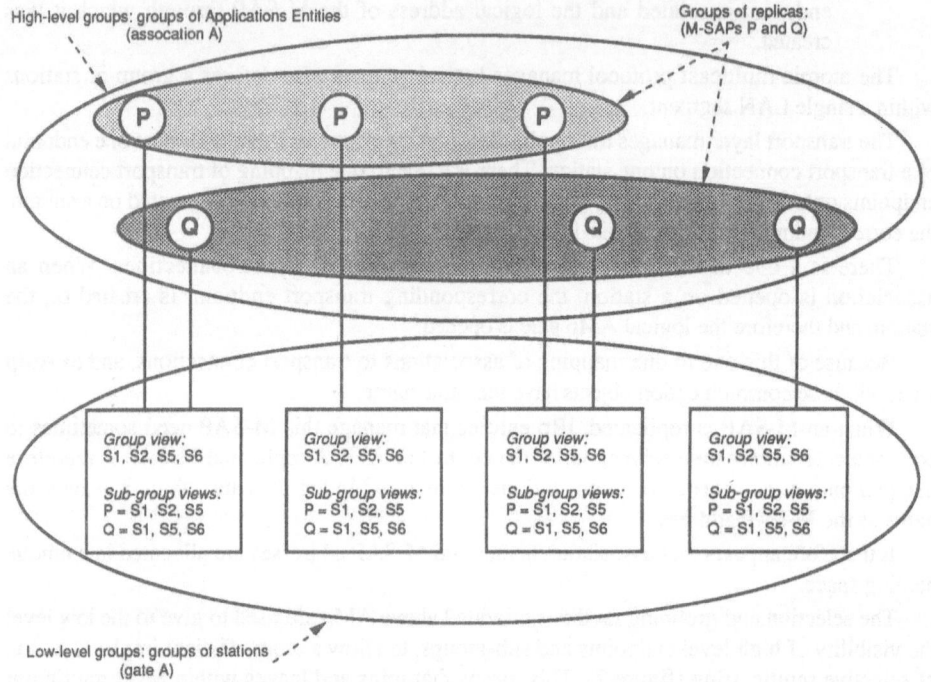

Fig. 7 - Communication Groups Mapping in MCS

8.2.1.1. The System Administration Space. This section aims to structure the functional view of system administration in a fault-tolerant open distributed system. Figure 8 shows the main dimensions that stretch the administration space and categorize the requirements for building an integrated administration system.

The dimension "Functions" corresponds to the management functional areas, which are used to structure the OSI-Management related ISO work. Of particular interest within the Delta-4 approach are fault and configuration management and to some extent the evaluation of the performance of a system, but the project also investigates specific topics on security management (see chapter 13).

A second dimension is dedicated to the components to which the management functions are applied. This is of course the MCP-stack itself. The fault tolerance approach based on replicated software components implies the need for fault and configuration management for *replicated* files and processes or general Deltase applications.

The different life cycle phases that a Delta-4 system may follow constitute a third dimension for categorising administrative requirements. There are design, planning, installation and commissioning phases, normal daily operation, and phases where the configuration changes with respect to station failures or system evolution.

8.2.1.2. Administrative Tasks. In the following a general view is given by defining three overall administrative tasks (ATs) along the life cycle phases of a Delta-4 system.

AT1: Planning and integration of redundancy and distribution;

AT2: Monitoring of system behaviour;

AT3: Fault treatment and maintenance

The illustration of the ATs is restricted here to the main topics of the Delta-4 approach. The overall management model and the implementation principles are however designed to provide support for all aspects of system administration.

The requirements of the Administrative Tasks identified below are concerned with the Delta-4 attributes "fault tolerance" and "distribution".

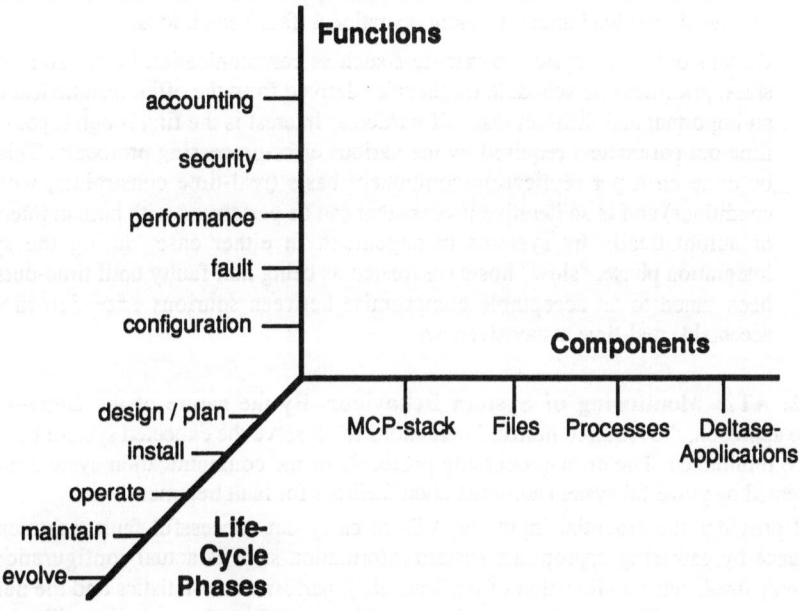

Fig. 8 - The System Administration Space

8.2.1.2.1. AT1: Planning and Integration of Redundancy and Distribution. The first step to configuring a distributed application is that of identifying application-specific characteristics such as:

- topological configuration (e.g., process control interface constraints);
- services to be made dependable (replicated software components);
- degree of possible redundancy (software component replication domains);
- error-processing modes dependent on host fault assumptions and required dependability attributes;
- contingency plans for application-specific reconfiguration strategies.

The system designer's experience is usually supported by simulation tools. Actual measured management information, derived from comparable configurations, can provide a quality of simulation significantly better than when using parameters derived only from the equipment's technical data.

The system integration phase in the real or a simulated application environment is used to validate the design decisions and to tune the system's operating parameters. In particular, administrative information and actions are required to support the following:

- Testing the fault tolerance mechanisms by using performable actions (on-line fault injection) to stimulate the use of available redundancy (passive replicas, cloning of further active replicas). Stimuli should cover all expected operational conditions such as failures of host and communication resources, buffer overflows and general overloading. This simulation of operational situations must also be carried out to demonstrate the system's capability for evolution.

- Performance evaluation of hosts and the communication system requires actions to initiate time-based measurement to obtain appropriate statistical data of a real or simulated workload under different operational (fault) conditions.

- Dimensioning the system parameters such as communication buffer and window sizes, priorities and schedule frequencies derived from the offered statistical data is an important and difficult task. Of particular interest is the first rough layout of the time-out parameters required by the various error-processing protocols. This must be done on a per replication-component basis (real-time constraints, workload conditions) and is an iterative process that can be performed with human interaction or automatically by systems management. In either case, during the system integration phase, "slow" hosts are treated as being non-faulty until time-outs have been tuned to an acceptable compromise between spurious error detection and acceptable real-time responsiveness.

8.2.1.2.2. AT2: Monitoring of System Behaviour. By the nature of the Delta-4 fault-tolerance approach, the required human interactions to preserve the expected system behaviour should be minimized. The error-processing protocols of the communication system must be supplemented by powerful system administration facilities for fault treatment.

AT2 provides the essential input for AT3 to carry out successful fault treatment and maintenance by gathering appropriate system information such as actual configuration data (redundancy used, current allocation of replicas, etc.), performance statistics and the numbers of (recovered) errors. The monitoring activity may be carried out by periodic polling, event-triggered polling, and/or by counting events such as error reports from the error-processing protocol entities.

Additional tests might be used to check the absence of faults within unused redundant system components. For example, a dual-ring LAN may require that the current inactive transmission direction be periodically tested in case a failure of the active one should occur.

8.2.1.2.3. AT3: Fault Treatment and Maintenance. The traditional "network management" view of AT3 is that of providing long-term management functions for system maintenance staff. Within a truly fault-tolerant distributed system environment this task is much more critical since it pertains to the prevention of serious consequences when faults accumulate. It should thus be automated as far as possible.

Fault treatment is achieved by supporting:

- fault diagnosis;
- fault passivation;
- system reconfiguration; and
- system maintenance.

Fault diagnosis is necessary to decide if a fault is permanent (e.g., judging the significance of time-out occurrence within the assumed and actual workload profile) and to assist in fault localization (e.g., by evaluation of error reports from error-processing protocol entities). If fault diagnosis should conclude that a permanent fault has occurred, then fault passivation must be carried out and system reconfiguration envisaged.

Fault passivation of fail-uncontrolled hosts must be done automatically by the integrated administration system, i.e., manual intervention should not be required. Note that fail-silent hosts, by definition, carry out automatic and autonomous fault passivation.

System reconfiguration can be envisaged if there are sufficient redundant resources. It entails re-allocation of the software component replicas that were resident on failed hosts to restore the level of redundancy required for the error-processing protocols to function correctly despite further faults. If re-allocation is not possible, then some software components may either have to be abandoned in favour of more critical ones. Alternatively, fault-tolerant operation is degraded to fail-safe operation to ensure safety and/or integrity of the distributed application(s).

Re-allocation of software component replicas is achieved by means of a cloning operation that creates a new replica on a specified node. Three sub-operations can be identified:

- Creation of a template of the software component at the new location. This can be done in advance of an actual cloning request according to application-specific contingency plans specified by AT1 (e.g., localization of passive replicas, designation of degraded modes,...).

- Creation of a copy of the component's persistent data or "state" at the new location.

- Activation of the new replica whilst ensuring replica-consistency. This involves the automatic management of the dynamic, configuration-dependent associations between replicated components.

Two techniques for cloning with different performance characteristics can be considered; they are termed *recursive transfer* and *snap-shot transfer*.

Recursive transfer requires the continuous identification and tagging of structures that are modified in the active replica(s) while a state transfer is attempted. The state transfer sub-operation is then repeated using only the tagged data. This is carried out recursively until no data structures are tagged during state transfer.

Snap-shot transfer involves the creation of a local copy (or "snap-shot") of the component's state on the host of the active replica(s). While this snap-shot is transferred to the new location, a log of messages sent to the software component is constructed (at the new location). When the snap-shot transfer is complete, the new replica processes the stored messages.

Other optimizations may be possible in the cloning operation; for instance the partitioning of the state of a software component into independently-lockable sub-structures.

The basic fault treatment administrative facilities outlined above are supplemented by facilities for configuration and maintenance management.

Configuration management facilities include remote initialization and loading of nodes, node passivation, version control, etc. These functions are necessary to assist in normal system evolution.

Maintenance management facilities are provided to minimize repair time. Such facilities include tools for post mortem analysis, remote access to the host's local operating system diagnostics and support for remote initialization or loading of off-line tests. In an ideal case, the maintenance staff is provided with information concerning failed boards (in stations) and failed transmission medium segments.

8.2.2. Structuring Concepts

This section presents a general structuring model for designing management systems (see §8.2.2.1) and its application to Delta-4. The model is supplemented by an enumeration of features of a formal management description language (see §8.2.2.2) that enables a specification of a management system structured according to the model's design guide-lines. The derivation of general architectural components from the model is given (see §8.2.2.3) and this forms the basis for the architecture of the Delta-4 management system (see §8.2.3). The relationship between the object model and the management of distributed applications is used as an example of the model's application (see §8.2.2.4).

8.2.2.1. Structuring Model. The inherent complexity of the various administrative tasks outlined in the previous section requires the development of a structuring model. This model should provide unified design principles and support a stepwise implementation strategy aimed to conform and coexist with related (draft) standards. This model should provide an abstraction of the system to be managed that is based on the viewpoint of a system administrator and which allows the system to be viewed coherently despite its complexity. The proposed model encompasses the following structuring principles:

1) manageable components;

2) management domains (or simply "domains").

8.2.2.1.1. Manageable Components. The basis for the model is to treat a system as consisting of a set of (interacting) typed components. A manageable system consists of a set of manageable components. These are an abstraction of normal components of some granularity with extensions relevant to management. In addition to the designed normal functionality, manageable components are characterized by their:

- static and dynamic management-related attributes such as version identification, operational state and parameters, error and performance-related statistical information, designed fault-tolerance behaviour, various management relationships to other manageable components etc.; and

- performable management operations such as create, delete, clone, reset, change state, get/set attributes or wait for events triggered by the manageable component.

Note: In the ISO/OSI Management standards, an abstraction of a physical or logical resource for managing this resource is called a managed object. An ISO managed object is defined by:

- the attributes visible at the managed object boundary;

- the management operations that can be applied to the attributes or the managed object itself;

- the behaviour exhibited by the managed object in response to management operations; and

- the notifications that can be emitted by the managed object.

A Delta-4 manageable component may be described as an ISO managed object. Thus, when the term managed object is used in the following to describe the management view of a Delta-4 resource, the ISO management view is applied with the above described properties.

Examples of manageable components in Delta-4 are stations, communication objects or software components. A survey of manageable components that are currently under implementation is given in §8.2.4. Note that software components (capsules) generated from Deltase objects form one type of manageable component (this is further illustrated in §8.2.2.4).

8.2.2.1.2. Domains. In the literature various definitions of a domain can be found. In [Sloman 1987] a domain identifies a sphere of influence of management. A more concrete definition can be found in [Robinson 1988]: here a domain identifies a "set of managed objects that share a common attribute; in particular, it is the set of managed objects to which the same management policy applies".

A common attribute of replicated objects in Delta-4 is that replicas can be dynamically placed on a given set of stations. Such a set of stations is termed a *replication domain*. Common management policies are applied to sets of replicated objects that have the same replication domain, for instance:

- restoration of the replication degree of objects which had a replica on a failed station by cloning them to stations offering spare redundancy; and

- migration of all objects from a station that is due for maintenance.

Various boundaries of domains can be considered, e.g., physical, organizational or security boundaries. For example, in a distributed system the components which conform to a communication standard form a domain. Manageable components encapsulating management information accessible through management operations may also be seen as a domain.

As illustrated in [Sloman 1987], it is possible to conceive of four relationships between domains: they can either be disjoint, interacting, overlapping or nested. These relationships indicate how the management entities that are responsible for the management of respective domains must cooperate to fulfil a common management policy.

8.2.2.2. Management Description Language (MDL). A management description language (MDL) can serve as a basis for a formal description of the structure and the functionality of a management system. In Delta-4, the development of such a language is under way; it should enable a system administration designer to specify formally the administration policy of a Delta-4 system.

The following language requirements have presently been identified:

- facilities to specify the domains and their interrelationships;

- facilities to specify the manageable components and their integrated management mechanisms;

- facilities to specify the transparency attributes of management services;

- facilities to specify decisions about the invocation of management services;

- facilities to specify possible system states — the system state is necessary to specify the conditions on which the above-mentioned decisions are based.

The current emphasis of our work on MDL is on the specification of the structure of the management system and the decision taking mechanisms. Structuring elements are derived from the structuring model given in the previous section; decisions are based on system monitoring.

The architectural components that make use of these specifications are not included in the language themselves. A classification of architectural components is given in the next section. A complete architecture includes more than this; it encompasses the set of architectural components as well as the set of relationships between these components according to a set of rules for their establishment [ANSA 1989]. The architecture of the Delta-4 management system is given in §8.2.3. Modelling the architecture in MDL is currently not an area of investigation.

8.2.2.3. Architectural Components. Having defined the structuring concepts and specification techniques, guide-lines for the development of the management system architecture can now be established.

Architectural components of a management system can be divided into two major classes:

1) domain managers;

2) (domain) manager applications.

8.2.2.3.1. Domain Managers. A domain manager can decide about, and carry out, management operations on the set of managed objects that constitute a domain.

A domain manager comprises a single domain manager process, or a set of peer domain manager processes that cooperate within the domain to carry out a domain-specific management task. A domain manager process may be replicated to meet the Delta-4 dependability requirements. However, the various interrelationships and interdependencies of domains mean that not all management tasks can be performed within a specific domain. Thus, inter-domain interaction is needed between domain manager processes across domain boundaries as well as intra-domain interaction between peer domain manager processes. Cooperation is carried out by a sequence of management interactions. For a particular interaction, a domain manager process may take either the role of an agent or a manager:

- A domain manager process taking the role of an *agent* (shortly: an agent) is responsible for performing the management operations upon managed objects within its domain, and for forwarding event notifications triggered by managed objects to a manager.

- A domain manager process taking the role of a manager (shortly: a *manager*) conveys management operations on managed objects to the responsible agent, and receives event notifications from agents.

Manager and agent roles are either statically or dynamically assigned to domain manager processes. In a static role assignment, one domain manager process takes the agent role and the other takes the manager role during the whole domain manager cooperation. In a dynamic role assignment, a domain manager process may take the agent role in one interaction and the manager role in a separate interaction during the period of the domain manager cooperation. The role assignment is dependent on the given relationship of the domains (figure 9) that the domain manager processes are part of:

a) For intra-domain cooperation between peer domain manager processes the roles are assigned dynamically (e.g., SMAP-SMAP cooperation, see §8.2.4.3.4).

b) There is no cooperation at all between the domain manager processes of two disjoint domains.

c) For cooperation between domain manager processes of two interacting domains the roles are assigned dynamically.

d) Two overlapping domains A and B have a domain intersection AB in which a common management task is to be performed. There are (at least) two approaches to manage the managed objects within AB:

    1. AB is itself regarded as a new domain. In this case, the cooperation between the domain manager processes that are responsible for the domain intersection is an intra-domain cooperation with dynamically assigned roles (cf. (a) above).

    2. Instead of regarding AB as a new domain, a new higher-level domain C is built that contains domain A as well as domain B (thus forming two nested domains, cf. (c) above). To synchronize management operations on the managed objects within AB, cooperation between the domain manager processes of domain C and those of A and B is necessary.

In such interactions, the domain manager processes of *C* always take the manager role, whereas those of *A* and *B* take the agent role. There is no direct cooperation between the domain manager processes of *A* and *B*.

e) Domain manager processes residing within nested domains interact using static role assignment where the domain manager processes of the containing domain take the manager role, and those of the contained domain take the agent role.

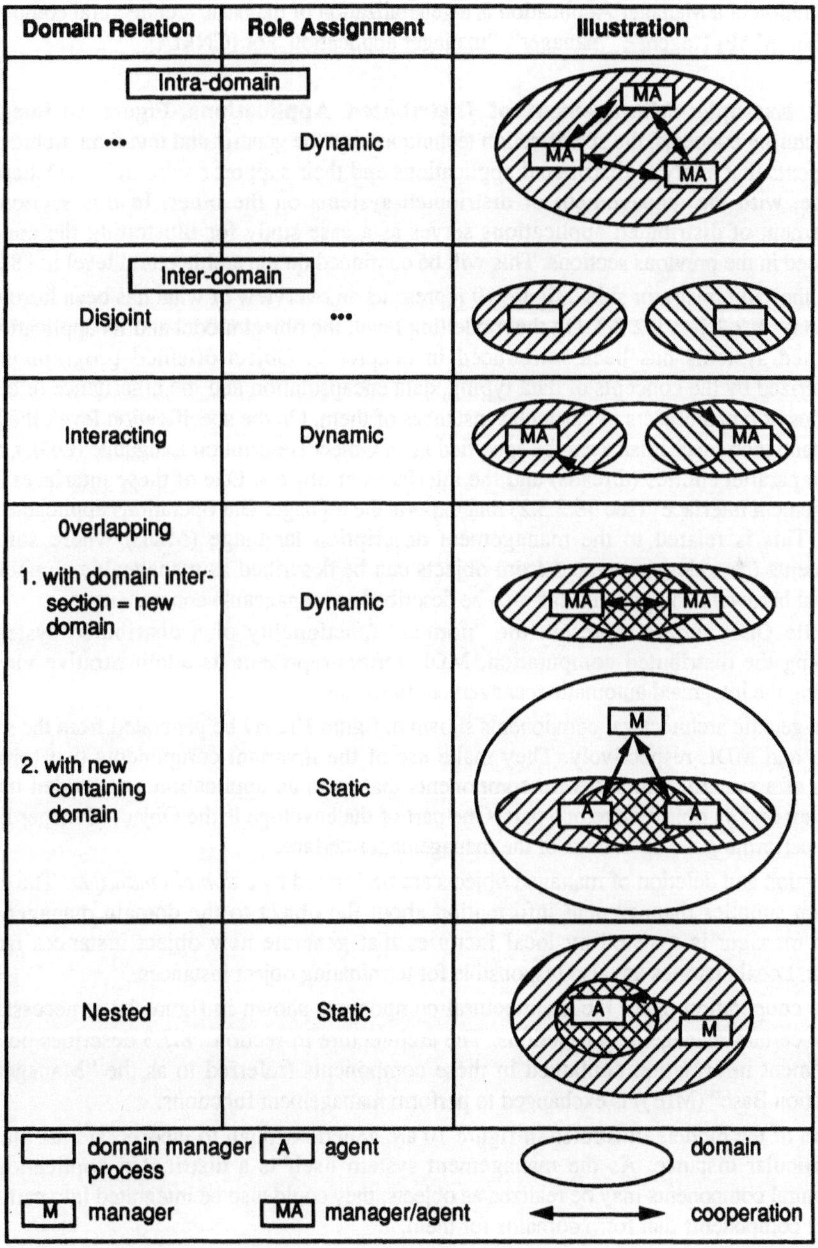

Domain Relation	Role Assignment	Illustration
Intra-domain		
•••	Dynamic	
Inter-domain		
Disjoint	•••	
Interacting	Dynamic	
Overlapping		
1. with domain intersection = new domain	Dynamic	
2. with new containing domain	Static	
Nested	Static	

☐ domain manager process	A agent	◯ domain
M manager	MA manager/agent	←→ cooperation

Fig. 9 - Manager/agent Role Assignment Dependent on Domain-Relationships

8.2.2.3.2. (Domain) Manager Applications. As outlined in the previous section, domain managers may not be able to carry out a particular (higher-level) common management policy without cooperating with domain managers of other domains. If the domain relationship between two domains M and A is of type "nested", and M contains A, then the domain manager of M is said to be a *"(Domain) Manager Application"* of the domain manager A. In this case, the domain manager processes of M always take the manager role when interacting with domain manager processes of A. Thus, the management policy is guided and synchronized by the (domain) manager application with the support of its agents in a containing domain. Note that this definition of a Manager Application is a generalization of the basic architectural components defined in [MAP] ("agent", "manager", "manager application") or [CNMA].

8.2.2.4. Example: Management of Distributed Applications. Figure 10 integrates the structuring concepts, the specification techniques and the generic and invariant architectural components that apply to distributed applications and their support environment (Deltase) on one side, with the management of distributed systems on the other. In this section, the management of distributed applications serves as a case study for illustrating the concepts introduced in the previous sections. This will be continued on the architectural level in §8.2.3.

On the "administrator side", figure 10 represents an overview of what has been introduced in sections 8.2.2.1 - 8.2.2.3. On the modelling level, the object model and its application for distributed systems has been introduced in chapter 7. Object-oriented programming is characterized by the concepts of data typing, data encapsulation and the inheritance of certain properties of typed objects to particular instances of them. On the specification level, the types and instantiation mechanisms can be specified in an Object Description Language (ODL), along with the parallel entities (threads) and the interfaces of objects. One of these interfaces is the "management interface" (see §8.2.3.2) that exports the management operations applicable to an object. This is related to the management description language (MDL) where software components (capsules) generated from objects can be described as manageable components (note that hardware components can also be described as manageable components).

While ODL scripts represent the "normal" functionality of a distributed system by describing the distributed computation, MDL scripts represent its administrative view by describing the integrated automatic management functions.

The generic architectural components shown in figure 10 may be generated from the scripts in ODL and MDL respectively. They make use of the invariant components that form the nucleus of a system. The software components that form an application are divided into an object part and an object envelope part. One part of the envelope is the Object Manager Entity (OME) performing the operations of the management interface.

Creation and deletion of managed objects are performed by a *domain manager*. The create operation supplies management information about the object to the domain manager. The domain manager is assisted by local factories that generate new object instances from a template. Local factories are also responsible for terminating object instances.

The cooperation of all the architectural components shown in figure 10 is necessary to perform certain management functions. The architecture in section 8.2.3 describes how the management information contained in these components (referred to as the "Management Information Base" (MIB)) is exchanged to perform management functions.

Both of the models illustrated in figure 10 are generic enough to incorporate the other as one particular instance. As the management system itself is a distributed application, its architectural components may be realized as objects; they could also be integrated into particular software components that form domains for them.

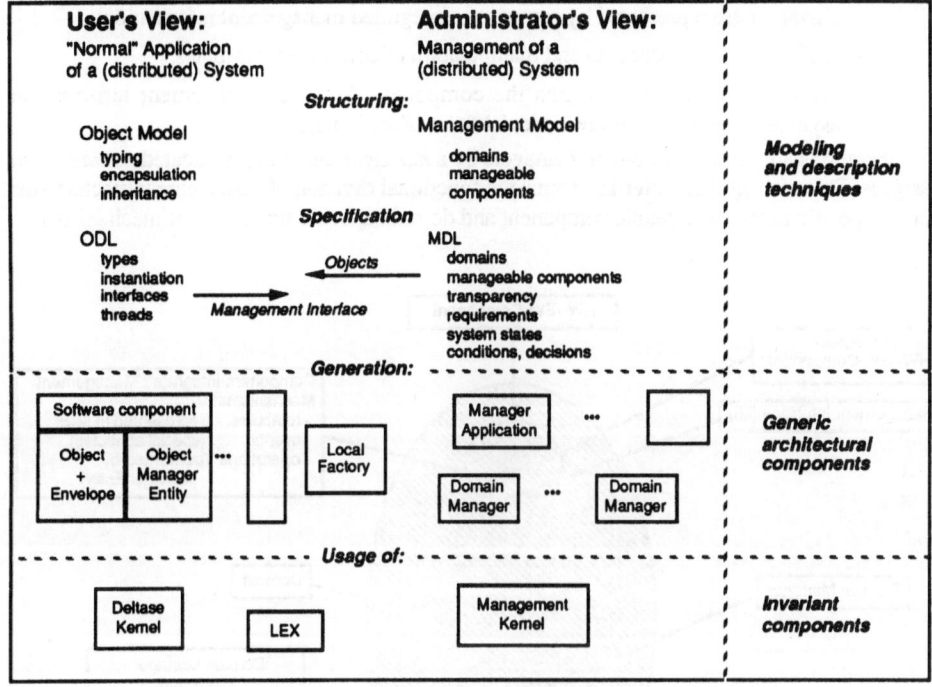

Fig. 10 - User's and Administrator's View of a Delta-4 System

8.2.3. Management System Architecture

The design of the Delta-4 OSA management system architecture in this section is described by:

- specifying the manageable components considered in Delta-4 OSA, including the management information, i.e., the representation of the attributes contained within the manageable components, and the management interface to the manageable components;

- specifying the domains considered in Delta-4 OSA, the respective domain managers, the management information contained in domain managers, the services and protocols for the exchange of management information, and the management services provided by domain managers;

- specifying the cooperation between manager applications, domain managers and manageable components that is necessary to perform management tasks.

8.2.3.1. Template Architecture. The generic architectural components of a management system as defined in section 8.2.2.3 must be correlated to get an architecture of the management system. These correlations are expressed in terms of service interfaces and cooperation types.

A manageable system consists of a set of *manageable components*. Manageable components are characterized by the fact that they possess integrated mechanisms for management purposes (figure 11). These must be specified and implemented in addition to the "normal" functionality of system components. These *integrated management mechanisms* comprise attributes, operations and events:

- *attributes* are represented by component integrated management information;
- *operations* allow access to and manipulation of component attributes;
- *events* are a means by which the component delivers management information asynchronously (events are a special form of attribute).

Operations are offered to domain managers by a *manager entity* at a dedicated *management interface*. This management interface forms the functional division of management mechanisms that are specific to the manageable component and domain-specific management mechanisms.

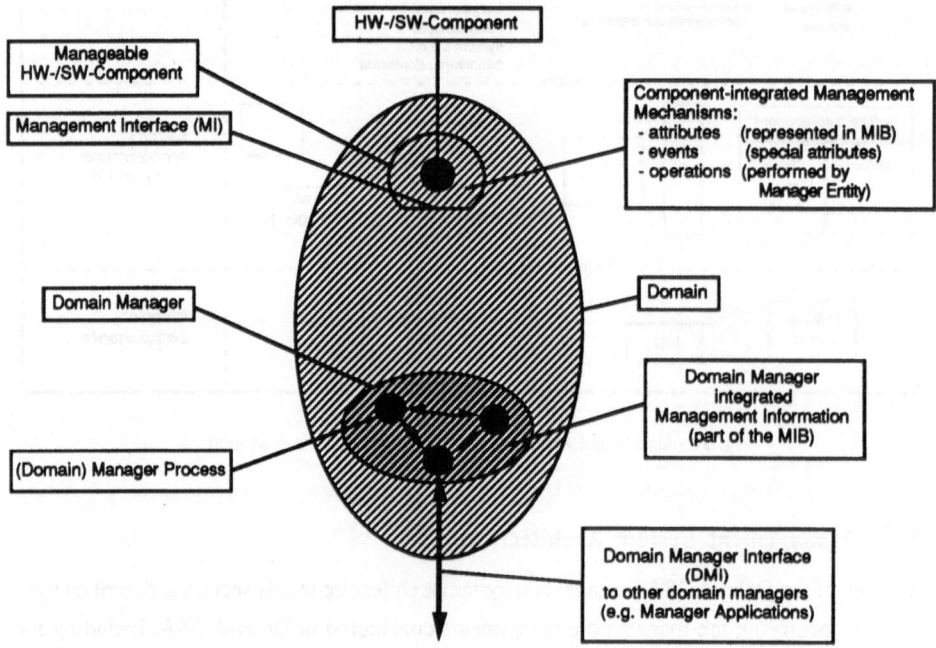

Fig. 11 - Manageable Components, Domain Managers and their Interfaces

The management tasks that are dedicated to a management domain are performed by a domain manager and may be categorized according to the five functional management areas (cf. §8.2.1.1). These tasks are offered to other domain managers at dedicated Domain Manager Interfaces (DMI) in form of management services:

- DMI_FM: DMI for fault management services;
- DMI_CM: DMI for configuration management services;
- DMI_PM: DMI for performance management services;
- DMI_AM: DMI for accounting management services;
- DMI_SM: DMI for security management services.

MCS supports these functional-area-specific management services in form of the Multipoint Common Management Information Service Element (M-CMISE). M-CMISE is part of the MCS application layer and offers basic multi-point services for the exchange of all types of management information between domain managers. The M-CMISE services enable the access to management information that is either encapsulated within the domain managers themselves or within the manageable components. In addition, they allow a domain manager to request other domain managers to perform a management operation upon manageable components and to send event notifications triggered by manageable components to other domain managers. As the M-CMISE services are mapped upon the MCS Session services, an M-CMISE service-user may also benefit from the multicast facilities of MCS. In ISO Systems Management, corresponding management services for the ISO/OSI bi-point environment are standardized in the application service element CMISE [ISO 9595].

Management information residing within the domain manager is information about the manageable component, e.g., the minimum, the current and the maximum number of replicas of the manageable component. On the other hand, management information integrated within the manageable component itself is information that is closely related to the "normal" functionality of the component, e.g., the context of a file in the case of a file server. This type of management information can be automatically generated from a dedicated MDL description (cf. §8.2.2.2) Both kinds of management information are conceptually summarised under the term *Management Information Base* (MIB) which is thus per definition distributed.

A *domain manager* may consist of a set of cooperating *domain manager processes*. Cooperation is needed for instance:

- to exchange management information between domain manager processes;
- to fulfil a common management task that was requested at a domain manager interface by a manager application; or
- to execute a distributed control or consensus algorithm.

8.2.3.2. Manageable Components

8.2.3.2.1. Objects. Objects that behave according to the strong object model, as described in chapter 7, encapsulate abstract data structures that can only be accessed and affected by operations via defined object interface(s).

This section discusses general aspects about managed objects as one type of manageable component. In subsequent sections these are applied to certain objects investigated in Delta-4.

a) Objects as Manageable Components

To be manageable, an object must have an *Object Management Interface* (OMI) at which management specific operations can be invoked. These allow access to *management information* that is encapsulated as object data within the object. The functional part within the object that performs the management operations is called an *Object Manager Entity* (OME). Thus the OMI and the OME, together with the object-integrated management information, comprise the additional part of an object that is necessary for the object to be manageable. Under the management view this kind of object is called a *managed object*.

b) Object-Integrated Management Information

Items of object-integrated management information are attributes that define static or dynamic properties of the object. The "object version identifier" is one example of a static object attribute. There are various types of dynamic attributes, e.g., counters, gauges, state information, logs, events, etc. Counters can be used for instance to indicate the number of invocations of "normal" object operations. Events can be seen as "special" attributes that are asynchronously generated by the OME if a defined object-internal situation occurs. They might be reported to the domain manager automatically or only following an explicit enquiry.

c) Interface to Domain Managers

The variety and complexity of management tasks require uniform basic architectural management mechanisms and a clear mapping of management functions onto the components of the management architecture. This is the major motivation for the identification of the *Object Management Interface* (OMI). It is situated between the *domain manager* and the (replicated) *managed object*. The OMI should provide a clear division between those management functions that are integrated within the managed object (and performed by the *object manager entity*), and those that are carried out by the domain manager. This functional division should be made both independently of the object type to be managed and of the domain manager type. As an object may reside within more than one domain, the OMI may be imported by several domain managers.

The OMI offers a variety of management operations to be performed on the object. On the one hand, their semantics should be generic enough to keep the number of management operations small, on the other hand they should cover all object types and their associated management information. Management operations are of two kinds: those which can be sent to an object to be applied to its attributes, and those which apply directly to an object itself.

Examples of the former are "get attribute" to read the current value of an attribute, or "set attribute" to modify the value of an attribute.

Examples of the latter are "create object" to direct the object manager entity to create an object, or "test object" to direct the OME to perform a specified test operation on the object.

8.2.3.2.2. Objects at the Application Level.
The architectural approach is applied to the following objects on the application level:

- *Processes:* Processes form the basis on which (existing) applications are built. In the present Delta-4 implementation, a Deltase capsule is represented by a UNIX process.
- *Files:* Global files may be useful in certain applications, so a server for managing replicated global files also been implemented.

For these components of different nature cloning techniques are investigated. Cloning of higher level objects, for instance file servers or databases, may be based on the cloning mechanisms of these components.

a) Application Objects as Manageable Components

For specifying the management of application objects the following approach is taken:

- the application objects are modelled as objects according to the strong object model;
- to describe these objects as managed objects, their integrated management information and the *object management interface* are specified.

The emphasis placed on the investigation of application objects as manageable components is to relinquish modifications in the *Local Executive Environment* (LEX). A precondition of this is that the object-integrated management information, which is obtained through services offered by the LEX, is sufficient to make the object manageable.

b) Application Object - Integrated Management Information

Management information related to application objects is represented in two ways:

- The *LEX* context: this is information that represents the state of the application object in a LEX-dependent way. This information may either be part of the application object (e.g., the loaded code of a process or the content of a file), or it may be established by the LEX when operations are applied to the application object (e.g., object descriptors). This information is not held within the application object, but is kept within the LEX, whence the term the "LEX context" of an application object.

- The *global* context: this is information that represents the state of the application object independent from the LEX. To generate this information, the LEX context must be transformed into an abstract notation. This is a precondition for domain managers to be able to interpret the context. If an application object is replicated, the global context is the same for all replicas, whereas the LEX context may have a different representation for different replicas. At cloning, the global context is transferred in voted packets over the network to the location of the new replica, where the LEX context is to be reconstructed from the global context.

c) Interface to Domain Managers

In Delta-4, domain managers are primarily investigated for the management of replication. A *replication domain manager* manages a set of objects; it determines the stations on which replicas of these objects should reside. To execute management tasks, it is assisted by *Object Manager Entities* (OMEs), which perform management operations on a particular object that is part of the domain. These operations are invoked by the domain manager. The assignment of management tasks to either the OME or the domain manager in Delta-4 is such that the OMEs:

- assist the domain manager in instantiation and termination of a replicated application object; and

- perform instantiation of new replicas (i.e., cloning) and termination of a replica on request from the domain manager.

Thus, the OMEs maintain both the LEX context and the global context of an application object; at cloning time, they transfer the global context to the location of the new replica.

8.2.3.2.3. Communication Object Management.
The management of communication systems and their communication objects is subject to standardization by ISO under the term OSI management. The concepts and the terminology that are elaborated in the OSI Management Framework [ISO 7498-4] and the Systems Management Overview [ISO 10040] fit naturally into the generic Delta-4 management model.

a) Communication Objects as Manageable Components

The management view of a communication system abstracts from the "normal" communication facilities and concentrates on those communication objects that need management support, and/or can supply management information. These communication objects are, for example, layer entities, protocol state machines, connections, service access points, or representations of physical devices.

To become manageable, communication objects are extended as described in §8.2.3.2.1 (a), i.e., management operations are invoked at a dedicated management interface to affect the communication object and its integrated management information.

Applying the OSI management terminology, a communication object that is regarded with such a management view is called a *managed object*. Instances of managed MCP objects that share the same management operations, attributes and notifications are said to be of the same *managed object class* [ISO 10165-1]. The set of managed object classes that have been defined for the MCP management is hierarchically organized. This hierarchy results in a so-called *registration tree* that is used for the identification of managed object classes. Figure 12 shows the current Delta-4 registration tree. One particular managed object class is identified by a path within this tree, e.g., the "MCP Session Endpoint" is reached by the path "Delta-4 1 5 2 1".

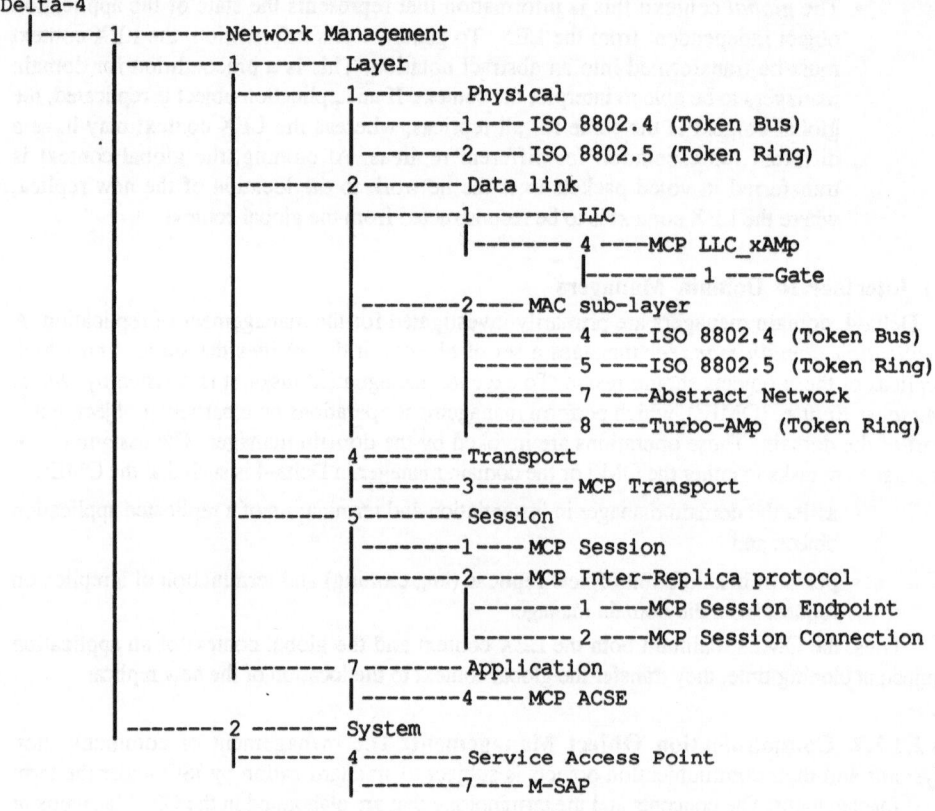

```
Delta-4
   |----- 1 --------Network Management
   |  --------1 ------- Layer
   |           --------- 1 -------- Physical
   |           |        --------1---- ISO 8802.4 (Token Bus)
   |           |        --------2---- ISO 8802.5 (Token Ring)
   |           --------- 2 -------- Data link
   |           |        --------1--------- LLC
   |           |        |--------- 4 ----MCP LLC_xAMp
   |           |        |--------- 1 ----Gate
   |           |        --------2---- MAC sub-layer
   |           |                --------- 4 ----ISO 8802.4 (Token Bus)
   |           |                --------- 5 ----ISO 8802.5 (Token Ring)
   |           |                --------- 7 ----Abstract Network
   |           |                --------- 8 ----Turbo-AMp (Token Ring)
   |           --------- 4 -------- Transport
   |           |        --------3-------- MCP Transport
   |           --------- 5 -------- Session
   |           |        --------1---- MCP Session
   |           |        --------2---- MCP Inter-Replica protocol
   |           |                --------- 1 ----MCP Session Endpoint
   |           |                --------- 2 ----MCP Session Connection
   |           --------- 7 -------- Application
   |           |        --------4---- MCP ACSE
   |  --------- 2 ------- System
   |           --------- 4 -------- Service Access Point
   |           |        --------7---- M-SAP
```

Fig. 12 - Delta-4 Managed Object Class Registration Tree

The current Delta-4 MCP managed object definitions, together with their attributes, events and actions, are specified within the MCP protocol specifications. Delta-4 *communication object management* has to manage MCP-specific communication objects to support fault-tolerant applications. Examples are:

- Delta-4 Multicast Service Access Points (M-SAPs), and
- Delta-4 Multipoint Associations.

These types of MCP communication objects are represented by two different managed objects. The first managed object represents the local-to-station management view of this MCP communication object. This managed object is locally kept on the station where the communication object exists. It contains, for example, management information that represents the local-to-station usage of the communication object. The second managed object represents pre-defined management information that is of global significance. The access to this managed object is essential for the instantiation of new instances of the communication object. It is therefore stored within the (replicated) "Global-MIB" that is itself a manageable component (cf. §8.2.3.2.4).

b) Communication Object Integrated Management Information

Management information that is integrated within managed communication objects is closely related to the communication activities of the communication objects. It consists of those types that are identified within §8.2.3.2.1 (b) and is specified in an abstract notation within a *managed object definition*. ISO is proposing such a notation in [ISO 10165-4].

Examples of Delta-4-specific management information within MCP are the attributes "number of session Protocol Data Units (PDUs) sent/received" of the managed object "MCP Session Entity", and the event "newStation" that is automatically generated by the SMAP (see below) when a new station has been attached to the network.

c) Interface to Domain Managers

All management operations upon layer-specific managed objects and their associated management information are performed by a *Layer Management Entity* (LME) that exists within each MCP layer. The interface type to this LME is the same for all layers and corresponds to a subset of the standardized object management interface as described in the management architecture. The domain manager process that accesses the communication object integrated management information through this interface is called the Systems Management Application Process (SMAP). The SMAP is described in section 8.2.4.3.1.

8.2.3.2.4. Global-MIB Management. The template architecture in figure 11 distinguishes two types of management information:

     a) component integrated management information;

     b) domain manager integrated management information.

There are several reasons why parts of the management information of type a) in real implementations are not encapsulated by the component; the LEX context of application objects is an example. In some cases, a practical solution can be to keep such management information within the domain manager.

As further explained in the implementation section (§8.2.4), there are also reasons for which parts of the management information of type b) are not encapsulated by the domain manager.

Management information that (in the implementation) is neither encapsulated by a component nor by a domain manager is assumed to be stored in an object called the *Global-MIB*. Conceptually, it is still seen as information of type a) or b) respectively.

a) The Global-MIB as a Manageable Component

As other objects, the Global-MIB is a type of manageable component and offers an *Object Management Interface* (OMI) to domain managers (see (c) below).

b) Global-MIB-Integrated Management Information

The statements of §8.2.3.2.1 (b) apply unchanged. As the Global-MIB needs management support (e.g., enhanced dependability), it has integrated management information. Two special situations may occur in practice:

    1) The Global-MIB's integrated management information (i.e., information of type a)) is stored in the Global-MIB itself. It can thus be accessed both via OMI and via normal object interfaces.

    2) The Global-MIB is managed by a domain manager that also has its integrated management information (i.e., information of type b)) stored in the Global-MIB.

c) Interface to Domain Managers

The statements of §8.2.3.2.1 (c) apply unchanged for the *Global-MIB* object. The functionality of the standardized OMI is sufficient to handle the two situations described above concerning the storage location of management information as well. However, there are further requirements for the design of domain managers managing the Global-MIB over and above those for domain managers concerned with ordinary objects. The domain primarily investigated is a replication domain for the Global-MIB (cf. §8.2.3.3).

8.2.3.3. Domains

8.2.3.3.1. Replication Domains.

A *replication domain* of an instance of a software component is the set of stations on which replicas of that instance may reside. A common management policy is applied to a set of software components that have the same replication domain. The architectural component that executes this policy is called the *replication domain manager*. It reacts on changes of the *station* or *node group* (cf. §6.8) that constitutes the replication domain, for instance, due to station failure or shut-down. It reconfigures *replica groups*, for instance, by cloning software components that had a replica on a failed station to another station offering spare redundancy. The reconfiguration strategy is not only dependent on this particular software component, but also on the *group of software components* that execute in the replication domain.

a) Domain Manager - Integrated Management Information

Domain specific management information is mainly the reconfiguration strategy and information about the objects and the stations of the domain that is necessary to execute the strategy, e.g.:

- the minimum, the current and the maximum number of replicas of an object;
- the desired and current location of replicas;
- the assigned communication objects, etc.;
- the maximum and current load of stations;
- the current state of stations (up or down, attached to the network), etc.

The update of this management information is in the responsibility of the manager of the communication domain (see next section); as the communication domain manager keeps this information in the Global-MIB, it can be retrieved from there. Significant changes of management information (e.g., station failure) are reported to the domain manager by spontaneous events.

b) Interface to Manager Applications

The replication domain manager offers a set of operations to its manager applications. Two different kinds of management operations can be distinguished:

- generic domain operations, independent of the domain type, e.g.:
  - *create/delete managed objects:* these services create/delete entries for objects in the domain managers information base;
  - *instantiate/terminate managed objects:* for instance, process instantiation/termination;
  - *list managed objects:* this service allows information about managed objects of the domain to be retrieved;
- operations that are specific to a replication domain, e.g.:

- *reconfiguration strategy:* this service supplies the domain manager with the reconfiguration strategy it should adopt for a subset of objects, which is to be applied if a certain situation should occur (e.g., station failure, station shutdown, etc.).

8.2.3.3.2. Interaction Manageable Component — Domain Manager. To perform its services, the domain manager makes use of the operations offered at the OMI of each application object.

At instantiation/termination of a process, the domain manager may be supported by the OME, which for instance performs some initialization and notifies the domain manager that all replicas have been instantiated/terminated successfully.

During cloning, the domain manager and the OME interact in the following way:

- Having decided that an object is to be cloned, the domain manager first instantiates a template of the object (and an OME for it) at the new location.

- When the new OME indicates that the (template) instantiation is complete, the domain manager instructs the OME of the already existing replicas to build and transfer the global context to the new OME.

- The new OME substitutes the current local context of the object with the one resulting from the received global context. It synchronizes the input and output message streams of the new replica with those of the existing replicas, while ensuring replica consistency. Having performed this sub-operation, the instantiated template has become a new replica, and the OMEs are now also replicas of each other.

- The OME responds to the domain manager to indicate that the context transfer is complete.

8.2.4. Implementation

A management system whose design is described in §8.2.3 is currently under implementation. This section gives a description of the architectural components and the managed objects chosen for that implementation.

8.2.4.1. A Replication Domain for UNIX System V Processes. The implementation comprises a prototype for the management of UNIX System V processes. Figure 13 illustrates the architectural components involved and how they interact.

Two replicated processes A and B are shown, which interact via an MCS (multi- or bi-point) association (other interaction, e.g., local inter-process communication, is disallowed to ensure replica determinism). These processes may include the Deltase run-time support system (Deltase/XEQ) which makes use of the MCS services offered on the host.

To make processes manageable, they include an *Object Manager Entity* (OME) as a library. The invocations of the MCS services are passed through the OME, which is thus able to control the complete input and output message stream of the process. Domain management is implemented by a *Replication Domain Manager* (RDM), which consists of one replicated domain manager process. Cooperation between RDM and OME is via a specific association, the RDM association.

To perform the instantiation/termination service, the RDM invokes via the RDM association the services offered by the *factories*. A *factory* is a manager of a domain nested within the replication domain; it is implemented as a non-replicated entity present on each host on which a process of the replication domain may have a replica; it uses the local UNIX fork/exec directives to instantiate a process.

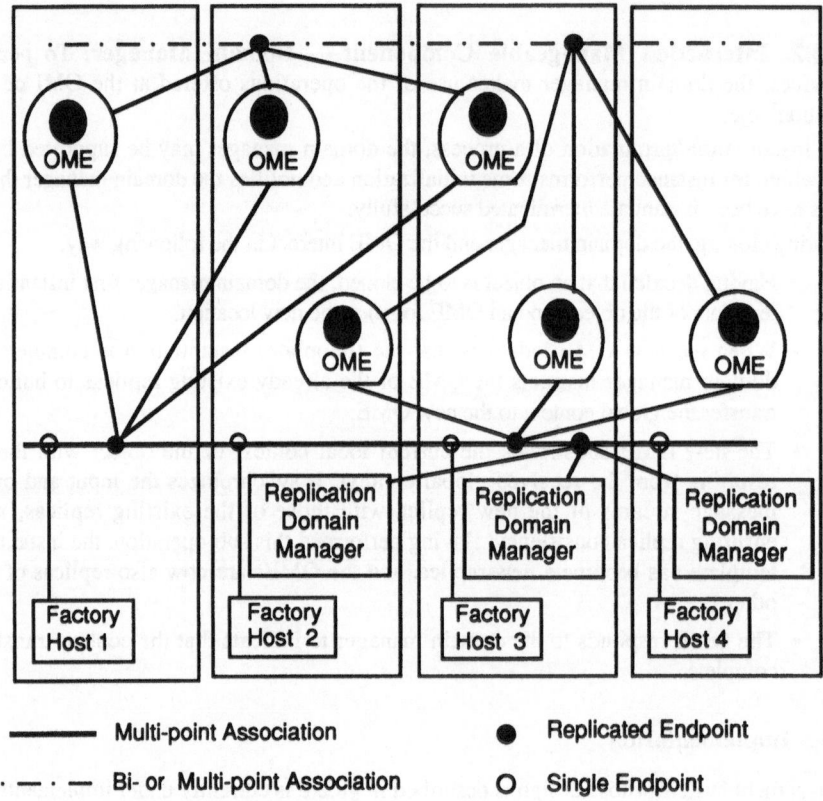

— Multi-point Association   ● Replicated Endpoint

·—·— Bi- or Multi-point Association   ○ Single Endpoint

Fig. 13 - Architectural Components for Process Management

8.2.4.1.1. The Replication Domain Manager (RDM). The Replication Domain Manager offers the generic services as described in section 8.2.3.3.1:

- Create Managed Objects;
- Instantiate/Terminate Managed Objects; and
- List Managed Objects.

There are additional services that are specific to a replication domain, e.g., the reconfiguration strategy service mentioned in section 8.2.3.3.1. A precondition for such a service is the definition of a Management Description Language (MDL, cf. §8.2.2.2), which enables the description of reconfiguration strategies. Certain strategies are directly implemented within the RDM.

An example of such a strategy is the RDM's reaction to station failure and station reentrance. Recognition of station failure/reentrance is in the scope of the *Communication Domain Manager* (CDM, cf. §8.2.3.3.2). If such an event occurs, the replication domain manager is notified. At station failure, it searches in its information base for processes that had a replica on the failed station, and updates the value of the attribute that determines the actual replication degree of this process. If the desired replication degree of a process (an attribute set by the Create Managed Objects service) is higher than the actual replication degree, the RDM searches for a "free" station, i.e., a station that does not hold a replica of this process. It instructs the local factory on that station to instantiate a template of the process, including an OME, and starts the cloning service provided by the OME. At station reentrance, the RDM again searches for processes that have fewer replicas than desired and instantiates replicas on the reentered station. In addition, all replicas that had earlier been cloned to another station, are moved back to the reentered station.

As the RDM itself must also be dependable, it is implemented as a replicated object. When a station fails, not only the application objects' degrees of replication are to be reestablished, but also that of the RDM. Thus, the RDM has a built-in OME that it instructs to clone itself.

8.2.4.1.2. The Object Manager Entity (OME).

The *Object Manager Entity* (OME) offers services at the *Object Manager Interface* (OMI) to the *Replication Domain Manager* (RDM):

- it assists the RDM in instantiation and termination of a replicated process;
- it performs instantiation of new replicas (i.e., cloning) and termination of a replica on request from the RDM.

The OME library not only supports cloning of actively replicated processes. Cloning is also useful for semi-active and passive replicas, for instance when a station is reinserted a new leader or primary may be required. However, the description here concentrates on cloning of active replicas.

As illustrated in figure 13, in a replicated process, each replica contains an OME library. The process and its OME use one replicated M-SAP. The process establishes endpoints on that M-SAP in order to communicate with other processes; in addition, the OME establishes an endpoint to the RDM association.

The OME acts as a replicated instance whose Finite State Machine (FSM) is always in the same state in each replica. Incoming events are generated:

a) when the process reaches an MCS communication statement (send/receive a message, establish/release an endpoint); and

b) when the OME receives a message, which is only possible when the process has reached a receive statement.

To preserve replica consistency, all incoming events are ordered identically at all replicas of the OME.

Outgoing events of the FSM depend on the current state. For instance, if the OME is executing a cloning protocol with several steps of context transfer, the process may reach a communication statement between two transfer steps, which may lead to a phase in which the process must be blocked.

The phases of the cloning protocol described below start from the assumption that the DM has created a template of the process with an OME library at the new location.

The new process's OME establishes the same M-SAP on the new station and receives all messages on the DM Association that the existing process/OME receives, but the new OME's FSM does not generate the same outgoing events. The RDM invokes the Instantiate Replica service of the OME.

Phase 1: *Reestablish Communication Context*

The context of all endpoints the process has currently established is transferred to the new location. The new OME re-connects to all associations on which the process has endpoints.

Finally, the input message streams of the existing replicas and the new replica are synchronized. Should the process try to establish/release an endpoint during further phases of the protocol, it is blocked.

Phase 2: *Take a Snapshot of the Computational Context:*
The LEX (UNIX) context of a process comprises:

- the data segment (initialized, uninitialized and dynamically allocated data);
- the stack segment, including the UNIX environment;
- the process registers.

The UNIX context is obtained by normal system calls. Modifications of the operating system kernel and libraries were not required.

Unlike checkpointing with passive replicas under a fail-silent regime (cf. section 6.6), cloning with active replicas requires that the snapshot be taken at the same execution checkpoint by each active replica. Furthermore, since the active replication technique is applied in a fail-uncontrolled environment, the context transfer messages must be able to be voted. This means that the context data in each active replica must be made bitwise identical; this is carried out in phase 3.

Phase 3: *Generate the Global Context:*
The system libraries (UNIX-, MCS-, Deltase- and OME libraries) may use local services that generate values that may be different between replicas. For example, local file descriptors, mailbox identifiers, context identifiers significant on the NAC, etc., can be found in the local context that are different for each replica. The OME builds the global context for which all local descriptors are equal in each replica, which is a precondition for starting phase 4.

Currently, application programmers must not use local resources for inter-process communication (signals, semaphores, pipes, shared memory, mailboxes). Only global files may be used (cf. §8.2.4.2). A reduction of these restrictions is planned such that local files may be used by the application.

Phase 4: *Transfer the Global Context:*
The global context is transferred in voted packets. During the transfer steps, the currently active replicas are allowed to continue execution; the new replica then avoids sending messages that have already been sent in the context transfer phase.

Phase 5: *Continuation at the Execution Checkpoint:*
Having received all context packets, the computational context of the new replica is substituted by the transferred context. The local resources needed by the communication and management entities (see Phase 3) are then established. The process now continues at the execution point at which the snapshot was taken. All replicas now synchronize their output message stream, i.e., the new replica is built and the RDM is informed.

8.2.4.2. A Replication Domain for UNIX System V Files. The architectural design outlined in section 8.2.3 has also been validated by a prototype for the management of replicated UNIX V Files. As in the case of the prototype for UNIX System V processes, the implementation architecture does not imply modifications to the UNIX kernel or libraries; the local UNIX file servers are used unchanged.

8.2.4.2.1. Management Services. As opposed to processes, files do not execute any computation and do not interact with each other. Operations for local file management (open/close, read/write, etc.) have defined semantics and are performed by a local file server.

To enable replication of files, the following support is provided:

- *Global file server:* this offers a subset of the UNIX System V file management operations. Applications using the global file server send requests for file operations as messages; if a requesting application is replicated, then its request-messages are voted.

- *Domain management:* The following generic domain management operations are provided:

  - *Create/delete managed object:* This delivers management attributes (e.g., degree of replication) to the RDM.

  - *List managed objects:* Instantiation/termination services are not provided; it is assumed that local copies of a file are present at each station as soon as the *create* service is invoked. The reconfiguration strategy for station failure/reentrance described in the previous section is also applied to files.

- *Object management:* Instantiation/termination services for new replicas of a file are provided.

8.2.4.2.2. Implementation. The functionality described above is implemented by a software component residing on each station of the replication domain. This software component is termed the *File Manager* (FM). The file manager offers service access by way of its *FM-Association* and comprises several sub-components, which are described below.

FM - Domain Manager (FM-DM)
This offers the domain management services listed above. It consists of one RDM Process with as many replicas as the replication domain has stations. A manager application of the FM-DM executes a reconfiguration strategy, which attempts to restore the replication degree of files in case of station failure and reestablishes the initial configuration when a station is reinserted. This manager application together with the FM-DM forms a *replication domain manager* (RDM).

FM - Application Entity (FM-AE)
The FM-AE offers the UNIX System V compatible remote access to replicated files. AEs are automatically instantiated by the replication domain manager when a new file is added to the domain (Create Service). This involves the generation of a file specific M-SAP and an endpoint on the FM-Association with the same number of replicas as the file.

FM - Object Manager Entities (FM-OME)
In a similar fashion to the AEs, OMEs are automatically instantiated by the DM when a new file is added to the domain (Create Service).

The object manager interface (OMI) between File-Manager-DM and the File-Manager-OMEs is an internal interface of the FM.

The file-cloning protocol starts with the generation of an OME and an AE (by the Domain Manager) and the establishment of the file's M-SAP and endpoint to the FM-Association on the new location. The new OME and AE from now on receive all messages on the FM-Association that the existing AE and OME receive. The OME then establishes the global context from the local context (as described in section   8.2.4.1).

The local context of a file comprises:

- the content of the file; and

- the file context maintained by the LEX, e.g., the file descriptors.

The global context is then transferred in voted packets. During the transfer steps, the services of the global file service continue to be provided. Having received all context packets, the content of the file is substituted and the file descriptors are reestablished at the new location (by local UNIX System V open calls). Finally, the output message stream of all file replicas is synchronized.

8.2.4.3. Management of the Multicast Communication Protocol Stack

8.2.4.3.1. The Systems Management Application Process (SMAP). The *Communication Domain Manager* (CDM) is the domain manager for the *Multicast Communication Protocol* (MCP) stack. The CDM is implemented as a set of cooperating domain manager processes, called *Systems Management Application Processes* (SMAP). As the set of MCP communication objects and their associated management information is not the same on each Delta-4 station, SMAPs are individual, non-replicated processes, and exist one per station. For dependability reasons, a SMAP mainly resides on the fail-silent *Network Attachment Controller* (NAC). The current structure and interfaces of a Delta-4 SMAP are shown in figure 14 and are further explained in the following sub-sections.

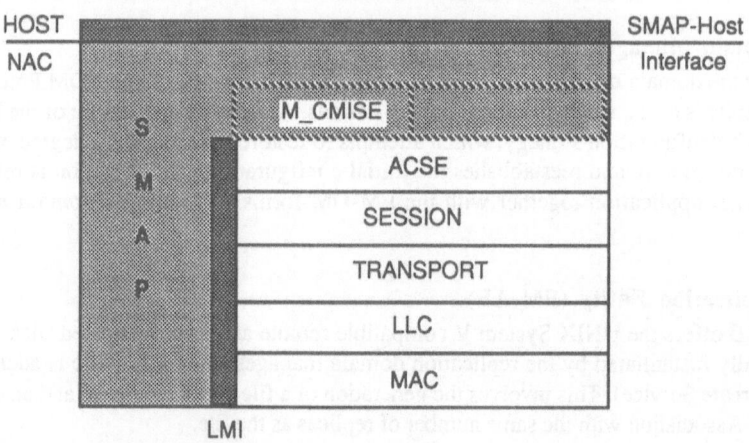

Fig. 14 - Structure and Interfaces of a SMAP

8.2.4.3.2. The SMAP-Host Interface. The SMAP-host interface allows management processes on the host to access the services of the SMAP. Whereas most of the SMAP software resides on the NAC, the implementation of the SMAP-host Interface also has some software components on the host. These are:

- two UNIX processes that handle the data transfer between host processes and the SMAP on the NAC (one for each direction);

- a set of C-language procedures that can be linked to any application process and allow the SMAP-host interface to appear as ordinary procedure calls to the application.

8.2.4.3.3. The Layer Management Interface. The *Layer Management Interface* (LMI) is the boundary between the SMAP and the MCP layers. The services that are offered at this interface are used by the SMAP to access the MCP managed objects. It currently offers the following services:

- LM_GET_VALUE to read management information;
- LM_SET_VALUE to write management information;
- LM_ACTION to require operations to be performed upon MCP managed objects;
- LM_EVENT_NOTIFY to report events spontaneously generated by MCP managed objects.

On the one hand, the LMI services are used by the SMAP to retrieve management information from the MCP layers. On the other hand, the MCP needs management support for "normal" MCP communication. This support is also given by the SMAP at the LMI. The following example illustrates the interactions between the SMAP and the MCP layers when an association endpoint is to be established:

When an application process issues an A_ASSOCIATE service request to the MCP ACSE sub-layer, the MCP ACSE generates an event at its local SMAP. This SMAP then checks if the association concerned is "open" ("open" means that all communication resources in layers 1-5 that enable the endpoint establishment on that association are allocated). If it is not, the concerned SMAP sends a request to the Global-MIB (cf. §8.2.4.4) to obtain the necessary parameters for the particular association (such as the services available on this association, the association context name or its amount of credit). Then the SMAP issues an action invocation to the LME of the MCP ACSE sub-layer to instruct it to continue with the A_ASSOCIATE service. The event and the action that follows it are transparent to the application (it is meaningless to perform the action alone, only the correct sequence leads to the desired result; generally, the correct sequence of such invocations is guaranteed by the SMAP).

8.2.4.3.4. Communication between SMAPs. SMAPs on different stations must communicate to perform their management task within the MCP communication domain. For this purpose they use the *Multipoint Common Management Information Service Element* (M-CMISE): this is an application service element within the application layer of MCP. M-CMISE is an extension of the ISO Common Management Information Service Element [ISO 9595] and provides generic multipoint services for the remote access to managed objects of all kinds.

As the SMAPs need dependable communication, M-CMISE is also implemented on the fail-silent NACs. It is mapped onto the MCP session services, and offers the MCP ACSE services or association control as pass-through services at the M-CMISE interface. Thus, management communication may also exploit the MCP multipoint communication facilities for its own purposes.

M-CMISE is used by the SMAP in the following way. The SMAPs use a single multipoint association that connects all SMAPs in the system. Each SMAP creates an endpoint on a dedicated multipoint association during its startup phase and keeps that endpoint throughout its lifetime. SMAP to SMAP communication then only requires a single M-CMISE service call in which the receiving SMAP is identified by its endpoint (or M-SAP address), thus saving the overhead of association endpoint establishment and release. Furthermore, by changing a parameter in the M-CMISE service call, a multicast or broadcast of a message from a SMAP to several or all SMAPs in the Delta-4 system is possible (which would be difficult using point-to-point connections).

8.2.4.3.5. Communication with the Global-MIB. Apart from the communication channel between the SMAPs, each SMAP communicates with the replicated Global-MIB. Communication between SMAPs and the Global-MIB is carried out by way of a dedicated multipoint association (the "Global-MIB-association") whose members are the Global-MIB and all Global-MIB users. The "Global-MIB-Manager" handles this communication and performs the required Global-MIB access protocol. See also §8.2.4.4.2 for a description of the Global-MIB contents and operations.

8.2.4.3.6. The SMAP during System Startup. The SMAPs play a special role during the startup of the communication software on a Delta-4 station. After the communication system and SMAP software has been loaded on the NAC and the NAC's processor has been started, the SMAP waits for a startup command from one of the SMAP-Host Interface processes on the host.

It then issues initialization action invocations to the communication layers, inserts the station into the network and allocates its own communication resources (and on "Global-MIB-stations" also allocates the communication resources of the Global-MIB).

Two different kinds of station startup are distinguished:

- startup during the initial startup of the Delta-4 system;
- startup of a station or restart of a repaired station when the rest of the Delta-4 system is already operational.

In the first case, the Global-MIB is started on a set of initial "Global-MIB-stations" during the startup phase. In the second case, the newly started station is always treated as a non-"Global-MIB-station" (however, the Global-MIB may be cloned to the new station after that station has become operational).

The Global-MIB is only accessible after completion of the startup phase of a station. Certain configuration parameters of a Delta-4 system must be known by the SMAP during this phase. They are generally specified at system generation time.

8.2.4.3.7. SMAP Support for Bi-Point Connections. Apart from Delta-4 multicast communication, the Delta-4 communication system also offers a standard ISO bi-point communication (with the extension of possibly replicated endpoints). Bi-point connections are mapped onto the Delta-4 communication resources of the lower layers (Transport to Physical) in the Session layer. This requires special support by the SMAP during connection establishment:

- Provide a unique global name for the communication resources "Transport Connection" and "Gate" per each bi-point connection. This is done by a request to a naming authority (a "global name server") which is integrated into the Global-MIB (cf. §8.2.4.4) and which manages a set of logical names.
- Ensure the allocation of the required communication resources (Transport connection, gate) on all requester and responder stations through a protocol between the involved SMAPs.
- Coordinate the replies from the responder stations (concerning the communication resource allocation) and instruct the local session entity to proceed with the bi-point connection establishment protocol, for which it needs the newly allocated communication resources (or to stop the protocol).

8.2.4.4. The Global-MIB. Management information that is stored in the Global-MIB is critical to the system's integrity and dependability; thus the Global-MIB must have a considerable degree of dependability. The implementation of the Global-MIB as a managed object comprises:

- an *Object Manager Entity* (OME) to manage information in the Global-MIB which is relevant to its replication and cloning; and

- a *replication domain manager* offering a service for the on-line cloning of new replicas of the Global-MIB.

The design of replication domain managers concerned with the Global-MIB has additional requirements compared with replication domain managers concerned with ordinary objects. In the current implementation, a dedicated domain for the replication of the Global-MIB is assumed.

8.2.4.4.1. The Global-MIB-Manager. The architecture for object management was outlined in §8.2.3.2.1. The *Object Manager Entity* (OME) of the Global-MIB is integrated into the *Global-MIB* object; the Global-MIB and OME software component is referred to as the "*Global-MIB-Manager*" (MIB-M). Similarly to the FM (cf. §8.2.4.2), the Global-MIB-Manager has an "application interface" and a "manager application interface". The Global-MIB Manager is implemented as a replicated manager process. Likewise, the replication domain manager for the Global-MIB is implemented as a set of replicated processes (the degree of replication and the location of these processes is the same as for the Global-MIB Manager).

8.2.4.4.2. Management Information Stored in the Global-MIB. This section gives an overview of the management information stored in the Global-MIB and how it is structured in the present implementation. Currently the Global-MIB contains "domain manager-integrated management information" for the CDM (cf. §8.2.4.3) and the FM (cf. §8.2.4.2).

1) CDM-integrated Management information
The "domain manager integrated management information" that is encapsulated by the CDM is described in §8.2.4.3. This information is local to a Delta-4 station and is accessed using the techniques described in section 8.2.4.3.4.

However, the Delta-4 fault-tolerance mechanisms require that parts of this management information be non-local to a Delta-4 station, i.e., the information that represents the attributes of the Delta-4-specific communication objects "multipoint association" and (replicated) "M-SAP". This information is held in the Global-MIB. The alternative having each SMAP on each station store its own copy of this information has been discarded because of:

- *Lack of memory space on the NAC:* The SMAPs reside on the NAC and would have to store the information in the NAC memory. The amount of information is not known beforehand but may be large. The Global-MIB is a replicated object on the host and can store its information on disc.

- *Consistency problems:* The SMAPs are "individuals" and do not act as replicas. It would have been necessary to implement a consensus protocol between the SMAPs without being able to use the Delta-4 Multicast Communication System facilities to their full extent. Using the replicated Global-MIB, which uses the multicast property and order guarantee of replicated Delta-4 endpoints, consistency among the Global-MIB replicas is preserved automatically.

2) Domain Manager information not stored in the Domain Managers

A domain manager may decide to use the services of the Global-MIB to store (part of) its global information. As an example the File Manager for replicated UNIX System V files stores information about file replicas in the Global-MIB. Manager applications of the FM make use of this information.

The information in the Global-MIB is organized as an abstract data type. The applicable operations are invoked at the application interface and are controlled by the MIB-M. The data is structured according to an identified set of manageable components, which are in the domains of the CDM and the FM, their attributes and the operations that can be performed on the attributes.

Currently, the Global-MIB defines 10 subtypes:

- delta_4_system;
- lan_segment;
- station;
- NAC;
- host;
- Multicast Service Access Point (M-SAP);
- multipoint association;
- replication_unit;
- file;
- replication_domain.

The first five data structures contain information about the configuration of the Delta-4 system and its major components, stations and LAN-segments, together with the primary communication objects that are to be managed: the multipoint associations and Multicast Service Access Points (M-SAPs).

The subtypes "replication_unit", "file" and "replication_domain" represent File Manager management information.

In general, attributes of any subtype may be:

- structural, reflecting the configuration of a Delta-4 System; or
- descriptive, holding information about manageable components or domains.

As an example, certain structural attributes show the decomposition of a Delta-4 system (figure 15).

A "delta_4_system" is composed of sets of "lan_segments", "stations", "replication_units", "M-SAPs" and "multipoint associations". A "lan_segment" is connected to a set of "stations". A "station" is composed of a "NAC" and a "host" and connected to a "lan_segment", etc.

An example of a descriptive attribute is the "lan_type" of a "lan_segment" (Token Ring, Token Bus, ...).

Note: Replicated M-SAPs and Multipoint Associations, as seen from Delta-4 management, incorporate two categories of management information: information pertaining to the managed object as a whole (e.g., its logical address), which is dealt with here, and information local to a Delta-4 station (e.g., the number of messages sent from a given Delta-4 station onto a multipoint association). This latter information corresponds to the ISO/OSI management information model, it is accessed using standard techniques: information exchange through M-CMISE, a service that uses an ISO/OSI addressing schema (object class registration tree and object instance containment tree [ISO 10165-1]). Figure 12 shows the object class registration tree used in Delta-4 for addressing local management information (which is specific for a given

Delta-4 station). Figure 15, on the other hand, illustrates attributes of the non-local management information that reflect structural relations between the managed objects. However, this is not used for addressing the information in the Global-MIB (an ISO-like addressing schema would be much too inefficient for the performance of the Global-MIB). As M-SAPs and Multipoint Associations have both categories of management information, local and non-local, they appear in both figures 12 and 15.

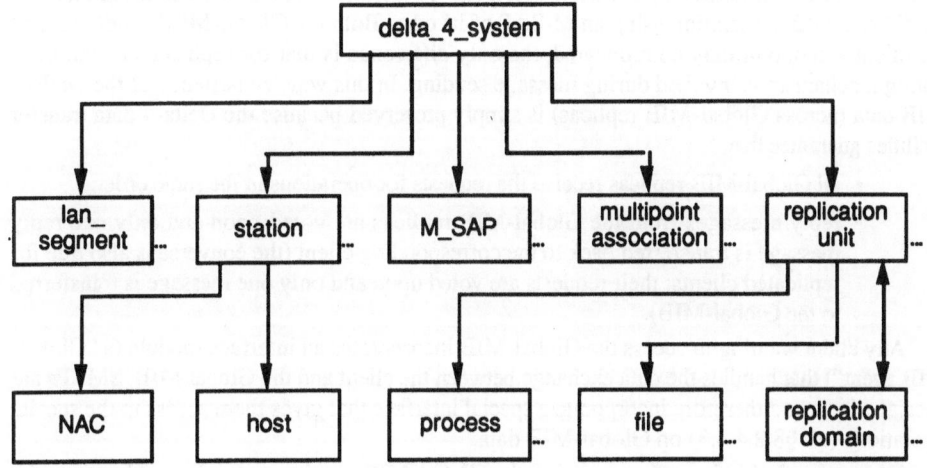

Fig. 15 - Structural Information about Manageable Components and Domains in the Global-MIB

8.2.4.4.3. Service Interface of the Global-MIB-Manager. As stated in §8.2.4.4.1, the *Global-MIB-Manager* offers services at an application interface and at a manager application interface. The application interface offers the normal object operations of the *Global-MIB* object. General operations offered at the application interface are:

- insertion or removal of Global-MIB entries, i.e., instances of data types representing manageable components;
- getting or setting values of single attributes in an instance of a manageable component;
- several operations for handling list type attributes;
- searching for a MIB entry of a given kind with a given name.

Special operations that may be used only by a *Systems Management Application Process* (SMAP) are:

- retrieval of all parameters needed for the opening of a given multipoint (or bi-point) association;
- retrieval of all parameters needed for the establishment of a given M-SAP;
- update of all concerned information in the case where a given multipoint (or bi-point) association has been opened or closed on a Delta-4 station;
- update of all concerned information in the case where a given M-SAP has been established or de-established on a Delta-4 station;
- preserve the Global-MIB database consistency and change NAC and host states in the case of a station failure or insertion of a new station into the system.

The manager application interface offers the services used by the Global-MIB domain manger for the on-line cloning of the *Global-MIB* object.

8.2.4.4.4. Service Protocol. Requests for operations on the Global-MIB and the results of such requests are transferred as messages via the Delta-4 multicast communication system. For this purpose, there is a dedicated multipoint association on which the Global-MIB owns an endpoint (characterized by the logical names of the association and the Global-MIB's M-SAP, which are both configuration parameters of a Delta-4 system). A Global-MIB client connects to the Global-MIB association using an M-SAP of its own. Both the Global-MIB's endpoint and the client's endpoint may be replicated, the only difference is that for replicated endpoints a voting mechanism is invoked during message sending. In this way, consistency of the Global-MIB data (across Global-MIB replicas) is simply preserved because the Delta-4 data transfer facilities guarantee that:

- all Global-MIB replicas receive the requests for operations in the same order;
- reply messages from the Global-MIB replicas are voted upon and only one reply message is transferred back to the corresponding client (the converse is also true for replicated clients: their requests are voted upon and only one message is transferred to the Global-MIB).

Any client wanting to access the Global-MIB incorporates an interface module (a "Global-MIB agent") that handles the data exchange between the client and the Global-MIB. SMAPs are special clients and therefore incorporate a special interface that gives them access to the special operations (cf. §8.2.4.4.3) on Global-MIB data.

The retrieval of information stored in the Global-MIB requires special considerations for system bootstrapping. No information stored in the Global-MIB is obtainable before the network is operational and the connections between clients and Global-MIB are established. Thus all information needed before that stage is configuration information of a Delta-4 network that must be fixed before network installation. Once the Delta-4 system is operational, its initial configuration may be changed dynamically.

Chapter 9

Extra Performance Architecture (XPA)

The OSA architecture offers a wide range of solutions for distributed fault tolerance. However, the conjunction of the attributes offered, especially the complexity inherent in its generality and openness, precludes its successful use in certain niche applications.

Indeed, for a number of critical applications, where high-performance, real-time and fault-tolerance are simultaneous goals, those major attributes of OSA become a constraint. The difficulty arises in the time domain, i.e., on the ability of offering assurances of timeliness, and on the user-perceived responsiveness and throughput of the system.

The Delta-4 *Extra Performance Architecture* (XPA) introduces mechanisms that support explicitly the requirements of real-time systems, with the introduction of priorities and deadlines, and the requirements of high-performance, with respect to both throughput and response. However, XPA will inherit as much of the OSA architecture as is possible within the constraints imposed by these requirements, and XPA and OSA will retain compatible interfaces as far as possible, see annexe H.

To meet performance and timeliness targets, XPA will inevitably lose in openness and generality. The functionality of the communication system is limited to that required to support XPA applications running under XPA-Deltase. Standards that cannot address real-time needs will not be used. There will be layer and service elimination in the communication protocols, and fail-silent hardware will be used to avoid the need for validation of computation results. Only homogeneous hosts will be accommodated, to avoid the performance overhead of conversion of data representation.

The XPA workpackage began much more recently than other work in Delta-4. This chapter therefore contains a synthesis of concepts and ideas that are less mature than those presented elsewhere. Since the present project is nearly completed at the time of writing; it is likely that a complete implementation of the concepts described will require further work.

9.1. Objectives and Definitions

The strategic objective of XPA is to give a Delta-4 solution to specialized application scenarios, where the OSA architecture proves to be inadequate. In Delta-4, these scenarios are found in the fields of real-time and high-performance systems. Since Delta-4 is aimed at providing distributed fault tolerance, technically, the XPA architecture has the difficult goal of combining:

- distribution;
- real-time;
- dependability;
- high performance.

Thus, XPA is required to provide timeliness assurances, by performing actions within bounded and known time intervals. The start and end of such an interval are called *liveline* and *deadline* in this chapter and chapter 5. All system support mechanisms, including communications, must have controlled latency and synchronism.

XPA is decoupled as far as possible from any particular approach to the issue of real-time scheduling mechanisms. The architecture is therefore open to whatever scheduling strategy a particular application requires, and can make use of whatever scheduling primitives are offered by the underlying LEX.

Each service of an XPA system is therefore assigned a *precedence* that provides a generic measure of both its importance and urgency. The precedence is an abstract data type that is interpreted by a system-dependent "Precedence Manager" into a set of scheduling parameters suitable for presentation to the underlying LEX scheduling mechanism.

The same precedence manager is used to interpret the precedence throughout a single XPA system, and schedulers of different resources, such as processor time and network bandwidth, use compatible strategies, in the following sense:

- XPA permits deterministic preemption of computation at predefined points a bounded execution distance apart (see section 9.4.3.2). Similarly, message transmission can be preempted in bounded time between individual frames. Whereas the LEX machinery for deciding between the instantaneous precedences of two computations might be complex and vary with time (e.g., [Jensen et al. 1985]), in the communication system precedence comparison can be simple, since the time granularity of communication is normally much shorter than that of computation. Indeed, a message carrying a typical RPC is normally contained in a single communication frame.

Thus the precedence manager exports an operation to tell an XPA resource allocator which of two precedences should be preferred. Precedences are inherited by all components on which a distributed computation depends, and the "inherit" operation is also exported by the precedence manager, since it is closely associated with the scheduling strategy.

In the XPA prototype, the LEX is the Real-Time UNIX developed in phase 1 of the Delta-4 project [Bond 1987, SVC200], and the precedence manager interprets the precedence as a "priority", or measure of importance, and a "targetline", or measure of urgency. In the context of this LEX, a service A is said to have higher precedence than a service B, if:

- either A has higher priority than B;
- or A has the same priority as B, and A has an earlier targetline.

In another implementation, the XPA precedence may be interpreted in a radically different way. For example, section 9.6.2.3.1 describes how a fixed cyclic schedule calculated at design-time can be encoded into an XPA precedence and reproduced at run-time by a suitable precedence manager and LEX.

The differing importance and urgency of services are recognised by assigning precedences to actions, and allowing high precedence actions to preempt low precedence ones. In a distributed system, the precedence concept should of course be propagated throughout the information path. In consequence, high-precedence messages in the communication system should also be allowed to overtake low-precedence ones.

Distributed fault tolerance in Delta-4 relies on message passing. The order and agreement attributes needed to preserve consistency of replicas should be guaranteed by the distribution support (group management and communications).

A distributed real-time system requires the maintenance of a distributed time service, which provides an abstraction of global time. This is required to ensure *participant and replica agreement on time*, for example: to trigger actions at pre-determined instants; to recognise when events occurred; or to provide replicas of a component with a common knowledge about time, a general condition for replica group determinism in real-time. Both an internal and an external time reference may need to be provided. The internal time is provided through a global clock approximation, maintained by a distributed time server. The external time consists of one of the standards of time, whichever is necessary for the particular use of a given XPA system.

A Delta-4 XPA sub-system is able to coexist and inter-work with Delta-4 OSA sub-systems. Application objects are portable between XPA and OSA systems; in both cases the application objects are supported in Deltase envelopes or *capsules*. XPA Deltase supplies the same services as OSA Deltase, with some variations to support real-time and high-performance applications (see section §9.4).

Disclaimers: XPA is aimed at providing support for application in which the cost of failure — including the failure to meet real-time constraints — is measured in financial terms rather than in terms of potential loss of human lives, i.e., XPA is *not* aimed at safety-critical applications although, like OSA, it may be usefully applied to safety-related applications (see annexe A).

Fault-tolerant sensing and actuating are not directly addressed within the present phase of the Delta-4 project. Input/output implications and requirements are however discussed in a Delta-4 context in chapter 12.

9.2. Overview

This section surveys the issues treated in the present chapter.

9.2.1. Real-Time and Performance

One major objective of XPA is to support real-time applications, which are defined as applications that are able to offer an assurance concerning the timeliness of service provision.

XPA in fact aims to support both hard real-time applications, which are deemed to have failed if they miss their deadlines, and soft real-time applications, which may miss their deadlines in some circumstances. An XPA system may also support non-real-time applications, which have no timing constraints.

XPA is also addressing applications where high performance is required, not only locally to hosts, but in the distributed environment. Special attention is thus given to the responsiveness of the system to distributed actions, optimization of the distributed user-to-user data paths, and development of efficient group communication protocols.

9.2.2. Dependability and Computational Models

The fact that distributed fault-tolerance is associated with real-time, also has implications in the dependability techniques to be used in XPA.

In XPA, fail-silent hosts provide the basis for error detection and replication provides the basis for error recovery.

Consequently, backup replicas can either be active or passive. The possible models differ in the role and degree of activity they assign to each replica. The requirements of XPA were in the origin of a new dependability model in Delta-4: the *leader-follower*, or *semi-active replication* model. The rationale and advantages of this model for real-time fault tolerance are discussed in chapter 6.

Computation in XPA relies on the XPA version of the Delta-4 applications support environment, XPA Deltase, which differs from OSA Deltase in two main ways:

- Performance enhancements: XPA Deltase and the communication manager are combined in one library, which copies data directly between the application and the LAN message buffers, bypassing the LEX.

- Support for real-time: XPA Deltase includes a real-time local execution environment, and interprets the precedence of each computation in terms of suitable LEX scheduling parameters.

9.2.3. Support for Distribution

The original design of XPA envisaged the simplification of the interface between Deltase and the communication support, for performance. Deltase was collapsed onto the group communication layer, implying the introduction of new communication services that led to the extended AMp (xAMp), and of an intermediate harmonising layer — the *group management layer (GM)*. The architecture is shown in figure 1.

The *Group Manager* (GM) is a distributed object, represented locally on every node of an XPA system. In fact, the group management entities are pseudo-objects, since they present an object interface to the Deltase world, but interact with the group communication layer as another layer above. The group manager is concerned with the management of groups of objects, and with the support of the distribution of such groups. The group manager incorporates knowledge of the different modes of replication (active, passive, semi-active, etc.), and based on that view, it is able to provide, transparently to the objects themselves, the appropriate level of support, using the xAMp.

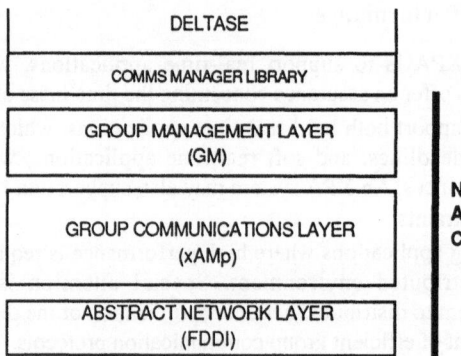

Fig. 1 - XPA System Architecture

The communication sub-system comprises the *abstract network layer*[1], and the *group communication layer*. As detailed in chapter 10, the implementation of the abstract network over different LANs allows the communication sub-system to be LAN-independent. The local area network used in the current XPA prototype is an ISO 8802/5 token-ring, featuring 4 Mbit/s. Taking into account the needs of XPA, both from the reliability and performance viewpoints, FDDI — a 100 Mbit/s high-speed, high-throughput fibre-optic LAN — is an envisaged alternative, whose advantages are discussed later in this chapter.

The group communication services of XPA are based on an extended version of AMp, xAMp (see chapter 10), which provides a set of primitives, offering extended group

---

[1]   We recall that "network" is not taken in the sense of layer 3 of OSI; it refers generically to the several communication infrastructures for xAMp.

management facilities and different tradeoffs between quality of service and efficiency. The services range from an *atomic* service to simpler and more efficient alternatives, down to multicast datagrams. The communication protocols will generally be executed on a special-purpose *Network Attachment Controller* (NAC) board. However, given that both host and NAC are fail-silent in XPA, it is also possible that the NAC be combined with the host itself.

The time service is provided at the xAMP layer, in order to improve precision.

9.2.4. XPA as an Integrated Machine

The previous sections have described the options available to the system designer in XPA. This section shows how the bricks just described interact, in one of the possible combinations to fulfil the real-time, dependability and performance objectives stated in the beginning, in the XPA prototype built. '

9.2.5. System Administration

System Administration (SA) is concerned with monitoring and managing changes in the operational status and degree of replication of objects, and changes in the system configuration.

This does not have to be implemented by a central "system administrator", or distinct *system administration subsystem*: system administration functions can sometimes be delegated to ordinary application objects and the support environments of the objects to be managed.

In Delta-4 XPA, because of its emphasis on high performance, there is an overall bias in favour of the local management of objects, and on delegated rather than centralized system administration. Objects are encouraged wherever possible to perform administrative operations themselves, when invoked by the appropriate manager, exploiting their self-knowledge to optimize performance.

9.3. Real-time and Performance

9.3.1. Introduction

One of the major objectives of Delta-4 is to support real-time applications. The techniques used to support real-time were developed first for XPA and in some cases depend on its restrictions, such as the assumption that hosts are fail-silent and have real-time local schedulers.

Real-time systems require predictably bounded performance. This is distinct from the other main objective of XPA, which is to support the highest achievable performance for all aspects of distributed computation.

Chapter 5 defines many of the concepts used in this chapter and explains the Delta-4 approach to real-time. In this chapter, section §9.3.3 discusses the requirements and computational philosophy of XPA real-time. The main techniques used to support real-time and high performance are described in sections §9.4, §9.5 and §9.6.

9.3.2. Markets Addressed

The market areas addressed by XPA were described in chapter 1, but a brief reminder at this point will assist understanding for readers who are more familiar with other real-time or high-performance market areas.

XPA is designed to address large dynamic systems in the process control and command and control areas. These systems are notable for the large, complex sets of assumptions that have to be made before it is possible to predict their behaviour during the design process.

For example, state-of-the-art process control systems typically contain hundreds of analogue loop controllers, each with analogue input sampling, control calculation and analogue outputs, and possibly thousands of digital inputs and outputs. Scan and response times may be of the order of milliseconds, and plant integrity may depend critically on reliable and regular execution of the scanning and control logic.

Such systems must support concurrent interaction with many operators. The operator interface has its own response time requirements in various conditions of use, and its quality can exert disproportionate influence on the commercial acceptability of an architecture.

Most process control events are periodic, but some aperiodic, and some "sporadic", the term used in [Burns 1990a] for events that are aperiodic, but have a minimum specifiable duration between their arrival times. We reserve the term "aperiodic" for cases where no such minimum duration can be specified.

Command and control systems contain large numbers of real-time events, periodic, sporadic and aperiodic. There are typically several modes of operation and several different priority levels. However generously the system is sized, peak activity levels depend on an uncontrollable environment, and may still cause a system overload.

9.3.3. XPA Real-Time

Chapter 5 contains a generic discussion of real-time concepts and requirements and defines many of the terms used in this chapter. This section is concerned with the more specific real-time requirements of the XPA prototype, nevertheless, for the sake of self-containment there may be some repetition of the material in chapter 5.

The *priority* of an activity is a measure of its criticality, i.e., the cost of not meeting its timeliness constraints. The highest priority is given to hard real-time components.

The *targetline* is the time at which the system designer aims to deliver the service; it lies between the liveline and deadline, which define a window within which service must be delivered. See chapter 5 for a generic discussion of these terms.

9.3.3.1. Hard Real-Time. A *hard real-time deadline* is one that must be met in order to avoid a costly timing failure. Such deadlines, the operational envelope in which they must be met, and any defensive behaviour must be captured in an adequate requirements specification, which must be unambiguous. This involves:

- Determining that there is sufficient system power. Worst-case component execution requirements and other resource usage must be determined. In particular, the system must have sufficient power, or "execution resource", and a means of allocating this power must be provided which has properties useable during the design process.

- Knowledge of execution times. Whatever events can occur, their worst-case timeliness consequences must be accounted for during system design. It follows that primitives that do not themselves give assurances of time bounds are unuseable as a basis from which hard real-time components can be built.

- Control over system component location. For example, [Burns 1990a] argues that migration of hard-real-time components cannot be permitted.

- Avoiding all deadlocks and bounding delays due to temporary resource conflicts (see section   9.3.3.4).

For example, the consequences of internal hardware faults must be bounded in time. Component failures will occur — they cannot be ignored in the design process. The consequence must be accounted for in specification, as a (specified worst-case) rate of occurrence, and perhaps in design, through the use of fault tolerance or recovery mechanisms.

If the time overhead for some such mechanism is not compatible with some computation's deadline requirements, then that computation cannot use the mechanism.

As another example, human operators are sometimes placed "in the loop" as a final resort to cover a possible emergency. Whereas it is obviously not possible to specify bounds to operator reaction times in such an open-ended situation, the information displayed and any commands issued must be processed by the system within specified time bounds so that it is not open to indictment as the cause of late response to the emergency.

9.3.3.2. Soft Real-Time. A soft real-time service has an identified timeliness requirement but no high cost is attached to a failure to meet it. An example is the updating of a display monitor that is intended to show the progress or present state of some production process. The majority of such information does not come into the category examined in the previous section, in that an operator will only be inconvenienced by an occasional delayed update. The quality of the interface is then measured in terms of its normal rather than its worst-case behaviour.

This is a very different world to that of hard real-time. Here, deadlines are regarded as desirable targets that it is not essential to meet on all occasions. The requirements specification must capture such possibilities. A consequence is that the system power can be less than implied by worst-case analysis. A system whose peak loads are many times normal loads, or only statistically predictable, might therefore be rendered economically viable.

Although this is distinct from hard real-time, it must be stressed that it is also quite distinct from the principles of "fair time sharing" employed elsewhere in the industry — notably in traditional mainframes and in communication systems, or the FIFO ordering used in batch control systems. In a "fair" system, when a new requirement arises, the execution resource is redivided; the new requirement gains a "fair" allocation of resources and every other requirement loses to a small extent. The effect is that if requirements arise arbitrarily, no assertions can be made about timeliness for any of them.

It is therefore necessary to be unfair, to give some requirements more importance than others. There may exist a spectrum of urgency to the different activities that arise. If not all outstanding activity can be fulfilled, that which is least important is postponed, but timeliness assertions can still be made about what is most important. This is one basis for controlled, or "graceful", degradation of behaviour.

9.3.3.3. Best-Effort. For soft real-time components such as the operator screen server, there are costs attached to each timing failure. The objective of a best-effort system is to minimize the probable cost function by achieving the lowest probability for the most costly failures.

The means of calculating the probable cost function for each proposed strategy must exist. This involves, as before:

- Determining that there is sufficient system power, this time as a factor directly determining the probability of meeting timeliness criteria.
- Knowledge of execution times, this time in terms of probability distributions rather than worst-cases. As discussed in chapter 5, we are usually obliged to build systems from microprocessors for which execution times are probabilistic due to internal pipelining and cacheing and the use of contention busses. Even were this not so, the complexity of operating systems, language compilers and realistic applications effectively prevents the construction of any rigorous timeliness proof, much less one short enough to be subjected to a meaningful peer review. Banning the use of unbounded constructs ([Puscher and Koza 1989] gives the example of unbounded loops) does not overcome these obstacles.

- Control over component location. If the pattern of external demands becomes such that certain nodes perform more than their share of high-urgency computation, some of this might not be completed whilst less urgent computation is completed elsewhere. This situation, if it persists, represents an unbalanced system. Components may be migrated to balance the urgency of processing achieved across nodes, providing care is taken to avoid migration-"thrashing".

- Avoiding all deadlocks (e.g., by always acquiring resources in the same order), and determining both the probability and the duration of any temporary resource conflicts (see section   9.3.3.4).

The method of devising schedules is important:

- Off-line schedulers (see chapter 5) do not provide direct support for the idea of preferring one computation to another.

- For on-line schedulers, there is a shift in emphasis. Some methods, like earliest deadline scheduling, can find a feasible schedule, in which all deadlines are met, if such a schedule exists. In a best-effort system, there may be no feasible schedule: such methods are then generally badly behaved.

- When not all deadlines can be met, the method must explicitly take account of which activities are preferred. For this, scheduling computations according to their priority is appropriate. It might be that under different modes of system operation, different priorities apply.

Note here the essential difference between use of priority and use of deadline. For feasible schedules, highest priority scheduling is not as well behaved as earliest deadline scheduling, which on one processor will find a feasible schedule if one exists. On the other hand, for unfeasible schedules, deadline scheduling is not as well-behaved as priority scheduling, which establishes a design preference over what is allowed to execute when not everything can be executed.

It follows that a best-effort system should rely primarily on highest priority scheduling, so that only lowest priority deadlines will be missed in a marginal overload. Only components of the same priority should be scheduled in the order of their deadlines or targetlines. For most priorities this will normally find a feasible schedule, but in a marginal overload some components of lowest priority start missing their deadlines. Since they are of lowest priority, the cost of missing their deadlines is minimal, and may be less than the cost of the extra system size needed to avoid such overloads.

9.3.3.4. Resource Conflicts. Resource conflicts do not cause unpredictable delays in simple static schedules, since the times when resources are required are known during design; the points of reservation and release are embedded in the static schedule.

However, in complex dynamic systems, where resources other than those for execution are being contended for and mutual exclusion must be practiced, there is no known generic mechanism that permits absolute (as opposed to probabilistic) assurances of timeliness to be made during the design process. In commercial systems, partial solutions have made use of application-specific knowledge, as discussed below.

If a highest precedence computation requires a resource, the current holder of the resource should start running with maximum precedence. The maximum delay is then the dedicated execution time before the holder releases the resource. This is a generic mechanism in that it is applicable to all types of resource. However, it works well only for the highest precedence computation. Furthermore, it does not resolve deadlocks; these must be avoided by application-level methods such as:

(a) declaring at the start of each transaction which resources may be acquired during the transaction [Eich 1988, Habermann 1969];

(b) restricting the order in which resources may be acquired [Kopetz and Schwabl 1989].

An alternative policy is to use a transactional model of computation and abort the holder of a resource required by a higher-precedence computation in much the same way as is proposed for deadlock recovery in [Moss 1981], through a sequence of messages from resource claimers to resource holders. The operation exported by the precedence manager to compare precedences is used to compare the precedence of a remote resource claimer with the local resource holder. This enables a shorter bound to be set for the highest precedence computation but leads to expensive re-execution of the aborted transaction; the timing properties are difficult to analyse, even probabilistically.

However, this mechanism will resolve cycles of deadlock, provided that precedence is always unique. To ensure this, one of the precedences must always be preferred, in a deterministic manner across replicas. For example, when importance and urgency are identical, precedence order may be resolved deterministically using unique computation identifiers.

The timing properties of these strategies are only calculable if applications hold resources for bounded periods, and only acceptable if these are short. This seems to be general; there is a similar requirement in static scheduling schemes[2], where components hold resources during pre-arranged parts of a static scheduling cycle (see section   5.2.4).

9.3.4. Performance

Homogeneity plays an important role in XPA performance: the communication system requires no presentation layer, and the use of nodes with similar power simplifies the synchronization of replicas. From an engineering viewpoint, the existence of only one hardware configuration optimizes performance enhancements.

The XPA communication sub-system is LAN-independent. LAN choice may then be conditioned, when possible, by performance improvement. Taking into account the needs of XPA, both from the reliability and performance viewpoints, FDDI — a 100 Mbit/s high-speed, high-throughput fibre-optic LAN — is an interesting alternative, in comparison with slower LANs, such as 4 or 16 Mbit/s token-rings or 5 or 10 Mbit/s token-busses.

The group communication service of XPA, xAMp, provides a set of primitives from an *atomic* service down to multicast datagrams. With this range of successively less costly services, the user incurs the minimum cost needed to provide a given functionality.

The communication protocols will generally be executed on a special-purpose *Network Attachment Controller* (NAC) board. However, given that both host and NAC are fail-silent in XPA, it is also possible that the NAC be combined with the host itself, yielding a simpler hardware configuration where data is moved around faster.

In fact, minimizing user-to-user transfer latencies also implies efficiency of the local executive and support services like buffer management. Mapping of user buffers into (and from) communication space (without copy) and scatter-gather DMA are desirable features.

The *Remote Procedure Call* (RPC) is one of several support mechanisms for distributed computation. This paradigm is one we wish to encourage designers to use. It underpins the

---

2   A resource required by more than one component can only be held by each component for part of the static scheduling cycle. There is a tradeoff between the time for which resources need to be held and the need for such cycles to be short enough to meet the latency bounds of response to rare but important events. Unless such events have the overhead of several adequate processing "slots" reserved (but normally unused) in every cycle, such response must be achieved through installing a new schedule at a predefined point in the current cycle. At this point the current cycle must not hold resources required by the new cycle.

support environment, and its performance in support of replicated interactions continues to be the subject of intensive optimization effort within the project. The end point of this effort is not yet in sight[3].

9.3.5. Timeliness

Timeliness, i.e., the correctness of the system in the time domain, is the distinctive issue in a real-time system. XPA is concerned with assuring the timeliness properties of the support environment, by bounding all system latencies and queueing times. To construct a system with timeliness attributes, other factors are also essential: system sizing, knowledge of component execution times, and control of component location (see section 9.4).

However, as explained in chapter 5, timeliness can only be guaranteed where circumstances and events fall within a predefined "operational envelope". If they ever fall outside that envelope, timing constraints may sometimes be violated, but the designer still has a responsibility to detect such violations and limit their consequences. A time service is thus mandatory. The timing requirements of XPA are discussed in section 9.5.

In XPA, a loss of timeliness is normally detected as a missed targetline, so it is the consequences of not meeting a targetline that need management. If the targetline is set before the deadline, then action taken at the targetline may yet avert a missed deadline. If the targetline is set equal to the deadline, then a missed targetline is a timing failure, but there may still be ways of minimising the resultant cost.

There is considerable variation in the types of response that it is possible to design into different computations:

- For some computations, a missed targetline may have no consequence; execution simply continues after the targetline has passed and the service is provided when possible. Even here, the missed targetline should be reported to a system administration module, which might log the event or increment a count of missed targetlines that could be used for assessing system performance or to support a load-balancing algorithm.

- For other computations, some local action by the component concerned might be appropriate. This might take the form of a change in the nature of the computation performed, or a change in the circumstances of execution:

- In some cases, such as that of figure 1 in chapter 5, the benefit/delivery-time graph shows that it is less costly never to deliver the service than to deliver it after the deadline. In this case the component should abandon its computation, but report the exceptional condition to its invoking client (which may be able to fulfil the required service in some other way).

- If it is a periodic process, it might cause the start of its next period to be omitted or delayed. This, whilst harmless in terms of its effect on other components competing for the same resources, must be assured not to lead to other sorts of instability, for instance if the component is in the control loop of an external process. Any replica of the component must act in a consistent way, so inter-replica negotiation may be needed.

- It might limit its own functionality in some application-specific way, such as rejecting rather than deferring outstanding requests. The global ability to tolerate

---

3 On the prototype, limitations are imposed by certain generations of VLSI medium access control chips, which appear to have been optimized for throughput rather than (as RPC demands) delay. Thus, figures of 3-4 ms for each network access have been measured on a particular Token Ring network interface chip. Such figures impose artificial limits that may not apply to the next generation chip or another medium.

such cavalier behaviour must exist. For example, a command and control system may reject older input data, but process newer input data that gives the recent position of an enemy aircraft.

- It might reduce the precision of some computation; indeed, it is quite reasonable to construct a certain class of components to refine a calculation until the targetline elapses (interpreted as "the deadline is about to elapse"), and then to deliver immediately the current "best estimate". An example is a chess-playing program. Such components exhibit hard behaviour.

There is such wide variation in possible responses that it is inappropriate to confine applications to a set of standard responses. When required, therefore, the support environment can inform the component of an elapsed targetline and allow it to interpret this as suits the application.

A system should normally be sized so that all hard real-time components can have worst-case execution time, and run right up to their targetlines, on every execution. So if a hard real-time component happens to terminate *before* its targetline, it may suspend until the targetline before starting its next task, without jeopardising the timeliness of its next task. By so doing, it releases resources to lower precedence components whose timeliness may be at risk in the current operating mode.

9.4. Dependability and Computational Models

9.4.1. Homogeneity

Chapter 7 discusses the use of heterogeneous hosts, languages and implementation conventions on Delta-4 OSA. The restriction to homogeneous hosts on XPA has the following advantages, from the computational and dependability viewpoints:

- *Performance:* the communication system requires no presentation layer, and performance enhancements need be developed for only one hardware configuration.

- *Synchronization:* it is important to bound the desynchronization between replicas, e.g., to maintain the accuracy of replica time (see section 9.6.6). This is easier if all nodes have the same power (see section 9.6.5.5).

- *Fail-silence:* XPA also requires fail-silent hosts, and it is easier to develop an acceptable degree of fail-silence for only one hardware configuration.

9.4.2. Replication Requirements

XPA requires a model of replication that supports real-time requirements such as bounded preemption and synchronization of replicas, and also the high-performance objectives of XPA, without compromising replica group determinism.

Several different models of replication are supported on Delta-4 (see chapter 6). They have different properties, and are not equally suitable for XPA. The differences are summarised below.

9.4.2.1. Active Replication.
Active replication is only used with deterministic software components, e.g., those structured according to the state machine model [Schneider 1990], so that the replicas pass through identical states without any external support apart from identically ordered inputs.

Active replication carries the minimum overhead in terms of inter-replica messages, but an active replica's input messages need to use the *atomic* service of xAMp, which is not as fast as its *reliable* service. Computation can proceed at the speed of the fastest replica, but if the fastest replica fails, there is a delay depending on the desynchronization of the replicas, (see section §9.6.5.5).

Active replication does not at present allow high precedence input messages to preempt low precedence messages or low precedence computation within bounded times. Because of these unbounded delays, it is unsuitable for supporting real-time computations of more than one precedence level. However, the *slotted* service of the extended AMp (xAMp, see chapter 10) could support negotiating active replicas, so that bounded preemption could then be achieved, if not as efficiently as with semi-active replicas.

9.4.2.2. Semi-Active Replication.
Semi-active replication supports non-deterministic software components, or components whose environments make opportunistic decisions about the preemption of messages or processes, in order to achieve bounded latencies for high precedence computation.

Messages to semi-active replicas can use the faster *reliable* service rather than the *atomic* service of xAMp (see chapter 10). The inter-replica messages cost less than those of passive replication, and can also be used to detect replica desynchronization and leader failure.

Semi-active replication can support preemption with bounded latencies. This is because a maximum precedence input message will raise the scheduling precedence of the destination leader to the maximum as soon as it enters the leader's environment. The leader must then run until the next preemption point, where the maximum precedence message preempts. The preemption latency is therefore bounded by the maximum execution time between consecutive preemption points, unless the leader fails.

If the leader fails, there is a longer but still boundable latency before a follower takes over. This depends on two other latencies:

- A follower can detect leader failure in a bounded time (see §9.6.5.6).

- The desynchronization between leader and follower can be bounded (see §9.6.5.5).

Because of this ability to bound latencies, semi-active replication can support hard real-time computation, as well as soft real-time and non-real-time components.

9.4.2.3. Passive Replication.
Passive replication, like semi-active replication, can support non-deterministic capsules, and opportunistic mechanisms such as re-ordering of input messages or preemption of processing. Messages to passive replicas are also able to use the faster *reliable* xAMp protocol.

Because only one of the replicas is running, passive replication may require less processing power than active or semi-active replication. However, this has to be set against the extra processing power used in the generation and storage of checkpoints, which also consume extra network bandwidth. The generation of checkpoints also tends to reduce the response time of the primary replica.

When failure of the primary is detected, a backup has to repeat processing from the last checkpoint. The delays associated with primary failure, namely replica takeover, are therefore larger than with the other models. However, they can still be bounded, provided the frequency of "I'm alive" messages and maximum reprocessing times are known.

Since preemption and replica takeover latencies can be bounded, passive replication can support hard real-time computation, provided the delays associated with checkpoint generation and backup reprocessing are acceptable.

9.4.2.4. Replica Determinism. Active replication unsupported by inter-replica negotiation is unable to support preemption, needed to bound the execution times of hard real-time components. The first XPA prototype will therefore make no use of active replication.

The choice between semi-active and passive replication depends on whether the extra processing power and network bandwidth required by checkpointing, and the extra delays when a primary passive replica fails, outweigh the costs of extra active replicas. This is most likely to happen with components whose checkpoints are large, frequent, or time-consuming to generate, e.g., some real-time databases. Since the XPA prototype may be required to support such components, it concentrates on semi-active replication.

On OSA, applications may be implemented on fail-silent or fail-uncontrolled hosts. If fail-uncontrolled hosts are used, the applications can only be made dependable by active models of replication, with validation of network messages. These models require replicas to behave deterministically, which means they must be structured as state machines.

On XPA, applications are always implemented on fail-silent hosts, and can therefore use semi-active and passive models of replication. Some of these models support non-deterministic applications by enabling the non-deterministic decisions of one fail-silent replica to be imposed on the other replicas. Consequently, XPA applications need not be structured as state machines.

9.4.3. Scheduling Principles

To summarise the main requirements of XPA, we wish to allow both hard and soft types of behaviour to coexist. That is:

- We wish to provide support mechanisms that allow a system designer to assert that a defined subset of required activity is assured to meet deadlines.

- We wish to offer the designer the possibility of another subset exhibiting "best-effort" behaviour, but this must not compromise the deadline assurance of the first subset.

- Finally, we wish to offer the opportunity for any spare resource to be occupied with non-real-time activity, as long as this does not compromise the other behaviour.

In consequence, a notion of *precedence* was introduced, with the possibility for higher precedence computations to *preempt* lower precedence ones. The next two sections discuss these issues.

9.4.3.1. Precedence. Since the XPA prototype is intended to support best-effort systems, it uses the Real-Time UNIX LEX, which takes the on-line approach to local scheduling, using the instantaneous "precedence" of computation. In this LEX, XPA precedence is interpreted as a combination of the priority and targetline. The coarse precedence of groups of components is established through their levels of priority, and within a priority level, the precedence of individual components is established by the component targetlines.

Precedence does not just apply to individual components, but is made to apply to the whole of a distributed computation. When a requirement for some real-time action arises, *all* components supporting the action must be conducted with an appropriate precedence. The precedence of the action is therefore propagated with the request for service from component to component. Some such mechanism is needed since it would be impossible to predict timing behaviour if a high-precedence client raised a request for service on a server that ran at low precedence. The arguments for precedence inheritance are discussed in [Bond et al. 1987, Sha et al. 1987].

A capsule may therefore receive different requests for service with different precedences. The XPA environment must support this in such a manner that the system designer can assert timeliness properties of the components concerned.

Thus, in an input queue of requests to a component, higher precedence requests must overtake lower precedence requests, or preempt their processing, otherwise the hard characteristics required for the highest precedence activity will be compromised. The ability to ignore the effect of queued low precedence computation in calculating the requirements for higher precedence computation is essential.

To support Delta-4 dependability models, the overtaking and preemption must be performed deterministically, since replicas must be certain to receive requests in the same order. To achieve this, some form of *negotiation* technique must be used. In XPA, the replication models supported are limited to those where such negotiation is inexpensive.

This question of precedence ordering of inputs has an interesting analogue in the form of a current debate in the Ada world. The language creates problems for designers of exactly this sort in the ordering of entry queues. When several rendezvous are possible, the Ada programmer may make no assumption about which will be selected first. This destroys the possibility of bounding the time it takes for the server to rendezvous with the highest-precedence waiting client. The Ada 9X committee are examining such shortcomings in the present language definition.

On Delta-4, we have experimented with a distributed Deltase/Ada system in which standard Ada programs are mapped onto Deltase capsules (see chapter 7). The communication between these programs is validated syntactically by standard Ada compilers, and semantically it is similar to Ada; however, at run-time these communications are managed in precedence-ordered queues.

9.4.3.2. Preemption. Preemption of independent processes running in separate address spaces can take place as soon as the higher precedence process is runnable and the lower precedence process has finished the current machine instruction. If the processes sometimes share address space, e.g., sometimes run in the Real-Time UNIX kernel, preemption cannot occur in the middle of logically atomic operations, so that the maximum preemption latency is longer than a machine instruction.

Preemption between threads in the same capsule, which share the same address space and global data, is also supported on XPA, because some concurrent activities do need to share global data. However, the thread preemption must take place at the same point in all replicas so that the their states do not diverge.

Thread preemption mechanisms for semi-active replicas are briefly described below. Leader replicas may be preempted at the next point in processing that can be identified in a notification (or mini-checkpoint) to their followers, which will then be preempted at the same point. Such points are either *receive points*, at which the capsule requests the next input, or *preemption points*, at which the thread declares its willingness either to be preempted or to continue processing.

As soon as a high precedence input is waiting to cause preemption, the destination process can be executed at the same high precedence. An input of highest precedence therefore waits no longer than the maximum dedicated processing interval between two preemption points.

The code executed at each preemption point is essentially as follows:

```
Preempt_point_no := Preempt_point_no + 1;
IF Preempt_soon_flag = TRUE
THEN (Determine whether preemption to occur here, etc.)
     (A follower may be instructed to suspend here)
ELSE (Continue execution of current thread)
```

This code could be invoked as a procedure in any procedural language. With some high-level languages, notably "C", a more efficient solution is available: the preemption point can take the form of in-line assembler code (see annexe I). The details of the preemption protocol are described in section   9.6.5.3.

9.4.4. Timeliness Enforcement

XPA requires a group of generation-time tools to support the following functions:

- Measurement of execution times and preemption latencies.
- Insertion of preemption points in applications.
- Design-time proving of timeliness properties.

Maximum task execution times can be calculated by source code analysis [Puscher and Koza 1989], provided the application obeys certain restrictions such as bounded loops. The PDCS project is developing a tool to do this [PDCS 1990]. Real-time applications that do not obey the required restrictions must be modified, either by the application programmer or perhaps automatically by a sophisticated tool.

Maximum execution times can also be measured by test execution on a dedicated processor with "worst case" input parameters. To determine the execution time of a section of code, calls to read a real-time clock have to be inserted at the beginning and end of it. This could be the first stage in the insertion of preemption points that are to be separated by a specified maximum execution time, so as to achieve a known and bounded preemption latency.

The insertion of preemption points should ideally be transparent to the applications programmer. To see why this may not always be possible, consider an application containing the following code:

```
A: X:= Global_Variable;
B:
C: Global_Variable := X + 1;
```

where X is a variable local to the current thread, but Global_Variable is accessible to all threads in the software component. If the tool inserts a preemption point at B, one thread can preempt another at this point, causing a double updating of Global_Variable, which may have undesirable effects. If the tool cannot place a preemption point at B, the required maximum preemption latency may be exceeded.

All accesses to global variables could be protected by mutual exclusion mechanisms such as Deltase semaphores. (Here the programmer is advised to use a standard mechanism that will be recognised by the tool.) At first, the tool could ask the programmer to insert the semaphore operations; in the future, they could be inserted transparently by a more sophisticated tool.

However, even if the programmer acquires a semaphore before line A and releases it after line C, a preemption point at B may still cause embarrassment, because the thread that preempts at B cannot acquire the semaphore if it needs to; so the effective maximum preemption latency may still be too great. The tool should thus plant preemption points before A and/or after C. If the processing between A and C exceeds the required maximum preemption latency, the programmer should be asked to recode it.

Another problem is that a single statement of high-level language code, e.g., a for-loop, may take longer to execute than the required maximum execution time between preemption points. Such a statement must be recoded, either by the programmer, or transparently by a more sophisticated version of the tool. For example, a long for-loop could be made into two for-loops, one "nested" in the other, with one preemption point in each iteration of the outer for-loop.

To prove at design time that a whole system, or hard real-time subsystem, will meet its timing constraints, it is necessary to measure the worst-case execution time of each component on each computing element and the maximum latency of each system operation. Also required is the following information:

- livelines and deadlines, or the means of calculating them, for each distributed computation;
- a definition of the operational envelope in which timeliness must be achieved.

A static allocation of the components to the available computing elements can then be determined, with redundancy sufficient to support the fault tolerance specified in the operational envelope. (Dynamic allocation of components to computing elements makes it more difficult to prove timeliness and is therefore confined to lower priority components whose timeliness need not be assured.)

Targetlines that will meet the timing constraints are then set for each component on each computing element and the (on-line or off-line) scheduling algorithm to be used is selected. To prove that an off-line scheduler can achieve the targetlines, the actual fixed schedules to be used for every possible task set are constructed. To prove that an on-line scheduler will find a feasible schedule, it may be sufficient to prove that a feasible schedule exists, or that the processor load will never exceed a certain level [Sha et al. 1988].

If the required timeliness cannot be proved, it is normally necessary to re-size the system and repeat the above calculations. Only in marginal cases will a change of scheduler render timeliness provable. On-line schedulers are more able to take advantage of the fact that worst cases do not normally coincide, but cannot assure timeliness more easily in the extreme situation when all worst cases do coincide.

9.5. Support for Distributed Real-Time Computing

A distributed system is formed by processors with no shared memory, no centralized control or clock, hosting participants in distributed activities, communicating only by exchanging messages.

In designing the distributed real-time computing support of an architecture like XPA, several fundamental issues deserve discussion. One of them is order: what is the role of order? The relative ordering of events and/or actions (which may be distributed) must be preserved, for the system to progress correctly. A range of design alternatives are available, from a freely interacting concurrent system, to a highly sequential and blocking one. Another issue is time: what is time needed for in a distributed real-time system? It is required to ensure *participants and replicas agree on the time the environment changed, or on the time to act upon it*. The last issue discussed in this section are the requirements put on the communication service, to support distributed real-time computing.

9.5.1. Order Requirements

Given that participants communicate by exchanging messages whose transmission delays are not negligible, the cause-effect relations between events lead to partial orderings [Lamport 1978].

One type of ordering defined in section 10.1.2, is the one established when one message departs from a site, and arrives at another site, before another message is sent; the two messages are then said to be in a *logical* order. Preservation of such an order is normally adequate for non-real-time systems [Birman and Joseph 1987].

If a component is actively replicated — e.g., for fault tolerance — another requirement, established in chapter 6, is that the group of replicas receive the same information in the same order. This should happen, regardless of the fact that the group of replicas, as a whole, is still subjected to the cause-effect relationships just described. We have seen that some replication techniques may relax these order and agreement requirements, but let us focus on the general case.

In real-time systems, there are circumstances where the logical order does not correctly represent causality. In those cases, implementations based on it produce *anomalous behaviour*, i.e., participants may observe an ordering that is not consistent with their perception of system state. These situations occur when there is information flowing between participants, which is not controlled by the ordering discipline. Examples:

- when participants exchange messages without using the ordering protocol;

- when the system is expected to interact with the real world, in terms of real time.

This latter case, typical of real-time systems, can be reduced to the first, the interactions between participants and the environment becoming the outside "messages".

Clearly, the logical order is not causal, for these cases. For example, when a participant sends a broadcast $m1$, then issues an RPC to another participant, which in turn sends a broadcast $m2$. If the RPC is made outside the ordering discipline, there may be no way of knowing that $m1 \not\!\!E\ m2$, and the resulting order may conflict with the view of the participants. The same may happen without using explicit messages. Physical feedback in a control and automation setting between two participants which otherwise communicate using a reliable broadcast protocol in a LAN, may produce similar effects. For example, suppose that a participant A performs an action, which generates a message $m1$ and an output that physically affects participant B. As a result, B generates a message $m2$ that, if only logically ordered, could be delivered at destinations before $m1$ — thus violating the real causal relationship established between them by the physical process. This may disturb distributed control algorithms programmed concurrently.

The solution for these situations consists in implementing *temporal order* of messages. In section §5.3, we saw it is possible to enforce temporal order with any synchronous communication system. Loose synchrony is the minimal condition for a reliable broadcast, so that correct behaviour be definable, in terms of real-time and ordering of events. This condition is relevant, since the basic communication protocols used in XPA are loosely-synchronous.

We have just discussed the ordering requirements of a general system. It may however be possible to reduce the generality of the distribution support, because certain assumptions about system ordering can be made. This is, of course, of particular interest in a highly optimized system such as XPA. Such assumptions, aimed at relaxing the default ordering and synchronism requirements of the system, may arise from:

- a particular knowledge of the applications, e.g., that concurrent senders are independent. A very immediate consequence is to reduce causality to individual senders, which yields the very well known FIFO order — in fact, a *temporal FIFO* order. Total order (see section 10.1.2) should still be provided when needed, to support replicas.

- a particular computational model semantics, e.g., that participant interactions are synchronous (RPC-like), and computations are single-threaded. This significantly reduces concurrency of the system; however, rendering it sequential simplifies the ordering requirements; as a matter of fact, causality follows quite simply the thread of control, as it moves from site to site.

- a particular replication model, e.g., the leader-follower. In this model, interactions are directed to a privileged replica (the leader). This replica represents the participant, in the system, and is in charge of ensuring the consistency of its replica group. Total order at system level can therefore be relaxed in this example, since it is ensured by a private protocol inside the replica group. Causal order, incidentally, restricted to FIFO in the computational model used, must still be respected.

The xAMp communication service ensures a variety of ordering properties, seeing to it that the system designer can use a variety of computational and replication models.

To ensure replica group determinism, a message multicast to a group of replicas must be received in a total order by all recipients. This is ensured by the *atomic* service of xAMp, or by the *reliable* service in conjunction with the semi-active or passive replication protocols.

Causal ordering (see section 10.1.2) of logically related interactions may be assumed by the application and is therefore supported by some xAMp services, e.g., the *atomic* service. It can also be ensured by the use of blocking RPCs, since the events in each logical activity are forced into a known sequence. Blocking RPCs are compatible with parallelism if the blocking is suffered by only one thread in a multi-threaded capsule.

Casts, i.e., messages without reply, are often used to represent external events within the system. Such casts may be generated in the order of event occurrence and the destination may be written to expect them in this order. If so, they should be given the same precedence (or the same priority plus targetlines in the order of generation), so that their relative order will not be changed by message preemption.

The leader-follower semantics is such that some interactions can do without any order: notifications contain enough information to ensure that they need not be delivered in FIFO order. For example, if the follower is waiting at preemption point N+1, the notification "Continue from N+K to N+L" cannot be obeyed until receipt of the one or more logically-earlier notifications that bridge the execution from N+1 to N+K (see section 9.6.5.3). One or more of these may contain decisions, e.g., "Continue from N to N+K, and then process input HIGH_PREC". Such earlier notifications must eventually be delivered, but not necessarily in FIFO order.

Section 9.6 describes the interaction model of then current XPA prototype, and gives examples of how ordering is relaxed, thus requiring more efficient communication primitives.

9.5.2. Timing Requirements

If we look at the temporal order issue under the general perspective of *participant agreement on time*, we will find two facets:

1) agreement on the time to trigger actions: synchronization of the concurrent progress of a system;

2) agreement on the time at which events occurred: distributed recording (logging) of events.

These requirements will depend on the anticipated range of applications. For example, in the computational part of a real-time system, the synchronism requirements of (1) — i.e., steadiness (bounded variation in the transmission times of different messages), and tightness (bounded variation in the times of delivery of one message at several recipients) — apply only indirectly to computation (see chapter 12), whereas they apply directly to interactions with the environment. Whereas we might say that the synchronism requirements of (1) depend on *application granularity*[4], those of (2) depend on the "environment granularity"[5], more accurately, the granularity with which we wish to perceive its evolution.

---

[4]   Granularity of an application is the minimum interval between any input event, and the subsequent response event.

Finally, replication for fault tolerance introduces the issue of *replica agreement on time*. This issue is a delicate one in real-time systems. The computation part is less demanding: whether for (1) or (2), solutions where agreement is slightly deferred, e.g., achieved by consensus, are acceptable. For example, a leader may notify its follower(s) that "the time of event E was T" to ensure replica agreement. Simplifying, there is not much difference between "the time is T" or "the time was T", provided that all replicas take the same "T". Fault-tolerant (i.e., replicated) I/O, however, requires a higher degree of synchronism. Non-replicated I/O already requires a certain accuracy towards real time; replicated I/O additionally requires precision among replicas. De-synchronism and loss of precision may lead to replica group non-determinism.

In conclusion, opportunities arise for relaxing the ordering requirements discussed above, and in consequence, the synchronism of communication and precision of the distributed time base. Those opportunities will be conditioned by the granularity and the real-action requirements of the envisaged applications.

An analysis of the various application domains for XPA, to discover relevant timing requirements, gives the following results:

Typical existing systems have required the time of the external world to be known only to the nearest second. This has conveniently been met through the use of radio broadcast time, received using compact specialized equipment at a single node and disseminated without any critical synchronization. If this service is temporarily lost, it has proved satisfactory to use internal dead-reckoning until the service is restored. As for internal time, most systems have relied on a centralized global time server (eventually derived from the source of external time). In normal conditions, such a system can yield low accuracy of the time read, due to delays. It is open to the unpleasant effects of arbitrary failure or unbounded dissemination time, unless a continuous check on the believability of the various time values is provided.

In XPA, it is expected to improve this situation. Within the system, there is a finer granularity called for, to distinguish the order of arrival of external and internal events such as alarm conditions. A typical current requirement is for about 10-50 ms granularity, for general events (local or remote). We hope in XPA to be able to determine event ordering to a finer granularity, even for distributed events. Internal clock synchronization is provided by the communication system, in order to achieve a time service of high precision (see section §9.6.6). This means bounding mutual deviations of the local clocks forming the global time approximation, to a close interval.

The clock synchronization protocol can in addition trigger actions at different sites within a defined time granularity, i.e., requirement (1) at the start of this section. Regarding requirement (2), the code that generates an XPA event when something happens in the external world (e.g., a LEX kernel interrupt routine) can be constructed to assign it a timestamp of the approximate global time of arrival.

Availability of good quality external time references (international time standards) will be considered (e.g., radio broadcast), when necessary for the relationship of the system with the external world. Given that this is not a general requirement, the time service will be independent of the approach to get external time. The latter may be injected in the system by one or more servers (fault-tolerant if necessary), having the required quality. Then, the external source may be used to *improve* accuracy of the approximate global time, i.e., keep the internal time as close as required to real time (typically 1 second).

---

5   Granularity of an environment is the minimum interval between any two physical events that must be perceived as being ordered (i.e., non-concurrent).

9.5.3. Communication Service Requirements

Basic concepts of real-time reliable multicasting have been discussed in chapter 5. Communication affects the timing of all distributed real-time activity, so the communication subsystem must respect the real-time constraints imposed by applications.

Previous material in this chapter has raised a number of requirements on communication, discussed below: bounded delivery *latency*; precedence (i.e., *urgency*) in the delivery of messages; *synchronism*.

A detailed description of the XPA communication system can be found in section 9.6.4.

9.5.3.1. Bounded Latency. Bounded delivery latency implies guaranteeing timeliness:

- of the underlying network, in the presence of disturbing factors such as *overload* or *faults*;
- of the communication protocols, whose executions must have a bounded and known termination time under specified fault scenarios; message delivery instants at several multicast recipients may differ, but only by a bounded and known amount.

To control frame delivery latency, the lower layers of the communication system should be engineered so that it is possible to parameterize the network size and load to the desired performance goals, if any guarantee is to be given on their achievement. Additionally, the operating conditions should be monitored so that error conditions can be detected and corrected. This includes both unsuitable loading situations and fault conditions, such as partitions, token loss, etc.

9.5.3.2. Urgency and Precedence of Messages. The requirement of precedence propagation implies that messages with different degrees of precedence receive privileged treatment, i.e., be delivered *ahead* of others. Despite such re-ordering, the following essential ordering properties are still assured by the replication and computational models used:

- total order of inputs to replicas (see chapter 6);
- sending order of messages with the same precedence and from the same source, e.g., checkpoints or notifications, and of causally-related blocking RPCs, whenever required.

Different urgencies imply distinguishing latency classes. In practice, there is a many-to-few mapping from (system-level) message precedence values to (communication-level) latency classes.

The urgency of a message should allow it to overtake other, less urgent messages, in the input queue of each destination access point. The overtaking time can be bounded for highest precedence messages at any particular destination. There could be similar overtaking in the output queue of a source access point, although this is not yet implemented.

In practice, high precedence messages often cannot overtake less urgent ones immediately, but they can do so after a bounded delay. If a low-precedence message is currently being transmitted when the high-precedence transmission is requested, the bounded delay will depend on the characteristics of the networks in use. For example, unless there is bandwidth sharing between latency classes, it will include the bounded time to complete the low-precedence transmission.

It is recommended that the lowest latency class on the network be reserved for hard real-time messages. These will then suffer only a bounded delay because of less critical messages. The delay which hard real-time messages cause each other must be bounded by system sizing and analysis at design time.

9.5.3.3. Synchronism. Real-time protocols must display synchronism, i.e., the duration of successive executions, as well as the skew of a protocol action at different sites, must be bounded (see chapter 5).

Tight synchrony, as implemented, for example, by protocols using approximately synchronized local clocks [Cristian et al. 1985], has the systematic performance overhead of having to take worst case communication delays into account. This would not be an extra overhead when proving the timing properties of a hard real-time subsystem, but is undesirable for best-effort or high-performance subsystems.

In a LAN, execution and inconsistency times may deviate considerably from their normal values, although the probability of these deviations occurring is normally very low. We plan to take advantage of this fact, for probabilistically tightening a loosely-synchronous protocol, since it behaves tightly most of the time. This would maintain recipients in synchrony in a best-effort manner, but of course, ordering could not be reliably ensured. In those cases where total order is needed (see section 10.1.2), another possibility is to build a higher-level tightly-synchronous clock-driven protocol on top of the clock-less one, to serve that part of the information flow requiring both total order and high synchronism. We believe both approaches cover the range of applications requiring performance and a high degree of synchronism.

9.6. XPA as an Integrated Machine

We now discuss how the computation and communication aspects of XPA are integrated to support real-time, high performance and dependability.

9.6.1. Prototype Architecture

In a bottom-up description (see figure 2), the XPA architecture is composed of the underlying LAN, on top of which the xAMp service is built. The group manager uses xAMp to supply the necessary distribution and replication management support to Deltase.

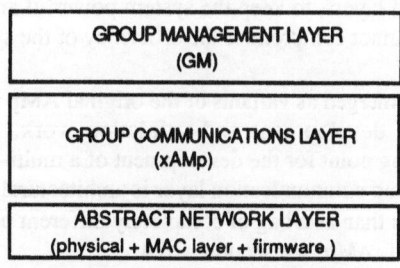

Fig. 2 - XPA Communication Stack

9.6.1.1. Choice of LANs for XPA. The LAN used in the current XPA prototype is a 4 Mbit/s token-ring, whose adapters were the ones available the earliest in the project. Note that thanks to the abstract network, XPA is virtually LAN independent. Still, its real-time guarantee and high performance aims, may condition the choice of LAN.

As an example, FDDI, which provides around 10 times the data rates of the other standard LANs used on the project to date (token-ring and token-bus), is under consideration, for future prototypes. It provides support for the implementation of urgent traffic classes (like token-bus), and has built-in medium fault tolerance, so that expected availability is very high (see section §15.4).

Although the sustained user-network data rate in FDDI is much higher, the objective in XPA, from a performance viewpoint, is achieving responsiveness not only to periodic but also bursty sends or receives. However, it has been shown, in section 10.7.2, that FDDI can enable increased responsiveness, reducing the end-to-end delay, and thence, the AMp primitives' execution times.

Compared with 8802/5 token-ring, FDDI has some speed enhancement measures like early token release and frame length limitation. A comparison with 8802/4 token-bus is more difficult, but these two LANs do have one thing in common: both access methods have a message priority mechanism based on bandwidth sharing by timed token priority classes. Although cable propagation velocity is basically the same, the transmission time of a frame in FDDI is about 30 times less that of slower LANs. For the same global load, medium length and number of stations, FDDI is capable of passing a much greater number of frames of the same octet length, per unit of time, than for example token-ring with which it has the most direct comparison. Added to the much larger number of nodes and distance allowed, FDDI also displays a better scaling capability. In the situations where frame dimension is not negligible, the throughput of FDDI becomes an important factor in the end-to-end delay.

9.6.1.2. Group Communication: xAMp. In the current OSA Delta-4 implementation the xAMp is part of an OSI-type protocol stack, with an inter-replica protocol implemented at the session layer, which also provides multipoint communication. In XPA, the application support environment (Deltase) is collapsed onto a group communication layer. A consequence of this is that the remaining layers must retain parts of the functionality of the removed layers. This is the reason for the introduction of new services in xAMp, and of the group manager as a harmonising layer.

As figure 2 shows, the XPA communication stack is based in a collapsed layering design where, to improve the efficiency of the whole stack, some of the layers present in the OSA architecture were removed. This means that the remaining layers must retain some of the functionality of the removed layers, to keep the system powerful and versatile. However, the increment of functionality cannot compromise the efficiency of the system, otherwise the result would be reversed.

The new services have emerged as variants of the original AMp primitive in order to profit from the previous design, development and validation work. Since the original AMp implementation was a starting point for the development of a multi-fold primitive, software is extensively re-used. The group communication layer is architecturally more elegant as a set of variants of the same protocol than as a bag of completely different protocols. For details about the extended service, dubbed *xAMp*, see section 10.4.

The group management services offered by the original AMp primitive were designed for the integration with an OSI-like stack, in the OSA architecture, where a single logical entity was the user of a gate group. With the original approach, where properties are only assured within a group, several high-level groups of entities were mapped on to a single AMp gate group. The mapping and multiplexing are done by some high-level entity, the session layer, with the original AMp. Evolution both in XPA and OSA has come to show that addressing support performed at low level can render those multiplexing and mapping functions more efficient. So xAMp should offer efficient primitives to ease the addressing of subgroups within one gate group and become the common communication support for both XPA and OSA. *Selective multicasting* is a key issue for this objective. With it, one can manage to address subsets of participants, e.g., separate between sender and receiver participants, in a dynamic, but user-friendly way. High performance address resolution is also obtained. These new features are extensively used in the XPA prototype.

Communication quality of service in XPA assumes a range from *atomic* to *datagram*, through variations of the same protocol, the xAMp, interpreted by a set of primitives, which we briefly recall: *bestEffortN*, *bestEffortTo*, *reliable*, *atomic* and *tight*. For details, the reader is referred to chapter 10.

9.6.1.3. Group Management.
Turning the xAMp into a super primitive would be a negation of the end-to-end argument [Saltzer et al. 1984], as much as just providing datagram quality would be a bad use of it. In fact, the argument is against providing more functionality than needed at a given level of a system, because this goes against optimizing efficiency. Nevertheless, if a class of applications requires a certain functionality (or quality of service), however complex it may be, the lower level should provide it, since it frees the user from programming it, and will probably have been optimized and widely tested. Going back to xAMp, since it has to support a diversity of applications, both in OSA and XPA, it provides a range of qualities of service, rather than a single one. The upper layer user will select the one that best suits its request.

The xAMp provides a comprehensive set of communication primitives and low-level group management tools, to support interactions between groups of components in a distributed environment. However, these interactions may assume a certain complexity, as is the case if components are replicated for fault-tolerance, making it worthwhile for them to be managed by a dedicated entity between XPA Deltase and the xAMp layer. To take advantage of the facilities provided by xAMp, in such operations as message reordering, overtaking, or reuse, it should share structures with xAMp. The *group management* layer is thus introduced as a harmonising layer between the xAMp layer and the communication manager of XPA Deltase, sharing structures with both of them.

9.6.1.4. Deltase for XPA.
XPA and OSA versions of Deltase (see chapter 7) present the same interface to application objects, except for the following extensions and restrictions for XPA:

- XPA applications can read and reset their precedence parameters, i.e., their priority and targetline (called "deadline" in the *Real-Time UNIX* documentation [Bond 1987, SVC200]), and the period of a periodic real-time task. However, only top-level client applications are expected to make use of these facilities; servers will normally inherit these parameters transparently.

- As explained in chapter 12, the management of *real actions* varies according to the model and degree of replication in use; yet it is an established ODP principle that replication should be transparent to the application. One way to achieve this is to perform the real actions in a transformer separate from the application.

  However, it may be that the application programmer himself wishes to supply a procedure that would perform the real action correctly in the simple non-replicated case. It may therefore be necessary to extend XPA Deltase so that it can call a real action procedure supplied by the application in a manner consistent with the actual replication details. For example, if the real action has already been performed by a leader replica, the Deltase environment of the follower would refrain from repeating it.

- XPA Deltase does not permit the recursive import of interfaces; i.e., a client may not import services from a server that also invokes the client. In the absence of this restriction, it would be more difficult to give timeliness assurances. Recursive import is also disallowed by some languages (e.g., Ada) but is permitted by OSA Deltase for reasons of openness.

However, the *implementation* of XPA Deltase differs quite radically from that of OSA Deltase in the following ways:

- Support for high performance: XPA Deltase copies network messages directly between the application address space and the message buffers in the network attachment controller, doing a minimum number of data copies, system calls and context switches. Details of these mechanisms are included in the *Delta-4 Implementation Guide* [Delta-4 1991]

- Support for real-time: XPA Deltase interfaces to a real-time local execution environment, which in the prototype is the *Real-Time UNIX* developed in the first phase of Delta-4, and to the *Collapsed-Layered Communication System* (CLCS, see section §9.6.4). XPA Deltase also propagates the XPA precedence from client to server, ensuring that both are scheduled accordingly. It supports the semi-active replication model, including its thread preemption mechanism (see section §9.4.3.2).

Good system generation tools are required to ensure that compatible versions of both application and system level objects are included in the same system build. This problem will arise more frequently during the development of the XPA prototype because the continuous effort to improve performance is likely to lead to several successive versions of important system interfaces, e.g., the interface between a capsule and the communication system.

Existing Deltase trading mechanisms and system generation practices will therefore be reviewed. If further mechanisms are required for XPA, consideration will be given to run-time checks in which the client presents the version number of the interface specification in its first invocation of the server. This is a marginal overhead that will detect any errors during system commissioning.

9.6.2. Scheduling

9.6.2.1. Scheduling Algorithms. As explained in section 9.1, XPA makes the minimum of assumptions about the scheduling strategy used by the underlying LEX. The consequences of using each of several common strategies in an XPA context are nevertheless of interest and are now examined.

Highest priority scheduling preemptively schedules components in the order of priority. Components that interact with the environment derive their priority from the cost of their potential timing failures, and other components inherit their priority from dependencies.

All XPA on-line schedulers use highest priority scheduling, because it is *stable*, i.e., during overload conditions, where the underlying LEX makes use of highest priority scheduling, the most critical components are still likely to meet their timing constraints although the least critical components may stop doing so. Refinements such as the ceiling protocol [Sha et al. 1987] prevent deadlocks and reduce delays to high-priority components, at the cost of blocking other components more often.

At each priority level, a scheduling algorithm can be used which discriminates between components of the same priority on the basis of their arrival time, targetline, deadline, slack time, period or some other principle. Different algorithms may be used at different priority levels, as recommended by the draft IEEE standard on real-time operating systems [IEEE P1003]. Only the most important algorithms are described below.

Earliest deadline (Targetline) scheduling preemptively schedules components in the order of their deadlines (targetlines).

Targetline inheritance is the policy of assigning a component's targetline to any component on which it depends, e.g., assigning to a server the targetline of its client. *Inherited targetline scheduling* is earliest targetline scheduling with such targetlines.

Targetline calculation is the policy of calculating a targetline for each execution on each computing element, by subtracting (estimated) subsequent execution times from the targetline of the whole distributed computation. For example, a server's targetline is equated to the targetline of its client minus the expected execution time in the client after the server replies. *Calculated targetline scheduling* is earliest targetline scheduling with such targetlines.

Inherited targetline scheduling meets all targetlines on a mono-processor system, if any schedule can achieve this [Halang 1986]. On a multi-processor system, this can only be achieved by NP-hard calculations (e.g., [Zhao et al. 1987]) and inherited targetline scheduling becomes non-optimal. Nevertheless, it performs well in multi-processor experiments [Sha et al. 1988] and is intuitively sensible, favouring those components that are part of the most urgent computations. Also it is simple and does not require accurate foreknowledge of execution times. It is therefore being prototyped in XPA.

Intuitively, calculated targetline scheduling derives component targetlines in a more realistic way and should therefore meet more targetlines in a distributed system. It also performs well in experiments [Sha et al. 1988]. It requires a knowledge of execution times, but this is needed anyway in a provable real-time system.

Several other scheduling algorithms could be used on XPA, but are not included in the prototype because of limited resources. Two examples are:

Rate-monotonic scheduling preemptively schedules independent periodic components in the order of increasing period [Liu and Layland 1973]. On a single processor, rate-monotonic scheduling completes every periodic activation within the period if the worst-case load is kept within certain bounds. It is stable if the more critical components are given the shorter periods.

Least slack time scheduling preemptively schedules components in the order of increasing slack time, (i.e., time till inherited targetline - remaining execution time). Intuitively, it is likely to have similar properties to calculated targetline scheduling.

9.6.2.2. The Prototype LEX

9.6.2.2.1. The Precedence Parameter.
A precedence is assigned to every distributed computation and communicated to every component of the computation. The precedence encodes the following information:

- The priority of the computation, i.e., whether it is hard or soft real-time or non-real-time computation, and how costly are the consequences of a timing fault.
- The targetline of the computation, which is normally derived from the deadline (and liveline, if any), or period for a periodic computation.
- A parameter whose interpretation depends on the scheduling algorithm used at this priority level, e.g., an estimate of the remaining execution time in the computation.

Every component of the computation (i.e., every message or process that can delay the computation for any reason) inherits the precedence and may not change the priority field, e.g., a component of a non-real-time computation may not claim that a hard real-time computation depends on it.

Each local scheduler preemptively schedules components in the order of their priorities and components of the same priority in an order that may vary from one priority level to another (see §9.6.2). The default is inherited targetline scheduling, i.e., components of the same priority are scheduled in the order of their targetlines. If two targetlines are equal or differ by less than the context switching time, the scheduler refrains from switching between the two components, which would increase the risk of missing targetlines.

9.6.2.2.2. Interpretation of the Targetline. If a deadline has been specified for a component, the targetline should normally be set at or before the deadline. If no deadline is specified, but the frequency of component activation is specified, targetlines should be set in a way that reflects this frequency, i.e.:

- Each periodic activation of a periodic process can be assigned a targetline equal to the start of the next period. If the periodic processing finishes before this targetline, the process suspends until the targetline — the targetline event unsuspends it and starts the next period. If the periodic activation finishes after the targetline, this is a warning of potential overload.

- A sporadic process, e.g., a process with a known minimum inter-arrival time, can be assigned a targetline equal to the earliest possible arrival of the next activation of the process. Again a missed targetline is a warning of potential overload.

9.6.2.2.3. Prototype Implementation. In practice the precedence parameters have to be converted into scheduling parameter(s) that are meaningful in each local execution environment. In the XPA prototype, this is done as follows:

- In the Real-Time UNIX developed in phase 1 of Delta-4, processes are scheduled in the order of their priority and processes of the same priority in the order of their "deadlines", which are set equal to their inherited targetlines.

- In VRTX32, which will be used in the prototype NACs, processes are scheduled in the order of a VRTX priority in the range 0 to 255, which encodes the priority and the most significant bits of the targetline.

- On a bus or LAN, messages typically belong to one of a few priority levels or latency classes. For example, only two latency classes may be used on the LAN, with only the hard real-time messages being assigned to the higher class.

The communication protocol software in the NAC will be organized on the basis of one thread per message to be transmitted. In the first prototype these threads are not prioritized, but it is intended that each thread should later derive its precedence from the message it manages, so that a high precedence output message overtakes concurrent low precedence output messages. (The group manager threads allow preemption only at certain points in processing to ensure protocol correctness.)

In the destination NAC, high precedence input messages may again overtake other messages; they may also preempt preemptible software components, provided the semi-active and passive replication models are used (see section   9.6.4).

9.6.2.3. Alternative LEX Mappings

9.6.2.3.1. Mapping of Off-Line onto On-Line Schedulers. If there is a static set of processes at one or more priority levels, an off-line scheduler might calculate a fixed cyclic schedule for them. It may be possible to communicate this schedule to an on-line scheduler by setting the precedence parameters to specify the times at which components are to run and the order in which components are to be preferred if they ever conflict for resources.

For example, if the static cyclic schedule is in fact the same as that which would be produced by a rate-monotonic scheduler (see above), the targetline can be set equal to the period, the third parameter of the precedence vector can encode the start time of the first periodic activation and a rate-monotonic scheduler can then reproduce the cyclic schedule planned at design-time by the off-line scheduler.

To reproduce a more complex static schedule, in which long-period processes sometimes preempt short-period ones, it would also be necessary to assign different priorities to the components. However, most static cyclic schedules can be encoded into a priority, period, start time vector for each component. (An exception would be any schedule that contains mutual preemption: A preempts B, which then preempts A.)

We can therefore envisage an XPA system in which the hard real-time components at the highest priority levels are executing according to a static schedule calculated at design time. The remaining system resources could be allocated to lower precedence components by a "real" on-line scheduler, calculating its schedules at run-time in the normal way.

9.6.2.3.2. Other Scheduling Strategies. The third parameter in the precedence vector is included so that XPA can remain open to other on-line scheduling strategies such as rate-monotonic and least slack time scheduling (see above). The detailed format and interpretation of the precedence are therefore hidden in a "precedence manager" object, which exports a fixed interface, e.g., will tell the caller which of two precedences is the higher, but which can be implemented in a number of different ways.

Various other scheduling strategies are described in the literature (e.g., [Johnson and Madison 1974, Minet and Sedillot 1987, Xu and Parnas 1990, Zhao et al. 1987]). Where possible, it is desirable to permit their use in XPA, although they will not be prototyped. The necessary restrictions are:

- All computations at the same priority level must compete using the same scheduling algorithm at each local scheduler.
- All local schedulers of the same resource, e.g., CPU time, should be identical on XPA. This makes it easier to predict timing properties, e.g., bound the desynchronization of replicas.
- Schedulers that perform extensive or NP-hard calculations are discouraged, as they tend to add to the problem they are trying to solve.

One reason for allowing the later introduction of other scheduling strategies is the wide variability of real-time system requirements. For example, a voice management system implemented by Ferranti has the following requirements:

- the start of service provision can be delayed if necessary, i.e., this is a "soft" requirement;
- but as soon as service commences, a fixed proportion of the processing power is required between the start and completion of each service, i.e., this is a "hard" requirement.

In this case there is no definable targetline, but the third parameter in the precedence vector could be used to hold the required proportion of processing power.

Non-real-time processes running at the lowest priority level may also have no "natural" targetline and may be given "fair shares" of any remaining resource. Alternatively, they may be assigned a targetline a fixed period after their activation times, leading to FIFO scheduling.

Figure 3 shows an example local scheduler that is using some of the above options. The arrows indicate the order in which the scheduler examines the tasks; it executes the first which is found to be immediately executable (i.e., not suspended waiting for any resources) and reexamines the list whenever a new task appears or becomes executable. There are four priority levels: one for the hard real-time tasks, two for soft real-time tasks of different criticality and one for non-real-time tasks. The hard real-time tasks are a static set, so they can be scheduled according to targetlines worked out by an off-line scheduler. The targetlines of the soft real-time tasks are equal to their deadlines; and the targetlines of the non-real-time tasks are derived from their activation times, so that they are scheduled like batch jobs.

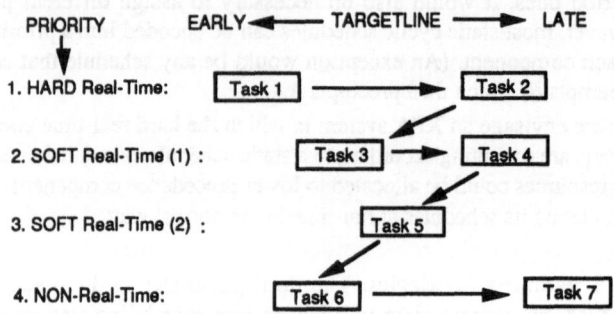

Fig. 3 - Scheduling Tasks of 4 Different Priorities

9.6.3. Relationship to System Administration

XPA system administration is described in section 9.5 and summarised in figure 7. System administration "events", such as missed targetlines, exhaustion of buffer space and changes in the replication status of the object are reported by the group manager to the Deltase communication manager. They are then handled by standard Deltase routines, unless the application code has substituted its own application-dependent handling routine.

Other "events" may take the form of UNIX signals, e.g., from asynchronous device drivers, but these too are handled transparently by Deltase unless the application explicitly substitutes some special handling routine.

All event handlers should preserve replica determinism (see section 9.7.6). A handler need not perform any local processing, but may simply report the event to the invoking client, as in the Ada exception model.

Application requests to read the time, or to be signalled at a particular time, are passed to the group manager and thence to the synchronized local clock, maintaining approximate global time, i.e., the local approximation of the common system time base (see [Kopetz and Ochsenreiter 1987]). For increased precision, clock synchronization is implemented within the communication system. Here the group manager is responsible for replica group determinism; e.g., having given a particular clock reading to a leader replica, it passes the same reading to the follower replicas.

9.6.4. Communication Issues

9.6.4.1. Communication System Design Aspects.
XPA uses a communication service (xAMp, see chapter 10) which provides *group communication*, i.e., which permits and encourages communication using addresses that can refer to more than one end-point or entity. Such communication capabilities are inherently useful in distributed systems and especially so where the use of component replication means that some kind of replicated communication must be provided either in the communication system itself or at the application level.

We chose to hide the complexities of using a particular group communication service (with all its various options and subtly differing QOSs) in a *Group Manager* (GM). This is a different entity from the layer in the communication service provider that implements group communication (xAMp). The group manager provides all the services that a user requires to use the xAMp.

This has turned out to be good decision, because the GM can encompass much of the implementation-dependent detail that is necessary for efficient communication within a node and between host and NAC. It also implements most of the leader-follower protocol, the rest being taken care of by a *host-NAC interface library* that straddles the gap between Deltase and the group manager and is responsible for hiding the complexities of the interface from everything else.

The xAMp was specified to be versatile, offering an easy way to address subgroups of participants and a multi-fold communication primitive offering different qualities of service.

The xAMp service is based on the gate group concept. We briefly review the concepts relevant for this discussion (detailed in chapter 10). Each gate group, or *gate*, possesses a logical identifier that is implicitly an address. A message sent to a given gate group will be delivered to all group members. Only a group can be addressed at send time. A message to be sent to two different groups must be sent twice. However, *selective addressing* allows the selection of a *subgroup* of a given group as the recipients of a message. In the xAMp, these subgroups are identified by lists of station names, but logical subgroup identifiers are constructed for higher-level references.

Within a gate group, a receive queue is associated with the gate at every group member. Messages sent using *atomic* or *tight* qualities of service are ordered in the receive queue such that total delivery order is provided to all members. Messages of the same precedence are inserted with FIFO order on the receive queue. However, the position of high precedence messages within the queue can be negotiated when the *tight* quality of service (QOS) is used.

Potential causal order is preserved for messages exchanged in the system, within the same domain of causality, which is identified by a label (*clabel*) independently of group membership. This service is provided by *atomic* or *tight* QOS. Messages of different precedence must use different *clabels*.

9.6.4.1.1. Mapping between Components and Gate Groups. We now discuss how high level components can be mapped on to gate groups. The discussion is needed since a designer using xAMp has several options that can be expressed by the following questions:

(1) May a component belong to several gate groups, or should the membership be restricted to a single group?

(2) May several components belong to the same gate group or should the group membership be confined to several replicas of the same component?

The way each question is solved by the designer has consequences in terms of the degree of parallelism, concurrency and order relations offered within the communication layer.

Question (2) concerns the *visibility* of multicast, also discussed in section 8.1. In the following discussion we assume that a component possesses a logical identification. A component can be replicated, having several replicas in different stations, but it is still the same logical component, since all replicas have the same identity. Communication with a group of these replicas is such that their existence is hidden from the higher MCS layers, thus it is termed *invisible multicast*. On the other hand, different components can be seen as forming a group of different cooperating entities. The leader-follower model is an example of such a component since each participant can be addressed individually (the leader, the first-follower, etc.). In this case, we have *visible multicast*. Participant visibility can go all the way up to the top system layers, if so wished.

To emphasise that components do not directly call xAMp, we will give different names to the primitives called by the high level components: when a component desires to receive the messages addressed to a logical name *lname* it will call an *OpenRec (lname)* primitive. To send messages addressed to a logical name *lname,* it will call an *OpenSend (lname)*.

9.6.4.1.2. Managing Concurrency. With xAMp communication primitives there are two different ways of achieving concurrency in the communication:

- The first is to associate different receive queues to the same component, letting the component open more than one gate. In this procedure, no causality relations can be assured between messages received through different gates.

- The second is to use *clabels* to establish causality paths while communicating through the same gate. This approach obliges the use of a single group, but it allows the users to incur the expense of achieving causal delivery only when required, without compromising efficiency.

The first option is appropriate for the model of causality based on RPC and leader-follower replication that has been described. The second option is made use of when other interaction constructs must be used; for instance where Casts must be received in potential causal order. These observations clarify the designer options to solve question (1).

9.6.4.1.3. Interactions between Single Components. Let us consider the case where components are only addressed individually and no identification is given to groups of components. If a message needs to be sent to several components, one copy is sent to each component. Components can still be replicated and replicas are still addressed transparently.

In such a case, a component can receive messages through a single gate and there is no need for several components to be mapped onto the same gate. There is a one-to-one relation between components and gates. Logical order of all messages exchanged (through the appropriate QOS) is preserved in the system. Since there is a direct mapping between components and gate groups, there is no need for any extra translation protocol.

With this configuration there is a direct mapping between an *lname* and *gate address*. The translation between *OpenRec* and *GateOpen* primitives and between *OpenSend* and *AttachGate* is also trivial.

9.6.4.1.4. Interactions between Groups of Components. Let us now consider the case where cooperating components need to be addressed both individually and as a group. Components will then be grouped and an identification will be given to each group.

If total order relations need to be verified among the messages exchanged by a set of components — and this includes both the ones directly addressed to a component and those addressed to one of the groups to which it belongs — all the messages must be received by each component through the same receive queue. This answers question (2) put earlier: group membership cannot be confined to several replicas of the same component since, when total order relations need to be preserved, cooperating components must be mapped onto the same group.

It will then be impossible to map directly component and group identifiers into gate group addresses since several logical names will be sharing the same gate address. So, some protocol must be added on top of the xAMp layer, executing the necessary mapping.

With this configuration, there are still two different strategies to disseminate the messages to the appropriate components. The first is similar to the one used in OSA before xAMp was available, where frames were addressed to all gate group members. At the upper layers the local existence of the logical name addressed was checked and the frame discarded or delivered to the addressed component. The major advantage of this method was that there was no need to run agreement protocols to open or close a logical name at a given station: as long as the gate group already existed the operation was purely local. The disadvantage was obvious: at the AMp layer the agreement was run among all the group members even when only a small subset needed to receive the frame. This was a severe performance penalty, so a second alternative was devised, as explained below.

Since xAMp provides a selective multicast service, which allows the frame to be addressed only to the relevant stations, we explored it to achieve better performance. If the sender is able to know, at transmission time, the identification of those stations where a given logical name exists, it can use the appropriate selective multicast field in any xAMp primitive to send the frame just to the relevant stations. There an extra lateral cost, since now an agreement protocol must be run to disseminate the information about open and close actions. However, since the environment changes are supposed to be relatively few in relation to the total number of messages exchanged, this can represent a substantial performance improvement.

9.6.4.1.5. The Use of Selective Multicast. We now summarise a simple protocol to support the use of selective multicast, when several components share the same group. An elaborated form of this mechanism was developed for the OSA architecture after the adoption of xAMp as a common component of the architecture.

On top of the xAMp layer the protocol keeps a table with two fields: *lname, listOfStations*. The table is just a conversion table, mapping logical names into the lists of stations where that logical name exists.

When a new station enters a group, and also to support the cloning protocol described in section 9.7.9, consistency must be maintained. All group members must have the same view of the table and a new station entering the group must obtain this view. The agreement protocol for consistent subgroup management depends on the *atomic* service of xAMp. Each time a component issues an *OpenRec* request, the protocol will broadcast to all the group members an *enter* message carrying the station identifier and the logical name to which the component becomes associated. When this message is received, a new entry in the table is created or, if the entry already exists, the station is added to the list. Each time a component closes the receive queue a *leave* message is propagated in a similar way, throughout all the group members.

When a message must be sent to a given component, the protocol just fetches the appropriate xAMp selective multicast list and requests the transmission to the xAMp layer.

9.6.4.2. Flow Control. Flow control is the problem of coping with limited buffer space. For hard real-time components, this can be achieved by sizing the system so that buffer space will never be exhausted. For other components, flow control can be achieved by blocking producers of messages when it threatens to be exhausted. Three types of component can be distinguished: hard real-time, soft real-time and non-real-time. In the XPA prototype, they are mapped onto different priority levels.

Note that the proposed mechanisms for flow control management have not yet been implemented in the first (1990) XPA prototype.

9.6.4.2.1. The Hard Real-Time Priority Group. Hard real-time components are assigned to the highest priority level. If we are to prove they meet their timing constraints, they must not be blocked for lack of buffer space.

The maximum buffer requirements of these components must be determined at design time and these buffers must be *reserved* for components in the hard real-time priority level. Components with a lower priority must *never* be allowed to use the reserved buffer space.

So a hard real-time component cannot be blocked for lack of buffer space, provided conditions remain within the operational envelope (see chapter 5). Outside this envelope, two things may happen:

- A hard real-time component may seize free space normally used by lower priority components.
- If all else fails, the producer component receives an event indication that offers it a chance to limit the damage in some application-dependent way. This is analogous to the missed targetline event and might be serviced by a similar event handler (see section 9.3.5).

Since hard real-time components are supposed always to have buffer space, inaccessibility for those components is not defined. In other words, it equals failure. In consequence, an xAMp message is never rejected due to inaccessibility.

9.6.4.2.2. The Soft Real-Time Priority Group.
Soft real-time components that are not part of the hard real-time subsystem may be denied buffer space even when conditions are within the normal operational envelope. However, it is important to make this an unusual event and therefore buffer space is reserved for the soft real-time priority levels, in the sense that non-real-time components may never use the reserved space.

When a soft real-time component cannot be given a buffer in its reserved space, two things may happen:

- A soft real-time component may seize free space normally used by non-real-time components.
- If all else fails, the producer component receives an event indication that offers it a chance to limit the damage in some application-dependent way. This is analogous to the missed targetline event and might be serviced by a similar event handler (see section 9.3.5).

When an xAMp message for a soft real-time component cannot be received in the destination NAC because of a shortage of soft real-time buffer space, the destination component is said to be inaccessible. The transmission may be retried later, but if this condition persists beyond pre-defined limits (duration, rate), it must be considered failed, either by itself, or through an action of system administration. When using the *reliable* QOS, that happens upon the first occurrence of inaccessibility, by definition. (In normal cases, the system administration module in the group manager is informed of the buffer shortage before it leads to such failure and may be able to avert failure by denying buffers to non-real-time components.)

9.6.4.2.3. The Non-Real-Time Priority Group.
Non-real-time components may not use the reserved buffer space of real-time components and the buffer space that they normally use may be seized by these higher priority components in exceptional conditions. It follows that all non-real-time components on an XPA system must expect to be blocked for lack of buffer space, or lack of other resources, and it may be difficult to predict their timing properties.

Although non-real-time producers may be blocked, it is not proposed that non-real-time messages should ever be discarded to release buffer space. This would lead to incorrect operation or abandoning of non-real-time components, which may in fact be as costly as a loss of timeliness in real-time components.

If there is no non-real-time buffer space for a non-real-time message to be received at a destination node, the destination component is said to be inaccessible. If the message was transmitted by *atomic* multicast, the multicast fails and may be retried after an interval. If the message was a *reliable* multicast, the destination component is deemed to have failed.

Non-real-time applications may, if required, be alerted by the missed-targetline or no-buffer-space events. An application might handle these events by explaining to a terminal user why it is delayed.

9.6.5. Computation Issues

9.6.5.1. Causality in Component Interactions. Deltase supports the Remote Procedure Call abstraction, in which object interactions are synchronous RPCs involving the passage of a global thread of control. Global threads are represented within the single address-space of a component replica by lightweight threads, which are scheduled deterministically, at well-defined points in the code. In figure 4, objects A, B, C, E and F are shown supporting multiple computations in this way.

Figure 4 also illustrates the facility that is provided to objects for a "family" of threads to come into being on behalf of the single global computational requirement (labelled iii in the diagram). Within object C, a "parent" thread has created "child" threads of control that subsequently die within that object, but in the interim are despatched through RPC to objects E and F. If the components supporting these are located on distinct hosts, simultaneous computational progress may be achieved; local "pseudo" parallelism can therefore give rise to distributed "true" parallelism.

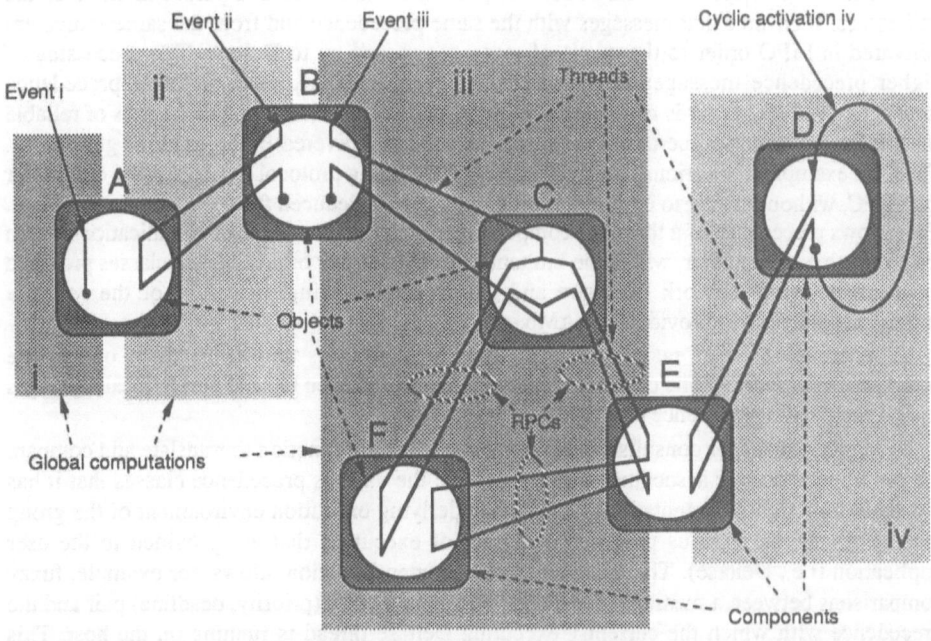

Fig. 4 - Distributed Computations

Apart from the purely within-component issue of birth and death, global threads that are "related" in this manner are causally-independent of each other during their life. In figure 4, this applies to those arriving at object F, which are just as independent of each other as those arriving at objects A, B and E. Causality follows each individual thread of control, as it animates different objects. Therefore, upon arrival at any one component, the messages that represent these threads, even from the same sending component, are necessarily independent if they are concurrent. They are viewed as "competing" and may be freely re-ordered by the component according to precedence. The machinery is described below.

9.6.5.2. Causality and Replica Determinism. XPA inherits the Delta-4 Open System Architecture lightweight thread scheduling. This is inherently replica-deterministic; it always takes place at well-defined points in code and the same data is used to take the same scheduling decision.

However, the group of replicas must reorder messages identically, and must permit these to sometimes cause preemption identically. The State-Machine approach to replica-determinism [Schneider 1990] used by the Delta-4 Open System Architecture cannot be used, since in effect it bans preemption. To overcome this, XPA has introduced the leader-follower, or semi-active replication model [Barrett et al. 1990].

Since stations are fail-silent, it is possible to permit all decisions on potentially non-deterministic behaviour to be taken by a privileged replica, the leader. Until it fails, this replica is in charge of ensuring the consistency of its replica group, both with respect to the order of receipt of messages that represent RPCs and with respect to any other potential causes of non-determinism.

9.6.5.3. Message Handling. The group manager takes messages from the network that are intended for recipients on its node, interprets their addresses and passes them on to the addressees. It ensures that messages with the same precedence and from the same source are delivered in FIFO order to the queue of messages passed up to Deltase. The processing of higher precedence messages is allowed to proceed at the expense of lower precedence messages, but FIFO order is maintained within a "precedence class", so that a series of reliable casts from the same source at the same precedence are delivered in the order of generation. Thus, for example, the various messages of the XPA cloning protocol can be sent by cast rather than RPC, without having to ensure that their order can be deduced from their internal content. This allows precedence of a thread of computation to migrate across the communication system from one object to another, within the limitations of the number of precedence classes provided by the underlying network hardware and the order-preserving properties of the *reliable* communication service provided by xAMp.

The group manager takes messages from local senders and sends them on via the communication service. The communication service is used even to send purely local messages so as to obtain the precedence properties that it provides.

The group manager consults a *precedence management* function to translate and compare the precedence classes associated with a message; the various precedence classes that it has available for its own computations (from the underlying execution environment of the group manager); and the various precedence classes of execution that are provided to the user application (i.e., Deltase). The precedence management function allows, for example, fuzzy comparisons between a message precedence that consists of a (priority, deadline) pair and the precedence with which the currently executing Deltase thread is running on the host. This comparison is then used to decide whether or not to interrupt the current Deltase processing in order to pass the (possibly more important) message to another thread or object. The comparison is not straightforward because it may, for example, use a cost-function which takes account of the time required to interrupt and schedule another thread. This may "cost" more to do, perhaps increasing the likelihood that one or both threads of computation may miss their deadlines, than simply letting the first thread run to completion, even though the new message has a slightly higher precedence when measured in the absolute sense.

As well as handling application messages, the group manager must also produce and process messages that are generated as part of the leader-follower protocol and administration messages controlling the creation, destruction and management of communication resources.

Concerning precedence management, it is mandatory for an RPC of highest precedence to cause bounded-latency preemption of lower precedence computation. The message must not only "jump the queue" on arrival; processing must begin at the same point in all replicas so that they do not diverge in state. Figure 5 illustrates the following solution to this problem:

Preemption points are small units of code, pre-installed in the object code a bounded execution distance apart (shown in figure 5 as small lines crossing the execution-path of the threads). Very few instructions are executed in the normal (non-preempting) path through a preemption point, which permits their distance apart to be small without significant overhead. A count is maintained to identify arrival at a preemption point; because of both deterministic lightweight thread scheduling and the machinery for handling asynchronous events about to be described, the same count (N-1, N, N+1, ... in figure 5) is assured to identify the same point in execution across replicas.

When the leader arrives at a preemption point, code examines a preemption flag and normally continues computation if it is not set; periodically, it sends "continue from N-K to N" instructions at such points (K is a constant and N-K is where the last instruction was issued). Such instructions consume very little network bandwidth and assist the follower to keep up or detect desynchronization. The set of such instructions provides a continuous description of execution requirements. When a follower arrives at such a point, the code compares its own preemption point count with the value supplied by the leader in the instruction currently being followed, and proceeds if it is less. When it arrives at an identified point, it obeys the leader's instruction to proceed or divert to accept a message (figure 5). Note that the leader must stay one instruction ahead; if this has not been sent, the follower blocks for a bounded time at the next preemption point beyond those for which it has received instructions.

When a message containing a request for service arrives at a station, its precedence is compared with that currently occupied by the destination replica (leader or follower), which is running with at least the precedence of its current computation. If the message precedence is higher, the replica is immediately given that precedence, and the preemption flag is set. If that precedence exceeds that of any other computation on that station, the replica gains exclusive access to the execution resource, and the execution distance to the next preemption point is then bounded in time.

The leader is diverted by this flag at its next preemption point, and will then select the highest precedence waiting message; this selection, together with the preemption point count when diverted, is sent to the followers. The service required is then invoked by running a local lightweight thread to temporarily represent the global thread in that replica, and will be identically invoked at the followers.

The group manager at a follower blocks the replica when it requires to read an input message until the corresponding notification has been received. Similarly when a replica requires to produce an output it is blocked until the notification confirming that the leader has sent it has been received. It would be possible to buffer output messages in followers, thus allowing computation to proceed, but for the moment we believe that the complications in message buffer management that this would entail do not justify the benefits to be gained. The argument is that the follower must be only a little way behind the leader, or the application would fail to meet its deadlines if the leader fails. There is thus very little to be gained by buffering output messages except programming complexity and probably reduced efficiency.

Follower replicas are also blocked by their group managers when they reach a preemption point for which a preemption notification has not been received from the leader. All such interactions between group manager and replicas (notification of reaching a preemption point; suspension/resumption of a replica, etc.) are handled by the *host-NAC interface library*. In the present implementation, this uses shared memory to reduce multi-processing context switches and a UNIX device driver for preemption of Deltase threads.

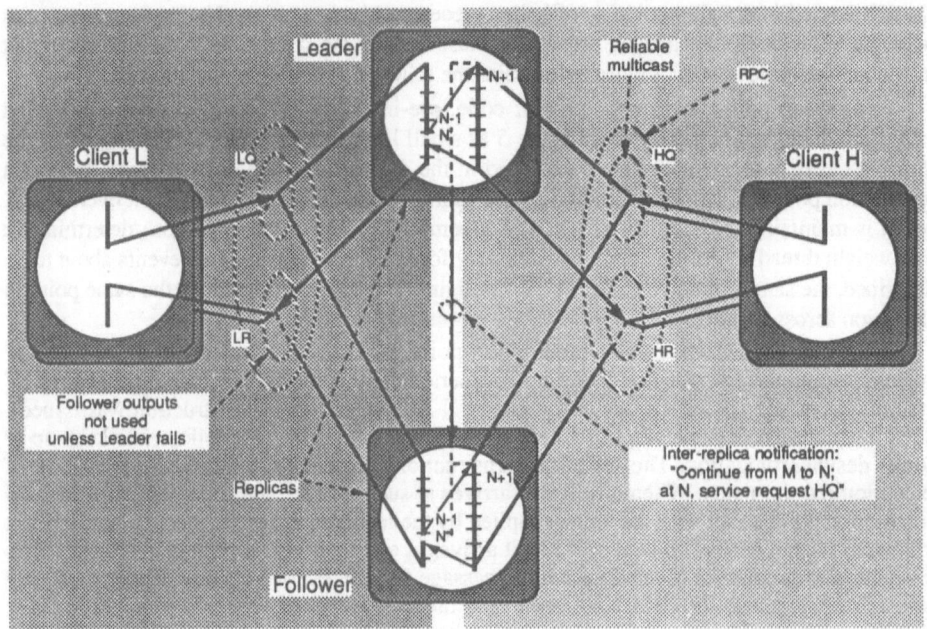

Fig. 5 - Replica Group Determinism, Leader-Follower Model

9.6.5.4. Supporting the Leader-Follower Protocol. The group manager relies on xAMp to inform it when a node containing a leader replica has failed, so that a follower must therefore become a leader. It is sufficient for the group manager of the follower that is about to become a leader to pass the information that the follower must become a leader in the normal flow of notifications to the follower. The follower can take appropriate action when it processes the notification (this is done within the host interface library). This assumes that correct ordering of messages and notifications is maintained within the group manager and that xAMp co-operates with the group manager by providing the indication that the leader node has failed in a suitable manner.

To support the leader-follower protocol, the group manager must support the following assumptions:

- total order of inputs to replicas;
- messages with the same precedence class do not overtake messages with the same or a higher class; i.e., FIFO — and therefore causal — ordering of messages with the same precedence and source is not violated;
- a similar consideration applies to notifications as messages.

The group membership management service for the *reliable* xAMp QOS must provide the following:

- notification of node failure (i.e., a group change indication) must be delivered to a node *after* any delivered messages from the failed node.

9.6.5.5. Replica Synchronization. With all forms of replication on XPA, the fastest replica will normally output messages to the network and to any non-replicated device. The role of the slower replica(s) is to provide a correct and timely service if the fastest replica fails. To provide this timely service, they must be synchronized with the fastest replica to within a bounded time.

Badly desynchronized replicas also present other problems: the slower replica may appear to have failed silently, or its input queue may overflow. When a follower replica reads the present time, it is given the value that was given earlier to the leader, to preserve replica group determinism; however this value is inaccurate unless replicas are well synchronized.

9.6.5.5.1. Detection of Desynchronization. One of the problems is to detect excessive desynchronization. Here a natural mechanism occurs as part of semi-active replication. A stream of notifications passes from leader to follower, indicating the results of non-deterministic decisions and successful transmission of the leader's output messages. Preemptible capsules send notifications of the form "Continue from N-K to N", with bounded inter-notification delays, which indicate that the leader recently passed preemption point N.

A follower's group manager can thus detect excess replica desynchronization. If the "Continue from N-K to N" notification is time-stamped T (in the approximate global time of the leader's node) and the follower has still not reached preemption point N at time $T+X$, (in the approximate global time of the follower's node), where X is an appropriate desynchronization event "trigger", the follower's GM raises the desynchronization event to both leader and follower.

This is a local event (see section 9.7.4.1) since it is handled differently at the two locations. The follower event handler reports the event to the local system administration object responsible for station management, which may merely log the event, or may initiate some long-term solution such as load-balancing. The leader event handler is discussed in the next section.

However, the follower may not only be desynchronized, but may have failed silently, e.g., because of a Heisenbug (software fault that manifests itself independently in different hosts) that causes it to fail at a point where the leader proceeded correctly. It is therefore necessary to define a time-out F larger than X and to raise the follower-failed event if the follower has still not reached preemption point N at time $T+F$. Such a follower is incapable of timely back-up of its leader, so it is correct to deem it failed, whatever the reason for its tardiness.

9.6.5.5.2. Synchronization Techniques. Two types of synchronization technique have been identified: lightweight probabilistic mechanisms that usually prevent the desynchronization exceeding X and, when the desynchronization is greater than X but less than F, a reliable protocol to keep the desynchronization below F unless the follower has failed.

Lightweight probabilistic mechanisms include:

1) Replicas should be allocated to hosts so that each follower (or slower replica) has as much processing power and other resources as its leader. For example, one host could support a group of leaders while an identical host supports the corresponding followers.

2) However, even with such a distribution of components, variations because of factors such as different disc interrupt times may cause some followers to finish early and others late, compared with the corresponding leaders. Those which finish earlier than the targetline, or which catch up with their leaders, should then suspend until the targetline or arrival of the leader's notification. This releases the spare resources for re-allocation to the late followers on the same node.

3) Synchronization of semi-active replicas can be optimized by conferring the role of leader on the slowest replica, i.e., on the replica that has fewest local resources such as processing power and buffer space. The followers will then complete most operations more quickly than the leader and cumulative desynchronization is unlikely.

4) A follower in fact requires slightly fewer resources than its leader, since it does not normally output network messages or perform real actions (see chapter 12). These factors help a follower keep up with its leader, although they are difficult to quantify.

5) When a high-precedence input message enters the input queue of either replica, it depends on the replica's current processing, which is therefore allowed to proceed at the high precedence of the pending input. If the follower is desynchronized, it will experience this "speed-up" effect for longer than the leader.

The protocol that ensures bounded desynchronization is as follows. The desynchronization event causes the leader and follower to switch roles, and the old leader then waits for the new leader (i.e., old follower) to catch up with it and send it an instruction. The new leader will do this unless it fails, which is detected when it reaches the failure desynchronization F (see above), if not sooner.

This strategy ensures resynchronization and tends to reduce the frequency of desynchronization events. Since the former leader is running on what was recently, and may still be, a less heavily-loaded host, it is likely to keep up with the new leader.

This strategy does not damage the timeliness properties of components where a design time proof of timeliness exists, since any such proof must apply to both replicas. However, in circumstances for which there is no such proof, the follower may miss a deadline that the leader achieves, so that role-switching can increase the probability of untimeliness. Some users may therefore prefer some other handling of the desynchronization event, and it may be necessary to offer applications a selection of desynchronization event handlers, plus a description of their properties.

Other synchronization mechanisms have been considered but rejected, e.g., it is undesirable to raise the precedence of a follower to speed it up, because this will adversely affect higher precedence components competing for the same resources. There could follow a useless "inflation" of many component precedences in an overloaded node. Before precedences are changed, there has to be a careful analysis to prove that the side-effects will be acceptable.

9.6.5.6. Latencies. As explained in section 9.5.2, a real-time system must bound all system latencies. This affects all parts of an XPA system, as can be illustrated by considering the various latencies that occur during a remote procedure call. Figure 6 shows the request phase of an RPC from a top-precedence client. Control paths are shown in solid lines and data paths in dashed lines.

The RPC starts in the client application. XPA Deltase copies its parameters directly into message buffers in the network attachment controller. For a top precedence application, which will not be preempted, this takes a bounded time, depending on the size of the parameters.

The client application then interrupts the NAC to draw attention to the output. There are then a bounded interrupt latency and context switching time and a bounded queuing time for a top precedence message, before the group manager processes the message.

On a token ring, there is a bounded latency before the top precedence message can be transmitted, since other stations may acquire the token first and transmit a bounded number of other messages, each of bounded length. Some LANs may entail unbounded latencies; a standard Ethernet, for example, may be unusable in a hard real-time context.

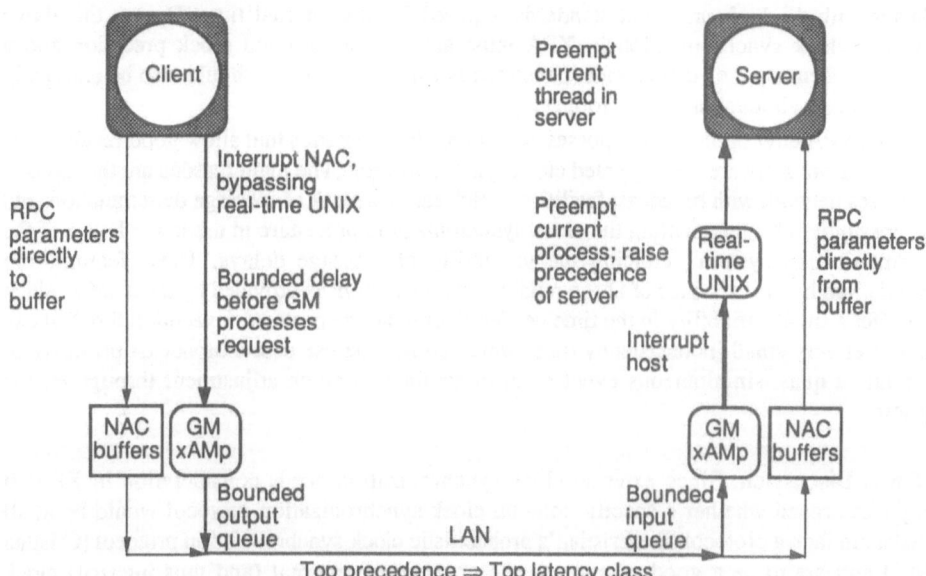

Fig. 6 - Request Phase of a Remote Procedure Call

Communication latency must be bounded, at least for messages that support hard real-time services. The basic XPA communication system provides a choice of communication services, with different tradeoffs between latency and reliability (see section   9.6.4). RPCs use the Reliable QOS, which involves no overheads to ensure total or causal order (see section   10.4).

When a top-precedence RPC request arrives at the destination NAC, the xAMp and group manager should process it in a bounded time. There are then more delays whilst the host is interrupted and its current process preempted. These are bounded under the Real-Time UNIX developed in Delta-4 Phase 1 [Bond 1987, SVC200]. The latencies involved in preempting a particular server and diverting it to the highest precedence thread, are bounded for the semi-active and passive models of replication.

However, in the worst case the fastest replica of the server fails. The latency that then occurs depends on failure detection (except for active replicas) and on the synchronization of replicas. Methods for bounding desynchronization are discussed in section   9.6.5.5. A follower can detect leader failure in a bounded time if the leader's environment generates an "I'm alive" notification to the follower whenever there has been no other leader-follower notification for a preset period. A similar technique can be used with passive replicas.

If this is a top precedence RPC, the server inherits its precedence and runs without being preempted until it replies. The reply message suffers similar delays to the request message.

9.6.6. The Time Service

In real-time computing applications such as those XPA will support, a facility to measure durations and the position of an event relative to the environment, in the metric of the external real time, is a major demand. Furthermore, in distributed real-time control systems, it is often required to measure the duration between two events that have been observed by two different nodes, or to specify actions to occur at several places in the system at a given absolute time (these operations may require a high degree of precision). It is also often required to establish comparisons with the real world time, with the assistance of some external time metric, not always with the high precision standards required for the internal time. Due to the above reasons, clock synchronization in XPA must always ensure good clock precision and a moderate accuracy, i.e., clocks must be internally synchronized, but should also be externally synchronized, although in a less stringent way.

The XPA environment itself possesses a set of characteristics that allow good results to be expected from software implemented clock synchronization. The main reasons are the use of a local area network with broadcast facilities, which eases the task of message dissemination, and the possibility of implementing the clock synchronization procedure in the lower levels of the communication system, reducing the variability of message delays. These features are materialized by the existence of cheap reliable broadcast services provided by the xAMp, which can offer a small variability in the time needed to execute the protocol (execution time). It can also offer very small inconsistency time, which allows the use of the broadcast primitive to simulate a quasi-simultaneous event or to disseminate a clock adjustment throughout the system.

9.6.6.1. Discussion. Since external clock synchronization needs consideration in XPA, it might be argued whether a specific internal clock synchronization protocol would be at all needed. In fact, a protocol like Cristian's probabilistic clock synchronization protocol [Cristian 1989] appears to be a good solution to the problem of external (and thus internal) clock synchronization.

However, we may cite two drawbacks of Cristian's protocol, as far as XPA is concerned:

First, unless the system designer is able to afford a great expense in traffic to achieve synchronization, the reading delay must be chosen with a value such that there is a good probability p of achieving the desired precision. So, although Cristian's protocol does not depend on the variability of the message delay, it depends, for practical implementations, on the distribution of these delays. In particular, it depends on how far the expected delay is from the minimum delay. If the sources of external real time are not assumed to be fail-silent, some kind of agreement must be reached before the value is read. This can take the average delay far from the minimum delay thus reducing the accuracy of the synchronization protocol.

Second, a direct use Cristian's probabilistic clock reading method, as suggested in [Cristian 1989], is extremely dependent on the availability of a source of external real time. Internal clock synchronization algorithms, in presence of any number of crash faults, will keep the surviving clocks synchronized (if faulty processors remain in the right proportions to correct ones), which may be enough for many applications. Given that XPA is supposed to support a number of applications, of varying needs, the cost of having an available and reliable external source [Cristian 1989] would be undesirable for some of them.

To avoid the total dependency on the presence of an external real time source, we propose a configuration where virtual clocks are both internally and externally synchronized.

To synchronize clocks internally, an approach similar to Babaoglu's clock synchronization procedure might then be envisaged [Babaoglu and Drummond 1987]. However, in XPA, group communication is achieved through the use of gates, reducing the number of *Full Message Exchange* rounds to those on behalf of applications using the *Multicast Group of Stations* (MGS). In Delta-4, MGS is mainly used by system administration. These applications are not expected to produce traffic with the frequency required to satisfy clock synchronization needs. Furthermore, to achieve *approximate timed simultaneity*, a need expected to be frequent in a real-time distributed system, an appropriate protocol must be run. The protocol proposed by Babaoglu (DemandAts) has many similarities with a clock synchronization protocol proposed by Srikanth and Toueg [Srikanth and Toueg 1987]. We analyse next this protocol and assess its adequacy for our purposes.

Srikanth's clock synchronization protocol possesses several advantages, namely: it achieves optimal accuracy; it is easily tuned to tolerate different kinds of faults; it is easy to implement. However, being a convergence non-averaging algorithm, the precision obtained depends on the worst-case message transit delays. This is a severe limitation that needed to be overcome to gain the maximum profit from the XPA architecture.

Kopetz's *Clock Synchronization Unit* (CSU) [Kopetz and Ochsenreiter 1987] could improve protocol performance and would reduce the computational cost of the clock synchronization protocol. However, the performance obtained by the CSU is not necessary in the envisaged applications. Furthermore, the CSU has the commercial drawback of being a single source component. The use of the CSU will thus be avoided.

The XPA clock synchronization protocol relies on the NAC fail-silent characteristics. However, the underlying physical clocks do not need to be fail-silent. When the physical clocks are assumed to exhibit an arbitrary behaviour, a majority of correct clocks will be required to run the protocol, i.e., a total of $2f+1$ clocks will be needed to tolerate f faults.

For internal clock synchronization in XPA we developed a clock synchronization algorithm that exploits the intrinsic characteristics of broadcast networks. The algorithm is based on a new variant of the well-known convergence non-averaging technique, dubbed *a posteriori agreement*. The precision achieved by the algorithm is not limited, as opposite to most of the published works, by the variability or worst case values of network access delays. Furthermore, our solution does not require the use of dedicated hardware.

The algorithm, that is described in detail in [Rodriques and Verissimo 1991], achieves clock synchronization using broadcast messages to generate simultaneous events in the system. In most existing broadcast local area networks, frame transmissions arrive almost simultaneously at the successfully receiving sites. This property is exploited by the a posteriori agreement clock synchronization to achieve precision. Note that occurrence of faults may prevent a given broadcast to generate a simultaneous event. We use acknowledgments to detect the effects of faults and several simultaneous events can be generated on a single resynchronization. The agreement for the appropriate simultaneous event — to be used as the base of a new virtual clock — is thus executed *after* the generation of the event, thus the name "a posteriori".

Virtual clocks provide a continuous function being obtained from the value of the local physical clock and the adjustments from internal and external synchronization. The adjustment term for external synchronization may remain null when not implemented. This allows the use of the clock synchronization service in XPA systems where no sources of "external" real time are available. Several solutions will be studied for external synchronization with different fault assumptions, since the use of a non-fail-silent source of real time may be interesting for economic reasons.

9.6.6.2. Primitives Provided. The aim of a clock synchronization service is to provide a synchronized virtual clock. The more important primitive is then *virtualTime*, which returns the value exhibited by the virtual clock. Since the virtual clock will be based on a local physical clock a primitive is provided to set its value during initialization (*setLocalClock*) and to read its value (*physicalTime*). The virtual clock is "started" with the *startVirtualClock* primitive, which initiates a clock locally and includes the node in the synchronization protocol.

9.7. System Administration

This section is concerned with areas of difference between system administration in OSA (OSA-SA), see section §8.2 and system administration in XPA (XPA-SA). XPA-SA provides an augmented subset of the services provided by OSA-SA. Although it is exclusively concerned with the leader-follower replication model for capsules and can take advantage of simplifications resulting from use of this single model, the administration consequences of the high performance and real-time features of XPA must be explicitly addressed. Thus, using the terminology of [Le Lann 1989], many of the additional XPA-SA services and mechanisms are concerned with the management of the time domain (timeliness properties) as well as the logical domain (correctness properties)

XPA-SA provides similar network management services to OSA-SA. These are fully described in section 8.2 and may be classified as the management of events that directly affect the logical domain, such as:

- the initial Delta-4 system configuration;
- system startup operation;
- station failures;
- operational changes to the system configuration.

XPA-SA additionally provides services to manage and attempt to minimize, the effects of faults and overloads in the time domain. Lack of resources (e.g., exhaustion of a finite number of computational threads, or message buffer space, or processing power) is manifested as failure to meet deadlines. The treatment within XPA-SA consequent upon the detection of such an event depends upon what sort of *event handler* has been provided for the object concerned; because of the significance of event handling in the context of the discussion of chapter 5, this forms a large part of the body of this section.

Some OSA-SA services are not relevant for XPA-SA or have a changed purpose. For instance, there is no need to collect statistics; nevertheless functions that permit pre-delivery tuning, verification of design-time deadline assertions and so on, will prove essential.

XPA-SA is biased towards logically distributed rather than logically centralized administration activity; separate concerns are where possible separately encapsulated. Such object-orientation is of particular importance in the target market areas, where individual projects are sometimes required to implement specialized administration paradigms. There is a twofold basis for this: substitution of individual system administration components and local-to-object thread-related event management.

System administration components are deliberately structured as a distributed federation to minimize the granularity of substitution. Individual system administration components may be substituted provided the subset of SA interfaces which are "presumed always to be present", together with the object design-level assumptions about their use, are preserved. Examples of such components are discussed in several sections below. The semantics associated with these interfaces may thus be extended, and in some cases reinterpreted, according to the needs arising in particular projects.

Designers and applications programmers are encouraged to take advantage of XPA mechanisms that permit events to be intercepted and managed within objects wherever this is appropriate. Application-specific knowledge may thus be brought to bear in order to achieve a necessary performance or timeliness requirement. Particularly in the case of a missed-deadline event, such knowledge and such local-to-object management may be essential to the design. Section 9.7.4 below discusses the event-handling mechanism chosen.

The processing required for the complete implementation of all SA functions may not be fully automated. The involvement of a human administrator will often be necessary in order to initiate and implement the system administration function.

As standard, a very small kernel of general-purpose functions is provided, largely derived from the equivalent work under OSA-SA. These constitute an object-library, implementing alternative application-level SA objects that provide interfaces "presumed always to be present" and a support-library of local-to-object SA event management mechanisms. Together, these provide administration paradigms of general value, which will be maintained and added to as additional paradigms are recognised and implemented. The library will always maintain upwards compatibility.

The original system configuration establishes the replication domains of each capsule as an ordered list of stations available for cloning. Replicas are instantiated on these stations in this order, up to the defined replication degree. The order of instantiation establishes an initial *"takeover order"* for replicas; which replica is the leader, which is the second (follower) and so on. When a station fails, promotion of followers is in this takeover order. The station chosen for instantiation of a new clone is the first free station on the list of the replication domain. The new clone is given the last position in the takeover order, now free as a result of promotion. Similarly, when a station is repaired and restored, it appears last on the ordered list.

The takeover order therefore evolves with time; after a number of node failures and repairs, the takeover order may bear no resemblance to that originally established.

In XPA, variants on this model may well be desirable in implementing a particular system; this can be achieved by replacement of selected SA components. For instance, it may be desirable to restore an initial configuration when this becomes possible through repair, perhaps to restore the set of assumptions taken during design with respect to timeliness.

9.7.1. The Structure of XPA-SA

Section §9.6.4 describes the collapsed-layered communication stack and group manager that replaces the OSA 7-layer communication stack and SMAP. This design is a consequence of the XPA performance and real-time requirements. The initial stages of fault tolerance are handled within the XPA group manager and the XPA-Deltase support environment by means of (error) event-handling code. All such event-handling code is considered to be part of XPA-SA and, indeed, code resident within the capsule takes part in the immediate handling of an event (e.g., follower promotion to leader, deadline-elapsed event handling). In OSA, care is taken to separate functionality in such a way that event handling is totally transparent to capsules that may execute on fail-uncontrolled hosts. In XPA, the capsules — executing on fail-silent hosts — can with advantage be given event-handling responsibility.

Fault-tolerance is achieved in two phases:

1) The *first phase* of fault-tolerance is error processing; failures of (fail-silent) servers/stations are detected (as time domain errors) and raised as events to the group manager and Deltase support environment in order to trigger immediate recovery. After server/station failure (leader or follower), an alternative follower replica is selected to take over the role of the failed replica.

   This immediate error processing is achieved on-the-fly within the group manager communication and Deltase libraries, transparently to the applications programmer. The objective is to complete the promotion of followers without compromising any deadlines currently in force.

2) The *second phase* of fault-tolerance is fault treatment, i.e., the restoration of the specified degree of redundancy through cloning, or load balancing by migration, of the capsules concerned onto another station within their replication domains.

   These actions are implemented as either local or global services embodied in system administration server objects with support mechanisms built into the Deltase envelope and the XPA group manager. In common with OSA-SA, these entities include a domain manager plus database, local factory, catalogue server, etc. One difference in XPA-SA is that those functions of the OSA *Management Information Base* (MIB) which are required in XPA have been absorbed into the domain manager database; there might otherwise have been a need for a protocol to maintain consistency between the two databases.

A general model of XPA-SA is shown in the data-flow diagram of figure 7.

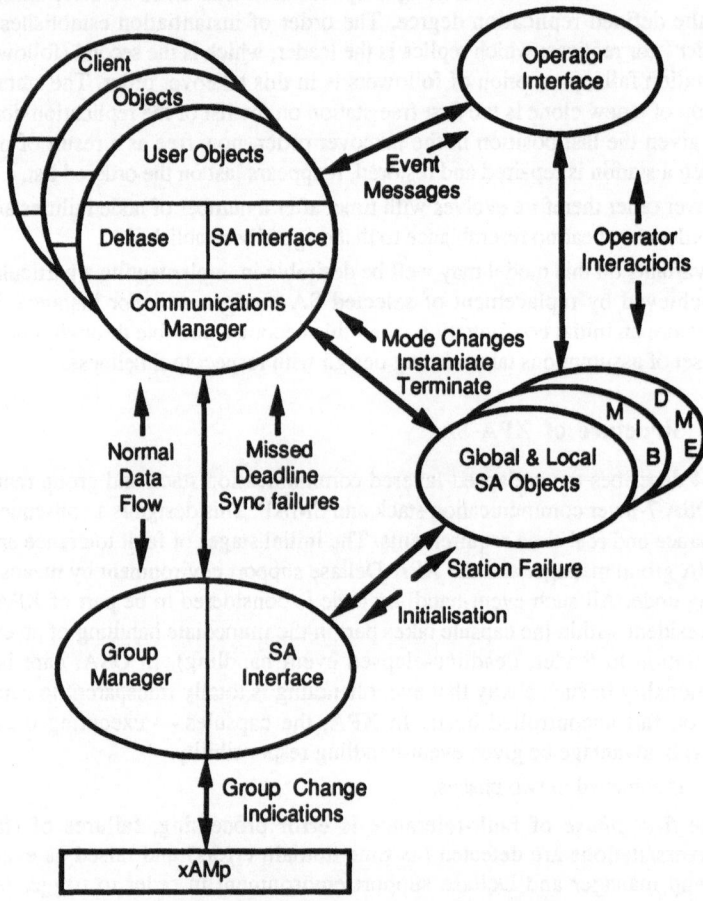

Fig. 7 - XPA System Administration Data Flow

9.7.2. SA Activities involving the XPA Group Manager

The group manager provides the interface between each capsule on the station and the xAMp communication system. The group managers on each station are not themselves replicas; they are replica coordination entities (cf. chapter 6). Each maintains its own local view of the identity and status (healthy, failed or quarantined, see section 9.7.8) of each station and communication group on the network, the identity and status (leader, follower, etc.) of each capsule on the stations and their replication domains.

The group manager on each station offers a management interface for processing requests for the instantiation and termination of the replica copies of each capsule on the system. The group managers ensure that each replica starts up in synchronism with the others and remains synchronized in the leader-follower role by the transmission of notification, acknowledgement and other messages.

The group manager is informed of communication group changes by the xAMp. Station failures are detected either as part of the normal message transmissions or by means of a multicast "Are You Alive?" request. The xAMp protocol for group change indications ensures that they are totally ordered at all recipients; see chapter 10.

All surviving GMs will therefore deterministically update their local opinion of the "takeover order" of each capsule. If the leader fails, the GM supporting the "first" follower will notify it to take over as new leader. (Note that the "takeover order" in XPA is maintained in the group managers. For leader-follower replicas in OSA, it is maintained in the Deltase envelopes of the capsules.)

This processing is carried out wholly by the system administration thread within each group manager, but the failed station event is also raised to an application-level system administration server, the XPA *Replication Domain Manager* (XPA-RDM).

The XPA-RDM is an XPA-Deltase object, unlike the OSA-RDM, and cloning of a new follower replica may be set in motion if this is required.

The group manager also detects most timing errors, excess desynchronism between leader and follower replicas and events caused by lack of resources, e.g., no thread or buffer space available for the onward transmission of a message. These are treated as fault events that may lead to failure unless the event handling can recover from them in time.

These events are handled as described in section 9.7.4 and passed to a local system administration object with an interface "presumed always to be present" for the purposes of event-logging. This local event logger is a prime candidate for evolution to allow presentation to a human administrator or an automated package capable of further action such as load balancing by replica migration.

9.7.3. SA Activities involving the Deltase Support Environment

The Deltase support envelope of each capsule, see figure 7, contains XPA-SA components and provides interfaces to the global and local XPA-SA server objects. Standard, or default, event handlers are provided in the Deltase envelope of each capsule, but provision is made to allow the use of event handlers within the application code which in turn may raise this or an alternative event with the client or an administration component (see section 9.7.4.4). The standard responses to timing failures and resource shortages are selected according to the real-time requirements of the capsule (see section 9.7.5).

The Deltase libraries also include the *Object Manager Entity* (OME) which, as for OSA-SA, is responsible for instantiating capsules under the control of the local factory in addition to handling capsule mode changes as described in section 9.7.8.

9.7.4. Event Management

In the context of XPA-SA, most events result from the detection of a failure to achieve either a correctness or a timeliness specification. This acts as a trigger for system administration event management action. Event management consists of any fault treatment and maintenance activities that enable the system to recover.

9.7.4.1. Event Types. The list of events that must be handled by XPA-SA is likely to change as XPA is developed. The following is a list of standard events handled by the prototype XPA-SA (note later discussion of the different implications of and treatment of events at leader and at follower):

- logical events:
  - station failure;
  - station overloaded for components of priority P or lower;
  - failure to find buffer space;
  - failure to find disc space or some other local resource;
  - failure to find a free processing thread[6];
- timing events:
  - missed deadline;
  - missed periodic activation;
  - excess desynchronization between leader and follower.

Any of these may occur on any station at any time whilst a Delta-4 system is operational. They are therefore "local events", affecting only one copy of a replicated capsule. Some initially local events may affect service provision. A failure to find a free processing thread ought to occur in all correct replicas. A deadline, if missed by a leader, will be missed by all. Where this occurs, a "replica event" is raised[7].

9.7.4.2. User Defined Events. These are defined in server interface definitions and by use of the "raise event" syntax of XPA-Deltase. An important example is conversion from another event type (e.g., deadline-elapsed) as the final act of the event-handler concerned. By this means, the event may be converted into some other event with greater semantic significance to its client (e.g., "cycle-abandoned"). Another example is some address-space-only computational condition (i.e., the classical "exception" concept). It could be useful for the programmer to abandon a computation for application-specific reasons and have nothing else to do but cause the client event-handler (if there is one) to run.

Note that all user-defined events are necessarily replica events; replica group determinism ensures that the computational circumstances that cause the event are encountered by all replicas. Replica transparency is preserved; the programmer never has to recognise or consider unwanted diversity.

9.7.4.3. Event Handling. The examples of real-time problems given in chapter 5 illustrate a tiny selection of the diverse requirements met in systems whose main characteristic is that their design must in some sense be assured to respect timeliness requirements. Generic mechanisms, such as are appropriate to incorporate into the fabric of the XPA support environment, must exhibit considerable adaptability and extensibility to be of value across a range of market requirements and to remain of value under rapid technological change. One of XPA's key mechanisms to address this requirement is that of event handling.

The main events of interest (consideration of which lead to the development of the model presented below) are those associated with timeliness — in particular, the elapse of deadlines. It is, however, noteworthy that the model appears to be applicable to other events and, in the spirit of an ODP recursively-defined architecture, allows a unified treatment of these, whether generated, or handled, at application-level or support-environment-level. Transparency is achieved by the provision of default event-handling.

---

6   Failure to find a free processing thread occurs when all threads are in use and an input tries to preempt a lower-precedence thread in the capsule.

7   A replica event is implemented like an input message: it is only passed to the follower at the same preemption point at which the corresponding event was passed to the leader, so the leader always sends an instruction for replica events.

System administration in XPA is considered to be part of the normal machinery available to fulfil application requirements. To achieve this, a mapping of events onto the computation models is desirable. The passage of control in the presence of an event must be defined; when an event is raised, an identified thread must be despatched. This may take a path through a series of objects; collectively, such objects make up XPA-SA. Some of these will have interfaces that are "presumed always to be present"; their implementation may, however, be standard or in some way system-specific. The path taken may therefore be more or less complex according to the particular system implementation choices taken.

In a symmetrical situation to that of user-defined events, user-defined handlers must be invoked only for replica events. Circumstances that are not replicated, such as a particular follower missing a deadline that the leader has met (and notified), are handled by invoking a local SA server ("presumed always to be present") that (at least) logs the occurrence. If the *leader* misses the deadline, then this will (as well as being locally logged) be notified to all followers by the normal XPA deterministic support mechanisms and will indeed result in invocation of all replica event handlers. Again, replica transparency is preserved; the programmer never has to recognise or consider unwanted diversity.

The event handling system must be sufficiently flexible to accommodate user-defined events and new requirements as they are identified, perhaps as a special case arising in a particular system implementation. It is not desirable to constrain the handling of events to a single outcome or even a small set of predefined outcomes.

9.7.4.4. A General Execution Model for Event Handling.

In an ideal world, the machinery underlying an abstraction would be totally transparent to programmers. However, an abstraction that is elegant to use is not always elegant in its implementation; abnormal as well as normal combinations of circumstances must be addressed in a manner appropriate to the system's target application.

A suitable application-level abstraction and a matching infrastructure are required. A computational model of administration events is needed that can map easily to programming languages. A programmer should be granted the ability to ignore particular (or even all) events and have them receive default handling; in turn, the default should be selectable by a project engineer from a set of standard handlers during system construction. There is a strong resemblance between the mechanism needed here and that which has proved successful in a quite different context; that of handling exceptional circumstances arising in a pure programming context. We therefore propose to introduce into Deltase-XPA a generalized exception-handler.

A possible execution model, which may be employed in implementing some of the ideas outlined above, is a distributed exception handling mechanism based upon that of the Ada programming language. This is concerned with the passage of control in response to an event that causes suspension of normal program execution. When an exception is raised in an Ada subprogram:

- If a local handler has been provided, this is executed in place of the remainder of the current subprogram. Where there is no handler, or where a further exception is raised within the handler, the exception is propagated to the point of call of the subprogram.

- If there is no handler provided by the calling subprogram, the remainder of its execution is also abandoned and the same exception is propagated (re-raised) to its point of call, the next subprogram outwards and so on until the body of the originating task is reached.

- If the body of the task (which can be thought of as the origin of the thread of control) provides no handler, then the task is terminated.

This model can be mapped onto a distributed environment. As always, a distinction is drawn between inter-object and intra-object behaviour; the former can be described without elaborating the details of the latter. Thus, within an object, the local passage of control as a result of an event depends on the language mapping.

The model may be extended in various ways. For instance, the exception propagation can generate an "audit trail". This includes a description of the path that otherwise would be lost. These parameters are important in the case of onward propagation through a series of objects that do not provide handlers for the exception concerned. Eventually the top-level administrative object is reached where, instead of a simple termination (of the relevant thread), a handler is eventually provided in the form of a human administrator as the only remaining basis for handling the exception. This human will certainly require the path parameters (and much other) information, as well as an interactive service that permits what is necessary in the way of delicate or drastic action.

Another possible enrichment of the exception model, where a (human or machine) exception handler can successfully correct the circumstances giving rise to the event, is to provide for the return of control so that execution continues. An example involving a human administrator might be circumstances that interfere with or augment the mechanism for cloning a new replica. It could be argued that all such circumstances amount to design faults in the cloning service, but this view fails to take account of the sheer complexity of possibilities that can arise and the advantage in terms of simplifying the machinery involved of including a human "in the loop", represented by an SA server. Once the human has successfully created additional disc space, or resolved local naming difficulties, or determined a suitable destination host, or installed additional required servers, or whatever, it should be possible for control to return first to the machine cloning service and, when this is complete, to any thread that was diverted from its normal activity to pass the exception.

An elapsed deadline event occurring at a capsule does not remove the need to receive any outstanding RPC responses. Although the servers concerned inherit the deadline as part of the precedence and so may also experience the elapsed deadline event, their threads of control are returned, as below, immediately or eventually, either as a raised event or as provision of service, depending on those servers' own local handlers.

The exact form of object behaviour in the presence of a "deadline-elapsed" event is decided by the programmer:

- If the programmer chooses not to provide an event handler, then a default is provided (selected on component generation) which abandons computation, marks any outstanding remote services as orphaned and then returns a "server-deadline-elapsed" event to the invoking client concerned (or "top-level-deadline-elapsed" to the XPA-RDM event-catcher, if the object is top-level).

- If the application programmer provides a handler, it is this that determines how the component behaves. The next section discusses how event handlers are mapped into the programmer's language.

9.7.4.5. Language Mapping of Event Handlers. It is desirable to make a distinction between events occurring at an object and events raised with an object by (remote or local) servers. The former invoke a handler, if provided, in a manner similar to UNIX signal handling in that the handler is run as a preemptive parameterless procedure (with the important difference that the preemption is at an XPA preemption point and is therefore deterministic across replicas). The latter involve the return of a thread of control and therefore constitute an abnormal return *from* a procedure for which a clear semantics must be established.

For any particular RPC (or LPC) and server-event, return values must take a state that is definable with respect to the interface concerned. Any return parameters expected may indeed be returned if the event is raised by server code that is able to set them; this is not mandatory but a matter for the language mapping established. If server code does not or cannot set the parameters, then the Deltase envelope, which unmarshals parameters from the message representing the returned thread of control, is able to construct agreed "null" values for the type concerned (typically all 0's). If the event is raised elsewhere than within the server, for instance by a group manager protocol detecting network partition, then no such parameters can be returned even though the message is identified with a thread of control; the Deltase envelope of the client again constructs "null" values. This permits control to return eventually to the statement immediately following the RPC (or LPC); before doing so, a handler is run.

The means of mapping this to a language depends on the facilities of that language. "EVENT" is typically an enumerated type, with the enumeration of standard events predefined and extensions as necessary for particular interfaces. An event is raised using a standard procedure exported by the Deltase envelope:

```
RaiseEvent (Cycle_Abandoned, String);
```

where "String" is a legible identifier of the source and reporting path of the event, so that the event can be understood and maybe managed by a human administrator.

The event handlers discussed here can then be achieved by use of "case" or "select" statements, or by introducing new keywords and associated syntax such as:

```
interface declaration:
(1) EVENT Cycle_Abandoned; assignment of event to handler:
(2) EVENT Cycle_Abandoned : Catch_All();
(3) EVENT Time_Out : {
              Do_Something; /* using local variables */
              RaiseEvent (Cycle_Abandoned, Local_Name);
            };
```

which are preprocessed to construct "case" or "select" statements. The latter solution is more elegant and transparent in that it allows automatic insertion of defaults to preserve the equivalent appearance of local and remote procedure calls.

This allows handlers to be context-independent or context-sensitive. A context-independent handler is a parameterless global procedure assigned to all occurrences of a server event on a particular interface by declaring the assignment at the point of import of service as in (2), or to selected occurrences of an event by declaring the assignment at the point of server invocation. A context-sensitive handler is assigned to an event by declaring the assignment and positioning its code at the point of server invocation (3).

A different syntax is used in the body of code to catch standard events such as time-out occurring at an object. At the start of service, the default handler is assigned to all standard events. At any point in the code, a new handler may be assigned to a standard event using the standard procedure AssignEvent; this handler then remains ready to catch that event on that thread until an alternative is assigned or the service completes:

```
AssignEvent (Time_Out, Catch_All);
```

At present, this is restricted to assignment to a parameterless global procedure. The more difficult issues of how to permit context-sensitive handling of events occurring at an object are the subject of ongoing study.

It is the nature of such handlers and their interaction with SA that determines the *run-time* behaviour of the component in terms of the categories of real-time. The programmer is offered an opportunity to decide whether a particular event is to be treated by object application code or by a system-administered strategy or by some combination of both. This is particularly important in the case of events such as "missed deadline", where the nature and specification of the application are what necessarily determines how this event is treated. At one extreme it is considered a failure and therefore treated according to Delta-4's model for failures, whereas at the other extreme it is merely a convenient prompt to the application of circumstances that must be addressed.

9.7.4.6. Mechanisms to Support Event Handling. The group manager raises all GM-generated events to the Deltase envelope of the affected capsule. In some cases, such as replica desynchronization, the event is raised to both leader and follower replicas. However, not all events are generated by GM — both application objects and their Deltase envelopes may also raise events.

The Deltase envelope contains a default handler for each event, which may just re-raise the event, replacing the reply to some RPC with a standard event message containing named path information.

9.7.4.6.1. Local Events. Local events are received as input messages or as error replies to LEX calls (e.g., no free disc space). If the group manager has a local event for a follower, it may create an input message accompanied by a special pseudo-notification ("receive this message and invoke local event handler at next preemption point; afterwards perform any replica action otherwise notified, to be performed at this preemption point"), or may raise a signal (e.g., in UNIX).

A local event handler may itself raise a replica event. For example, if a leader runs out of disc space, the event handler in the leader generates the "leader failed" replica event. Indeed, an event is transformed from a local to a replica event only if it occurs at the leader capsule's station. For certain local events, for example persistent desynchronization, a follower capsule raises a replica event by notifying the leader capsule.

9.7.4.6.2. Replica Events. The transformation from local to replica event is implemented by either the leader's GM or local event handler raising a high-precedence input message to the capsule, which will therefore be deterministically received by all replicas. Deltase accepts the replica event at a preemption point and decodes it. When the event is decoded, it may be found to be directed at a particular thread; Deltase then runs the appropriate event handler in the context of the affected thread. Some events, such as "leader failed", concern the whole capsule rather than a particular thread and these are handled in the Deltase envelope itself.

Some computational events, e.g., "floating point exception" and "illegal instruction" will affect all replicas and are inherently deterministic since they occur immediately after the offending instruction (e.g., in UNIX they are presented as signals that preempt the next instruction). There is therefore no need to use the XPA determinism arrangements to treat these as replica events. From the point of view of the application, they are presented and may be handled in the same way as events arriving as input messages.

9.7.5. Hard and Soft Real-Time

"Hard" is a design-time classification (see chapter 5). At run time, a timing failure of a "hard" capsule is assumed to be an indication that the external world behaviour is no longer within the operational envelope. This assumption recognises that the entire responsibility for assuring that hard real-time requirements are met depends upon the system design process; there is nothing that can be done at run-time to tolerate such design faults.

The design process must therefore consider the "worst case" scenario. The "worst case" scenario conceivable is when all stations are in a non-operational mode. This drastic (and highly improbable) condition is, however, not useable for design purposes. The operational envelope is a definition of some less extreme worst case scenario with acceptably small probability of being exceeded, say in the lifetime or mission time of the system.

For a system subject to unboundable demands from the environment, the operational envelope may be exceeded and hard deadlines may not be met. What happens to "hard" components under such circumstances is not defined by the classification but by the application programmer; a "hard" component may have a missed-deadline handler defined for such an eventuality[8]. According to the design strategy, the passage of the event through programmed or default handlers and administration components may cause any action up to and including a controlled emergency shut-down of the entire Delta-4 system.

The following separation of concerns is made: the system designer is concerned to assure that the missed-deadline handler will not be invoked within the operational envelope and therefore the application programmer can assume this to be the case.

A capsule is classified as soft real-time when real-time deadlines are regarded as desirable targets. Unlike hard capsules, timing faults do not conflict with the requirements specification and can be tolerated. Delay or loss of instances of service can be said to occur "gracefully"; that is, any such degradation of system performance is not catastrophic and can be arranged to occur in stages. As the load on an XPA system is increased, the services of lowest priority are the first to suffer timing faults and the most important and valuable services are the last.

The real-time type classifications, "hard" and "soft" actually represent examples of a continuous spectrum of design assumptions that effectively define a function representing the cost of inability to achieve computational timeliness. The following paragraphs discuss intermediate type classifications made possible by the characteristics of the leader-follower model under XPA.

Consider the detection of a missed-deadline within a single capsule replica. This is considered to be a local event but if the affected replica is the leader, it will also constitute a replica event; the service deadline has not been met and there is reason to propagate this event to a client (and perhaps ultimately the top-level object, according to the mechanism described in the exception model). If the replica that misses the deadline is not the leader but a follower, then the ability of the system to meet service deadlines will be at risk if the leader should fail.

After leader failure, the follower should normally be able to perform the low-level protocol to become leader prior to deadline. The follower's missed deadline is evidence that if the leader fails, this may no longer be possible.

A strategy that cannot be used during the design process to achieve absolute assurance of hard behaviour, but may offer a high probability that such behaviour will nevertheless be exhibited, is as follows:

The follower's missed deadline causes SA to declare that the follower has failed because of station overload rather than station failure. This is then handled by cloning or migration of the capsule to a less loaded station, so as to reduce the risk of a future actual missed deadline. Note that there are several small contributions that must be taken into account when determining the

---

8 If the "hard" deadline is also "critical" (in that if it is missed then the consequential damage is incommensurate with the benefits of the service provided by the system as a whole) then there may be no action that is reasonable to take in that eventuality or there may be no action that limits the scale or level of damage. To a first approximation, however, "criticality" is only compatible with "boundability"; only when the system is subject to boundable demands can the design process give an assurance that the catastrophe will never occur. Given such an assurance, a run-time missed-deadline handler is irrelevant.

probability that this will succeed. There may be pathogenic coincidences; the leader may fail coincidentally with the follower first missing its deadline, there may be no lighter-loaded station, there may be a failure during migration, etc. Even if such a calculation results in a probability that is close to 1 over the lifetime of the system, it is still possible for a pathogenic case to occur. This is very similar to the case discussed earlier of the possibility of all stations failing simultaneously.

9.7.6. Real-Time Events and Replica Group Determinism

Real-time events need to be handled efficiently, but without compromising replica group determinism. The implications of this are discussed below.

Section 9.3.5 lists some possible responses to a missed deadline event. The handler of a local event must not create replica divergence, or raise the event to a client, since the local event may yet be masked by the normal behaviour of other replicas of the server. The local event is therefore handled in the Deltase envelope and the handler runs in a diverted thread in a manner analogous to that of a UNIX signal handler (it cannot be given its own thread since the free thread resources in different replicas would then be different). The context of this handler is severely limited; it is, in effect, necessary to treat the execution as occurring in a separate virtual machine (although this may occupy part of the address space of the capsule). All interactions between this handler and elsewhere must be through remote service invocation unless or until drastic action is necessary, after which the capsule ceases to be a replica and terminates. Local events are normally handled by reporting them to the local SA event logging object, which may merely record the event, or may initiate some recovery or reconfiguration, or appeal to some human administrator. The local event handler may contain no preemption points of its own. Missed-deadline events are handled by the thread that missed the deadline, at a precedence higher than that thread, to ensure the event will preempt the thread. The missed deadline event can be a replica event, i.e., all replicas have missed the deadline, or a local event, i.e., only the local replica is known to have missed the deadline.

Missed-deadline events in a component that is known to be non-replicated cannot compromise replica group determinism. A Deltase handler might take advantage of this, but a handler in the application code itself should not assume any particular state of replication.

Missed-deadline events at the leader are necessarily transformed into replica events and are handled as follows. When the deadline elapses according to the local NAC clock (which is approximately synchronized with other local clocks, see section 9.6.6), the group manager local to the leader sends a replica event to all replicas. This is an input message of higher precedence than the thread that has missed the deadline, to ensure that thread will be preempted.

If a high-precedence thread and a low-precedence thread in the same component both miss a deadline at the same time, the high-precedence missed-deadline handler runs first in all replicas. As with the processing of normal input messages, the low-precedence thread cannot delay the high-precedence activity and replica group determinism is preserved.

The NAC clock is managed by a NAC clock server that maintains an ordered list of deadlines and the threads that are to be informed when these deadlines elapse. When a thread meets its deadline, it either terminates or sets a new deadline and the old deadline is removed from the list. A missed-deadline event may be produced in the NAC whilst the thread meets the deadline in the host; in this case, the event is discarded.

The handler can report the missed deadline to the local SA logger. This provides another example of the intended project engineering flexibility of the event handling model. If, in an XPA system where such mechanisms are important, the logger is constructed to further raise the local event to a special SA component[9] which is constructed to manage a load-balancing strategy, this might respond to the fact that a soft real-time follower has missed a deadline which its leader has met by migrating the follower to an underloaded node.

The missed deadline event is typical of the real-time events, signals, or interrupts that should be reported to XPA software components without compromising replica group determinism. One of the above methods of preserving replica consistency should be selected in each case. The need to preserve replica consistency can increase the latencies of event reporting or preemption handling. In [Le Lann 1989], it is argued that this normally has small real costs; so long as latencies are small compared with computation times, even a large increase in latency causes only a small increase in required processing power. Nevertheless, XPA seeks to minimize all such latencies and to ensure they have known bounds so that required processing power can be calculated.

9.7.7. XPA System Startup

This section discusses the initial startup of an XPA system. When it is switched on, an XPA system becomes fully operational after a sequence of steps, each of which depends only on its predecessors:

1) Initialization of GM on each station. This includes the initialization of message buffers for each group of priority levels to manage flow control and credit allocation (see section   9.6.4.2).

2) Creation of the GM group

3) Activation of Deltase on every station. This will lead to creation of the catalogue group or groups and RDM group, as well as local factories on each station, as with OSA. (The functions of the OSA MIB are fulfilled on XPA by the RDM database.)

4) A startup file contains instructions interpreted by a system startup component that invoke the necessary services on the RDM, and thence the local factories, to instantiate Deltase capsules, again as on OSA. The GM is also instructed to create communication facilities for each capsule.

5) Exports and imports are made via the trader, which uses the (possibly federated) catalogue (normally this is transparent to application code), again as on OSA.

6) Application-code startup activity, e.g., testing availability of and initialising I/O, setting up application-specific initial conditions within servers through invoking their services, etc. During this phase, timeliness properties cannot be assured of capsules; this is a convention that must be taken account of by the programmer.

7) After they have initialized, each component individually awaits initialization of its imported servers. A top-level component also raises its precedence.

8) Normal run-time operation. Servers await invocations of service, top-level components are autonomous (and usually cyclic). This is the phase with which the design-time assurances of timeliness are concerned.

In order to avoid individual entry to normal run-time operation causing conflict with its own timeliness and server availability requirements, the order of exit from startup is important and is controlled as follows:

Under OSA, if a client object invokes service on an imported server interface during the client's startup activity, then it will be blocked until the server object has completed its own

---

9   Special components invoked by variants of standard SA components are themselves considered to be part of SA.

startup activity and is able to receive and service requests. This characteristic is made use of in XPA to assure that all of a client's servers are able to perform in a timely manner before the client itself enters normal run-time operation. This is done by introducing, for XPA, a "dummy" server invocation, recognised by the server envelope as being part of its clients' initialization and immediately returned, rather than being treated as a normal invocation requiring the allocation of a thread, or as an interfacing error requiring the raising of an event.

Each client must, after completing its startup activity but before entering normal operation, invoke this "dummy" service on all imported servers. This is done in the envelope and therefore is transparent to application code. In the case of a client that is itself a server, this is done within the envelope immediately prior to first waiting for service invocation.

In the case of a top-level client (i.e., one that does not export its service) the initialization code is in effect long-lived. Before entering the portion of this code considered to represent normal run-time operation, a real-time top-level client must raise its precedence, e.g., to a value determined in the object's environment string or the system startup file. A standard (envelope) procedure is invoked by application code to enter normal operation that both raises precedence and the necessary "dummy" service invocations. A non real-time top-level client need not raise its precedence or issue "dummy" service invocations, so need not call this procedure.

Since all objects during initialization have non-real-time precedence, then top-level clients completing initialization and gaining their normal precedence are in a better position to meet deadlines (i.e., have less competition) than under normal global run-time conditions. The above mechanism ensures that all their servers (to whatever degree removed) have preceded them into normal run-time operation. Therefore conflict with timeliness is avoided during the period of transition.

One restriction imposed by this mechanism is to prevent the mutual import of interfaces that would otherwise permit recursive interaction between objects. This is, in principle, permitted by OSA Deltase, but some language mappings (e.g., Ada) already impose this restriction. The restriction is not considered serious in XPA; recursive interactions would make timeliness predictions extremely difficult.

The group manager activities are initiated by a command from each host. (1) includes such operations as the initialization of the pool of free message buffers in the NAC. (2) and (3) involve distributed protocols described in the *Delta-4 Implementation Guide* [Delta-4 1990]; in (2) the group managers may discover that not all the expected stations are operational.

All capsules are generated with a connection to the local trader and binder; after stage (3), the local traders can access the catalogue, so that stage (5) becomes possible. Thus far, XPA startup differs little from OSA startup in that equivalent structures are created in an equivalent way; e.g., the GM group, used in the detection of station failure, is equivalent to the SMAP group.

However, XPA has to impose restrictions on the later startup of hard real-time activities at run-time. This is because the timeliness of hard real-time components is normally assured as part of the design process; therefore, to change the configuration of these components at run-time invalidates this assurance. All hard real-time activities must therefore be instantiated during the startup sequence. Soft real-time components may be instantiated later, but the appropriate SA service is only made available to authorized humans, e.g., a "real-time administrator". For non-real-time components, the service can be less restricted.

9.7.8. Maintenance/Fault Diagnosis

Facilities are needed to enable the human operator to start up a new station, or to re-introduce a station that has failed, into the Delta-4 communication network after it has been repaired. The new configuration will now support the cloning of those user capsules whose replication domains include the new station so as to restore the required level of capsule replica redundancy to pre-fault conditions. It is necessary to ensure that this is carried out in a controlled manner for two reasons:

- The cloning of a number of different user capsules to the new station must be scheduled to minimize the additional loading of the existing stations.

- The fault may not, in fact, have been discovered and repaired whilst off-line so that a restored station, or an individual capsule, may show the fault again either immediately or after a short period of operation. The effects of this on the rest of the Delta-4 system must be minimized.

An interesting method of controlling the re-introduction of the capsules on a restored station that will be studied for inclusion in the prototype XPA-SA is described below:

A policy might be that, on a new station or perhaps a station on which maintenance action has not revealed a suspected fault, none of the objects is trusted (they are prevented from becoming leader) until a period of "quarantine" (a mode, held as a node- or object-level status) has elapsed. During this period, behaviour comparison with the trusted leader is desirable, to discover whether the fault is still exhibited. The duration of this quarantine period, and the testing that is possible, are entirely under human control. Any necessary promotion of a quarantine object to leader, arising as a result of the failure of other nodes, is explicitly detected as part of the failure event handling. This is another example of an event appropriate to present to a human administrator "server" as described in section 9.7.4.4, in this case to decide whether to lift the quarantine or take some other action such as creating or migrating an additional replica.

Methods of comparison of the outputs of a suspect capsule replica against those of a healthy replica will be studied on the XPA prototype and will require additional services within the group manager and Deltase.

9.7.9. Cloning in XPA

Cloning of Software Components in XPA makes use of the same architectural components as cloning in OSA: that is the Object Manager Entity (OME), the Replication Domain Manager (RDM) and the factories, and their relationship is that described in section §8.2.4. These components for XPA differ in detail from those in OSA, due to the different communication system to which they interface, and the use of the leader-follower model of replication. This section describes the differences in implementation of cloning between OSA and XPA.

Phase 1: Reestablish communication context.

In OSA, messages queued in the NAC for the Software Component before and during this phase, and which will not arrive at the new replica, are always received by the Software Component before phase 2 starts, due to the ordering properties of the Atomic service of the xAMp (see section 10.4). In XPA, which uses the xAMp Reliable service, this cannot be guaranteed. Therefore in XPA there is an extra phase during which messages queued at the leader replica are transferred to the new replica by the group manager, and duplicate messages at the new replica are discarded. This new phase is carried out in parallel with phases 2, 3 and 4, but must be completed before phase 5 can commence.

Phase 2: Take a snapshot of the computational context.

This phase is essentially the same as in OSA.

Phase 3: Generate the global context.

In XPA, only the leader replica sends the checkpoint data (no voting takes place on checkpoint packets) and therefore the OME does not have to ensure identity of replicated checkpoint data before starting phase 4.

Phase 4: Transfer the global context.

This phase is essentially the same as in OSA.

Phase 5: Continuation at the execution checkpoint.

In XPA, before the new replica can continue, the OME and group manager must discard instructions generated and associated messages received during phase 1, i.e., messages which form part of the snapshot taken in phase 2. The instructions to discard are those whose preemption point counter is less than the value included in the snapshot.

Chapter 10

The Atomic Multicast protocol (AMp)

The utility of reliable group, or *multicast*, communication protocols was recognised in section §6.9. The rationale behind the desired qualities of service was also discussed, i.e., which properties the service implemented by these protocols should offer, to support distributed fault-tolerance. We will have the opportunity to formalize those properties and to present the relevant protocols, in this chapter.

In building a reliable communication service to support distribution in Delta-4, we were faced with three concerns: (i) network design, in order that it would be as portable as possible; (ii) architecture design, so as to harmonize the use of standard components with the procurement of adequate levels of fault tolerance; (iii) protocol design, in order to take advantage of the network architecture, both in terms of performance and reliability.

These concerns were materialized in *AMp*, the Atomic Multicast protocol. The present chapter concentrates on the design of the overall architecture supporting the AMp service, and the description of the protocols. We begin by presenting a model for the kind of problems Delta-4 communications wish to solve, and we formally describe the required properties of AMp. However, since no service can be more reliable than the system supplying it, the dependability model of the communications architecture deserved particular attention: section §10.3 identifies the building blocks of the architecture; determines their individual behaviour and the way they interact with one another; establishes a system fault model, and finally, discusses the measures to ensure system dependability. Two aspects of relevance are the use of *self-checking* components and of a *non-replicated* LAN, and their implications on the fault model, i.e., the fact that a "nice" behaviour can be expected, and that time domain redundancy must be used, to recover from transmission errors. Sections 10.5 and 10.6 deal with the provision of an *abstract network* service by the channel[1], and of a group communications service, by a protocol built on top of the abstract network.

A note about the framework concerning AMp, in the project: protocol design has been discussed in several papers, where variants of the basic multicast protocol were presented: some, specifically for given local area networks, namely the token-ring and the token-bus networks, require hardware modifications [Guérin et al. 1985, Veríssimo et al. 1987]; the other version is a software implementable protocol, which is LAN independent [Veríssimo et al. 1989], and which is currently ported to the various LANs supported in the project (token-ring, token-bus, FDDI). Detail about these various implementations is given in §10.8.

The implementation specifications of both these versions are fairly complex, recommending the use of semi-automated methods for validation. In consequence, a fault injection campaign is in course, with the aim of forecasting faults and assisting in their removal, with the help of a

---

[1] In the remainder of the chapter, the word *network* will be used not to mean layer 3 of OSI, but rather the *networking infrastructure*, that is, the entities concerned with providing the abstract network service: the LAN Medium Access Control (MAC) sub-layer and Physical layer, and the relevant firmware.

specialized tool [Arlat et al. 1990]. The software variant was formally specified, in Estelle, and is being formally validated [Baptista et al. 1990]. Detail about this work is given in chapter 15.

10.1. Notions about Reliable Group Communication

A reliable broadcast protocol is a protocol that fulfils a set of safety and liveness and/or timeliness properties, in the form of agreement, order and synchronism paradigms. If the addressing modes supported by the protocol concern subsets of participants (multicast), rather than *all* the participants (broadcast), then it is called a *reliable group communication protocol*. Delta-4 is concerned with the latter.

10.1.1. Agreement

Distributed *agreement* in the presence of faults has been the subject of a number of publications. Useful paradigms have been identified, among which the well-known Byzantine Agreement [Lamport et al. 1982]. Informally, it seeks to make a participant disseminate a value to all the other participants, so that those who are correct, accept the same value; if the sender is correct, the value accepted has to be the one it sent, otherwise it is some default value. The problem as originally equated, is concerned with a specialized, phased-execution environment in which faulty participants can behave arbitrarily. To ensure that participants behave consistently, forms of agreement such as those found in *atomic* broadcast protocols specify, alternatively, that a message may or may not be delivered, but in affirmative case, it is delivered to all intended recipients. The word "intended" is used to underline that the specification may not include all participants, but only a *group* of them, or refer to relaxed forms of agreement, such as majority, at-least-N, etc.

In general, the conditions for achieving distributed agreement can be equated in terms of *agreement* and *validity* properties [Perry and Toueg 1986]. The strongest form of agreement is unanimity:

> **Unanimity**: Any message delivered to a recipient, is delivered to all correct recipients.

The ancillary *validity* properties specify the conditions in which agreement is, or is not, performed. A normally necessary validity condition called non-triviality, specifies that any message received is a "useful" message, i.e., not forged or spontaneously generated, not a pre-agreed message, etc.:

> **Non-triviality**: Any message delivered, was sent by a correct participant.

In systems where components exhibit fail-silent behaviour (see section 6.2), it is possible to define accurately situations where components may temporarily *refrain* from providing service, without that having to be necessarily considered a failure. We call this concept *inaccessibility* and it is detailed in section 10.3; the resulting validity property is therefore:

> **Accessibility**: Any message delivered, was delivered to a recipient correct and accessible for that message.

Given that a reliable group communication protocol is supposed to actually deliver messages, it is necessary to specify the (hopefully rare) conditions when the message may not be delivered without that constituting a failure:

> **Delivery**: Any message is delivered, unless the sender fails, or some recipient(s) is(are) inaccessible.

A practical example of the use of inaccessibility in Delta-4 is to represent buffer-full conditions at recipients: this situation — under limits — is one valid situation for the protocol not to deliver a message. The other one concerns sender failure: logically, it is rather irrelevant whether a failure occurred right before a sender sent a message, or right after it did it, let alone during the execution of the protocol. What is important is that recipients perceive whatever happens consistently — this is guaranteed if the protocol secures the unanimity property.

The unanimity property has a cost that may be unnecessary in some situations. For instance, queries to a group of replicas need only to reach one of replicas, or a quorum of them, it does not matter exactly which. Depending on the validity condition, i.e., whether the message must always reach N recipients, or whether it may not reach all of them if a failure occurs (e.g., the sender), an *at least N* or a *best effort N* agreement semantics is obtained, respectively. For completeness, the situation where $N = 0$ is the well-known *datagram* semantics.

> **At-least-N**: Any message delivered to a recipient, is delivered to at least N correct recipients.

> **Best-effort-N**: Any message delivered to a recipient, is delivered to at least N correct recipients, in absence of faults.

A slightly different but also useful semantics, is the one where the set of recipients is a *named* subset of recipients. Consider the example of passive or semi-active replication schemes, where there is a privileged participant, i.e., the primary or leader replica, or the example of a cooperating activity where there is a coordinator. For correctness reasons, it is mandatory that any message arrives at that privileged participant; for performance reasons, it may be interesting that the message also arrives at the other participants, should they need it, taking advantage of the multicasting facilities[2]. The considerations made about validity conditions are still relevant, and the following properties are then obtained:

> **At-least-To**: Given a set P_{to} of recipients, any message delivered to a recipient, is delivered to all correct recipients in P_{to}.

> **Best-effort-To**: Given a set P_{to} of recipients, any message delivered to a recipient, is delivered to all correct recipients in P_{to}, in absence of faults.

10.1.2. Order

10.1.2.1. Partial and Total Orders. In a distributed system, so that participants coordinate their actions in a decentralized way, they must perceive how the system evolves, i.e., the order in which actions and events take place. Each participant will observe system evolution, as a sequence of events, which may not be the same for all participants, due to the relativity of their positions (space-time view). In other words, the best that can be achieved is a *partial ordering* on events [Lamport and Schneider 1985].

The cause-effect relation is the natural partial ordering of events in a system. Consider two events a and b occurring at different sites of a distributed system. The event a and b can be ordered if an information departing from the site where a occurred, arrives at the site of b before b occurs. The event a is said to *precede* event b, $a \rightarrow b$, in those conditions. Given the space-time relation, it may occur that neither of them can cause the other, in which case they are *concurrent*, i.e., $(a \rightarrow b \text{ or } b \rightarrow a)$, and may be ordered in any way. For practical reasons, protocols that seek to respect the causal order, normally do so by guaranteeing a "precedes" order, i.e., that events are *potentially causally* ordered. So, a potential causal (causal, for simplicity) order in communication is defined as:

> **Causal Order**: If any two messages, delivered to any correct recipients, are not concurrent, they are delivered to each recipient in their "precedes" order. If the messages are concurrent, their order of delivery is undefined.

---

2   Although, in case of error, it can be forwarded by the privileged participant.

There are essentially two ways of implementing the "precedes" relation. If participants can be made to only exchange information by sending and receiving messages, they can only define causality relations through those messages. Messages ordered in this way are said to be in *logical* order. Lamport proposed such an implementation, using logical timestamps [Lamport 1978], and Birman later gave an implementation using message piggybacking [Birman and Joseph 1987]:

> **Logical Order:** A message m_1 is said to logically precede message m_2 if: m_1 is sent before m_2, by the same participant **or** if m_1 is accepted by the sender of m_2 before it sends m_2 **or** there is a chain of such recipients and senders linking m_1 to m_2.

For clarity, it is to be noted that the order obtained is thus a *logical (potential) causal order*.

There are however circumstances where the logical order does not correctly represent causality. This can occur when interactions have to take place in real-time (see §9.5.1 for details). The solution for these situations consists in implementing potential causality with a technique based on *temporal* order of messages, for example, by locally time-stamping them, from synchronized physical clocks [Lamport 1984].

> **Temporal Order:** A message m_1 is said to temporally precede message m_2 if: m_1 and m_2 are sent by the same or any two participants, respectively at *real times* $t1$ and $t2$, and $t2-t1 > \delta t$.

The variable δt is introduced to achieve an implementation-independent definition of the relationship defined above — which we shall call δt-*precedence*. Given that message transmission speed is not invariant in the systems dealt with, messages do not arrive naturally in order. One solution is to transport precedence (space-time) relations to the time domain, establish a discrete quantifier, δt, to account for the influence of space[3] and order them based in their time differences of module δt. Thus, δt is the granularity with which it is possible or desired to distinguish orderings between messages. This way, the *temporal (potential) causal order* definition obtained is implementation independent: δt is just the minimum real time difference for the order between two messages to be recognizable, by a given communication system. One way to enforce it is the one mentioned above: using approximately synchronized real-time clocks. The clock precision will correspond to δt. A number of methods exist to maintain local clocks approximately synchronized [Cristian 1989, Schneider 1986, Srikanth and Toueg 1987].

Clearly, not all events that are potentially causally related are actually causally related. Distinguishing this implies that the ordering discipline acquires some specific knowledge. In consequence, when serving a particular application semantics, or computational model, or replication technique, chances are that the causal order may be relaxed. These situations are detailed in section §9.5.1, since they form a practical example of how orderings can be relaxed.

The most obvious example is that of a causal order degenerating to *FIFO* (first-in-first-out) order, if senders are all concurrent, for example, if requests from different clients commute, which is a frequent situation in some distributed client-server applications:

> **FIFO Order:** If any two messages, delivered to any correct recipients, were sent by the same participant, they are delivered in the order sent. If the messages were not sent by the same sender, their order of delivery is undefined.

Given the orthogonality of properties, if messages exchanged by a group are all commutative, order may be completely relaxed, while maintaining other useful properties of the

---

3 A distance over a velocity.

protocol. An example may be a replicated state machine application where all requests commute [Schneider 1990].

Regardless of the way in which participants causally relate themselves, if a participant is actively replicated, there is the requirement (already discussed in section 6.5), that the replicas process the same messages in the same order. One way to do this is to ensure that they receive *all* of them in the *same* order. So, to start with, messages should be **totally** ordered, as opposed to partially:

> **Total order:** Any two messages delivered to any correct recipients, are delivered in the same order to those recipients.

Then, to receive all messages, the replicas also require unanimity, an agreement property. The combination of total order with unanimity, yields what is called an *atomic multicast* protocol. In other words, a message is either delivered to all recipients, or not at all. Any two delivered messages are seen in the same order by all recipients [Cristian et al. 1985].

As said before, in other forms of replication, namely semi-active, there is a privileged replica, the leader, which performs ordering operations. In this case, a simpler protocol can be used, simply providing unanimity, and no order or at most FIFO order. It is called a *reliable multicast* protocol in Delta-4.

10.1.2.2. Incomplete and Complete Orders. Given that maintaining a globally consistent view of system events by participants obliges the properties of the service to be valid system-wide [Chang and Maxemchuck 1984, Cristian et al. 1985], and given that a significant part of those events are not related to one another, it will be advantageous, from a point of view of performance and simplicity, to ensure consistency only between *related* participants instead of *all* participants in the system. Birman has studied the problem in [Birman and Joseph 1987]. The ISIS CBCAST uses a labelling method controlled by the high-level user, which allows the paths of causality inside the system to be traced. In consequence, it only orders the potentially causal relationships that are significant (for the application). This means restricting the universe of observation, to a subset of the messages exchanged by all participants, or to the messages exchanged by a subset of all participants — in short, by means of incomplete orders (as opposed to complete):

> **Incomplete Order:** An order is incomplete, if the set of messages under the ordering discipline is not the set of all messages delivered.

Being complete or incomplete is orthogonal to being total or partial. "Partial" has been used to name both complete and incomplete orderings in previous work: using clabels in the CBCAST protocol or establishing orderings in the conversation groups in [Peterson et al. 1989] are ways of implementing incomplete orderings. Notwithstanding the fact that the user may end up perceiving a single order, either from using a protocol providing a complete partial order, or several protocols supplying incomplete orders, the fundamental difference is that partial ordering results from an *a posteriori* observation on the way the system evolved. This obliges all events in the system (messages in this case) to obey the discipline defined. On the contrary, incomplete orders are obtained after *a priori* precluding relationships between separate flows of information. Groups are a way of structuring a system in order to reason in terms of incomplete orders.

10.1.3. Synchronism

From our viewpoint, a synchronous protocol is one that has bounded and known execution times. This is a mandatory property for real-time operation and when it necessary to establish temporal order. The reader may find a more detailed discussion of synchronism in chapter 5.

10.2.   Related Work

Architecturally, existing systems which provide reliable broadcast services, belong to two major groups: (i) problem-oriented, closed solutions, with dedicated hardware and software, normally designed around Byzantine agreement protocols and networks with multiple redundant message-passing links [Babaoglu and Drummond 1985, Cristian et al. 1985]; (ii) high-level network-independent solutions over standard systems (e.g. UNIX, Ethernet), which are open, but whose achievable performance and dependability are limited by network independence [Birman and Joseph 1987, Garcia-Molina et al. 1988].

Few works exist, however, which directly use standard local area networks as a low level solution [Cart et al. 1987, Chang and Maxemchuck 1984]. This is a fundamental aspect of the Delta-4 architecture. Another fundamental aspect is the preference for local area networks. LANs are a standard means of communication, and components are largely available. Additionally, they have architecture and technology attributes that can be used for improved performance and dependability.

Disadvantages of LAN based solutions are the limited scope of the low-level approach, and the scale problem. Generally, LAN dimension is somewhat limited both in distance and number of nodes[4]. Network-independent approaches scale better than LAN based ones. However, they will hardly be able to take advantage of the optimizations achievable by low-level data-link solutions. On the other hand, the data link is sufficiently low-level to allow several options for upper layers: OSI-like multipoint stacks, such as the Delta-4 MCS architecture, described in §8.1; transfer layers [Chesson 1988], architectures compacting the network and transport layers and providing optimized access to and from user space, very suitable for high-performance real-time; or collapsed application support environments, such as the Delta-4 XPA communications architecture, described in §9.6.4. Importantly, high-performance, hard real-time or highly fault-tolerant applications, may take advantage of the optimized and controlled environment yielded by a single LAN used in a closed fashion.

Further to that, our system, unlike the protocols in [Babaoglu and Drummond 1985, Cristian et al. 1985], does not use clocks; it relates more directly with other *clock-less* approaches [Birman and Joseph 1987, Cart et al. 1987, Navaratnam et al. 1988]. However, among other differences, we have studied the capability of addressing real-time applications, by using techniques to enforce known and bounded execution times. This clock-less approach trades the predictability of clock-based protocols, for faster termination in absence of errors.

In [Babaoglu and Drummond 1985], a phased execution Byzantine protocol is presented, using exact clock synchronization and multiple LAN channels. The protocol family of [Cristian et al. 1985] is diffusion-based. It relaxes the clock assumption to approximate synchronization, but requires all processes to participate in the protocol and delays termination to a worst case time Δ. Δ depends on network parameters and clock precision.

Chang describes an asynchronous atomic broadcast protocol that provides a global order; requests pass through a centralized token holder to be ordered. To tolerate failures of the token site, it is rotated; in consequence, a message is only guaranteed to be committed by all recipients, after two token rotations, introducing a significant latency, which is not bounded a priori. The work of Navaratnam is based on the approach taken by Chang.

Two works on reliable *group* communication — i.e., multicast instead of broadcast — use piggy-backing to establish incomplete causal orderings, materialized in [Birman and Joseph 1987] by piggy-backing messages and using *clabels* in the CBCAST protocol, and in [Peterson

---

4   Although emerging fibre-optic standards like FDDI, feature up to 1000 nodes, in 100-200 Km, added to an improved *bit error rate*, typically 10^{-14}.

et al. 1989] by piggy-backing references in *conversations* in the Psynch primitive. These orders are partial: only subsets of messages are bound to be ordered so they may be satisfied by several orderings. Birman also provides another primitive, the ABCAST, that enforces a total but not causal order. The AMp provides an ordering that is both total and causal. It uses properties of the underlying network, to achieve it at low cost, in comparison to the alternative methods, based on explicit sequencers [Chang and Maxemchuck 1984, Navaratnam et al. 1988], logical clocks [Cart et al. 1987, Lamport 1978], or piggybacking [Birman and Joseph 1987], involving significant context exchange to enforce logical ordering of messages.

Our approach takes advantage of the properties of broadcast LANs and, such as in [Cart et al. 1987, Chang and Maxemchuck 1984], it integrates communication layer error processing in the reliable broadcast layer. Additionally, it integrates participant management with communication. Participant management includes failure detection, which is normally performed by a centralized monitor, elected or selected in some way [Birman and Joseph 1987, Chang and Maxemchuck 1984, Navaratnam et al. 1988]. However, instead of a preexistent monitor [Birman and Joseph 1987, Chang and Maxemchuck 1984, Navaratnam et al. 1988], the monitor is only elected when needed, on a contention basis. The information needed for recovery actions is very little, making the whole process of election, investigation and recovery, reasonably fast. Failures trigger reconfiguration, which can be a complex process, if histories of past and pending transmissions are to be kept [Chang and Maxemchuck 1984]. In our protocol, any node can become monitor, but the history is reduced to the last accepted message, per sender node.

Approaches using space redundancy, like the architectures proposed in [Babaoglu and Drummond 1985] or in [Cristian et al. 1985], assume replication of the message passing layer. Our architecture uses standard LANs, and space redundancy only exists in the physical layer[5], to resist permanent medium failures.

The V-kernel group IPC [Cheriton and Zwaenepoel 1985] is worth mentioning. It differs from AMp because, while very efficient in multicasting, it does so at the cost of other attributes: faulty process behaviour is not handled, agreement semantics is at-least-one and order is not provided. The kind of attributes of reliable group communication just discussed would have to be built on top of V [Navaratnam et al. 1988].

10.3.   System Architecture

This section identifies the building blocks of the Delta-4 communication system architecture and explains the relevant fault model. It should be borne in mind that standard LANs are to be used in order to gain in openness and portability.

The characteristics of the Delta-4 communications architecture, i.e., not using a global clock for communication, and not having space redundancy (single logical LAN), raise two interesting problems, related with its real-time capabilities: how to maintain nodes interconnected in the presence of channel faults and how to achieve synchronism in protocol execution. The first is treated in this section, whereas the second is postponed until section §10.7.3.

10.3.1. Fail-Controlled or Fail-Arbitrary

System modelling and relevant assumptions have been discussed in chapter 6. Let us recapitulate some concepts, and introduce others, which will support the explanations of the

---

5   I.e., the design of the protocol does not take into account the possible replication of the physical media — if redundant *physical* media exist, they are treated as a single *logical* message-passing link.

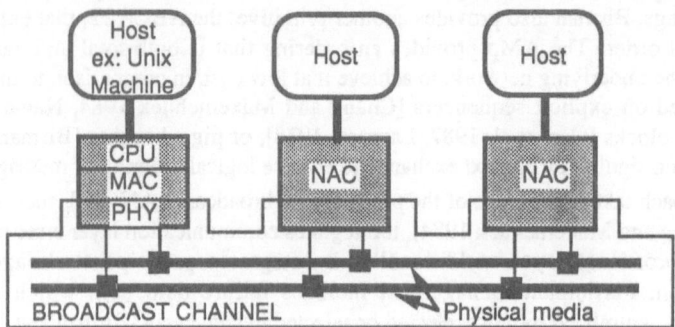

Fig. 1 - Local Computer Broadcast Network

following sections. In the system model used, components interact exclusively through input and output ports, or *service access points*, *delivering* messages to one another. A component produces system errors because it *fails* to follow its service specification. Although each component may be regarded itself as a system, at the level of abstraction of figure 1, where a Delta-4 local computer network is depicted, the system components are: a broadcast channel (possibly composed of several redundant physical media), interconnecting several network attachment controllers (NACs). These serve the computing units (Hosts). Each host-NAC set is a *node*.

The aim of this coarse granularity is trying to decouple errors in the interactions between components, from errors inside components, in order to obtain practical results concerning system design. In fact, the former (*system errors*), are visible outside component boundaries: they are the subject of the system error processing measures that will be discussed throughout the text. The latter are *component* errors: they should propagate to the outside only in accordance with the admissible faulty behaviour specified for the component. The separation between component and system errors yields a well-founded *fail-controlled* system: given the universe of all possible faults, components only display a subset of that universe; the remainder of the universe is supposed to have a negligible probability of occurrence. From now on, when just mentioning *errors*, we mean system errors.

In essence, system errors are caused by component failures. From now on, when just mentioning *failures*, we mean component failures. Component failure modes in the *value* domain are avoided, in our model, in the interest of building an efficient protocol; it has been shown that the comparative cost of coping with them in reliable broadcast protocols is high [Cristian et al. 1985], in relation to time domain errors, the most general of which are timing errors. This work is concerned with *timing* and *omission* errors.

Omission errors in communication may have many origins: mechanical defects in a cable or electromagnetic interference may garble a passing frame; a modem loosing synchrony; a receiver overrun or a transmitter under-run, etc. These component failure modes sometimes occur in bursts. We call *Omission degree*, *Od*, the number of consecutive omissions produced by a component.

Certain kinds of components may temporarily *refrain* from providing service, without that having to be necessarily considered a failure. That state, that we call *inaccessibility*, is definable, if: (i) it is made known to the client of the component; (ii) inaccessibility limits (duration, rate) are part of the component specification; (iii) violation of those limits implies permanent failure of the component. This attribute will assist the definition of timeliness properties of the service.

10.3.2. NAC Fail-Silence Assumption

The last section discussed the conditions for obtaining a well-founded *fail-controlled* system. Restrictions on the behaviour of the NACs and the Channel will be imposed so that errors do not propagate to the outside of the communication system. Assumptions like *fail-silence* are the most restrictive type (see chapter 6): a component delivers messages correctly (as specified), until it stops functioning (after its first failure). In essence, this means the system always exchanges correct messages on behalf of the hosts. In consequence, systems built to this assumption are: simple, provided the measures taken to justify the fail-silent assumption are also simple; and efficient, because they do not have to take into account processing of errors like timing or value. So, we begin with the working hypothesis that components are *fail-silent*, and that a single component may fail during a *protocol phase*[6].

However, the broadcast channel is equivalent to a single LAN. There is no space redundancy at the message level and consequently, transients in the medium[7] during transmission are unavoidable; they cause omission errors, and have the same impact as if caused by the NAC. So, the model will be weakened in order to take a certain omission degree (Od) into account. The failure-of-a-single-component assumption then means:

Failure of a single component:

- During a protocol phase, any errors result from failures in the same component: it may produce *multiple* errors ($Od = j$, j integer), or it may fail permanently, by silencing.

Forgetting for a moment the particular NAC interconnection method, i.e., blaming the NACs for any errors, including those of the network, the overall behaviour of a fail-silent NAC is defined by:

Fail-silent NAC:

- **Fs1:** A NAC may omit to deliver messages to other NACs; if k is the allowed omission degree *(Od)*, then a NAC with $Od > k$ fails permanently (remains silent).

- **Fs2:** All the messages a NAC does deliver, are delivered correctly.

Although having some similarity with the fail-stop processor approach [Schneider 1984], rather than a processor, the NAC is a communication component that provides a reliable communication service to the hosts (processors), which they may use to implement fault-tolerant computing [Powell et al. 1988]. Certain errors are allowed, but when the pre-defined behaviour criteria are violated, there is an abrupt transition to a permanent failure state, by silencing (Fs1). This decoupling yields a simple design of the NAC, using self-checking by duplication and comparison. Additionally, unlike fail-stop processors, it does not require stable storage since all NAC information is volatile. Furthermore, the failure of a NAC is detected not through a local readable failed predicate (which would require the NACs to be fail-operational) but indirectly, through a distributed monitor function.

Now, we proceed by correctly identifying in the model the causes of temporary errors. Remembering that there is a network underlying the NACs, those errors will arise both in the NAC and in the medium. We are going to rework the model to represent these errors in a uniform way, given the broadcast nature of the channel.

---

[6] A **phase** is a well-delimited portion of a protocol execution, which is a containment domain for error detection. A protocol may have several phases.

[7] The medium is the passive part (cabling) of the channel.

Observing the internal structure of a real NAC (in the left of figure 1): the operative part — processor, memory, etc. — dubbed CPU, runs the protocols software; the MAC contains the LAN specific medium access protocols — normally cast in communications VLSI — being a frame-level part[8]; the PHY part — containing interface components, like modems, codecs, amplifiers, etc. — is the bit-level, electrical signalling part. PHY is the omission error prone part, so system structure is modified, by moving component boundaries, to include PHY in the channel, as shown in the figure. In consequence, the portion of the physical NAC that must exhibit fail-silence includes the CPU and MAC parts: it executes the communications software. The remainder of the parts, encapsulated by the thin line in the figure, compose the channel. Clearly, if "channel" is substituted for "NAC" in Fs1 and Fs2, we have the behaviour required of the channel.

10.3.3. Channel

The broadcast channel is itself formed by several sub-components recalled here: (active) receiving and transmitting parts (the physical layer (PHY) entities of the NACs) and the (passive) medium. Interaction-wise, the channel offers a pair of ports (input and output) at each node.

The remarks made in the beginning of §10.3.2 should now be apparent: the duplication shown in figure 1, is concerned with providing high availability of the channel. Please remark that the dual cables shown are only symbolic. They may be switch-over buses, or reconfiguring dual rings, depending on the LAN being considered. MAC receives from a single medium at a time (from the part of the channel that is currently selected for reception), so if a transmission error occurs, the frame has to be retransmitted, because it is lost. Then, taking our failure assumptions into account, the faulty behaviour of a channel during a limited number of frame deliveries — a *protocol phase* — will be the following: omission errors only; up to k consecutive errors, in the same outputs, for deliveries coming from any input (receiver or medium failures); up to k successive errors, in any output, for deliveries coming from a single input[9] (transmitter failures). The issue is further discussed in section   10.5, where the behaviour of the channel with competing transmissions from several AMp groups is characterized.

The major consequence of this observation of channel behaviour is property Pn3, in table 3 (Pn3 is discussed in the next section), allowing to establish the foundation of our *bounded omission degree technique:*

- Pn3, in short: if $(k+1)$ series of N transmissions are made, then in at least one series, all N transmissions are indicated in all destinations;

- the protocol is resilient to temporary omission failures, provided that, during each protocol phase, they are produced by one single component, with a bounded omission degree of value k ($Od \leq k$) and k is known during the life-time of the system.

In certain kinds of networks, the channel may become temporarily *inaccessible* (e.g., upon token loss recovery or medium reconfiguration). Although with a very low rate of occurrence, duration may be large, compared to normal frame delivery delay. We specify:

---

8   The protocol offers a message-level delivery service to the user. A whole protocol execution is composed of several network-level *frame* deliveries.

9   Successive does not mean consecutive: consecutive input failures may be interleaved with other transmissions.

Channel inaccessibility limits:

- A channel, during a whole *protocol execution*, has at most one inaccessibility period, whose duration is bounded to a known limit T_{ina}, for all envisaged cases.

In conclusion, this section stated the behaviour required of the different components of the local computer network, in order to meet assumption Fs. Next, it must be discussed how to enforce it: the upper part of the NAC will not be addressed here (see chapter 11); channel implementation issues are discussed in the following sections.

10.3.4. The Extended AMp or xAMp

The *Atomic Multicast protocol* (AMp) is the basic communication primitive of Delta-4, on which the various fault tolerance mechanisms rely. The AMp service offers some of the properties identified earlier as being useful in this context. The strongest quality of service (QOS) possible in this architecture, *atomic multicast*, whose properties are highlighted by arrows in table 1 below, formed the basis of the design of AMp. This was the only QOS available during the early phases of the project. The need, discussed earlier in this chapter, for a range of QOSs to obtain the best match between performance and functionality, led to the multi-service *xAMp*. This is an extension of the original AMp, where the atomic QOS is retained, and the additional QOSs derived from the main core of the protocol. Discussions about architecture, dependability and performance are thus made around this core implementing atomic multicast. We will be using AMp and xAMp interchangeably, except where specifically noted. The complete set of properties offered by xAMp is enumerated in table 1.

10.3.5. xAMp Execution Model

A generic communication system should support several applications, possibly disjoint; the entities running them are *fault-tolerant participant groups*. Those groups should operate with parallelism and independence. For example, several independent fault-tolerant groups in the same system, or several groups working in parallel for the same fault-tolerant application, such as groups of replicated clients, accessing the same group of replicated servers. That support should additionally allow for dynamic evolution of groups, if possible in a location independent manner.

This may be established by isolating sets of participants. Let us say that participants group themselves in *universes*. Membership of participants in the different universes may overlap, nest, or be completely disjoint. The word universe was chosen to signal that what happens inside a set of participants is *a priori* independent from the rest of the system, i.e., in other disjoint universes. Each universe is identified by a designation e: U_e.

Let us define how participant interactions are supported, and which properties they should observe. Participants interact through the messages they send one another; they are disseminated, universe wide, rather than system wide (i.e., *multicasted*). The attachment of a participant to a universe is thus materialized by an entity used to send and receive messages — a communication *gate*, with the universe designation — and membership, at a given time, is given by the group of gates of that designation that exist in the system. There is a one to one mapping between a node (S_u), a gate $(G_{e,u})$ and a participant $(P_{i,u})$[10], i.e., there is a single point of access to a universe, in each node, owned by a single participant. A protocol (the AMp) controls all the messages exchanged between participants, ensuring that the sets of received messages — the receive queues at each gate, $M_{e,u}$ — observe certain order,

---

[10] It is recalled that the objective is a low-level primitive. Multiplexing of a single gate by several high-level entities is likely to be performed by a participant that is an upper layer protocol, a service element, etc.; applications or processes access the group through that participant.

Table 1 - AMp Properties

("⇨" designates properties of the *atomic* QOS)

- **Addressing**

  - **Pa1** — *Selective addressing*: The recipients of any message are identified by a pair (g, sl), where g is a group identification and sl is a selective address (a list of physical addresses).

⇨   - **Pa2** — *Logical addressing*: For each group g there is a mapping between g and an address Ag, such that Ag allows all correct members of g to be addressed without the knowledge by the sender of their number or physical identification.

- **Agreement**

⇨   - **Pa3** — *Unanimity*: Any message delivered to a participant, is delivered to all correct participants.

  - **Pa4** — *At-least-N*: Any message delivered to a participant, is delivered to at least N participants.

  - **Pa4.1** — *At-least-to*: Given a subset Pto of the participants, any message delivered to a participant, is delivered to all correct participants in Pto.

  - **Pa5** — *Best-effort-N*: Any message delivered to a participant, in absence of sender failure, is delivered to at least N participants.

  - **Pa5.1** — *Best-effort-to*: Given a subset Pto of the participants, in absence of sender failure, any message delivered to a participant, is delivered to all correct participants in Pto.

- **Validity**

⇨   - **Pa6** — *Non-triviality*: Any message delivered, was sent by a correct participant.

⇨   - **Pa7** — *Accessibility*: Any message delivered, was delivered to a participant correct and accessible for that message.

⇨   - **Pa8** — *Delivery*: Any message is delivered, unless the sender fails, or some participant(s) is(are) inaccessible.

- **Order**

⇨   - **Pa9** — *Total order*: Any two messages delivered to any correct recipients, are delivered in the same order to those recipients.

  - **Pa10** — *Causal order*: If any two messages, delivered to any correct recipients, have the same *clabel*, they are delivered by their logical order of precedence.

  - **Pa11** — *FIFO order*: If any two messages from the same participant, with the same *clabel*, are delivered to a correct participant, they are delivered in the order that they were sent.

- **Synchronism**

⇨   - **Pa12** — The time (T_e) between any AMp service invocation by a correct participant and the subsequent indication at any correct and accessible recipient is known and bounded.

- **Consistent Group View**

⇨   - **Pa13** — Given any receive ordering observed by the participants of a group, each change to group membership is indicated, in a total order, to all correct group participants.

agreement and synchronism properties. Once this is ensured, the consistency rules for each universe or group of participants can be defined. Location transparency is achieved by mapping the universe designation onto a logical address, which is used as a location independent message destination address. This feature is obtained efficiently by using hardware level LAN multicast addresses.

A *correct* participant in a group is a pair $(P_{i,u}, G_{e,u})$, operating according to its specification. So, not only must the participant itself be operating, i.e., timely servicing its input queue $M_{e,u}$, but also the gate must exist. Gate closure implies participant failure, and vice-versa.

The execution view of the system is the one shown in figure 2, where the concepts presented in §6.9.7 appear and one can see how the relevant services map onto the various components. Bottom-up: the *network* service, provided by the channel and all MAC parts, implements the interconnection between protocol entities. In fact, all happens as if there were several virtual broadcast channels (one per group) over a single physical broadcast channel, as shown in the figure. Each group communicates on a virtual channel, through instantiations of the *xAMp* in every node where a gate of the group exists. The xAMp runs on the CPU part, and provides a service to *participants* $P_{i,u}$, residing at nodes S_u. These participants may use more than one gate (e.g., $P_{1,2}$ is in groups e and f).

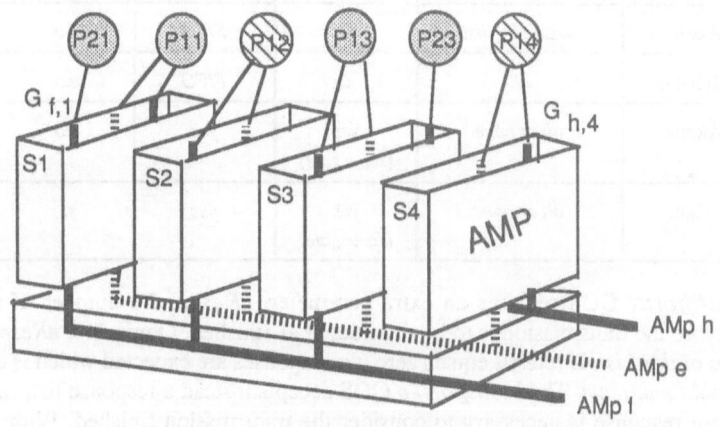

Fig. 2 - Execution View of the System

10.4.   Summary of xAMp Services

The properties of the various xAMp qualities of service are summarised in table 2. These qualities of service are available through a multi-fold *amp.request* primitive. The relevant primitives have the following format:

```
res = amp.request ( QOS, groupId, destList, frame,
                    prio, {QOS dependent params} );
amp.indication ( groupId, frame, priority, nSlots );
getSlots.request ( groupId, nSlots );
```

As far as group management is concerned, in addition to the *gateOpen* and *gateClose* primitives, offered by the basic AMp protocol, two new primitives are included: *AttachGate* and *DetachGate*, that allow non-group members to communicate with a group.

In the *amp.request* primitive, *QOS* stands for quality of service, which can be one of the following: *bestEffortN*, *bestEffortTo*, *atLeastN*, *atLeastTo*, *reliable*, *atomic*, and *tight*. They derive from a common AMp core, and each of them offers an incremental quality of service at the price of an incremental loss of performance. All primitives share the same three parameters: *groupId* (the gate identifier), *destList* (the selective list of the destination stations), *frame* (the frame to be sent), and *priority*. All the primitives are then *selective* since only a subset of all the group members needs to be addressed.

Table 2 - Summary of xAMp Service Properties

Quality of service	Agreement	Total order	Causal order per clabel	Rx queue re-ordering
bestEffortN	*best effort to N*	*no*	FIFO	*no*
bestEffortTo	*best effort to list*	*no*	FIFO	*no*
atLeastN	*assured to N*	*no*	FIFO	*no*
atLeastTo	*assured to list*	*no*	FIFO	*no*
Reliable	*all*	*no*	FIFO	*no*
Atomic	*all or none*	*yes* *(same gate)*	*yes*	*no*
Tight	*all or none*	*yes* *(same gate)*	*yes*	*yes*

The *bestEffortN* QOS requires an extra parameter, *nResps*, the number of responses necessary before the transmission is to be considered as finished. Obviously, *nResps* must be less than size of *destList*. If *nResps* equals zero, no responses are expected which is equivalent to an *unreliable multicast*. The *bestEffortTo* QOS accepts instead a response list, *respList*, of stations whose response is necessary to consider the transmission finished. With these two qualities of service, agreement is relaxed. The frame is retransmitted in order to obtain a positive acknowledge from at least *n/ respList* of the addressed stations. Accessibility constraints are not tested and no assurance of delivery is provided when the sender fails. The "at-least" qualities of service enforce stronger agreement, assuring that a given sub-set or number of the participants will receive the frame, even if the sender fails. As with best effort QOSs, both *atLeastN* and *atLeastTo* primitives are available. The *reliable* quality of service is a particular case of *atLeastTo*, where delivery is assured to all the addressed participants.

The frames sent through *bestEffortN*, *bestEffortTo*, *atLeastN*, *atLeastTo* or *reliable* QOS are delivered to the user as soon as they are received. No effort is made to assure that the frame is totally ordered in relation with other frames sent through this or through any other QOS. Also, no effort is made to avoid the violations of potential relations of causal order. Only FIFO order is guaranteed, between frames with the same *clabel*.

The *atomic* QOS is the basic AMp quality of service, with "slotted" messages and providing incomplete orderings through the use of *clabels*. As in the basic AMp, frames are, upon reception, always inserted at the end of the receive queue. The *tight* QOS provides the same service but allows the negotiation of the final position of the frame in the receive queue. The frame can be inserted between two pending frames or even between the remaining slots of a frame being consumed. In addition, the *tight* QOS offers a reduced *inconsistency* time.

With these last two QOSs, accessibility is tested in all destinations. The frame is only delivered if all destinations are accessible to received it. If a frame is delivered to at least one participant, it is delivered to the participants in all the addressed stations. Frames sent through *atomic* or *tight* QOS are totally ordered in relation to other frames sent in the same group. Potential relations of causal order are preserved at the communication level between all frames carrying the same clabel.

Frames sent through *atomic* or *tight* QOS can be associated with several "slots". When $nSlots > 1$ a descriptor of the frame is kept at the head of the receive queue. No other frame with ordering constraints is delivered until $nSlots - 1$ slots are removed using the *getSlots.request* primitive. Since high priority frames can negotiate their place within the remaining slots, a new frame can be received as the result of getSlots operation.

10.5.   The Abstract Network

In the design of AMp, it was tried to take advantage of LANs, but the network interface, although LAN oriented, has a general set of properties, being in essence, LAN independent. That independence was limited to guaranteeing those properties to be fulfilled by a set of existing LANs, namely: 8802-4 token-bus [ISO 8802-4], 8802-5 token-ring [ISO 8802-5] and FDDI [ISO 9314]. In the criteria of choice, relevant factors were: standardization, existence as industrial products, with availability of (possibly second-sourced) VLSI, possibility of implementing media redundancy, inherent real-time and reliability attributes.

In consequence, the *abstract network* service interface formulates, in a way usable by the AMp, a set of helpful properties (table 3) that are typical of LANs, along with the reliability attributes of the channel, discussed in the last section. At each node, there is a pair of service access points to the network, one for transmitting frames (source), the other for receiving them (destination). These access points will be used by the xAMp machines. The properties are defined in terms of a network delivering each frame, transmitted at a source, to all destinations. Pn1, Pn2, Pn3, Pn6 satisfy the channel model developed in §10.3. Pn4 and Pn5 are the foundation of the ordering attributes of the protocol. Using LAN terminology, the abstract network implementation will comprise functions of the PHY and MAC layer communication and management entities, complemented with the necessary hardware and/or software.

Pn1 and Pn2 impose detection of value domain errors, in a broadcast. This derives directly from the CRC protection mechanism used in LANs, and has its coverage. Pn4 and Pn5 can also be provided by LANs.

Behaviour in the time domain is defined by Pn3 and Pn6. Let us define a protocol phase to be composed of up to t series of up to N broadcasts from N different stations. In consequence, Pn3, the *bounded omission degree* property, guarantees that at least one fault-less series of N broadcasts is obtained in any protocol phase, provided that $t \geq k+1$, where k is the allowed omission degree. This holds even if omission errors are not consecutive in the network[11]; the property also holds for any virtual channel used by the AMp groups, replacing N by the group dimension. It is easy to verify that a single group of the maximum dimension, N, yields the worst case scenario; smaller groups in competition for the channel have a more favourable situation, because they "share" channel errors.

---

[11] In that case, they must, by assumption, originate in the same component. Reliability of individual components has thus to be such that the single failure assumption holds during the worst case duration of a phase: $(k+1)N$ transmissions.

Table 3 - Network Properties

> - **Pn1** — *Broadcast*: Destinations receiving an uncorrupted frame transmission, receive the same frame.
>
> - **Pn2** — *Error detection*: Destinations detect any corruption by the network in a locally received frame.
>
> - **Pn3** — *Bounded omission degree*: In a network with N nodes, in a known interval, corresponding to $(k+1)$ series of unordered transmissions, such that each of the N access points transmits one frame per series, all transmissions are indicated in all destination access points, in at least one series.
>
> - **Pn4** — *Full duplex*: Indication, at a destination access point, of frame reception, during transmission by the local source access point, may be provided, on request.
>
> - **Pn5** — *Network order*: Any two frames indicated in two different destination access points, are indicated in the same order.
>
> - **Pn6** — *Bounded transmission delay*: Every frame queued at a source access point, is transmitted by the network within a bounded delay $T_{ina}+T_{td}$.

Acceptable coverage of the bounded omission degree assumption can be enforced through redundancy in the physical and "medium" layers. An example of an implementation for a dual cable token-bus LAN, allowing real-time switch-over between media, is described in detail in [Veríssimo 1988]. On the other hand, when the channel deviates from the behaviour postulated, it fails permanently. A distributed failure detection and fault treatment mechanism for a redundant LAN channel, which acts upon each node's opinion of the channel state, is also given in [Veríssimo 1988].

Pn6 depends on the particular network, its sizing, parameterizing and loading conditions, which must be known, in order to calculate T_{td}. The value of T_{ina} depends on the network alone and can be predicted for a set of known local area networks. According to the definition of inaccessibility, exceeding T_{ina} implies permanent failure.

To end with, it must be underlined that, in a sense, the *abstract network* extends the concept of LLC — Logical Link Control sub-layer — the LAN independent sub-layer of the IEEE, and later ISO, 802 standard. Historically, the LLC was intended to hide the differences between LANs, multiplex the broadcast channel access, and provide some additional reliability to the MAC datagram service. The abstract network concept is innovative in that, besides offering a message (service data unit) delivery service, it makes visible a set of functional properties that are common to LANs in general. It is believed that the principle can be used to optimize design of protocols intended to work on LANs.

10.6.   Two-Phase Accept Protocol

Without restriction on the kind of faults allowed, the cost in traffic and time to achieve agreement is rather high. In consequence, the Delta-4 approach relies on fail-silent property of the attachment controllers, to centralize protocol execution. The protocol, forming the core of xAMp, that implements the atomic multicast quality of service, is a *two-phase accept* protocol. Its operation resembles that of a commit protocol, in that the *sender* coordinates the protocol: it sends a message, implicitly *querying* about the possibility of its acceptance, to which recipients reply (dissemination phase). In the second phase (decision phase), the sender checks whether *responses* are all affirmative, in which case it issues an *accept* — or *reject*, if otherwise. In the event of sender failure, protocol execution is carried on by a termination protocol. However, in this case, although Pa3 is always respected, delivery is no longer ensured (see Pa8) for that execution. This core protocol is formally described in annexe J; together with the formal presentation, the outline of a correctness demonstration is also given.

Two protocol variants have been derived from that description, differing in whether mutual exclusion is ensured between executions, or not:

- The *token-based* protocol assumes that there is *at most one message transmission in course* in a group, at any time, in the whole network. For an efficient implementation, such a token should be managed by hardware, otherwise execution times increase and the forced serialization becomes a performance problem. The token may be the LAN token itself, in token-based LANs. Such a token implementation is discussed in §10.8.1. In fact, these hardware based variants collapse the AMp and the abstract network in a single modified MAC layer, since they are based on changing the standard 802.x machinery.

- Avoiding hardware modifications, the *token-less* variant is discussed in §10.8.2, allowing *several concurrent* message transmissions, for different groups, and *several competitive* message transmissions, for the same group. That variant relies on an abstract network implementation as discussed in the last section.

An informal explanation of the protocol implementing the atomic multicast quality of service is now given. Since some functionality is variant dependent, we base ourselves on the token-less variant in what follows, in order to cover all the relevant details.

10.6.1. Protocol Structure

Each *gate*, the entity used by a participant to communicate, uses an instantiation of the AMp machinery. This comprises a local *GroupMonitor* agent, which participates in error recovery and fault treatment procedures, and two context structures, the *GroupView* and the *ReceiveQueue*, containing, respectively, the group composition and the frames received for that group. A station may belong to any number of gate groups and their number in the LAN is only bounded by implementation limits. Users (mapped on gates) join and leave a group at any station, through local gate opening or closing operations. Due to the nature of the communication architecture, joins and leaves are not truly independent of communication, such as found in [Cristian et al. 1986]. Instead, those actions are performed in a privileged state, which temporarily obliges all participants to synchronize. This ensures that the group views in all gate group members are updated consistently, in relation to the ordered flow of information (§10.6.5). However, the operation is perceived as being dynamic in most situations. The group join/leave protocols in [Birman and Joseph 1987] work similarly.

Error detection is done on a transfer-by-transfer basis, and relies on consistency of the group view by each member. The minimum information needed is a *concise GroupView*, which contains only the number of group members. The concise view is used to detect station failures or undelivered frames (omission errors). For instance, if an emitter requests acknowledgment to a frame, it can compare the number of responses received with its view to detect the presence of an error. However, to allow fast identification of failed stations, the permanent maintenance of a complete list of member identifiers would be desirable: a so-called *extended GroupView* was implemented[12].

---

12 Given that the approach is very expensive in terms of storage, a compression technique was used.

10.6.2. Assumptions

We proceed by describing some assumptions that support protocol operation, followed by the description of operation itself. The set of assumptions that guarantee correctness of operation of the token-less protocol implementation is presented below.

Assumptions:

    A1   There are at most k message transmission pending from each node, at any time.

    A2   There may be several concurrent transmissions in course in the network.

    A3   There may be several competitive transmissions in course, in the same group.

    A4   From each node, at any time, there may be only one transmission in course associated with a given *clabel*.

    A5   The sender positively confirms that all correct participants receive a decision, if it is *reject*.

    A6   A transmission, once started, executes atomically, i.e., it is not preempted by other emitting actions, for example, from the GroupMonitor.

Assumption A2 is the source of external parallelism in AMp: unlike other approaches providing global order [Chang and Maxemchuck 1984, Cristian et al. 1985], AMp enforces *incomplete* orderings and in consequence several concurrent executions run simultaneously. On the other hand, assumption A3 allows internal parallelism in a simple way. Group members just run transmissions competitively, in a fully decentralized fashion. There is no coordinator to achieve order, unlike the work in [Chang and Maxemchuck 1984]; several transmissions may be initiated simultaneously.

Order and agreement are achieved by the protocol, based on the network properties, and the error detection and recovery mechanisms provided, which rely on Assumption A1. In fact, this assumption reflects a subtle restriction to parallelism, which maintains protocol simplicity, namely, in obtaining order properties. Together with Assumption A5, it also allows safe use of an error recovery algorithm detailed later in the text, which uses no context about previous transmissions. In consequence, maintenance of histories or lists of significant dimension often found in other approaches is avoided; this greatly simplifies recovery and makes the monitor operate very efficiently.

10.6.3. Atomic Multicast Transmission

A multicast transmission is performed by a protocol entity called the *Emitter Machine*. Assumption A1 limits the number of simultaneous transmission from each node, so there is a limited number of Emitter Machines available at each node. Emitter Machines are locally identified by an integer, s, in the range *1* to k. Thus, every Emitter Machine in the network can be identified by a pair *(n,s)*, where n is the node id. In the following text, an Emitter Machine will be simply designated as *"the sender"*.

An atomic multicast transmission is initiated by the protocol coordinator, the sender (E), by sending a multicast frame containing the message. The *Dissemination* phase (figure 3) then proceeds as follows:

    • After transmission, E will expect a number of responses indicated by its group view, within a predefined response time (TwaitResponse). When all responses arrive or TwaitResponse has elapsed, they are analysed and if some recipient cannot accept the frame, decision = *reject*.

- Normally, responses are of "can accept" type, meaning recipients are accessible; then, if all recipients responded, according to the sender GroupView, decision = *accept*. If there are responses missing, the data frame is retransmitted.

- If some station does not answer within the retry mechanism, it is considered failed. However, the execution proceeds, allowing timely termination: an *accept* decision may be sent if all the remaining stations can accept the message. Stations considered as having failed are removed from the group view, by the Group Monitor.

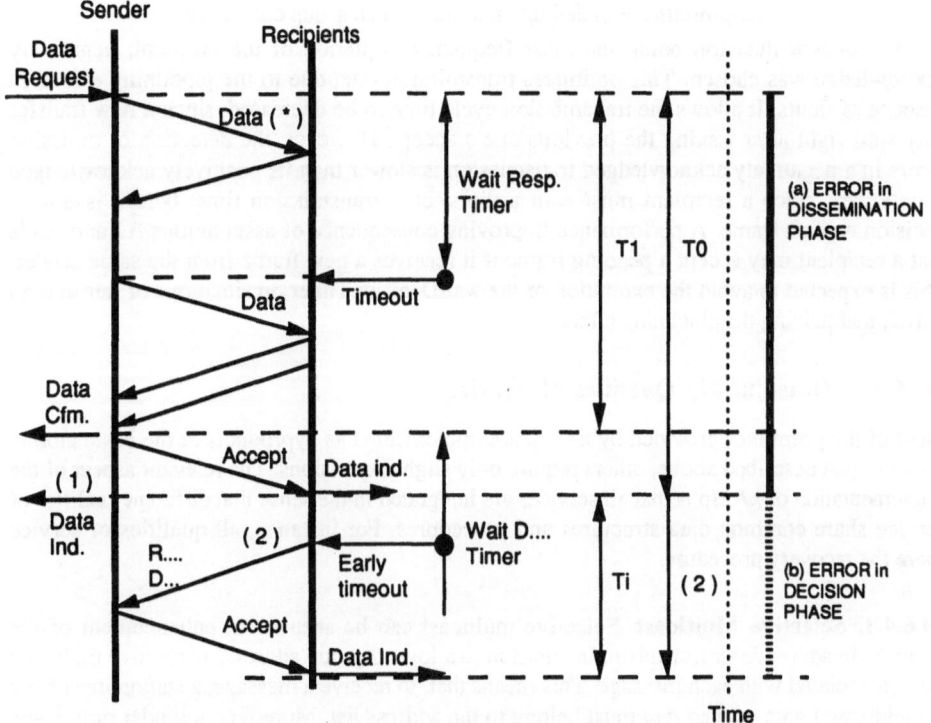

Fig. 3 - Protocol Timing

The *decision* phase is implemented in the following way:

- The *reject* frames always require response (assumption A5). A station that does not answer within the retry limit, is considered failed.

- The *accept* frame, on the other hand, does not require response, in the interest of improved performance. A time-out mechanism, at the recipients, covers omission errors in the transmission of a decision: after receiving an information frame and responding, a timer is started with a predefined TwaitDecision time. If no decision is received within this time, a recipient requests the decision from the sender (figure 3). Safety of this method is based on a simple algorithm, relying on assumptions A1 and A5:

  - all participants log the sequence number of the last message delivered (*lastDeliv* for sender) or accepted (*lastAccept* for recipients);

- omitted decisions can then be recovered very simply: a recipient requests a missing decision, and the sender may respond *accept* if *lastDeliv* has a higher sequence number; else, it is not yet finished with processing it.

- Note that in case of reject, a sender only starts a new transmission after ensuring that all the group members received the reject (assumption A5). So, when a sender receives such a decision request it can answer with an accept without any knowledge of the past, or proceed, if it was still processing that frame. The recipients will retransmit the decision request, until the retry limit is exceeded. When that happens, the sender is considered failed and the GroupMonitor is called upon, to reestablish group coherence.

The *accept* decision being the most frequent completion of the protocol, negatively acknowledge was chosen. This optimizes transmission rate, due to the pipelining effect, in absence of faults. It allows the transmission cycle time to be decreased, since a new transfer may start right after issuing the previous one's accept. However, the detection of omission errors in a negatively acknowledged transmission is slower than its positively acknowledged counterpart, since a recipient must wait a worst case transmission time, before issuing a decision request frame. A performance improving consequence of assumptions A1 and A5 is that a recipient may accept a pending frame if it receives a new frame from the same sender. This is expected to avoid the expiration of the waitDecisionTimer, in situations of fair to high traffic, maintaining the pipelining effect.

10.6.4. Other xAMp Qualities of Service

Most of the primitives provided by xAMp are implemented as byproducts of the basic atomic protocol just described above, others require only slight expansions. The relevant aspect of the implementation of xAMp is that all services are integrated in the sense that different qualities of service share common data structures and procedures. For instance, all qualities of service share the recovery procedure.

10.6.4.1. Selective Multicast. Selective multicast can be seen as an enhancement of the basic AMp addressing mechanism. In addition to a logical group address, a selective multicast list is associated with each message. This means that, to receive a message, a station must have the addressed gate opened *and* must belong to the address list. Moreover, a sender only waits for responses from stations in the selective address list. This excludes non addressed stations from the multicast transfer. The *abstract network* recognises selective address lists in order to perform the address recognition task in the lower layers of the architecture and, whenever possible, by the underlying hardware.

10.6.4.2. Exploiting AMp Message Delivery Procedures. The frame delivery procedure of the basic AMp, TxwResp, is sketched out in annexe J. Its properties, that were illustrated in Lemma *1, can* be stated informally in the following way: when a sender executes TxwResp, frame delivery is assured for all the intended recipients if and only if the sender remains correct during the execution of the primitive.

This procedure is exploited in xAMp to obtain the *bestEffortN*, *bestEffortTo*, *atLeastN*, *atLeastTo*, and *reliable* qualities of service. The *bestEffortN* and *bestEffortTo* QOSs are direct sub-products of the *TxwResp* procedure. The "at-least" qualities of service require extra work since delivery must be assured even when the sender fails.

In the basic AMp primitive a frame is only accepted after being received by all group members. If the sender fails before unanimity is reached the frame is rejected. This means that, in case of failure, the recovery procedures do not need to re-transmit the frame, only to disseminate a *Reject* decision. Now we want to develop a primitive where a frame is accepted as soon as it is received, before the unanimity had been assured, avoiding the use of the two-phase accept protocol. However, since the sender may fail before termination of the TxwResp procedure, only those correct stations who did receive the message can assure protocol termination. So, to offer this quality of service, every receiver must be ready to act in the role of the sender, retransmitting the frame until the protocol terminates. The *"at-least"* qualities of service are then implemented as follows:

The sender transmits the frame and executes the TxwResp procedure. When a recipient receives a message, an Emitter Machine is activated at that node, also executing TxwResp, to control the termination of the protocol. Emitter Machines, when activated at the recipient side, omit the first step of the TxwResp procedure and start immediately collecting acknowledgments, avoiding unnecessary retransmission of the data message.

10.6.4.3. Exploiting Two-Phase Accept. The two-phase accept protocol can be enhanced to allow the sender, which is acting as a coordinator, to obtain the state of the recipients receive queues. If the access points to the receive queues are locked until the frame is accepted, the sender may choose an appropriate insertion point for high priority messages, such that the total delivery order is not violated. The Accept would then disseminate the position of the high priority message in the queue (possible within the "slots" of a message being consumed). This mechanism is the basis for the negotiation protocol available in the *tight* quality of service.

10.6.4.4. Using Responses to Exchange Group Views. The algorithms described until now rely on the knowledge of the group membership to assure protocol properties. This conflicts with an xAMp goal: the ability the allow non-group members to send messages to a given group. For xAMp a solution is envisaged which consists of inserting the group view in the response frames, using a similar mechanism to the one currently used in the AMp *GetView* procedure. This will allow the sender to obtain the group view *during* message transfer, without significant loss in performance.

10.6.5. Distributed Group Monitor Function

The *Group Monitor function* executes, under a privileged state, critical activities relevant to correct operation of the protocol. Namely, it maintains consistency of the GroupView, recovering from station failures. Additionally, it runs the termination protocol in case of sender failure. It also controls group membership: joins and leaves from an MGateGroup require activation of the GroupMonitor so that all GroupViews change consistently.

The distributed Group Monitor function relies on information provided by the various local GroupMonitors of a group. It may be invoked by several groups simultaneously, executing with total independence from the monitors of other groups. The local GroupMonitors are normally inactive. So to speak, a GroupMonitor only exists when needed. At that time, an election is performed, if there is more than one contender, information is gathered from the local entities if necessary and a decision is made and disseminated to group members. If an Active Monitor fails, it is replaced by another GM who detects the failure. The procedure is recursive.

10.6.6. Active Monitor Election

The need for group monitor intervention can be detected simultaneously at two or more group members so it is to be expected that several GroupMonitors (GM) will compete for the activity.

To ensure that only one monitor becomes active, the first frame sent by a candidate carries the Suspension attribute, which suspends multicast traffic on the MGateGroup. Other candidate GMs will find the traffic suspended and return to the Standby State.

However, this simple solution must be improved to avoid deadlock and contention. Deadlock occurs when an Active Monitor fails, leaving the traffic suspended. Contention can occur in the presence of an omission fault during the competition for the activity if two different monitors lock different subsets of the recipients.

These two problems are solved with a mechanism based on the association of a value, the suspension level, to the suspension attribute. When a receiver is suspended it stores the suspension level associated with the suspending frame; this value is the Current Suspension level. If another frame, with the Suspension attribute, is received during the suspended state, this frame is rejected unless it carries a suspension level higher than the current one. In this case, its sender will become the new Active Monitor, preempting the old one (and the current suspension level is updated). Deadlock and contention avoidance with the suspension level mechanism are now explained.

To solve the deadlock problem a timer is started when the traffic is suspended. Whenever the Active Monitor (AM) sends a new frame the timer is restarted. If the AM remains silent for a long time, the timer will expire and the failure of the AM is assumed. The GroupMonitor who first detects the failure becomes active, incrementing the suspension level. The contention problem is also easily overcome: when one monitor detects contention (receiving some responses reporting traffic suspension and others acknowledging its frame) it retransmits the frame incrementing the suspension level. In both situations, the new frame with a higher suspension level than the current one will be accepted by all the suspended participants, establishing consensus on who is the new GroupMonitor.

10.6.7. Handling of Failed Stations

Whenever a station fails, the GroupMonitor function is invoked, to reestablish group coherence. The GroupMonitor winning the activity must, if needed, finish the transmission interrupted by the failure and disseminate a new group view. To accomplish this objective, the monitor executes in two phases (StepOne and StepTwo).

These phases include the identification of failed stations, search for the presence of pending messages from failed emitters, decision to *accept* or *reject* those messages and finally the dissemination of the new group view.

The decision process for the frames pending from failed emitters is the most difficult step of all monitor actions. The Monitor must ensure that the pending transmissions are finished correctly. This means that the Monitor must investigate if the message had been accepted by any member of the multicast group and if so, all the other members must also accept it. If none of the group members had accepted the message, it can be rejected.

This action can only succeed if the recipients are able to provide some knowledge that can map onto past transmissions, since many other messages may have been received (from other emitters) prior to detection of the failure. Since no context is kept about previous transmissions (see §10.6.2), an indirect information will help solve the problem. Each station keeps a table with the Multicast Data Number (MDN) of the last message accepted from each sender in the LAN[13] (StationMdnTable[14]). The active monitor reads its contents concerning the failed

---

[13] From assumption A1 there are at most k senders at each node, thus only $k*N*sizeof\ (MDN)\ bytes$ need to be reserved for monitor action (where N is the number of nodes in the system).

[14] Note that a station may belong to many gate groups and more than one group may keep information about the same station. Thus, to save both execution time and storage, a structure containing information about all the stations with AMp capability is created in every station: the *StationMdnTable*.

station(s), on all group members, chooses the highest MDN value for each sender and disseminates pairs *senderId/MDN*, during the next phase. These search for the presence of a message in the receive queue, sent by the failed senders. If found, should its MDN be lower or equal to the received MDN, it is accepted, else it is rejected.

The first phase (StepOne) covers the identification of the failed stations, the search for pending messages and, finally, the investigation of the *StationMdnTable*. The investigation frame carries the identification of the failed stations. The responses will carry a list of triplets containing the id of the failed station, the content of the StationMdnTable for that station and a Boolean stating if there is any pending message in the receive queue sent by that station. After this, a second phase (StepTwo) is performed, including the dissemination of the decision for the pending frames and group view. An exceptionDecision frame simply contains pairs *senderId/MDN* where the MDN is the highest MDN (for that sender) received in the first phase. The recipients will use this value and the decision algorithm presented above, to finish pending transmissions. The first frame sent in the monitor action, always carries the suspend attribute, while the last one carries the resume attribute.

10.6.8. Joining and Leaving the Group

There are two control frames sent to change the group membership, inserting a participant in or removing it from a gate group. These frames, called respectively *OpenGate* and *CloseGate*, are sent in response to a local request by a participant, to join or leave a given MGateGroup. Since these two frames change the GroupView, they may interfere with other transmissions in course. So the traffic is temporarily flushed and suspended, prior to the processing of a Gate frame. In a normal situation, this gap is hardly perceived by group users.

The *CloseGate* action is very simple. A frame with the identification of the participant to be removed is sent with the Flush and Suspend attributes. If the traffic was not already suspended by another participant, the end of the current message transmission is awaited for, before a *close accept* frame resuming the traffic is sent. When the accept frame is received, each participant changes its GroupView and group activity is restarted.

The *OpenGate* action needs an extra step since the participant desiring to enter a multicast group must, prior to sending the Gate frame, obtain a view of the group. The OpenGate action is then started with an GetView frame, which is sent with the Suspend and Flush attributes as above: this ensures correctness of the View obtained. If no response is received to the GetView investigation, the frame is retransmitted. If silence persists after $k+1$ tries — allowed omission degree plus one — the requesting participant assumes it is the first member of the MGateGroup and initializes his GroupView. If the GetView investigation obtains a response, it is followed by the *OpenGate* frame, which is retransmitted if needed. When it is acknowledged by all the participants, an *accept* frame is sent. At this moment, the GroupViews are changed, inserting the new participant in the group. Traffic is resumed.

10.7. Performance and Real-Time

Definitions about real-time communications protocols have been given in chapter 5. Requirements for XPA high-performance real-time communications, have been stated in chapter 9. This section addresses the derivation of the execution time expression, calculation of execution times under several scenarios and, finally, the demonstration of the existence of a known upper bound for the execution time, necessary for synchronous operation. The framework covered by the present section is mainly that of high-performance, fault-tolerant, real-time systems. These issues are addressed in chapter 9. The relevant scope in communications is that of a service with timeliness guarantees — among other attributes — specially important for XPA (see section 9.6.4).

10.7.1. Atomic QOS Execution Time

To quantify the duration of an execution of AMp, one has first to observe that it depends very much on the LAN used, and the particular protocol implementation. The following observations will be made on the token-less protocol implemented in firmware, as described in [Veríssimo et al. 1989]. The atomic QOS, being the most complex and covering many aspects of weaker qualities of service, was chosen for this example. The execution time (T_e) expression[15] for the atomic service, taking the possible errors into account, is given in table 4. Observe that the protocol is structured in transmission-with-response series of the several frame types, $t_{xwr}(FR)$, where FR is: message msg; decision DEC; request for decision $reqDEC$; monitor action $Step1$ and $Step2$. These alternate with several processing or waiting steps, accounted for by CPU times (t_{pr}) and timers $(t_{waitResponse}, t_{waitDecision}$ and $t_{earlyDecision})$. The transmission-with-response is itself composed of datagram transmissions and processing. Its duration highly depends on the target network and NAC performance parameters.

The sender uses timer $t_{waitResponse}$, after receiving confirmation of transmission by the network, to control recipients activity. Each recipient uses his timer $t_{waitDecision}$, to control sender activity. In fact, it is a two-shot timer, to optimize recovery in case of omissions, with a first time-out given by $t_{earlyDecision}$.

The variables O_i take values according to the allowed error scenarios, and the execution time expression changes accordingly. For example, O_3 accounts for the existence of errors in transmission of the decision frame. If errors occur, the term on $\overline{O}_3$ in the expression is replaced by the one on O_3.

10.7.2. Performance

This section deals with the performance implications of supporting distributed applications, with reliable broadcast protocols. In general-purpose computing systems, most of the time domain requirements are of the *on-line*, or *soft real-time* kind. That is, applications require responsiveness, fastest possible reaction and a probabilistic treatment of worst-case response times. To encourage utilization of reliable broadcast protocols in such applications, it is mandatory that the above-mentioned benefits in quality of service are not considered too costly in performance, by the user(s).

From the performance viewpoint, there are three questions in the design of reliable broadcast protocols, which influence the final result: (i) which fault model; (ii) what level in the communication stack; (iii) which network? It is assumed that the communication system is *fail-silent*. The protocol was designed both to run on top of LANs, i.e., at the data link level, and not to depend on a particular LAN.

There is clearly a difference between LANs with moderate date rates, from 4-16 Mb/s, and LANs with high data rates, of the order of 100 Mb/s. So, to predict performance of AMp, we will concentrate in one example of the lower class, namely a 10 Mb/s 8802-4 token-bus. Second, it is analysed whether AMp performance will benefit from migrating to a higher throughput LAN, such as the 100 Mb/s FDDI ring.

The scenario defined is a small cell network for real-time manufacturing control, in an industrial environment: a 500m network with 32 stations. The performance assessment will be

---

[15] We recall that the execution time is the time between the send request primitive and the issuing of the last receive indication for that message.

Table 4 - Temporal Expression for AMp Execution Time (T_e), with Several Error Scenarios *(token-less variant*

	All situations
T_e	$(O_1+1).t_{xwResp}\,(msg\,)+$ $\overline{O}_5.\{\,t_{pr} + (O_2+1).t_{xwResp}\,(DEC\,) + t_{td}\,\} \;+$ $O_5.(\text{-}\;t_{td} + t_{AM}\,) + O_6.t_{ina}$
	Active Monitor action — after sender failure
t_{AM}	$(O_3 + 1).t_{xwResp}\,(INV\,) + t_{pr} \;+$ $(O_4 + 1).t_{xwResp}\,(DEC\,) + t_{td}$
	Variables for error situations
O_1	Number of Dissemination errors ($\leq k$)
O_2	Number of Decision errors ($\leq k$)
$O_{3,4}$	Number of Monitor action errors ($\leq k$)
O_5	Sender Failure after Dissemination: true — O_5=1; false — O_5=0
O_6	Inaccessible Network: true — O_6=1; false — O_6=0

based on an evaluation of the execution time, T_e. Values will be extracted from the expression in table 4. The various parameters will be quantified taking into account the specific LAN, the scenario, and assuming a well-engineered implementation on a high performance NAC. This supposes a very efficient local executive, very powerful CPU and fast data paths. The objective of this "optimistic" approach is to show the possibilities of the architecture and protocol, rather than those of a given implementation, and mainly, to appreciate the comparative behaviour between different LAN ports.

Real implementations will hardly combine all of these favourable attributes. A normal AMp implementation will run at about four times the optimum-case execution times quoted.

Some example situations were extracted, as a function of message length, number of group participants, and several typical error situations. The relevant values are presented for T_e of AMp on a token-bus LAN, in table 5.

With the purpose of comparison with the token-bus LAN, predictions for an FDDI network with the same dimensions are then made. The channel rate, which was 10Mb/s, becomes 100Mb/s. The predictions for T_e in the same situations as done for token-bus are repeated in table 6.

These results are detailed in [Veríssimo and Rodrigues 1990]. The use of FDDI, with a 100Mb/s rate, seems to be advantageous: it is shown that for small messages not only AMp throughput increases, a natural consequence, but also *speed*, measured in duration of single AMp executions[16]. This fact is of importance, since it has been recognised that communication

---

[16] In essence this is due to the following facts: (i) the increased available bandwidth, which reduces the impact of protocol frames in channel utilization; (ii) the increased speed, because of the shorter rotation and transmission times, for the same load condition; (iii) the assumed low values for processing times, which are achievable for a low-level implementation.

speed is the dominating requirement for distributed computing. On the other hand, it shows a way of using technology to improve performance without compromising portability. While keeping a neat, independent interface in the LAN world, something can be done to increase performance, by merely changing LAN.

Table 5 - Execution Time Predictions, T_e (ms) (AMp on TB)

T_e	80 octets		$n = 6$	
	$n = 6$	$n = 12$	320 oct.	1280 oct.
no faults	2.7	3.1	2.8	3.6
1 om. f. diss.	4.5	5.1	4.7	5.5
1 om. f. dec.	5.5	6.2	5.7	6.5
k om. f. diss.	6.4	7.2	6.6	7.4
sender fail..	13	14	13	14

Table 6 - Execution Time Predictions, T_e (ms) (AMp on FDDI)

T_e	80 octets		$n = 6$	
	$n = 6$	$n = 12$	320 oct.	1280 oct.
no faults	0.72	0.94	0.73	0.81
1 om. f. diss.	1.0	1.3	1.0	1.1
1 om. f. dec.	1.15	1.4	1.2	1.3
k om. f. diss.	1.3	1.6	1.3	1.4
sender fail..	2.4	3.0	2.4	2.5

10.7.3. Synchronism

One of the questions raised in the beginning of this chapter was how to achieve synchronism with a clock-less protocol, given that classical approaches to synchronous protocols are clock-based [Babaoglu and Drummond 1985, Cristian et al. 1985]. Synchronism is taken in the sense of the existence of a known time bound for all executions.

A fundamental issue about synchronism is the way timing errors are treated, i.e., if there can be, for example, delivery of messages outside the bound, and if so, how is this event treated. Timing errors are avoided by imposing a performance specification on the NAC. However, load variability imposes a significant amount of head-room in that specification, since worst case delay situations are far from normal ones. Since the execution proceeds in transmission-with-response series, late deliveries are detected as not belonging to the present series and can be rejected. This is equivalent to transforming a timing error into an omission error, reflected in the omission degree: a NAC doing more than k successive timing errors is failed. This way, sporadic timing errors are allowed, and thus the performance specification can be tightened.

The main issues about the clock-less AMp structure concerning synchronous operation are the following:

- achieve and determine upper bounds on frame delivery delays by the abstract network, in the presence of overload and faults (T_{td} and T_{ina}, discussed in §10.5);

- impose a performance specification on the NAC hardware and software (CPU, kernel, etc.) in order that processing times of the protocol actions be bounded and known for the specified worst case traffic scenario;

- structure the protocol in phases, so that an execution predictably has a bounded number of phases; clearly delimit phases, in what concerns error detection and recovery (omission and timing), and permanent failure detection;

- structure each phase as a series of timed-out transmissions-with-response, so that it can be decomposed in time, in a sequence of frame deliveries and protocol actions as specified above, thus having a known duration bound.

With these measures, the AMp execution time expression of table 4 is bounded to a known value (and in consequence, T_i, once $T_i \le T_e$). Given the assumptions made in §10.3, namely the single failure assumption, only some error combinations are allowed. To determine $T_e = T_{emax} - T_{emin}$, those error scenarios have to be exercised in the expression to compute T_{emax} and T_{emin}. The existence of the bound ultimately means the AMp is *synchronous* in the sense of property Pa12 in table 1.

10.8.   Implementation Issues

Two implementation approaches are possible, as already seen. One consists of embedding the atomic multicast mechanisms in the Medium Access Control (MAC) sub-layer, by intervention in the MAC state machines. The second approach consists of implementing atomic multicast on top of the exposed MAC interface of a LAN VLSI, while still presenting the same extended interface to the user (figure 4).

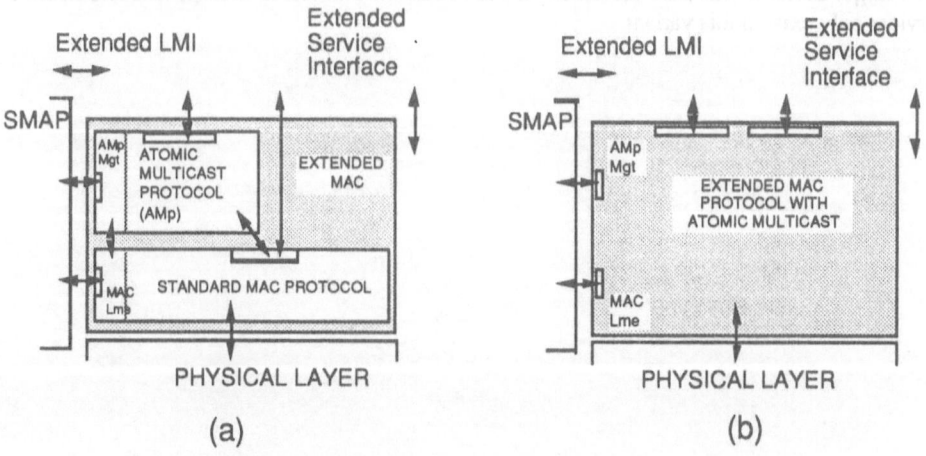

Fig. 4 - Functional Decomposition of the Alternative AMp Implementations

10.8.1.   The Token-Based Protocols

Hardware support was considered mandatory to achieve an efficient solution for the token-based protocols in Delta-4. The approach followed consisted of designing a superset of the MAC sub-layer, in order to include AMp services. The neat interface between the *abstract network* and the *AMp* itself disappears, since they are compacted in a "super"-MAC. Consequently, the protocol becomes LAN dependent.

A theoretical study of the approach for token-bus was made, whereas a prototype for token-ring is actually being built in Delta-4: it is dubbed "Turbo-AMp" and the protocol was formally specified and validated. It is a superset of the standard, with the existing services plus the AMp functionalities (figure 4b). Turbo-AMp only implements the atomic multicast service. Although implying modified VLSI (for example an ASIC), it is an interesting approach for providing a standard LAN with atomic multicast capability. Intervention in the state machines allows establishing a clear protocol execution containment domain, through *token hold control*. This hardware token[17] yields an efficient implementation of a variant of the two-phase accept protocol of figure 1, annexe J. A high degree of *synchrony* and low *error latency* are given by a transmission with multiple acknowledgment mechanism, based on on-the-fly operations on the passing frame.

10.8.2.   The Token-Less Protocol

The token-less variant implementation, shown in figure 4a, is less efficient, yet readily implementable in software or firmware. The protocol implementation is detailed in [Veríssimo et al. 1989]. The main points are: higher parallelism, portability (LAN independence by means of the abstract network); evolvability (being software-based). The protocol particularities, with regard to the core protocol in annexe J, are the non acknowledged decision and the group monitor actions. The protocol was described informally in section   10.6.

In consequence, the token-less variant is a *generic* protocol, LAN independent, and ported to the various LANs in Delta-4, just requiring an implementation of the abstract network on each target LAN. It was also the basis for the evolution that led to xAMp, whose extended services only exist in this variant.

---

[17] Which is the same token as the LAN itself.

Chapter 11

Fail-Silent Hardware for Distributed Systems

The architectural frameworks assumed in this chapter are those of OSA and XPA, and are summarised below:

a) distributed computations are assumed to be structured as software components communicating via messages;

b) XPA: software components execute on fail-controlled nodes with the *fail-silent* property: a node either functions according to the specification or stops functioning;

OSA: the fail-silent property is not essential for nodes, so software components can execute on ordinary (potentially) fail-uncontrolled nodes; however, all the protocols for message passing are executed on fail-silent hardware (the Network Attachment Controllers, NACs);

c) nodes communicate with each other through redundant communication networks;

d) software components can be replicated on distinct nodes for increased reliability; the degree of replication (if any) for a software component will be determined by the failure characteristic of the underlying nodes: K+1 replicas can tolerate up to K replica failures if the nodes are assumed to be fail-silent, whilst 3K+1 replicas are needed if the nodes are assumed to be fail-uncontrolled.

This chapter concentrates on the issues concerning the design and implementation of fail-silent hardware (nodes and NACs). The basic idea is to replicate processing on two processors that can check each other's performance to form a self-checking processor pair. Since a NAC is really a processor with a network interface, we will begin our discussion by considering how a general purpose, fail-uncontrolled processor can be made into its fail-silent counterpart (a fail-silent node); later on we will discuss some specific details of the fail-silent NAC design. We discuss and evaluate two different approaches to the construction of fail-silent nodes.

11.1. Fail-Silent Node Models

11.1.1. General

Individual hardware components are not inherently fail-silent. For some devices, simple models of their correct behaviour exist, e.g., memory devices should output exactly the data that was originally input to them. In these cases, faults can easily be identified by the addition to the data of redundant information, e.g., parity bits or CRCs, which can be checked when the data is output. However for complex devices, for which there is no simple correlation between their inputs and subsequent outputs, e.g., microprocessors, the easiest method of adding redundancy is to duplicate the device and compare the outputs of the two devices.

Fail-silent nodes have been used widely, for example in commercial transaction processing systems (e.g., [Bernstein 1988]). Such nodes have been designed with the assistance of specialized comparator hardware and clock circuits. A common (reliable) clock source is used for driving a pair of processors that execute in lock-step, with the outputs compared by a (reliable) comparator; no output is produced, once a disagreement is detected by the comparator. We term a node designed this way to be a *hard fail-silent node*. An alternative design being developed within the Delta-4 project requires the processors of a node to execute clock synchronization and order protocols "to keep in step". A node implemented according to this design will be termed a *soft fail-silent node*, since no special clock or comparator circuits are employed. Such nodes can provide an attractive alternative to hard fail-silent nodes since no special hardware assistance is required. Furthermore, as the principles behind the protocols do not change, the protocol software can be easily ported to any pair of processors (including the ones expected to be available in future). In essence, hard fail-silence is implemented in hardware and is transparent to software so that standard software will run. Soft fail-silence is implemented in software and is transparent to hardware so that standard hardware may be used. Note that since only two processors are used within a node to check on each other, the fail-silent characteristics of a node can be guaranteed only if no more than one processor within a node is faulty.

Intuitively, fail-silent behaviour ought to mean that a node never generates an erroneous output, i.e., the node can only either generate correct outputs or remain silent. However, this is impossible to implement since output messages take a finite time to transmit, and a fault may occur leading to an error during the transmission of a message. A definition of fail-silence must include the case where a message receiver rejects such erroneous messages. Thus a two-processor node will be said to exhibit fail-silent behaviour in the following sense: the outputs produced by it (if any) are either valid messages or *detectably invalid* messages; this behaviour is guaranteed so long as no more than one processor in the node fails.

11.1.2. Hard Fail-Silent Nodes

In a hard fail-silent node, duplicated microprocessors are closely coupled and run in micro-synchronization. Each component is initialized to an identical state and then performs identical actions on identical data on a clock by clock basis, so that on every clock cycle the data output by the components is identical. The principles underlying the node architecture can be explained easily by examining figure 1. Since the data streams to be compared are in exact lock-step, a simple hardware comparator can be used to check that the data streams are identical and to prevent any outputs once a discrepancy is detected. Although two replicas of software are actually running, because they are micro-synchronized and compared by hardware, the software takes no part in the replication and comparison process and is "unaware" of it. Thus the replication is transparent and imposes no requirements on the behaviour of software. Indeed, a hard fail-silent environment may be produced such that software is wholly unable to distinguish it from a standard non-duplicated environment. When hard fail-silence techniques are used, the correct and erroneous message sets sent over the network are distinguished by the fact that the only erroneous messages that can be sent are incomplete correct messages, since the occurrence of a fault during the transmission of a message can stop transmission within one clock cycle. Such incomplete messages are easily identified by the receiver since they will contravene the lowest levels of network protocols.

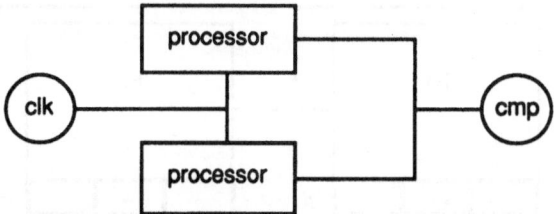

Fig. 1 - A Hard Fail-Silent Node using Duplicated Processors

11.1.3. Soft Fail-Silent Nodes

In the case of hard fail-silent nodes, there is no problem of replica group determinism other than the hardware design issue of ensuring identical results of response to external asynchronous events, so for this purpose replication has no consequences to software design. In contrast, soft fail-silence nodes cannot be expected to run arbitrarily structured programs and preserve replica determinacy. In essence, programs have to be structured as state machines [Schneider 1990]. We assume, as stated before, that distributed computations have been structured as a number of software components (referred to as processes in the rest of this chapter) that interact only by way of messages. A process has an *input port* through which it receives all the messages directed to it and an *output port* through which it can send messages to other processes. This simple model is sufficiently general in that other models, such as clients and servers interacting through remote procedure calls, or objects communicating by messages can be seen as special cases (although, certain enhancements to the model are possible, and will be discussed subsequently). We assume that computations performed by processes are *deterministic* so that if all the correctly functioning replicas of a process have identical initial states then they will continue to produce identical responses to incoming messages provided the messages are processed in an identical order.

The overall node architecture is shown in figure 2. Each of the two processors (P_1, P_2) has network interfaces (n_1, n_2) for inter-node communication over (redundant) networks; in addition, the processors are internally connected by a communication link, l, for intra-node communication needed for clock synchronization and order protocols. Each non-faulty processor in a node is assumed to be able to *sign* a message it sends by affixing the message with its (the processor's) unforgeable signature; it is also assumed to be able to *authenticate* any received message, thereby detect any attempts to corrupt the message. Digital signature based techniques [Rivest et al. 1978] can be relied upon to provide such functionality.

It is necessary that the replicas of computational processes on processors within a node select identical messages for processing, to ensure that they produce identical outputs. Identical message selection can be guaranteed by maintaining identical ordering of messages at input ports and ensuring that processes pick up messages at the head of their respective input ports. An *order protocol* is then required to ensure identical ordering if both processors are non-faulty.

An implementation of this order protocol will require that the clocks of both the processors of a node are synchronized such that the measurable difference between readings of clocks at any instant is bounded by a known constant ε. Algorithms for achieving this abstraction exist (see [Halpern et al. 1984, Lamport and Melliar-Smith 1985]). Communication for clock synchronization and ordering takes place by way of the internal link. Given that the clocks of both the processors are synchronized, the order protocol can be implemented using a version of the signed message algorithm for Byzantine agreement [Lamport et al. 1982]. As there are only two processors, the protocol is particularly simple, since it is expected to work only in the

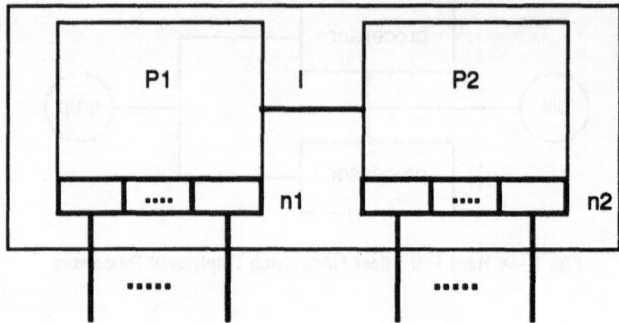

Fig. 2 - Overall Node Architecture

absence of any failures. Essentially, an *order process* of a processor stamps a message to be ordered with its local clock reading. A copy of the time-stamped message is signed and sent over the link to the *order process* of the other processor in the node. If T is the timestamp of the message received from or sent to the order process of the other processor, then the message becomes *stable* at local clock time $T+d+\varepsilon$ where d is the maximum transmission time taken for a time-stamped message to travel from one order process to another order process over the link. A message with timestamp T will be said to become stable, if no message with timestamp $T_1<T$ will be received by an order process. Stable messages are queued at the relevant input ports in the increasing timestamp order (with care being taken not to queue a stable message, if its replica has already been queued).

Each non-faulty processor of a node has five processes:

a) **Sender Process**: this process takes the messages produced by the computational processes of that processor, signs them and sends them via the link to the neighbour processor of the node for comparison.

b) **Comparator Process**: this process authenticates all incoming messages from the neighbouring processor; an authenticated message (m_i) is compared with its counterpart produced locally. If the comparison succeeds, the authenticated message m_i is counter signed (by considering the first signature as a part of the message) and this double signed message is handed over to a *transmitter process* for network delivery to destination nodes. A message that cannot be compared because its counterpart does not arrive within a time-out period or a comparison that detects a disagreement indicates a failure. Once a failure is indicated, the comparator process stops, which results in no further double signed message being produced from that node.

c) **Transmitter Process**: this process is responsible for sending the doubly signed messages to destination nodes.

d) **Receiver Process**: this process accepts authentic messages for processing from the network, discarding any duplicates; such valid messages are sent to the local *order process*.

e) **Order Process**: this process executes the order protocol (mentioned earlier) with its counterpart in the other processor and attempts to construct identical queues of valid messages for processing by the computational processes.

A correctly functioning node will generate two identical copies of its output messages. A receiver process at a node will discard any duplicate messages received over the network. When the comparator process of a non-faulty processor in a node detects a failure and therefore stops, no new double signed messages can be emitted by the node; any messages coming from this node that are not double signed will be found to be unauthentic at the receiving nodes. Any authentic but old messages from a faulty node will also be discarded by the receiving nodes as replicas of already received messages.

11.2. A Comparative Evaluation

11.2.1. General

Certain general observations regarding the two architectures can be made at the outset. The main advantages of the soft fail-silent nodes are that: (i) technology upgrades appear to be easy; since the principles behind the protocols do not change, the protocol software can be easily ported to any pair of processors (including the ones expected to be available in future); (ii) the second advantage is that since the replicated computations are loosely synchronized, the architecture is likely to be capable of detecting common mode transient failures. This is because transients are unlikely to affect the computations on the processor pairs in an identical fashion. The advantages of the hard fail-silent nodes are: (i) the ability to execute arbitrary programs that are not necessarily deterministic — this is because the common clock source ensures lock-step synchronized execution; and (ii) better performance than the software-based approach since there is no need for protocols for message ordering.

The disadvantages of the two approaches are very much the converse of the above characteristics. For the software fail-silent nodes, these are: (i) the restriction that application programs need to be deterministic; in particular this requires that potential sources of non-determinism, such as, interrupts and time-outs have to be treated as messages, thus requiring careful treatment; and (ii) concern over the performance of the node due to overheads of the various protocols described above. For the hardware constructed fail-silent nodes, the disadvantages are that every new microprocessor architecture is likely to require substantial design overheads and that tightly synchronized processors may not be resilient to common mode transient failures. Furthermore, lock-step synchronization at very high clock speeds (50 to 100 MHz) may well turn out to be difficult to achieve. We next discuss several specific issues in detail.

11.2.2. Replication

Figure 3 shows a typical fail-uncontrolled node, with constituent components. For this discussion, we will use the term *network interface* (NI) to refer to a simple interface to the network, such that most of the communications protocol is executed by the host processor. The term *network attachment controller* (NAC) will be used to refer to the device with a processor and NI, with the capability of executing virtually all the communication protocols without involving the host processor. A logical way of constructing the fail-silent version of the above node would be to duplicate all the above components. In this subsection we will examine the basic issues concerned with replication: not surprisingly, they are hardware related for the hard fail-silent nodes and software related for the soft fail-silent nodes.

11.2.2.1. Hard Fail-Silent Nodes. One important design consideration is to minimize the need for specially designed interface components. If duplicate busses are employed then this requirement cannot be met, making this form of replication unattractive. On the other hand, many busses (e.g., VME) do not employ any redundancies in their data and address paths, so such busses must be considered potentially fail-uncontrolled. Figures 4 and 5 show two designs based on the assumption that busses are fail-uncontrolled, and not replicated (bold

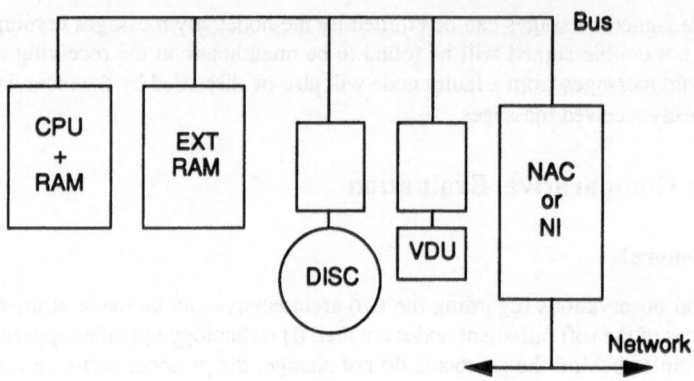

Fig. 3 - Typical Fail-Uncontrolled Node

boxes represent fail-silent components). We note that RAM has been duplicated, rather than being interfaced directly to the bus (a RAM with error detection capability can still be vulnerable on a fail-uncontrolled bus). The common clock that is necessary within a fail-silent component can constitute a single point of potentially uncontrolled failure unless it is itself implemented to be fault-tolerant (see, e.g., [Moreira de Souza and Peixoto Paz 1975, Moreira de Souza et al. 1976]). Also, even though comparators may be built that are self-checking (see, e.g., [Wakerly 1978]), the final comparator output checker can also constitute a single point of fail-uncontrolled behaviour.

Data held on non-replicated discs needs to be protected using checksums or similar forms of redundancy, which must be generated and checked within fail-silent environments (CPUs, figure 4 and in addition NACs, figure 5). Similarly, messages to be transmitted over the network must contain redundant information (checksums, CRCs) generated and checked within fail-silent environments.

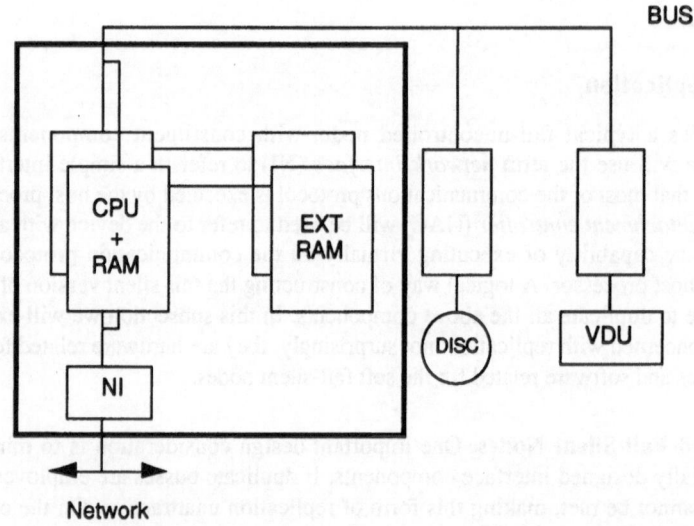

Fig. 4 - Hard Fail-Silent Node *(design 1)*

Methods of handling the interactions between two asynchronous clock domains, each run in lock-step synchronization, are well-understood. Applying such methods requires access to and special treatment of the interface between the domains. The interfaces between asynchronous clock domains are often buried deep within such VLSI chips and so are quite inaccessible. Fortunately, clock signals of processors are accessible, so processors can be replicated. The particular case of concern in this regard is the VLSI Network Interface controller chip, whose clock signals are not accessible, making it impossible to duplicate. An alternative would then be that the chip be discarded and several boards-full of MSI and LSI components be substituted. Such a development is not considered practical, given the over-riding desire of using standard VLSI components. Thus the sensible approach is not to duplicate the NI, but to regard it as a part of the network, which is treated as fail-uncontrolled. For this reason, only a single NI (which is potentially fail uncontrolled) is shown in figure 4. In the similar fashion, the two NACs will share some common interface circuits (a single fail-uncontrolled NI).

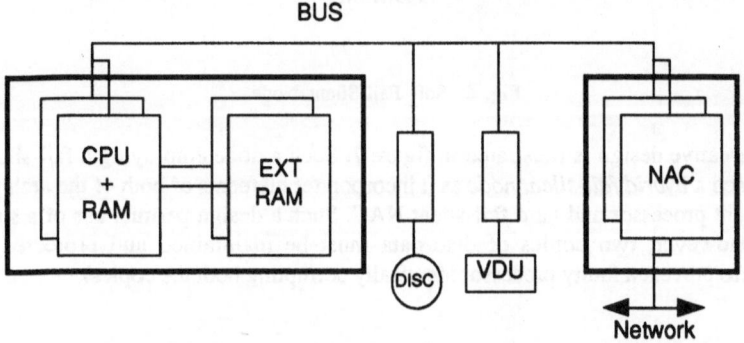

Fig. 5 - Hard Fail-Silent Node *(design 2)*

11.2.2.2. Soft Fail-Silent Nodes. In this architecture, two fail uncontrolled nodes (figure 3) can be duplicated as illustrated in figure 6; the node can be built entirely from standard components. It must be stressed however that the guarantee of the fail-silent behaviour rests on the coverage achieved by the signature and authentication mechanisms used. Of course, as observed before, such nodes, unlike the hard fail-silent nodes, cannot be expected to execute arbitrary "black box" software; rather software modules must be structured as state machines (section 11.1). Most practical systems require some additional functionality in their concurrent processing model than that assumed in section 11.1. Typically, this will include the capabilities of selecting a message within a given time interval (waiting for a message with a time-out), selecting the highest priority message from the set of available messages, and responding to interrupts coming from "alarm messages". Naturally, protocols to enforce identical replica behaviour will be necessary to deal with the treatment of such interrupts, priority messages and time-outs; some work in this direction is discussed at length in [Tully and Shrivastava 1990]. We mention it here to indicate that the soft fail-silent architecture presented here can be adapted to incorporate these functionalities. However, it must be stated that performance implications are not fully well understood and that no operational experience is yet available (a prototype node is currently being implemented [Shrivastava et al. 1991]).

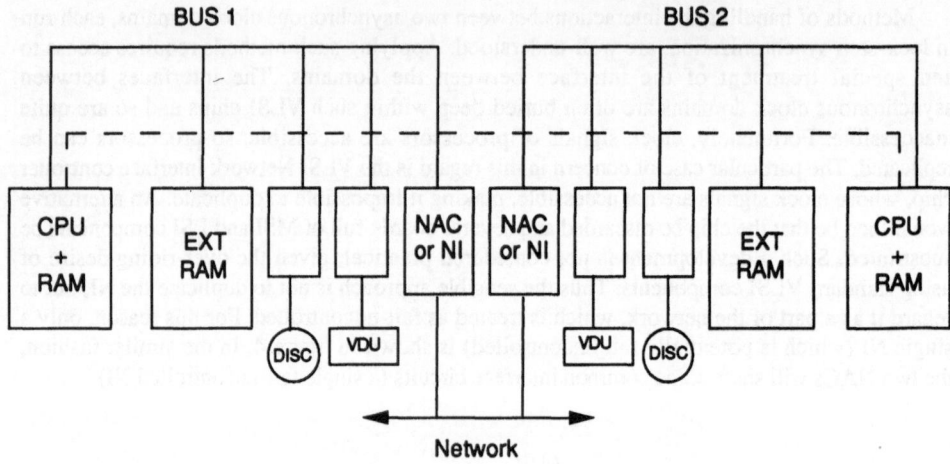

Fig. 6 - Soft Fail-Silent Node

An alternative design is illustrated in figure 7. Such a node employing a fail-silent NAC will be termed a *hybrid fail-silent node* as it incorporates elements of both of the architectures: soft fail-silent processor and hard fail-silent NAC. Such a design permits use of a single bus and disc. However, two copies of disc data must be maintained and protected against corruption (to prevent a faulty processor identically corrupting both the copies).

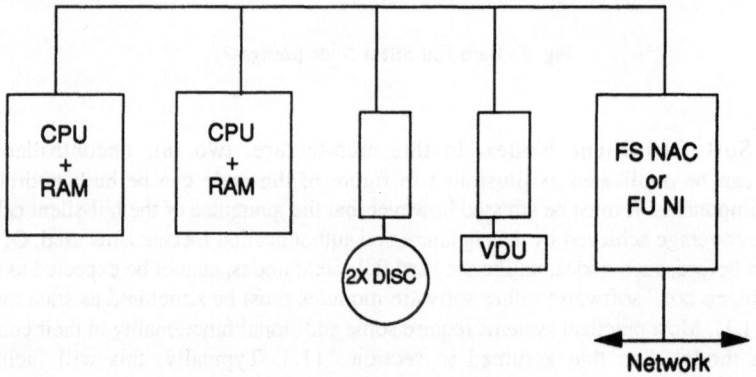

Fig. 7 - Hybrid Fail-Silent Node

11.2.3. Discs

The disc drives are one of the most expensive components of a node, and require particular attention. Various techniques are described in the literature to achieve very reliable but expensive fault-tolerant disc subsystems; typically this involves mirrored discs dual-ported to dual disc channel cards, with special attention paid to live-unplugging of busses for maintenance purposes. Since we are only looking for fail-silent properties rather than fault-masking or recovery, the problem is simplified. With much less cost and performance penalties, this can be met by added redundancy without full duplication. If standard discs and channels are to be used, the redundancy must, however, be added and checked within a (hard or soft) fail-silent environment elsewhere in the node. In the case of hard fail-silent nodes, this means that disc drivers resident in the fail-silent host must add and check the redundancy information of all data held on disc that could affect the correctness of computation leading to output messages. Only one copy of each data item is held. Soft fail-silent nodes really need duplicated discs, unless the hybrid arrangement is used which will require some form of data protection technique.

11.2.4. NIs and NACs

In Delta-4 XPA, the fail-silent host assumption should lead to two simplifications: (i) communication protocols (e.g., for reliable multicast) may reside in either the fail-silent host or in the fail-silent NAC; and (ii) there is no need for voting protocols for managing replicas. As a consequence of (i), the use of simple NIs, with protocols implemented within the device drivers running on fail-silent hosts becomes an attractive design option. Work on the design of protocols for soft fail-silent nodes with NIs has been reported in [Ezhilchelvan and Shrivastava 1991].

We now discuss some specific design issues for hard fail-silent NACs that have been constructed for use within OSA. The NAC and the host (which could be either fail-uncontrolled or fail-silent) share a bus (the VME bus, in the particular design under discussion).The main system components of a NAC are: (i) a NAC processor plus RAM pair, acting as a hard fail-silent node (where all the communication system protocols are run); and (ii) a NI with communications RAM (for reasons discussed earlier, the NI's RAM remains non-replicated). A fail-silent NAC must be able to cope with potentially uncontrolled failures in the host, the bus connecting the host and NAC, parts of the NAC associated with the interface or accessible across the bus, the NI and the network. In the case of a fail-uncontrolled host, errors cannot be detected by the host itself as any checksums or similar methods used to validate data might themselves be incorrect. Therefore any checking must be performed by comparing messages generated by replicated software components running on distinct hosts. This checking is carried out in the communication system by voting on active replicas' messages. This can only detect faults that are transmitted on the network and it is possible to have long periods of latency before faults are detected.

For fail-silent hosts, the majority of hardware faults will be detected by the host itself. However, the data that is transmitted from the host can be corrupted. Since this can happen as soon as the data leaves the fail-silent environment, it is necessary to generate some form of checksum within the fail-silent environment if voting on active replicas is not performed. This checksum should not be a simple adding of bits since it will be transmitted over a large area of fail-uncontrolled environment before its validity is checked and therefore a combination of errors could lead to the checksum appearing valid. Any data arriving at the host should be checked for validity. The data should be copied into the fail-silent environment and the checksum should then be used to verify the data. The checksum can then be discarded if required.

The bus must always be considered as fail-uncontrolled if it is not replicated and no checksum or parity is automatically generated (this is the case in the design being considered). Even if the bus itself could be considered as reliable, the interface at either end certainly cannot. Therefore, as mentioned above, a checksum of some sort must be generated from within the fail-silent environment on the host or NAC if the data is being transmitted from the fail-silent part of the NAC. This checksum must be checked wherever the data is used.

Data that is transmitted to the NAC across the bus can be placed in either the communications RAM (non-replicated) or the processor RAM (replicated). Both the host and the network interface may perform direct memory accesses into the RAM of the fail-silent NAC. This allows messages to be passed without extra copying processes between protected and unprotected memory areas. This means that a fault in either the host or the network interface could cause corruption of data or program within the NAC. To prevent this occurring, while allowing messages to be placed in the RAM and then protected whilst being processed, a dynamically controlled protection scheme is provided which allows the duplicated processors to control the protection on their respective RAMs. Each environment has two masks that may only be altered by the NAC processor in the associated environment. One mask in each environment prevents the host from writing to the RAM, the other prevents the network interface from writing to the RAM. Each bit in each mask protects 128 bytes in the RAM, so that the NAC software can allocate buffers to each interface in granules of 128 bytes. An attempt to write to a protected granule causes an interrupt to both NAC processors so that the NAC software is aware that the host or the network interface may be faulty. No write occurs, but the access terminates normally. This protection mechanism can be tested at system start up, and therefore it would require failure of the independent protection mechanisms on both RAMs before a fault in the host or network interface could cause a RAM corruption that might lead to a malicious fault.

Since these masks are the only means of protection from an arbitrarily faulty host, they are replicated. This should prevent a failed host from transmitting apparently correct messages over the network. It has already been stated that any data transmitted over the bus needs to be checked. If the data is to be altered by the NAC, it is essential that software on the NAC checks the validity of the data before altering it. If data is altered the NAC must also generate a new checksum in order for the data to be recognised as valid at its destination. This checksum must be generated from within the fail-silent environment since if it is not, the data that is being read cannot be relied upon and therefore a valid checksum could be generated for incorrect data. Wherever possible, to reduce processing and transfer overheads, data received from the host is not altered on the NAC, but additional data is added around the original data, i.e., an envelope is created which encapsulates the original data. The additional data should have an additional checksum or checksums added and should be produced within the fail-silent environment and should not rely on any of the original data that has not been validated. This will ensure that the initial checksum generated in the fail silent part of the host is preserved and therefore the encapsulated data can still be verified against the original checksum.

Any data that is received by the NAC over the network must carry out a check on the envelope around the initial data and make sure that no corruption of data has occurred in transmission. The envelope should then be stripped from the original data and this data transmitted to the host. Any data that is to be transmitted over the network should already have had checksums added within the fail-silent part of the NAC and this can therefore be checked at the destination NAC. The network interface is effectively treated as part of the network and therefore liable to data corruption.

The design just discussed has illustrated the practical considerations that go into the construction of fail-silent hardware, particularly when some of the components cannot be duplicated.

11.3. Concluding Remarks

In addition to meeting technical criteria, to be marketable the system implementation must also have other attributes.

a) **Visibility:** The fault tolerance of the system must be made visible and comprehensible to the customer.

b) **Credibility:** The dependability mechanisms must be credible to customers. The mechanisms must also be developed and tested to high quality control standards; it is expected that it will be necessary to offer such evidence to potential customers. The fault injection testing of NACs (see section   15.3) has been performed with this requirement in mind.

c) **Cost-effectiveness:** The perceived gains in dependability must outweigh the increased cost and performance penalties of a fault tolerant system, and do so at least as well as competitive methods (see also section   15.4, where dependability evaluation is discussed).

In this respect, both hard and soft fail-silent approaches provide visible and credible solutions; redundancy within a node is visible, and its exploitation can be explained to non-expert customers. The ability to run third party software, e.g., concurrent database systems with non-deterministic behaviour, with minimum performance penalty makes hard fail-silent nodes cost-effective. Factors working against such nodes are that there could be market resistance from customers to the use of non-standard processor and NAC boards and that new processors require considerable re-design effort. In contrast, the performance of a soft fail-silent node can never be made as good as its hard fail-silent counterpart; furthermore, not every third party software can be made to run on such nodes without any changes. Factors working in the favour of a soft fail-silent node, as observed earlier, are that they can be built entirely using standard hardware components and no major re-design effort is required for new processors. Given these observations, we can only conclude at this stage that both types of nodes have a role to play in dependable distributed systems.

Chapter 12

Input/Output: Interfacing the Real World

Consider a (chemical, nuclear,...) process plant controlled by a computer system. The design of such a process plant includes global dependability analysis by process experts, for whom the plant is a system of which the computer control is a subsystem. This dependability analysis may range from cost-effectiveness considerations to safety assurance through hazard analysis. The dependability models and mechanisms presented in other chapters describe how such a subsystem may contain and recover from its own misbehaviour.

The connection between the computer system and the plant has significance as a subsystem in its own right. The issue of interest for this chapter is to consider means whereby the dependability of this subsystem may be commensurate with, and contribute to, the dependability properties of the whole plant. At some stage, the process experts must consider whether the dependability properties required of the plant itself permit a design in which, for instance, a single piece of wire emerging from the dependable computer control subsystem is the only output controlling a crucial valve or pump or heating element. There are many obvious risks associated with such a design, which need not be enumerated here.

If such risks are unacceptable, the engineers concerned may require the dependability model to extend its boundaries beyond those of the control subsystem into the mechanics of the plant itself. This could entail, perhaps, such well-established "mechanical voting" techniques as on/off valves or other equipment being duplicated or tripled or even quadrupled; a common example is a series/parallel arrangement of valves and pipes (see figure 1).

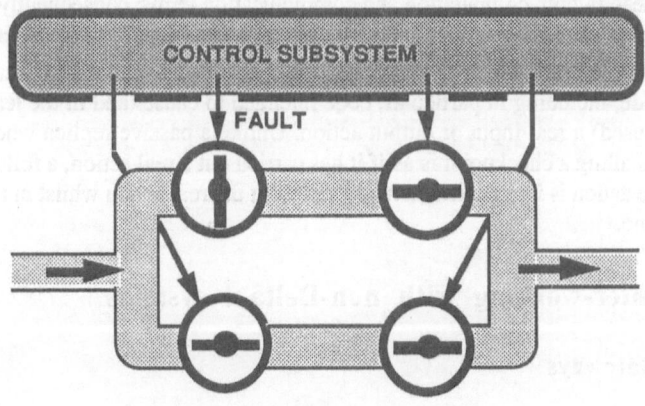

Fig. 1 - Example of "Mechanical" Voting

We are therefore concerned with input/output operations that cross the boundary of a Delta-4 system and interact with, or observe, entities in the enclosing environment that are not subject to Delta-4 models or assumptions. These entities may be complex and based on computers (commonly called "intelligent"); an elaborate example and special case of which is that of an external networked system, which is discussed in section 12.1 below. Alternatively, these entities may be primitive electromechanical devices (commonly called "dumb"), as in the above example; the rest of the chapter is concerned with various examples of this case. While the subject falls outside the remit of the Delta-4 project itself, it is worthy of discussion at length, since any system based on Delta-4 must find solutions and there are indeed opportunities inherent in the architecture that should be identified.

In this chapter, semi-active replication (see section 6.7) is revisited many times; the model seems to be well-suited to solving many problems of interaction between Delta-4 and the outside world. Two major distinctions between the Delta-4 models of semi-active and active replication are of particular importance in this chapter and so are summarised here:

- Active replication does not require a host to exhibit the fail-silence property (indeed, the discussion is made under the assumption that the host is not fail-silent) whereas semi-active replication, like passive replication, does require the host to be fail-silent.

- Whereas active replicas have a symmetrical relationship and during execution there is no assurance as to which replica will perform some computational step first, semi-active replicas are like passive replicas in having an asymmetry in their relationship both in terms of authority (which replica is "in charge" of computational progress and decision making) and time.

There are also two major distinctions between the Delta-4 models of semi-active and passive replication of particular importance in this chapter; again, these are summarised here:

- In passive replication, the computational overhead of taking, and the message size of transmitting, checkpoints is such that there is good reason to limit their rate. This is reflected in the discussion below, where a complex of output activity is assumed to occur between two checkpoints. In semi-active replication, the construct which "corresponds" to checkpoint taking and sending is action notification, which is cheap in both computation and communication terms; consequently it is appropriate to associate such notification with each individual input or output action.

- The construct which "corresponds" to checkpoint installation is actual execution of code, including in particular, code intended to cause (and in the leader, having just caused) a real input or output action. Unlike a passive replica whose new state on installing a checkpoint is as-if it has carried out a real action, a follower will, unless the action is intercepted, actually carry out the real action whilst in transit to the new state.

12.1. Inter-working with non-Delta-4 systems

12.1.1. Gateways

The MCS bi-point model (see section 8.1.4.1) offers services that are compatible with ISO standards and that allow standard ISO applications to be ported at very low cost in an MCS environment. However Delta-4 systems have their own addressing schemes up to the highest level of the OSI model: the inter-working with non-Delta-4 applications therefore needs a gateway function that must be performed within the application layer. Inter-working is illustrated in the following figures. Figure 2 shows inter-working between applications running in a Delta-4 environment and a separate ISO environment. Figure 3 illustrates the point that this inter-working may take place on the same LAN because of the compatibility with standards that has been maintained at the low layers of MCS.

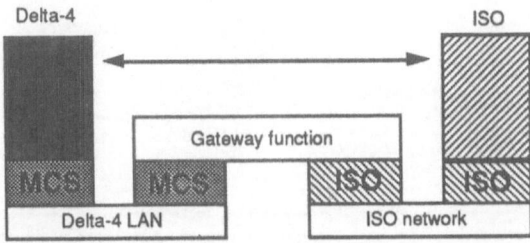

Fig. 2 - Inter-working between Applications running in Delta-4 Environment and ISO Environment

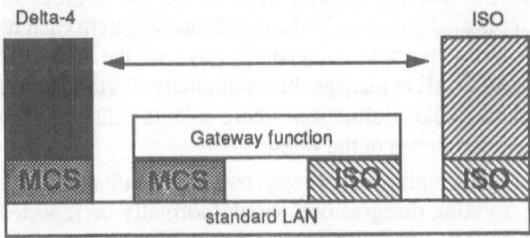

Fig. 3 - Co-existence and Inter-working of Delta-4 Environment and ISO Environment on a Single LAN

A gateway is thus an application-layer relay that provides a migration path for inter-working with other network architectures. Such a gateway interfaces both the Delta-4 world and the outside world. However, dependability cannot be conferred on this interface through use of Delta-4 techniques alone to replicate the gateway function on several nodes. Although these techniques offer transparent redundancy of the gateway function within the Delta-4 system, such transparency is not obtained for the non-Delta-4 applications outside the Delta-4 system. Although, later in this chapter, proposals are discussed whereby Delta-4 can contribute to the provision of dependable interfacing and the replication of primitive sensors or actuators, the provision of fault-tolerance to an ISO based application is quite another matter. This section will therefore not discuss how fault-tolerance might be achieved in the non-Delta-4 world, but will instead examine how Delta-4 techniques might help to preserve the availability of the interface between the two worlds. With this in mind, we now examine two ways of implementing such a gateway, based first on the active replica technique and second on the semi-active replica technique.

12.1.2. Replicated Gateway with Active Replication

In this approach, the gateway function is actively replicated with respect to its interactions within Delta-4. Each time the gateway is used to interwork with a non-Delta-4 (e.g., ISO) Application Entity (AE), each replica of the gateway function must establish an ISO connection to this AE. The ISO AE must explicitly manage the consequences to itself of this redundancy. It will receive several copies of each message from the Delta-4 world, and must separately send copies of messages destined for the Delta-4 world on several links. An example of such inter-working is shown in figure 4:

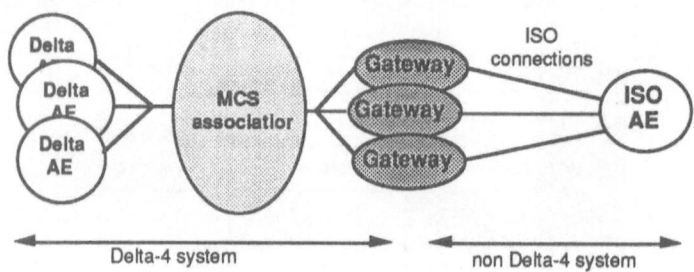

Fig. 4 - Gateway with Active Replication

In the figure, a replicated Delta-4 Application Entity inter-works with an ISO AE through a replicated gateway. In the Delta-4 system, MCS hides the degrees of replication of the replicated AE and the replicated gateway. In the non-Delta-4 system each replica of the gateway function must establish an ISO point-to-point connection with the ISO AE. Some software must therefore be added to the ISO AE to manage this multiplicity. Furthermore, the gateway must be carefully programmed to avoid a situation where events related to the ISO connections compromise the replica determinism of the gateway.

On the Delta-4 side of the replicated gateway, output validation will be carried out by Delta-4 voting mechanisms; a voting disagreement would normally be reported to Delta-4 System Administration by MCS. Such a disagreement could equally be due to a fault occurring somewhere on the ISO world; either in the communication on one of the ISO connections, or in the host environment where the ISO AE executes. The consequence is that the Delta-4 voting mechanism in the gateway can help to discover some misbehaviour outside the Delta-4 world. This must be balanced against the fact that fault treatment (diagnosis and repair) can become very complex, especially where no proper error containment domains have been constructed in the non-Delta-4 world. Whilst it is possible to arrange for the Delta-4 error-detection mechanisms to provide input to suitable fault-management applications that exist or could be created in the non-Delta-4 world, this approach should be seen as offering improved availability of the gateway rather than error detection.

Reconfiguration of the Delta-4 side is automatically performed in the Delta-4 system. On the non-Delta-4 side, disconnection of the failed replica and eventual reconnection of a new replica must be explicitly programmed. When a new replica of the gateway is installed, some actions must be performed on reconnection on both sides of the ISO connections to synchronize the gateway replicas with respect to ISO communication. In short, this solution implies that much of the machinery transparently provided by MCS in the Delta-4 system must be explicitly reproduced above ISO point-to-point communication in the non-Delta-4 system. Moreover, this management must be added to any ISO AE that is intended to inter-work with the Delta-4 system. This approach is therefore rather expensive and should be reserved for a few particular cases that require a high availability and fast recovery of the gateway function.

12.1.3. Replicated Gateway with Semi-Active Replication

The following approach presents a lower cost in terms of functionality having to be added to non-Delta-4 components. The model used will be recognised as corresponding to that of semi-active, or leader-follower, replication (see the introduction above). Although support machinery for this model has been developed within Delta-4, the description given here is of what is necessarily an application-level implementation.

At any one time, only one gateway replica establishes a connection to the ISO AE, as shown in figure 5:

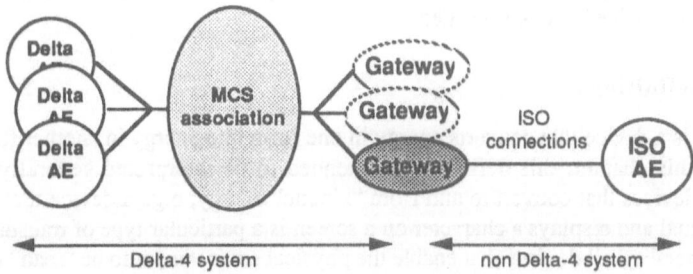

Fig. 5 - Gateway with Semi-Active Replication

All gateway replicas are programmed so as to appear to be true replicas when viewed from the Delta-4 system. However, they do not behave identically with respect to communication with non-Delta-4 AEs; in this direction, only one replica is seen as "active". Although all replicas receive the Delta-4 messages to be forwarded to the ISO AEs, only the "ISO-active" (leader) replica forwards them. The other replicas (the followers) instead queue them temporarily; when the leader receives confirmation of delivery from the destination ISO AE, it sends a checkpoint that allows the followers to delete the corresponding message from their queues. In the other direction, only the leader sends the messages it receives from an ISO AE by way of MCS. Such messages can be sent by using the Leader-Follower support of MCS, which allows the leader to send in the same atomic communication a message to a destination together with a checkpoint to the followers. If the leader gateway replica fails, all replicas are informed and a new gateway replica will take the role.

From the ISO AE's point of view, this solution does not totally hide the redundancy of the gateway. Some management of reconnection is required when the "active" gateway replica fails, with controls to ensure that messages are not lost or duplicated on the ISO connections when disconnection and reconnection occur. The Leader-Follower replication technique requires a fail-silence assumption on the hosts on which the gateway function resides. In the case of a gateway to an external world, this assumption is implicitly extended to this whole world. The disadvantage of this model is therefore that it does not provide any help for the detection of errors coming from the outside world; in the absence of application-level safeguards, such errors may propagate into the Delta-4 system.

The advantage of this approach is that it offers improved availability of the gateway function, with a minimum of essential additions on the non-Delta-4 side. It is therefore the recommended way to implement gateways between Delta-4 and non-Delta-4 systems if the risk of error propagation mentioned above is considered acceptable.

12.2.   Inter-working with the Physical Environment

We will now turn attention to "dumb" peripherals, which are the normally understood domain of Input-Output. Several of the issues of the previous section are related to or recognisable in what follows. Moreover, some of the approaches identified below appear general enough to apply equally to communications systems, microprocessor-controlled peripherals and the like.

There are two major divisions to the discussion, for which the issues involved are very different. The first, for which Delta-4 is able to offer significant support, is concerned with the acquisition of information from the environment; the so-called "sensor" problem dealt with under section 12.2.2. The second is the converse, the so-called "actuator" problem discussed under section 12.2.3; a more difficult issue, in which particular cases are straightforward but explicit support outside Delta-4 is needed.

12.2.1. Definitions

A transducer is a device that converts energy in one form into energy in another form. For the purposes of this chapter, this definition is intended to be interpreted generally and is not restricted to devices that convert to and from "kinetic" energy; e.g., a device that receives an electronic signal and displays a character on a screen is a particular type of transducer. To the necessary generality, all devices that enable the physical environment to be "read" or "written" by a computing system are transducers.

However, the terms input transducer and output transducer are used in this chapter to refer to devices (normally electronic) which are connected (e.g., by cabling) to the external environment but are physically inseparable from and so must be considered part of the host: examples are analogue-to-digital converters and output driver transistors. The devices in the external environment to which such transducers are connected, whilst themselves transducers according to the above definition, are distinguished respectively as sensors in the input case and actuators in the output case.

12.2.2. Input Sensors

Sensors (of temperature, flow, pressure, torsion, weight, level, position, direction, torque, etc.) are in many cases notoriously unreliable devices. To perform dependable computation on input from a basically unreliable sensor is to risk encountering the "garbage-in, garbage-out" phenomenon. Consequently, mechanisms for achieving and managing sensor and access mechanism redundancy, to ensure consistency and tolerate faults, are important. Below, we will evolve models that might be appropriate and coexist naturally with a Delta-4 system.

Three cases should be distinguished:

- A single sensor accessible from a single host (and therefore in the same fault containment domain as the host),
- A single sensor accessible from several hosts (via separate paths as shown in figures 6 and 7) (sensors and hosts are in different fault containment domains),
- Multiple (redundant) sensors each attached to different hosts.

12.2.2.1. Non-Redundant Sensors Accessible from a Single Host. The access mechanism is to perform non-replicated parameter value acquisition (including conversion, and other preprocessing if convenient). At the end of a (hopefully small) non-replicated chain of hardware and software components outside the domain of Delta-4, a single software component in effect "represents" the sensor within the Delta-4 system. There may indeed only be one identifiable acquisition component, though this is unlikely if, for example, Fieldbus-based I/O equipment is used. This chain of acquisition components up to and including the Delta-4 component is therefore collectively referred to in this chapter as the representative of the sensor.

Note the similarity between a representative and a "transformer" as described in chapter 7; it may indeed be convenient to implement such representatives as Deltase transformers.

The structure of a single representative accessing a single sensor can be considered as a logical subsystem. The representative is the only source of sensor values to the rest of the system. It can nevertheless interact with true replicated components, the characteristic of such interaction being assured by Delta-4 mechanisms to have the required property that all replicas receive identical values at identical points in computation. As an example, consider replicated Deltase client objects making an RPC to a non-replicated sensor-representing server/transformer.

If a host fails, the sensors attached to that host are lost. We now explore dependability first through representative redundancy and second through sensor redundancy.

12.2.2.2. Non-Redundant Sensors Accessible from Several Hosts. The next refinement is to consider the case where the reliability of a single sensor is considered sufficient, but the dependability of a single representative is inadequate (or thought to be a weak-link in the "dependability chain"). Can the representative be replicated?

To achieve representative redundancy, it is necessary for the sensor to be physically connected to a suitable input transducer on each station on which replicas of the representative might reside, i.e., the relevant replication domain. A point of concern is whether there are input transducer failure modes that are common-mode (for example, can an analogue or digital input transducer short-circuit the input wire). However, unlike the complementary case of outputs which is discussed below, there is little evidence that this is a serious problem.

In section 12.2.2.1, it is pointed out that within a Delta-4 system, replicas are assured by Delta-4 mechanisms to acquire identical values from a representative. However, interaction with the external world is outside the scope of Delta-4 mechanisms; replicated representatives under active or semi-active models of replication have no such automatic assurance of receiving exactly the same data when reading an input value from a sensor. If each of a set of active or semi-active replica representatives is allowed to interrogate an external sensor independently and directly, it will do so at a slightly different time. Even if perfect synchronism could be assured, it is unlikely that two replicas will receive bit-for-bit identity in a parameter value provided by the configuration given in figure 6, which shows normal analogue input preprocessing (analogue-to-digital conversion followed by linearization and normalization, though the latter two processes are normally deterministic and so do not compound the argument).

Even in the simple case where replica determinism is assured by computational-path-identity (see section 6.4.1), active replication of *representatives* is generally not possible. However, we now show that it is straightforward under the passive-replication model, and is possible with support from the application support environment under the semi-active replication model (see figure 7). Note that although figure 7 shows a similar physical configuration to figure 6, the replication models concerned are based on fail-silent nodes.

Under the passive-replication model, checkpoints are taken each time the sensor value is provided to other computation. Under the semi-active replication model, a somewhat different approach must be adopted needing environment support. When a sensor value is requested by a replica, this request must be treated differently depending whether the replica is a leader or follower. Only the leader actually collects values from a sensor. This value is then immediately sent by the support environment to the followers, as described in chapter 9 for propagating a consistent opinion on time. As in that case, the local environment then supplies the propagated value when a follower requests the same reading.

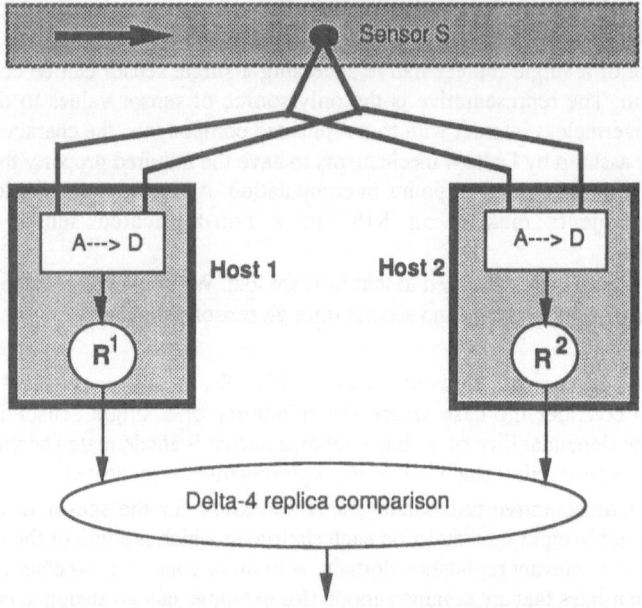

Fig. 6 - Single Sensor read by Two Active Replicas

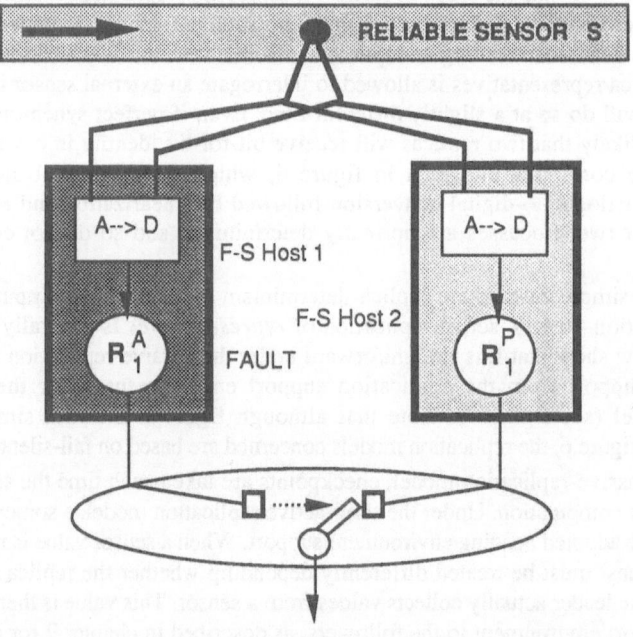

Fig. 7 - Single Sensor read by Two Passive or Semi-Active Replicas

By this means, the principles of semi-active replication are met; exactly the same reading is provided to each replica, even though the internally-generated requests are separated in time by a delay up to the maximum latency between the leader and its slowest follower. In the case of a failure of a leader after a sensor reading has been made but before this reading has been sent to the followers, the reading has not yet influenced the computational progress of the object. Providing the sensor is, as we are here assuming, a "pure" input device, the reading can safely be discarded and another one taken by the new leader replica.

Section 7.2.1 discusses various Delta-4 transparency mechanisms and the means by which these are managed. The model described here is one in which at least part of the distinct behaviour required of a leader replica and a follower replica could conceivably be provided, with assistance from libraries and tools, in a programmer-transparent way. Generic support mechanisms require that suitable abstractions be identified which nevertheless permit the many different actuator interfaces which might be encountered to be constructed. A less ambitious approach is to construct support libraries, language mappings and tools for each interface of interest.

A point of concern is that both passive and semi-active replication require a station to be fail-silent and the above discussion assumes that this property extends to the input transducers and all other components of the representative. This raises the design issue as to how to construct, for example, fail-silent analogue-to-digital converters.[1]

12.2.2.3. Sensor Redundancy. Consider now a set of sensors each intended to measure the same plant parameter (the flow in a pipe, the direction of motion,...) and each of which is connected to a (different) single station on which a single representative executes.[2]

The redundant sensor model that develops naturally from the above starting point is termed the "Rivals" model, in recognition of the fact that different sensors intended to measure the same parameter are not, and cannot be modelled as, true replicas. Even when correctly working, they will not provide bit-for-bit identical results other than by coincidence; this may or may not happen often, but cannot be guaranteed. On the contrary, they should only be thought of as providing similar results, where the degree of similarity is sensor-representation-specific[3] and subject to the laws of physics and perhaps statistical characterization. Nevertheless, to permit proper design, a formal definition of the (highly application-specific) meaning of this "similarity" must be captured in an application requirements specification.[4]

Judging what is to be considered a consensus value in the presence of sensor faults for the parameter concerned is necessarily application-specific (indeed, derivation of consensus even in

---

1 In this particular case, a design approach may be based on the so-called "inverse" method: a D/A converter is used to convert the digital value back to an analogue value that is then compared with the input value by an analogue comparator (with appropriate hysteresis). However, the analogue comparator (the need for which arises because of the approach taken; it is not an "inherently required component" of the conversion function) must be designed so that its failure leads to silence. The requirement has been changed and perhaps simplified, but not solved. It can be solved; the point is, however, that the design issues involved in any such exercise must be fully confronted.

2 More complex possibilities, where sensors are redundant and each sensor has replicated representatives, are ignored in this discussion for simplicity. In the event that such a design approach proves to be justified, the principles described here may "simply" be combined.

3 There may not even be type-consistency: the model given here can cope with, for example, one sensor representation of "direction" as integer degrees $[0..359]$, another as floating-point radians $[-\pi..+\pi]$, and a third as an unnormalized fixed-point Cartesian coordinate (X,Y)

4 The fact that identity is not sought between rival sensor representatives permits a spectrum of possibilities. There is not only no need for sensors to be of identical type but, in fact, the representatives might work in very different ways. For example, one "opinion" on the flow rate in some pipe might be derived directly by means of a flow sensor, and a rival "opinion" might be derived indirectly by calculation of the rate of change of level in a destination vessel.

the absence of sensor faults is highly application-specific) and a suitable algorithm should also be captured in the application requirements specification.[5] Well-known examples are: calculate mean, find median, discard extremes, etc., often these are used in combination and iteratively.

The provision of such application-specific algorithms falls outside the domain of standard support environment mechanisms; thus, in addition to the representatives, an application component that provides the consensus algorithm is required. The crucial observation that we wish to make here to complete this description of the basic model is that *the consensus component may be replicated* (see figure 8). The replication model is not constrained; active, semi-active, or passive consensus components are all possible.

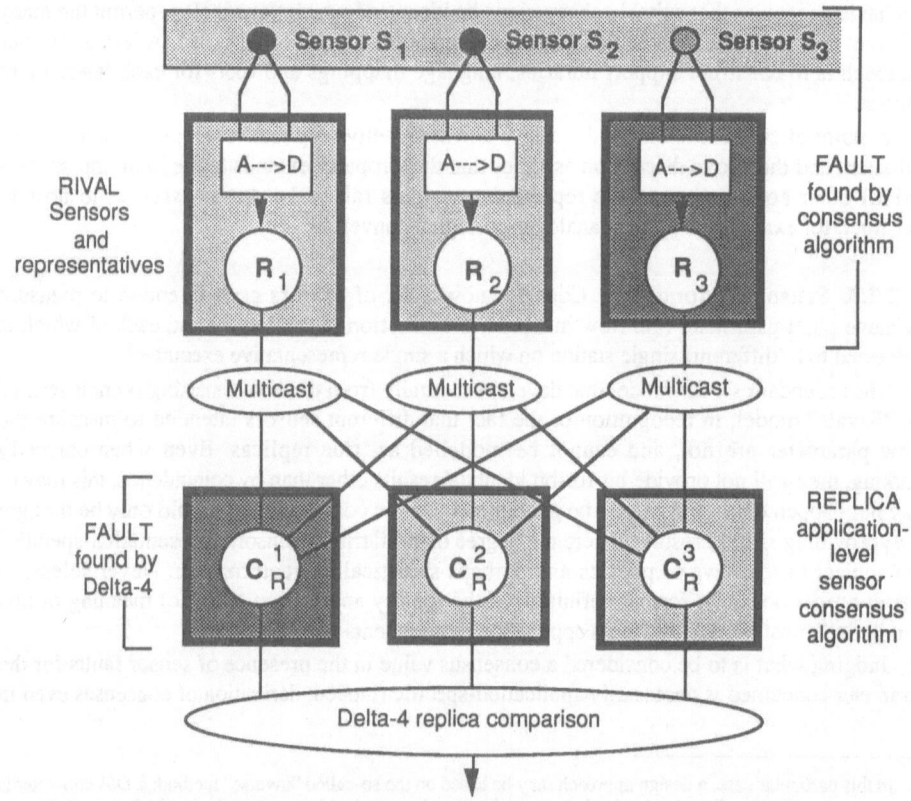

Fig. 8 - The "Rivals" Configuration of Multiple Sensors and Representatives

Note that, unlike replicas where failure is a managed event, detected and reported to system administration by MCS voting algorithms, representative/sensor combinations are found to have failed by the consensus algorithm. In order for such a discovery of failure to be treated as a managed event, application level consensus code must invoke, and so needs an interface to, system administration (see sections 8.2 and 9.5).

There are various means by which the apparent performance limitations of the model may be overcome, should they prove to be significant for some application. First, if a parameter is

---

[5]   The specification must, of course, also define the type of the result!

needed only in one replicated computation, there is no need for the consensus algorithm to be implemented in a separate address-space (component) to that performing that computation (Deltase offers a good way of still maintaining structural separation of function, described in chapter 7). The same might apply to the calculation of an indirect "opinion", as in the flow example above. Second, there may be no need for the sensor representative to be constructed to only acquire, convert, linearize and normalize a single value when prompted to do so by an RPC. Such components may instead use local cyclic activity or respond to externally generated events, in consequence maintaining values ready-transformed and date-stamped, for immediate response to RPC. Third, it is probable that a single component will represent a multiplicity of sensors and may indeed be constructed with a database-like interface capable of more complex local processing.

12.2.3. Outputs to Actuators

12.2.3.1. Interfacing Issues. We first consider problems associated with the design of output-transducer-to-actuator interfaces and the nature of the output transducers and the actuators themselves.

As with inputs, Delta-4 voting mechanisms play no part in assuring the correctness of outputs, even though these are generated by replicas. If the host concerned is fail-uncontrolled, the output that appears at its interface may be incorrect. Even if the host concerned is in all other respects fail-silent, it is difficult even to define a meaning for this property with respect to the type of output signal with which this chapter is primarily concerned. For example, a digital output might have two active states, driving current and not driving current. Neither state corresponds to "silence"; indeed there is no obvious state that can generally fulfil this role, although mechanisms which isolate such outputs may be effective in some cases. For a standard 4-20 mA analogue output to a valve position actuator, there is again no condition of the output that is in any satisfactory sense "silent". Preserving the last correct output value may also meet the requirements of some cases but is not generally acceptable.[6]

Application code does not act upon the external environment. Instead, the action is carried out through the intermediary of special hardware components that we refer to as *output transducers* and perhaps through support environment software components that we refer to as *output transducer drivers*. Such intermediary components are conventionally designed to offer the programmer a useful procedural abstraction of the output action concerned. Assuming there are no design errors, when a procedure is called, an effect occurs at the interface between the host and the external world that can be directly interpreted as corresponding to the intended external result.

Where greater dependability is required than is provided by the single actuator representative, there are three cases of connection of replicated representatives to actuators to consider.

- "Simple" actuators, with no special provision for the connection of multiple output transducers.
- "Compound" configurations of "simple" actuators.
- Actuators with multiple interfaces.

---

6   Some valve position actuators have a mechanism to ignore outputs which demand a change exceeding some percentage of the total output travel. For these, "silence" could be approximated by issuing a demand to go to one limit; the output is ignored unless the actuator is already close to this position. However, this interpretation is with respect to actuator behaviour on transducer failure. In contrast, replicate transducer takeover on transducer failure might require "silence" to be an open-circuit. Such a condition may not be satisfactory to the actuator during the interim period, or as a safe state if there is no more redundancy. Actuator silence is not (generally) transducer silence.

The first case is discussed in the following section 12.2.3.1.1, where the general concept of actuator representatives is also introduced. The other two cases have sufficient commonality to be discussed together in the subsequent section 12.2.3.1.4.

12.2.3.1.1. Actuator "Representatives". In the highly specialized world of industrial actuator technology, it is rarely the case that an actuator is designed with replication in mind. For the task concerned, no special provision has been made for the connection of more than one output transducer. The actuator possesses, in effect, a single interface. The system must offer such "simple" actuators a single, validated version of the output in the form required, masking, if necessary, any time skew between different replicas.

If the dependability consequences are acceptable, a single synchronized version of the output may be achieved by constructing a single active intermediate software component driving a single output transducer. Validation is then performed by Delta-4 mechanisms on an internal-to-system representation of the output, e.g., a message specifying a new value. This intermediate object is, in effect, representing the actuator within the Delta-4 system; a dual of the situation described earlier for sensor representatives. This can be expected to be a common configuration and would typically be based upon a simple conventional Fieldbus link between a Delta-4 system and an "intelligent" peripheral.

Where the reliability of the single actuator is considered sufficient, but the dependability of the single representative is not, it is necessary to physically connect multiple transducers to the actuator interface in such a way that *from the point of view of the actuator interface*, there is only one active transducer. See figure 9, which shows two examples of this. The first is a simple configuration of a pair of driver transistors each able to cause a simple valve actuator to go to the "on" position. The second is the very common traditional *changeover* system, where devices such as solid-state switches or even electromechanical relays are used as a basis for isolating an output transducer. Requirements are imposed on the replica driving of node output transducers, and upon the nature of the interconnections between the output transducers and the "simple" actuator.

12.2.3.1.2. Node Requirements. A single output at a time is naturally achieved using passive replication, but semi-active replication can also be given this property if, at the follower host interface, a call for output is prevented from having an effect. When an output procedure is called by application code on some host, an effect normally occurs at the host output transducer that corresponds to the intended external result. If a more elaborate output transducer driver is constructed such that, although all replicas are active in that they execute the output call, then the effect only occurs at one host transducer; at the other replicas' hosts, the transducer drivers "intercept" the call for action.

Such a mechanism introduces a bias into the status of the replica that is "really" performing the output, so naturally belongs with replication models where such a bias is already evident and managed. Semi-active replication has such a naturally-occurring bias.

In both passive and semi-active cases, output fault containment is required. Since, under these models, the node involved is fail-silent, it is tempting to appeal to the properties of fail-silence. When a node fails its output transducers are assured to be disabled in whatever way is appropriate for that class of output. As pointed out at the start of section 12.2.3, this is highly application- and transducer-specific.

12.2.3.1.3. Interconnection Requirements. The major issue in interconnecting multiple output transducers and "simple" actuators is that of common-mode failures.

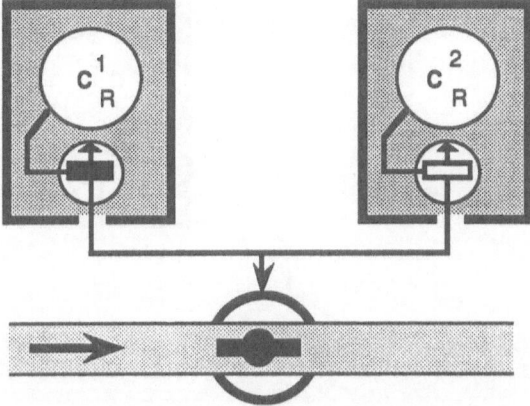

An on/off valve driven by two output driver transistors

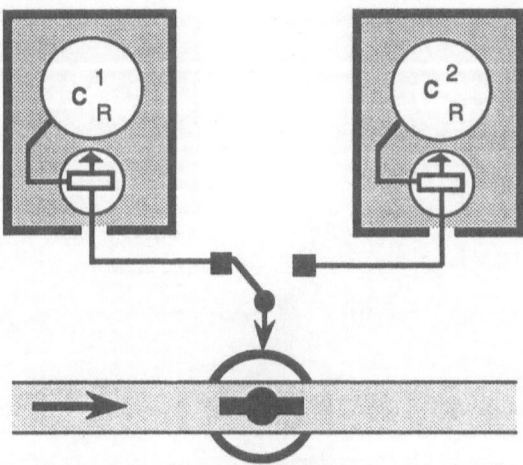

A simple mechanical output "changeover" system

Fig. 9 - Single Actuator Controllable from Multiple Sources

In the first example of figure 9, if the driver transistors are required to carry heavy current, they possess an unfortunate probability of failing into a short circuit condition, as a molten lump of silicon; the consequences are shown in the first example of figure 10). The valve is permanently "on" until the interface is disconnected.

In the second example of figure 9, if the basis for isolating an output transducer is inappropriate, the condition illustrated in the second example of figure 10 can arise. Although, in the configuration shown, the fault remains latent, there is no obvious means of detecting that the fault exists, the intended output transducer redundancy is not available, and there is a difficulty in replacing the faulty component on-line.

A fault in inter-station signal earthing can also be very difficult to detect. When a transducer fails, a current path for its newly-activated replica might not be through the intended route and so cause all sorts of problems, from excess noise-sensitivity to activating mains safety cutoffs.

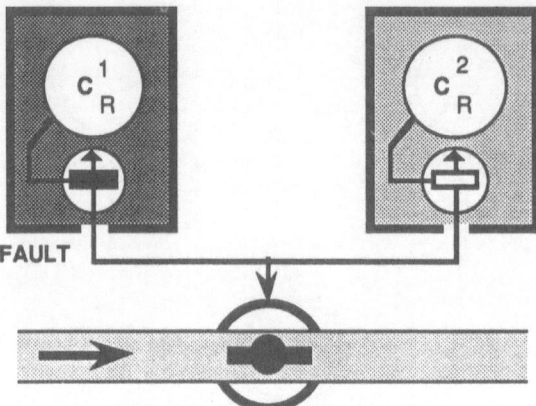

Output transistor fails short-circuit

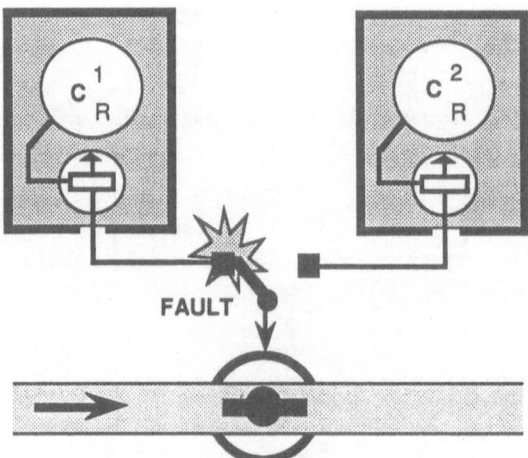

Relay contact welds fails stuck-left

Fig. 10 - Common-Mode Failures

To summarise, the physical connections to the actuator from each of the nodes on which the intermediate voting object might reside must be made in such a way that there is no interactive interference (e.g., from output feedback sensors or unexpected current paths) and that no arbitrary failure of the output transducer on one node can prevent another output transducer from fulfilling the output requirement, within the necessary level of coverage.

Primitive output transducers rarely exhibit benign failure modes. Through feedback, the local fault-detection of each node concerned may be extended to the output transducer and perhaps beyond (the condition of the output is sensed and compared with the desired output). In the case of passive replication, the present primary replica, or in the case of semi-active replication, the present leader, or in the case of active replication, some nominated replica, is the master and hence connected to the peripheral device. Under as many conditions of failure as possible some form of isolation device should disconnect the output.

Due care must be paid to the peculiar properties of isolation devices. For relays, this includes such design issues as "wetting" current and contact material migration, the significant chance of a contact welding shut, the very limited number and rate of operations, the long (relative to electronic switching) changeover times, open-before-close vs. close-before-open designs, vibration-sensitivity, etc. However, isolation thresholds are better than with solid-state switches, which can be important where cable runs are long, equipment is on different phases of mains or handles high voltages, earth-return-current design issues are troublesome and so on.

12.2.3.1.4. Voting Actuators. Where an actuator is explicitly intended to be used with replicated outputs, care is taken to provide for more than one interface, to isolate faults in these interfaces from each other, and to provide a solution to interactions between these interfaces, based on some model for how these interfaces are to be used and how and what failures are tolerated. Such actuators, and configurations of simpler actuators that exhibit similar properties, are now discussed. An example of a compound configuration of simple actuators, introduced at the start of this chapter, is the series/parallel combination of on-off valves.

Active replicas will each cause a version of some output to appear at the several respective host output transducers. We shall refer to these as "replica outputs". An actuator that is capable of accepting several loosely-synchronous outputs from such replicas and majority-voting on them locally to determine the action to be carried out, may use three or more of such replica outputs directly. Suppliers of computer control equipment may be required by the plant contractor to provide three or four replica outputs, to be voted externally by independently-supplied mechanisms of this variety.

Delta-4 provides an appropriate means to construct the active replicas and therefore the replica outputs (see figure 11), given that the degree of desynchronization experienced by active replicas can be assured to be within the bounds specified for the voting mechanism.

Where a difference in outputs occurs, one of the following conditions is the cause:

1) An unmanaged opportunity for application non-determinism
2) Excess desynchronism (beyond some specified limit)
3) Computational subsystem failure
4) Output transducer failure

The first two cases are, respectively, (1) an application design and (2) an application/support system interaction design problem.

The voting actuator is irrelevant in the presence of such problems, which are not covered by any replication model. The system is rendered unable to provide the required service by any such event. The voting actuator gives additional and valuable coverage against cases (3) and (4). It is therefore consistent with and supports the fail-uncontrolled host active replication model.

We now consider fail-silent host models. In the semi-active case, majority voting implies a leader and at least two followers. Outputs generated by the followers must not be intercepted, but must arrive at and be taken account of by the voting actuator. From the actuator point of view, follower failure and leader failure are equivalent. Indeed, the actuator "sees" no significant distinction between active and semi-active replication, other than that one output (that of the leader) will always precede the others.[7]

---

[7] In XPA, the leader precedes by at least a bounded amount (minimum leader-to-follower desynchronisation) and with a larger overall bound(maximum leader-to-follower desynchronisation). Since actuator voting requires the outputs from the leader and at least one follower to arrive, the time bound for the voted result is the difference between the two; this is more tightly bounded than are the replicates themselves.

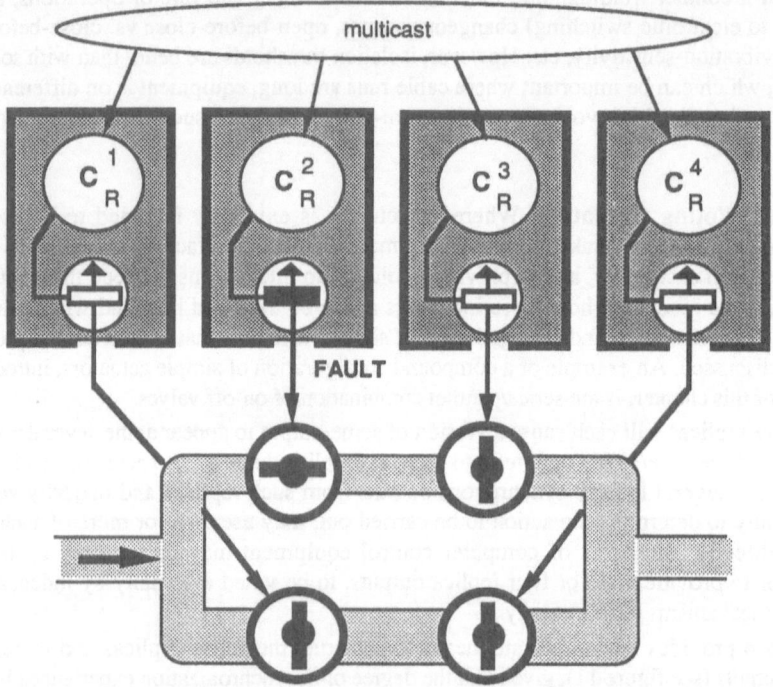

Fig. 11 - Active Replicas controlling a "Mechanical" Voting Actuator

A point made earlier in section   12.2.3 was the difficulty in extending the notion of "silence" to output transducers. However, with voting actuators, if this variety of host fault does not result in "silence", this can be tolerated. How far inside the host can this need for silence be relaxed in the presence of a voting actuator? Again, consider the possible causes of a difference in outputs. This might result from either (1) or (2) as above, for which the same conclusions apply in the semi-active case, or:

  5) Computational subsystem lack of coverage (failure to exhibit fail-silence)

  6) Output transducer lack of coverage (failure to exhibit fail-silence)

The voting actuator gives additional and valuable coverage against (5) and (6) which may otherwise be difficult to handle. The technique described in this section is therefore a means of tolerating fail-silent support environment lack of coverage in the particular case where this leads to output failure, and of relaxing the fail-silence requirement of semi-active replication with respect to the transducer hardware itself.

12.2.3.2. Output Error Detection and Recovery

12.2.3.2.1. Imperfect Output Error Containment. The discussion of "simple" actuators above assumes that fault-containment at an output transducer has been perfectly achieved. This may not be the case, e.g., an output transducer may not fail cleanly and a transient pulse might occur which causes the external world to transit to a different state to that represented in the control system, or it might not be known whether a particular output has or has not occurred at the point of failure. An approach to solving such problems is to use feedback or application knowledge to establish consistency between the control system and the outside world, and then to resolve any difference between the discovered state and the desired state of the outside world. This approach, which also applies to faults arising with the actuators themselves, is discussed in the following sections.

Under passive replication, host hardware faults are tolerated by backward recovery to a previous valid state and subsequent re-execution (see chapter 6). This concept is difficult to extend to the real world, since restoring a previous state will not, in general, be possible. The real world can only roll forward; real actions cannot by any single mechanism or principle be rendered "undone". Once cash has been released into a cash terminal customer's hand, it cannot be taken back. Once a valve has opened and a chemical reaction has begun, it cannot be reversed. In the general case, devising a way whereby a future state of the global system is arranged to coincide with a previous state is perhaps impossible; instead, an acceptable future state must be reached.

For a cash terminal, the requirement might be to complete, if possible, the transaction correctly. Presumably cash should indeed be released to this customer if the transaction has proceeded to the point where it is mechanically ready for release. Can the system discover what amount has already been released? If this proves impossible, a secondary wish might be to limit the damage, perhaps in terms of financial loss or reputation.

Similarly, for the process plant example the requirement might be to complete the reaction correctly. The amount of reagent might be determined as a function of flow rate and time; what was the flow? When was the valve opened? Has a quantity limit or critical time been passed? If so, a secondary requirement might be to limit the damage, perhaps in terms of environment or plant down-time.

To deal with this type of situation, the notion of *forward error recovery* (see chapter 4) is used, and is discussed in the next section.

12.2.3.2.2. Computation Issues. With simple output actions, if no checkpoint is received, a real action may not have occurred at all, or may have been carried out by the primary replica which immediately failed prior to propagating the fact to the rest of the system. If the checkpoint is sent before the real action, this action might or might not be carried out by the primary replica, but the checkpoint information is that it will be. Rolling forward a computer control system that has experienced failure may therefore cause an output either in the first case to be repeated, or in the second case never to occur. Propagation of information in two physically disjoint directions (out to the real world, and internally to other computation) from a single sequential computation cannot be atomic.

With compound output actions, where several "primitive" real actions take place between checkpoints, the problem is one of dealing with the consequences of failures that occur while the compound action is actually being carried out. These consequences may be varied. The failure may have occurred immediately after an object took its first checkpoint, in which case the compound action will not yet have begun. It may have occurred immediately before the final checkpoint, in which case the compound action will be complete. Alternatively, it may have occurred somewhere in between these points, in which case the compound action may have been partially carried out.

Here, as in the cases examined in the previous section, the solution is to use *forward error recovery*. This involves executing application-level recovery mechanisms specific to the external system involved. The first requirement is to establish consistency between internal opinion of external state and actual external state; i.e., to achieve a "known" state. One way to achieve this is to correct the internal state, for instance by interrogating the environment to determine the actual situation: *has the cash been released? Has the valve been opened?* This requires that arrangements are made in the design — i.e., the necessary sensors are included — to obtain all relevant information. If complete interrogation is impractical, it may be possible to make use of assumptions (e.g., the laws of Physics) which allow some properties of the environment to be inferred: *what is the temperature in the reaction vessel? How much reagent has been used?* Another, more interactive, possibility is to change the external environment so as to progressively reduce the uncertainty, i.e., carry out some initial compensatory actions that are guaranteed or can be reasonably assumed to place the environment and the system in a compatible state: *shut off reagent flow, flood reaction vessel with neutralizer*.

From the known state determined by such activity, the necessary actions must be applied to complete, or safely terminate and possibly restart the action anew (the nearest equivalent to the concept of "abort" in a roll-forward world): *dump neutralized reaction products for later analysis and recovery, scrub reaction vessel, restart with new reagents* (see figure 12).

12.2.3.2.3. At-Least-Once Semantics. The discussion so far has established that, in the general case of computer subsystems linked to non-dependable I/O subsystems, under host failure conditions:

- Where the semi-active or active/passive models are used, since output and notification/checkpoint cannot be atomically associated, a new leader or new primary replica cannot simply continue, but must pursue a forward error recovery procedure.

- Where the active model is used, there is no built-in mechanism to synchronize or bind input/output actions with computation, and, if the failed host is the one responsible for output, then an event must be raised to cause all replicas to pursue a forward error recovery procedure.

However, in the particular case of *idempotent* output actions, i.e., actions that do not cause incremental change and can therefore be repeated (for example, a valve position demand), application-specific forward error recovery procedures can be avoided if at-least-once semantics can be assured. In section   12.2.3.2.2 above, it is shown that if a checkpoint is sent before a real action, it is not known whether this action takes place or not. However, if, on takeover, a new primary performs the real action, then at-least-once semantics are achieved. Similarly, in the semi-active case, if a notification is sent before a real action is requested, then the new leader can achieve at-least-once semantics in the manner described below.

In the general discussion of forward error recovery above, the granularity of this recovery is assumed to be coarse. If at-least-once semantics are to be effective, the granularity must be made sufficiently fine for idempotence to be practical in the context of a dynamic controlled process. If the implementation involves an excessive inter-command latency as regards changing the valve setting, then the plant may not be controllable, or may move to some state that requires a full forward error recovery. Reducing the granularity ensures that the possible discrepancy between plant and control system is correspondingly reduced.

Semi-active replication allows short inter-notification latency because of the small network bandwidth consumed by such notification (see chapter 9). Thus, although the concepts discussed below can also be applied in principle to passive replication, such an approach would not be practical.

In what follows, a real output is assumed to be a primitive single-step or atomic transducer event at the leader's host. The last decision that is known to have been taken by a failed leader is defined in its last notification. We need to ensure that a new leader is not diverted to a different execution path prior to completing the real actions previously requested by the ex-leader.[8]

---

[8]   A new leader can receive an external event and therefore be diverted to a different execution path at any subsequent decision point. As long as there are no externally visible consequences, it does not matter if the original leader had previously passed this point without diversion. Real-world actions are, however, such

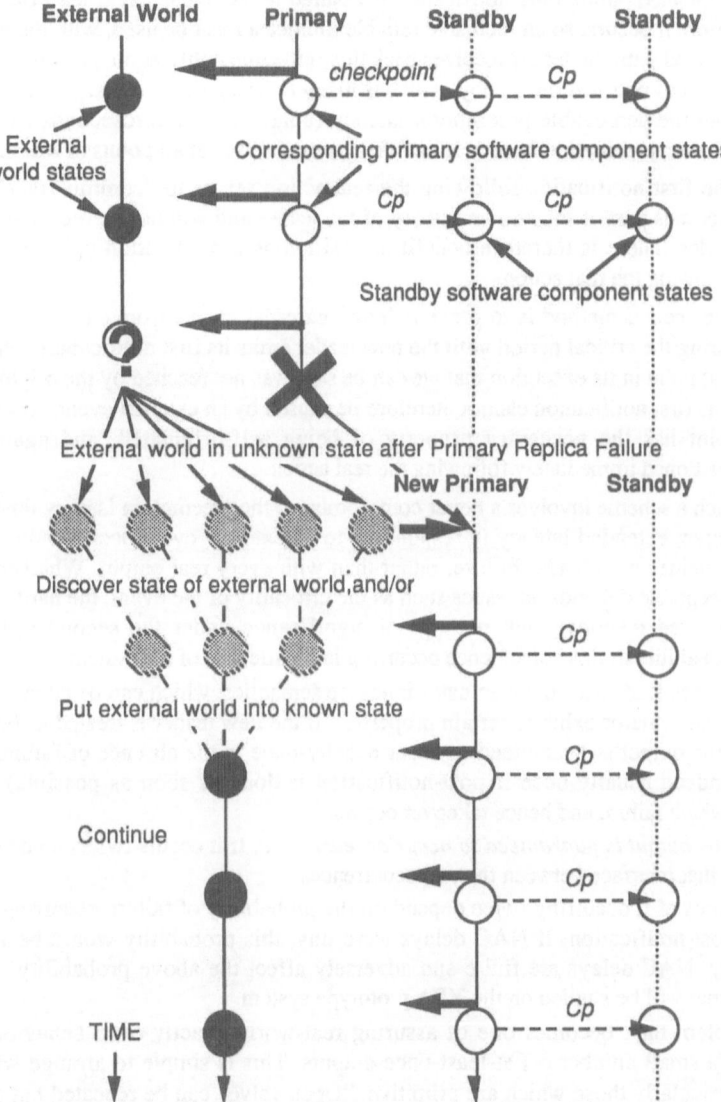

Fig. 12 - Forward Error Recovery

There are two methods by which this may be achieved. Both make use of the concept of a "notification point". This is a point at which, unlike a normal decision point, notification always occurs. A notification point may or may not also be a decision point.

- The first method is to locate such a notification point before the real action, with no decision points between this point and the real action. The real action can still be

---

externally visible consequences; to be assured that a different execution path does not "orphan" an action in this way, the new leader must not be allowed to divert execution at any decision points encountered between the last point notified by the previous leader and the point of action.

"orphaned" unless the notification is assured to have reached at least one follower before it occurs, so an inclusive reliable multicast must be used, with the real action blocked until the leader receives back this self-same notification. Note that this delay is that of notification latency, which is likely to be significantly larger (milliseconds) than the achievable preemption latency (hundreds of microseconds). Under this scheme, preemption latency is therefore compromised at all points of real action.

The first notification following the real action serves to "commit" that action; it becomes part of the known history of the leader and will not be repeated by a new leader. There is therefore benefit in positioning a notification point immediately following the real action.

- The second method is to prevent "new" external events from causing preemption during the critical period until the new leader emits its first notification, which is the first point in its execution that we can be sure was not reached by the original leader. This first notification cannot therefore be caused by an external event. A notification point has the necessary property of being self-originated, and again is best positioned immediately following the real action.

  Such a scheme involves a better compromise to the preemption latency time than the above; extended latency in responding to an external event occurs only when this coincides with leader failure, rather than with every real output. Whether either is acceptable depends on issues such as the criticality of the event, the hardness of the required response and, perhaps of significance under the second scheme, the probability of this coincidence occurring in the lifetime of the system.

Both schemes lead to a simple actuator interface semantics, which can be taken advantage of providing the actuator exhibits certain properties. If the new leader is treated as being so in all respects, the output is guaranteed to occur exactly-once in the absence of failure, and at-least-once (indeed usually-once if post-notification is done as soon as possible) for each occasion on which failure and hence takeover occurs.

That is, the *output is guaranteed to occur at-least-once*. If it occurs twice, no other output will occur on that interface between the two occurrences.

The chances of it occurring twice depend on the probability of failure occurring between action and post-notification. If NAC delays were tiny, this probability would be very low. Unfortunately, NAC delays are finite and adversely affect the above probability. This is a design issue that will be studied on the XPA prototype system.

The problem then becomes one of assuring real-world exactly-once behaviour in the presence of (a small number of) at-least-once outputs. This is simple to arrange with many actuators, particularly those which are primitive. "Open valve" can be repeated but the valve opens once; repeated demands are, in effect, ignored. This is a canonical example of a very general solution that can be applied to more complex actuators, given that there is some control over their design. This solution is to construct these as (mechanical, hydraulic, electromechanical, electronic,...) state machines, which, again in effect, ignore all but an expected subset of the universe of possible outputs when in each state; each acceptable output causes transit to a new state for which the same holds. The at-least-once variety of output repetition is tolerated by arranging to ignore any repetition of an immediately previous output; if repetition of action is required, it can be arranged by introducing intermediate outputs causing transit to specially-introduced intermediate states.

If the actuator is "given" but does not have such properties, there is a problem with this form of replication. Nevertheless, some quite complex cases are amenable to the approach outlined here. Consider the following "intelligent" actuator: A riveting robot has operations *step left, right, up, down, in, out*, etc., and also *fire, reload* rivet (it is immaterial here whether these commands are pulses on individual control lines, simple codes sent over a serial or parallel interface, or something altogether more sophisticated). *Fire* is at-least-once in that it will not succeed unless *reload* has been issued since the previous *fire* command. The same is true of *reload* with respect to *fire*; these operations naturally alternate between a pair of states. *Stepping* is incremental, but since the steps are tiny and the position is sensed (as part of a closed control loop) before a *fire* command is issued, an occasional error of one additional *step* is naturally tolerated by the control philosophy.

12.3.   Summary

This chapter has argued that extending the boundaries of dependability to encompass sensors and actuators must be managed in an application-specific way. Dependable input-output design is not amenable to generic support, although there are some cases where particular modules may be constructed to be consistent with a generic architecture; examples are indeed available from some manufacturers.

We do not consider it sensible to construct an extension to the Delta-4 environment to encompass the provision of dependable input-output. Much of the limited generic support that can sensibly be offered by an environment like Delta-4 is already present or planned. Where it is not (such as in the area of intercepting follower output calls), it will take experiment to establish the ideal mechanism and to create the appropriate tools. This work, whilst not inconsistent with Delta-4, has separate concerns to the central thrust of the present project, so should be pursued in another context, such as productization of the Delta-4 architecture.

Chapter 13
Security

Since Delta-4 is open and compatible with current operating systems and applications, the basic security is provided by the local operating system, e.g., UNIX. However, in some distributed contexts, this basic security is not sufficient, e.g., because some individual sites or operators cannot be trusted, and/or because the information processed by the distributed system is very sensitive. This chapter presents some solutions to provide secure distributed application services, such as a file archiving service, and to manage the security of a Delta-4 system. The solutions are such that an intrusion into a part of the distributed system will not endanger its security, i.e., they are *intrusion-tolerant* solutions [Deswarte et al. 1991].

Computing system dependability has been classically considered as being the ability of a computing system to deal with faults that are, implicitly, seen as being accidental. Security, i.e., the ability of a computing system to deal with intentional attacks, was historically a somewhat later worry than "classical" dependability. Such intentional attacks are becoming more and more numerous and subsequent losses are ever-increasing, causing growing problems in computing centres, banking systems, etc. This is particularly true in large network environments. The numerous cases of computer fraud that are related in the newspapers underline the fact that although quite many techniques for achieving security have been developed, the problems of computing system security have not really been satisfactorily solved.

Until now, security and dependability have had their own meetings, terminologies, standardization committees... and very few research groups work on both aspects. The quite evident relationship that exists between security and dependability has not been clearly formalized. Nevertheless, recent attempts at relating these two domains tend to consider dependability as a general concept that embraces security, reliability and safety as different attributes. This viewpoint is not only conceptually interesting but should lead to the application of concepts originally intended for accidental faults to intentional faults, and vice versa. This approach implies the designing and implementation of secure and reliable systems using homogeneous solutions rather than viewing security and reliability as two opposing requirements.

Dependability can be more rigorously defined as that property of a computing system that allows reliance to be justifiably placed on the service it delivers (cf. chapter 4). This general but precise definition does not need any specification of the impairments or faults that may lead to a system failure. These faults may be either accidental ("classical" dependability) or intentional (security). This is why dependability has to be considered as a general concept, involving all kinds of faults. Reliability and safety may also require security in the sense that an intentional action may alter the mechanisms intended, for example, to ensure system reliability. It is also true that security requires reliability because an accidental fault in a security mechanism may allow intrusions that are usually not possible. In fact, it is not possible to integrate one aspect into the other as is sometimes attempted, but on the contrary these concepts should be considered at the same level with tight relations linking one with each other. Figure 1 shows a conceptual organization of dependability as a generic concept (cf. chapter 4).

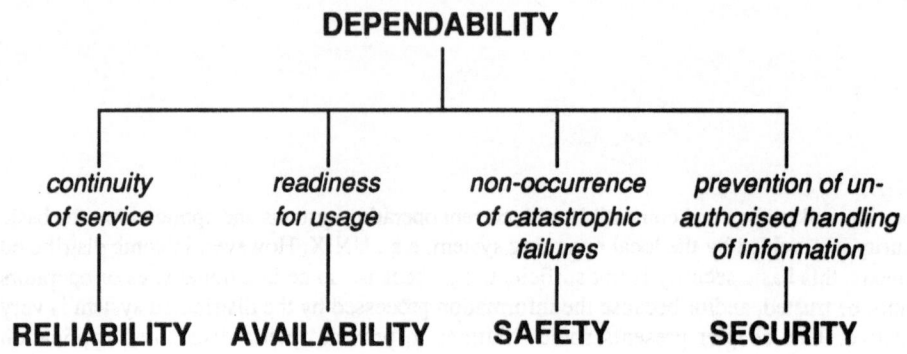

Fig. 1 - Perceptive Attributes of Dependability

Impairments to security come primarily from intentional faults, which may be either design faults or interaction faults. Intentional design faults (usually called *malicious logic*) are introduced into the system so that it delivers a service that is not as described in the specifications in order to mislead the user about the operations performed by the system. These illicit functions are often carried out without being detected by the user who may continue to work with a corrupted system for a long time before discovering the faults. Interaction faults are performed while the system is already in use, taking advantage of the weaknesses in the security mechanisms or policy. They are usually called intrusions. Both classes of intentional faults may cause a failure of the system. Their effects depend on the efficiency of the mechanisms designed to ensure security.

13.1. Principles of Intrusion-Tolerance

13.1.1. General Concept

An *intrusion* can be defined as an *intentional operational external fault* (cf. chapter 4). Different types of intrusions can be classified according to who makes the intrusion:

- It can be somebody outside the system who tries to access it. This is the most well known kind of intruder, but not the most important. In this case the intruder has to *bypass* physical, procedural and logical protections.
- The second kind of intruder is a user of the system who tries to access information or services for which he has no right of access. The intruder tries to *extend his privileges*. This is the most common intrusion. The intruder has "only" to bypass the logical protection.
- The third — and the most dangerous — type of intruder is a security administrator who *uses his rights* to perform illegitimate actions. In this later case, the administrator has enough access-rights to do these actions but, according to the security policy, is not supposed to do them.

Intrusions can be treated with the same means as for other faults, i.e., by means of fault-avoidance and fault-tolerance. Intrusion avoidance techniques are the most common in secure systems and in particular are the basis of the notion of *trust*. For instance, when the *"Orange*

Book" [DoD 5200.28] states that a reference monitor must be *tamperproof*, it means that you must prevent intrusions into it. In such systems, the protection mechanisms must prevent unauthorized actions. If an intruder succeeds in bypassing these protections, the security of the system is no longer ensured. If a security administrator decides to carry out illegal actions, there is no logical protection to prevent him from doing so. He could only be detected by using an intrusion-detection model [Denning 1986] which is able to detect intrusions by monitoring audit records for abnormal patterns of system usage.

The principles of intrusion tolerance are different [Deswarte et al. 1991, Fraga 1985, Fray et al. 1986]. The system tolerates a bounded number of misuses. If one or more intruders bypass the protection mechanisms and if the number of misuses is less than a given threshold, the security properties of the system (confidentiality, integrity and availability) are still ensured.

Three types of intrusion tolerance can be formulated:

- for *confidentiality*: read access to a subset of confidential data gives no information about the data,
- for *integrity*: the change of a subset of data does not change the data perceived by legitimate users,
- for *availability*: the change or deletion of a subset of data or of a server does not produce a denial of service to legitimate users.

For each property, a tolerance threshold is defined. If the reading, modification or destruction is done on a part D' of data/server D such that |D'| (size of D') is less than the threshold, the properties are always verified (figure 2).

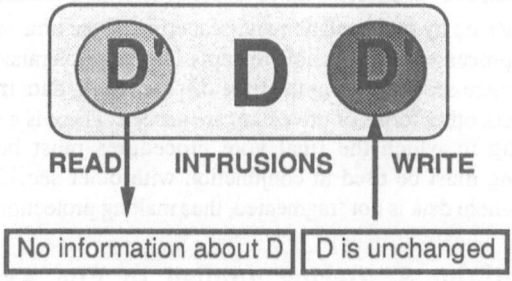

Fig. 2 - Intrusion Tolerance

13.1.2. Fragmentation-and-Scattering

Fragmentation-and-scattering is an intrusion-tolerance method that can be compared with redundancy in a classical fault-tolerance context. Redundancy is used to ensure that the occurrence of a fault in one copy will be of no consequence because this fault will not appear in the other copies. Fragmentation consists of defining different fragments of the data to ensure that once isolated, every fragment is of no interest due to the lack of sufficient information. Scattering refers to the way by which each fragment is isolated from the others. Once this operation is carried out, an intrusion into a part of the system has no consequence due to the lack of significant information involved in the fragments to which the intruder has access. The number of intrusions that can be tolerated without delivery of significant information is dependent on the way the operations are defined. This number is a parameter that can be chosen to determine the tradeoff between security and performance.

Fragmentation may be carried out in many ways, with or without a secret parameter (fragmentation key). It can be performed with the help of other techniques such as cryptography to increase the level of security. It may also be performed with various granularities of the data such as bit level or byte level. The granularity of fragmentation directly influences the security and the performance of the method.

Scattering may be carried out in different ways, depending on which intrusions one wants to tolerate. *Geographical scattering* may be used for both communications and data storage. One example of using different communication links for sending different fragments has been proposed for meshed networks [Koga et al. 1982]. As another example, Rutledge discusses the use of different telephone lines to transmit the different fragments to the receiver [Rutledge 1987]. Geographical scattering for data storage may take place in a network environment where several archive sites are available. This solution will be described later.

Another form of scattering is *temporal scattering*: fragments are sent over a communication channel at different and unpredictable moments. In this case, fragments coming from different sources must be the same length in order to hide their origin.

Frequential or spread-spectrum scattering consists of the use of different frequencies to transmit the fragments in radio-communication applications or in broad-band local area networks.

Yet another scattering technique is *privilege scattering*, which is a way to implement the separation of duties proposed by the Clark-Wilson policy [Clark and Wilson 1987]. With this technique, certain sensitive operations need the cooperation of several users to be performed.

The main idea that governs all forms of scattering is the difficulty that an intruder would have in retrieving all the information. A successful intruder would have to carry out several intrusions in different places or at different times or frequencies in a consistent manner.

Data cannot be protected by fragmentation-and-scattering all the time since it must exist as a whole when it is to be processed. It is possible to apply intrusion-tolerance for communication security or for data storage security but at the time of processing, data must be reassembled. Once data is reassembled, other forms of protection are needed. There is a similar problem with classical error-masking in which the final vote procedures must be trusted. Although fragmentation-scattering must be used in conjunction with other security means, it greatly restrains the locations where data is not fragmented, thus making protection easier.

13.2.   Fragmentation-Scattering applied to File Archiving

13.2.1.   Overall Framework

In Delta-4, a *secure data archiving system* has been designed to deal with different types of intrusions that could occur either in the storage devices (archive sites) or in the communication channel. Data security is provided by intrusion-tolerance and more precisely by geographical fragmentation-scattering (see figure 3).

The basis of the fragmentation and scattering technique is to cut every sensitive file into several fragments in such a way that one or more fragments (where the number of fragments is less than the total number of fragments) are not sufficient to reconstitute the file. These fragments are then stored in geographically distributed archive sites. An intruder accessing some sites cannot obtain all the fragments of a given file unless he has almost overall control of the complete distributed system. On the other hand, to ensure availability, several copies of each fragment are stored on different archive sites. This service thus meets all three security requirements: confidentiality, integrity and availability. Confidentiality and integrity are directly provided by fragmentation-scattering while availability is obtained by the replication of fragmented data. This section, after a presentation of the environment needed for implementing

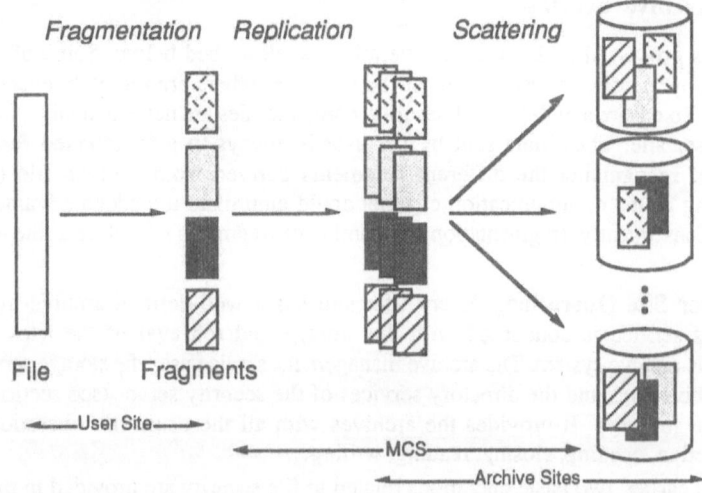

Fig. 3 - General Principle of Fragmentation-Scattering Applied to File Archiving

fragmentation-scattering, gives the basis of the different operations that will take place to secure sensitive data.

To describe the file archiving service, only two kinds of sites are considered, interconnected by the multipoint communication system (figure 4):

- *user sites*, which are personal workstations that constitute the physical domains of their users,

- *archive sites*, designed for long-term data storage, while the user of the data is not logged-in,

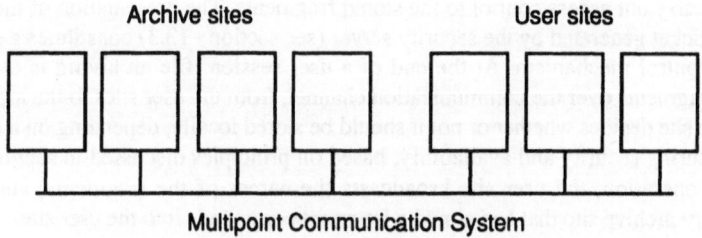

Fig. 4 - System Environment

A user has full authority over the objects stored in the machine he/she is working on. During the user session (log-in time), nobody else can work on the same machine. Key management and access right management will be addressed in the next section (see section §13.3).

13.2.2.   Archive Service

An archive is processed under a set of operations as described below. Some of the archive management operations are carried out in the user site, others are remotely executed by the archive sites. To ensure a high level of security, complete files are never available, except in the connected-user site. Thus, data sent by the user is always in a fragmented form with no possibility of recognising the different fragments derived from a given file (otherwise, eavesdropping of the communication channel could annihilate the added advantages of the technique). Consequently, fragmentation and naming of fragments take place at the user site.

13.2.2.1. User Site Operations. Every user site has a well-defined archive-management service. This service is concerned with the storage and retrieval of the files within the distributed file archive system. The archive management-service uses the storage services of the different archive sites and the directory services of the security server (see section   13.3) to build the file structure. It provides the archives with all the usual file operations such as creation, deletion, opening, closing, reading, writing...

As stated earlier, two basic operations related to file security are provided in the user site: fragmentation and naming of the fragments. The fragmentation operation uses a fragmentation key that is distributed by the security server. This operation is based on fast and simple algorithms that give flexible access to any file whilst ensuring a high level of security due to the scattering of information. The names given to the different fragments are generated by cryptographic methods using the fragmentation key, such that no information can be derived from these names. The naming is carried out in such a way that fragments have a unique identifier, derived from the fragmentation key, the name of the file and some other parameters. During the read operation, the original file is reconstituted with the same key that was used for fragmentation.

13.2.2.2. Archive Site Operations. A storage service is provided by the archive sites. The operations of these sites correspond to simple space allocations on the physical storage devices and data transfers between the storage device and the network. These storage operations are only available to the file-management service embedded in the user sites. The archive sites carry out access control to the stored fragments. The presentation of the fragment names and a ticket generated by the security server (see section   13.3) constitutes a capability-type access control mechanism. At the end of a user session, file archiving is executed by sending the fragments over the communication channel, from the user sites to the archive sites. Every archive site decides whether or not it should be stored locally, depending on a distributed algorithm ensuring security and availability, based on principles discussed in section   13.2.5. For the read operation, the user site broadcasts the names of the fragments, and for each fragment, every archive site that had stored a fragment copy sends it to the user site.

Each archive site acts as a file server that would only store fixed length files, with a "flat directory" structure. The operations managed by these archive sites are fragment reading, fragment writing, and fragment deletion. Only fragment names are visible to the archive site; thus, it cannot determine where a fragment comes from or to which file a fragment belongs.

13.2.3.   File Access Session

The aim of this section is to clarify the different actions needed to access a file either for writing or reading from a user site, where the user is already logged-in, and thus recognised as authorized.

13.2.3.1. Write Operation. The write operation is achieved with the following steps:

W1: The user request for file opening with write-access is transmitted to the security server, which checks the access rights according to the archive descriptor.

W2: If this user has write-access rights to this file, the fragmentation key is transmitted by the security server to the user site.

W3: The file is fragmented and the fragments are named, using the fragmentation key in the user site.

W4: The fragments are broadcasted from the user site to the archive classes in a random order (temporal scattering). An intruder who is eavesdropping on the communication channel or who controls an archive site does not know the order in which the fragments have been sent. The intruder would have to run a large number of trials to reconstitute a file page (see §13.2.4.2).

W5: Each archive site that receives a fragment decides whether it should be stored locally depending on a specific, pseudo-random, distributed algorithm that takes into account the relative available storage space on each archive site (the need to take into account the available space at each site is necessary to maintain a good balance among the different archive sites).

W6: The user-site copy of the file must be deleted.

13.2.3.2. Read Operation. The read operation entails a similar sequence:

R1: The user request for file opening with read-access is transmitted to the security server, which checks the access rights according to the archive descriptor.

R2: If this user has read-access rights to this file, the fragmentation key is transmitted by the security server to the user site thus allowing re-computation of the names of the fragments.

R3: Fragment requests (transmitted in a random order) are broadcast to the archive sites that send the fragments back to the user site.

R4: Once all the fragments have been received, the file is reassembled at the user site.

The following sections will discuss more accurately the way in which both fragmentation and scattering have been implemented.

13.2.4. Fragmentation Principles

A general approach is proposed for the fragmentation operation. This operation consists of defining all the fragments of a file. The file may be of any length and of any type. The fragmentation operation must ensure that, once the fragments are isolated, no information can be obtained from them. Moreover, it must be impossible to guess their origin, this requirement implying that all the fragments from a given file must be of fixed length and that their names do not allow any information to be deduced. Finally, an important requirement concerns data integrity: modification of a fragment must be easily detected at the time of reuse.

13.2.4.1. Partitioning. A method has to be defined that is suitable for producing identical-length fragments from files with very different lengths. The solution proposed is to first cut every file into pages of fixed size (partitioning). Every page may then be fragmented into an identical number of fragments. The files are padded out to reach a size equal to a multiple of a page size (figure 5). All the fragments so obtained will have the same length, which may be equal to that of a packet sent on the communication channel, or a quantum in the mass storage, for example. These choices may also improve the speed of access to information. Another advantage is that one does not need to get the whole file: pages can be retrieved independently. So, a user does not need to reassemble a whole file if he only needs a single page. A signature, built as a cryptographic checksum, is added to each page; this signature is checked by the read operation to verify the integrity of the page.

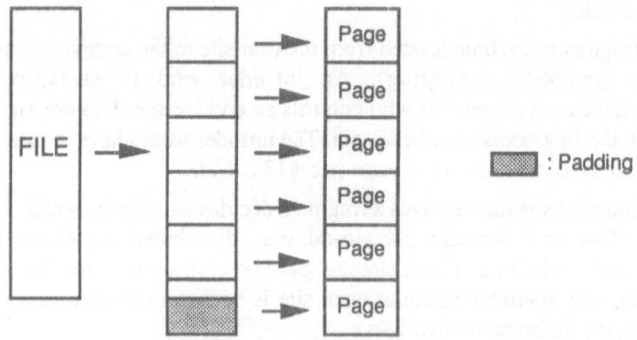

Fig. 5 - A File Partitioned into Pages

   The mean overhead due to padding information is half a page, and is of course a large overhead for small files. The shorter a page is, the smaller the overhead is, but the longer the management time, mainly due to the fragment storage time.

13.2.4.2. Fragmentation. A very simple fragmentation scheme is to scatter the data symbols among the fragments, without any real ciphering. However, in this case, an intruder having several fragments could possibly derive the whole page by guessing the missing data. At the very least, he could obtain a good idea of the symbol vocabulary used in the non-fragmented file. Cryptographic methods must thus be used in conjunction with the fragmentation. The question that one may then ask is: why use fragmentation since cryptographic techniques must still be employed? There are two good reasons:

- first, the geographical scattering of fragments makes theft of individual storage media of no avail to the intruder — even if he possesses the cipher key,
- second, the added security of scattering means that the ciphers employed can be much simpler and thus faster than conventional ones.

   One may imagine different schemes to realize ciphering and fragmentation together. Our choice consists of the fragmentation of ciphered pages (figure 6). Each page is first ciphered and the fragments are obtained from this ciphered page. The fragmentation itself uses a fixed scheme wherein each successive quantum of data is put into one of the fragments according to a fixed distribution that does not depend on the key. This operation leads to a fine-grain scattering of the data among all the fragments.

13.2.4.3. Choice of an Appropriate Cipher. To make it as difficult as possible for an intruder to decipher an individual fragment or sub-set of fragments, it is preferable to choose a cipher scheme that makes the ciphertext of each data-quantum, and thus each fragment, dependent on the others. This may be achieved by using a stream cipher. With such a cipher, the key used to cipher a quantum of plaintext changes for each quantum: the preceding ciphered

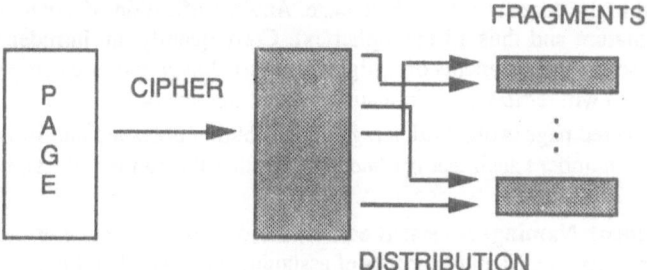

Fig. 6 - Ciphering/Fragmentation

text is necessary to determine the following plaintext. This introduces secrecy if any quantum of text is missing.

For this application, the stream cipher is based on a random number generator (RNG) chosen to make the ciphering very fast. The keystream is generated by two 32 bit random number generators (figure 7). Before each page ciphering, both RNGs are initialized with the two halves of the 64 bit key K (used as the seeds of the RNGs), then step by step the first one is reinitialized by the ciphertext quanta. The two results are exclusive-ored to generate the keystream K_i. Then, this keystream is exclusive-ored with 32 bit quanta of the plaintext. The first RNG is only used as a one-way-function applied to the previous ciphered quantum. The second RNG is equivalent to a large sized Linear Feedback Shift Register. Thus, the keystream K_i depends on the properties of the RNGs and on the previous ciphered text.

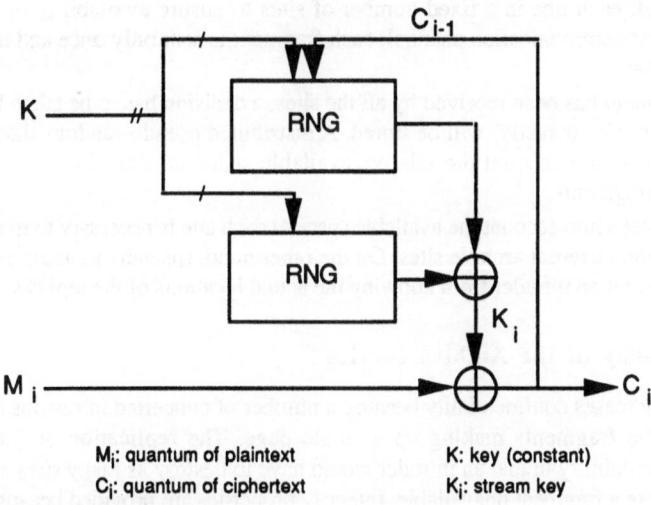

M_i: quantum of plaintext   K: key (constant)
C_i: quantum of ciphertext K_i: stream key

Fig. 7 - Random Number Generator Stream Cipher

This method enables the integrity of the fragments to be checked, because if one quantum of the ciphertext is modified, all the following deciphered text is changed. However, if one of the last plaintext quanta is modified, only the last ciphertext quanta will be changed. It is more interesting that the modification of any quantum changes all the ciphertext. To do so, the beginning of the plaintext is constituted by a signature issued by a hash (cryptographic) function applied to the content of the whole page. Any modification of a byte of the plaintext changes the signature and thus all the ciphertext. Consequently, an intruder who observes successive releases of the fragments of a slightly modified page will see completely different fragments, and then will retrieve no information on the fragment order.

Once the ciphered page is obtained, a regular distribution is carried out: fragment number j receives the bytes number i such that $j=i \ (mod \ N)$, N being the number of fragments.

13.2.4.4. Fragment Naming. As stated earlier, the operation of naming the fragments is carried out on the user site. Naming consists of assigning a unique identifier to every fragment; this unique identifier is derived from the fragmentation key, the name of the file, the index of the page and the index of the fragment. The naming algorithm is based on one-way cryptographic functions such that no information concerning one fragment can be derived from its name.

13.2.5. Scattering Principles

The fragment write, read or delete requests that are transmitted by the user site to the fragment server sites, are sent in a random order for each page. This means that if an intruder is eavesdropping on the network or controls a fragment server site, he can receive all the fragments of a given page, but he is not able to ascertain the order in which he has to select the fragments before attempting a cryptanalysis of the page. For instance, if a page is cut in 16 fragments, an intruder may need to attempt about $16!/2 \approx 10^{13}$ trials before finding the correct arrangement. Thus, the confidentiality depends more on this random order than on the efficiency of the cipher.

Once the file is fragmented in the user site, the fragments are broadcasted to the archive sites to be stored, each one in a fixed number of sites to ensure availability by redundancy. Using a broadcast communication channel, each fragment is sent only once and is received by all the archive sites.

Once a fragment has been received by all the sites, a decision has to be taken by these sites to ensure that R copies (exactly) will be stored. A distributed pseudo-random algorithm is then required that takes into account the relative available space at each site to decide the final locations of the fragments.

The need to take into account the available space at each site is necessary to maintain a good balance among the different archive sites. On the other hand, (pseudo-)random behaviour can be applied to prevent an intruder from knowing the actual locations of the replicas.

13.2.6. Security of the Archive Service

Scattering increases confidentiality because a number of concerted intrusions are necessary to retrieve all the fragments making up a single page. The replication of fragments also increases data availability in that an intruder would have to destroy as many sites as the number of replicas to make a fragment unavailable. Integrity properties are provided because an intruder would have to modify all the replicas and so must carry out several intrusions. Moreover, it is very unlikely that an intruder could modify even one byte of a fragment without modifying the cryptographic signature of the page.

Other techniques can be used to implement a secure file archiving server. The first one consists of ciphering the file on the user site, and storing several copies of the ciphered file on different archive servers. In that case, confidentiality relies on the efficiency of the cipher algorithm. An intruder who is eavesdropping on the network or who gets a copy of the ciphered file (e.g., a file server back-up tape) can take all the time and all the computer power he wants to cryptanalyse the file; with one intrusion, he gets all the information he needs. With the fragmentation-scattering technique, the intruder who gets all the fragments of a page, would have to try 10^{13} arrangements before cryptanalysing the ciphered page. That means that the cipher can be $\approx 10^{13}$ times less strong, and can be much faster. For integrity, the two techniques are comparable. For availability, the two techniques are equivalent for accidental server failures. However, the fragmentation-replication-scattering technique is less robust against simultaneous destruction of storage servers: if an intruder is able to destroy R (out of N) fragment server sites at the same time, he will make many more files unavailable than if he destroys R ciphered file servers. The overhead of the two techniques is equivalent for the communications and the storage space, but fragmentation-scattering can be made much less CPU-consuming than the ciphered file approach.

Another technique has been proposed by Rabin for fault-tolerant file servers: the "Information Dispersal" approach [Rabin 1989]. This technique consists of coding the file with a special error-correcting code and then splitting the coded file in n pieces, such that m out of n of these pieces are sufficient to rebuild the complete file. The n file pieces are stored by different storage servers. The coded file is longer than the original file, but the redundancy is much smaller than replication with $(n - m + 1)$ copies, for nearly the same availability and integrity. This technique is much more CPU consuming than the fragmentation and it does not ensure file confidentiality: to prevent information disclosure to eavesdroppers and storage server intruders, the file has to be ciphered before coding. To summarise, the storage and communication overhead is much less important in the Rabin's technique, but it consumes much more CPU time.

13.3.   Intrusion-Tolerant Security Server

This section describes the Delta-4 approach to the distributed system security management, by means of a distributed, intrusion-tolerant security server.

13.3.1. Basic Principles

13.3.1.1. Centralized versus Distributed Security Management. Most of the currently developed secure systems are based on paradigms such as the *access control matrix*, *reference monitor*, *security kernel* or *trusted computing base concepts*. These concepts are essentially centralized, in order to keep their implementation simple and verifiable. Such a centralized approach is inconsistent with the distribution, local autonomy and concurrency that distributed systems are supposed to provide.

Nevertheless, this approach has been maintained for the design of some distributed systems in which all the security relevant functions have been centralized in a specific site that is a mandatory mediator within any interaction. For instance, in the Secure File System proposed for the Newcastle Connection [Rushby and Randell 1983], a specialized, trusted site called "secure file manager", implements a "file access reference monitor" that is responsible for the enforcement of the mandatory access control policy. The throughput of such an approach is very low because all communications between a subject and an object must be relayed by the reference monitor site. This site is a "hard core" for both security and availability of the whole distributed system.

Another approach is proposed by the "Red Book" (Trusted Network Interpretation of the Trusted Computer System Evaluation Criteria [NCSC 1987b]), in which a Network Trusted Computing Base (NTCB) is composed of a set of cooperating Trusted Computing Bases (TCBs). Within each site of the distributed system, a local TCB is responsible for the authentication of local users and access control to local objects. For the accesses from local subjects to remote objects, the local TCB must cooperate with the remote TCBs responsible for the objects. The enforcement of the authorization policy is based on the cooperation between the TCBs, which must trust each other. This approach is unsuitable for heterogeneous, open distributed systems. Moreover, a successful intrusion of a local TCB can endanger the security of the whole distributed system. Such a case has to be considered seriously since, with current workstations, it is easy for a local user to obtain complete local control (e.g., as superuser).

For open distributed systems, yet another approach is to give the responsibility of user authentication to a trusted site, the *"authentication server"*, and to give the responsibility of authorization to the servers responsible for object management. An example of this approach is the Kerberos system [Steiner et al. 1988]. In such systems, security depends on the trusted servers and on the trustworthiness of the people in charge of these servers. For instance, in the Kerberos authentication servers, all the user passwords are recorded in a plain text file, which makes this file and this site good targets for intrusions and the administrators of this site good targets for bribery.

The approach described here consists of giving the responsibility of nearly all the security functions to a set of specialized sites so as to realize a *distributed security server*. While this global distributed security server can be *trusted*, it is not necessary to trust any of its components or any individual site administrator: such a security server is *intrusion-tolerant*. Since no individual site has to be trusted, such an approach is compatible with the heterogeneity of open distributed systems, such as Delta-4.

13.3.1.2. Security Server Requirements. The objectives of the intrusion-tolerant approach are:

- openness, no specific hardware.
- reduction of TCB (by intrusion tolerance).
- modular security requirements.

In Delta-4, the security view of the network defines three kinds of sites (figure 8):

- *User sites* are untrusted computers where users can log in. The local security is ensured by users.
- *Security sites* are computers providing security services: registration, identification and authentication, authorization, sensitive information management, audit and recovery services.
- *Particular servers* whose access needs to be secured. In the current system, this is the case of the Archive service located on Archive sites.

The fault assumptions for security sites are:

- The probability of more than one intrusion before detection and recovery is small.
- The probability of an intrusion into a site is independent of the previous intrusion(s) in other site(s).

The characteristics of the intrusion-tolerant security approach are:

- An intrusion or a misuse on one security site is immediately masked and has no consequence on the service and on its properties.
- If errors occur on some security sites before recovery, the number that will be masked depends on the services and their properties (confidentiality, integrity, availability). The service performances can be degraded.

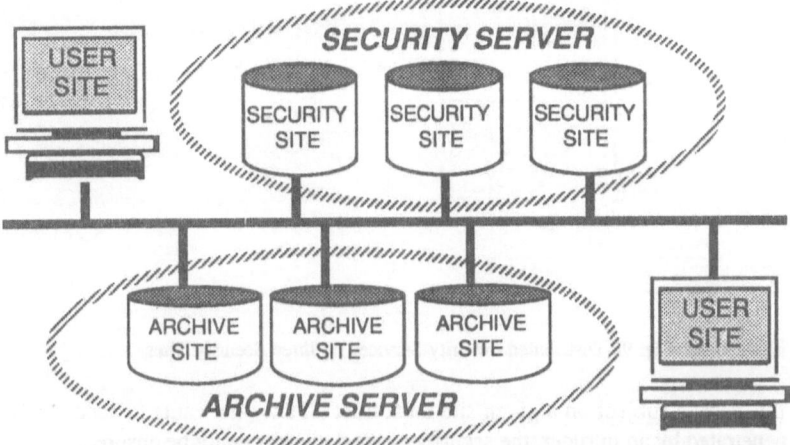

Fig. 8 - The Different Types of Sites of the Network

13.3.1.3. Intrusion Tolerance by Distribution. The fragmentation-and-scattering technique has shown that distribution can be exploited for intrusion-tolerance. Other intrusion-tolerance mechanisms have been proposed such as some cryptographic tools like the Shamir's threshold scheme [Shamir 1979]. The data is split in "shadows", each shadow being stored on one security site. To build the data you only need a sufficient number of shadows called the threshold. If you do not have enough shadows, you cannot build the initial data. The same scheme can ensure availability and integrity.

Among all the information managed by the security server, some data is not confidential; so it is possible to replicate such data on each security site. Data is *shared* or *replicated* according to its confidentiality.

Another important point is the prevention of denial of service. In this case, it is not just data but a service that has to be protected. The server thus has to be replicated on each security site in order to prevent denial of service in case of security site unavailability. However the different sites cannot take certain decisions independently. They must agree by communicating data and local decisions. To ensure the last property, the servers must obey two implementation principles: *replication* and *agreement* (figure 9).

The distribution of the security service permits a geographic distribution of the security sites. This makes the intruder's task more difficult since even if he succeeds in accessing one site, it will be more difficult for him to access other sites if they are not in the same place. It would indeed be a pity to distribute programs and data to perform logical intrusion tolerance and not to provide geographic distribution to assist physical intrusion tolerance.

13.3.1.4. Distributed TCB. In the Red Book architecture, each computer includes a part called the Trusted Computing Base (TCB) including the Trusted Interface Unit (TIU) and the Network Trusted Computing Base (NTCB) partition. The TIU assumes only communications security whereas the NTCB partition implements the local part of the network authorization policy. This NTCB partition must have a very high physical and logical protection. Moreover all sites have to trust one another. On each computer, accesses are mediated by the local TCB that is the implementation of the *local* authorization policy. The *global* authorization policy is implemented in the set of NTCB partitions. When all computers are connected by a network and a subject wants to access a remote object, the NTCB partitions communicate with each

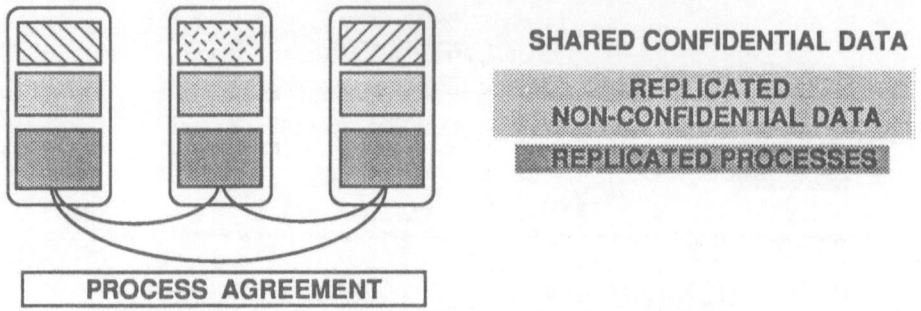

Fig. 9 - Distributed Security Services on three Security Sites

other. In this case, a subject on a given site must trust all sites he wants to access. If one site has been penetrated by an intruder, the security of the network cannot be ensured.

In the Kerberos architecture, the greater part of the TCB is within the security server. Each server also has an large TCB, which carries out authorization operations. However if an intruder succeeds in penetrating a server, he cannot access the other ones. The consequences of an intrusion are thus limited. The only site you really have to trust to ensure global security is the security server site. However, in this case a single point of failure exists.

In the intrusion-tolerant approach, there is no local "Trusted" Computing Base on the security sites. Only the set of security sites is globally trusted (figure 10). There is also a small local TCB on the user sites and the secured servers. The servers themselves, for instance the Archive Service, can have a distributed TCB. In this case, the only single point of failure is on the user site. This can be minimized if the user site is considered as a one-user computer when a user accesses a secured service. If this were not so, back-doors would always exist in the local protections between users.

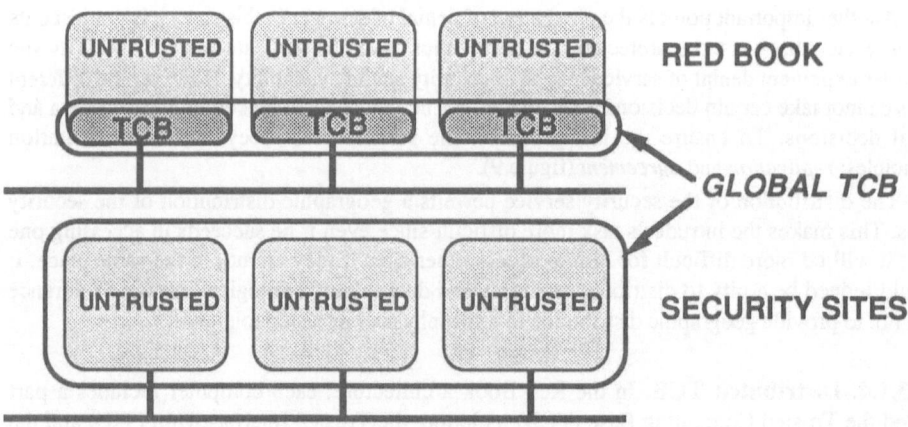

Fig. 10 - Red Book versus Security Sites TCB

The security server is trusted, but not the different computers. The security service is considered as one global server. If an intrusion in one computer is successful in a classical architecture with a local TCB, the security of the full system is no longer verified. On the contrary, if protections of one security site in the intrusion-tolerant approach are bypassed, the security of the global system is maintained. These differences come from three different points of view about security and networks:

- In the Red Book, a network is only seen as a communication channel (low layers) between centralized systems. A subject, or an object, is located on one site and cannot be shared between several sites. In the Delta-4 architecture, the network is a support for distributed applications. A subject or an object can be shared on several sites.

- In the Red Book, sites are time-sharing systems with several users working on the same host. These computers are controlled by a system administrator. In Kerberos, sites are workstations under the control of one local user. In a secure open distributed system, whenever possible, the use of one host by several users during access to a secured service should be prohibited.

- The Red Book architecture is a support for DoD policy, a multilevel mandatory policy where confidentiality is the most important property to be ensured. All sites thus need to have a large trusted part. It is impractical to implement this policy in the Delta-4 system (see section 13.4), mainly because trusted paths between all user sites would be needed.

13.3.2. The Security Services

The different services that must be provided by the security sites are registration, authentication, authorization, sensitive data management, audit and recovery services.

13.3.2.1. User Registration Service. Registration of a new user of the system must be done under the control of the security administrator on each authentication site. The user is registered on the first site by the first security administrator, on the second site by the second administrator and so on. On each site, the identity of the user is stored with some authenticator that will be used by the authentication process to verify the claimed identity. The authenticator can be a password, a secret permanent key or a public key. Except if public keys are used, different authenticators will be stored on the different authentication sites for the same user. This can be considered as an implementation of some *separation of duty*: one or a minority of security administrators cannot register an illegitimate user and cannot prevent the registration of a legitimate user on a majority of authentication sites. Moreover, if such malicious administrators try to use local information stored on their sites to impersonate a registered user, they will fail because they will not be authenticated by the other sites.

13.3.2.2. User Authentication Service. When a registered user wants to access remote servers from a workstation, he has to be authenticated by the authentication server: an authentication protocol has to be run between the user site and the security sites. This protocol is composed of three phases (see figure 11). In the first phase, the user site attempts an independent authentication with each of the security sites. This authentication can be based on classical authentication algorithms (see [Lamport 1981, Needham and Schroeder 1978],...) but with different authenticators (or different challenges with public key systems) for the different security sites. In this first phase, each security site independently decides for itself if the authentication attempt succeeds or not. During the second phase, each security site broadcasts its own decision to all the other security sites, and receives the decisions of these sites. In the third phase, according to the majority of all the received decisions and its own decision, the security site authorizes (or not) the session for the user, and confirms any session authorization by sending a session key to the user site (or a ticket containing the session key, depending on the authorization algorithm that has been selected). A different session key is randomly generated by each security site. This session key or ticket will be used by the user site to authenticate its requests in the authorization process (see section   13.3.2.3).

In this protocol, the majority voting on the different authentication decisions enables the tolerance of accidental faults affecting the registration data stored on the various security sites or communication faults during the protocol. It also allows tolerance of intrusions into a minority of security sites, which could lead to false local decisions. Moreover, different session keys are generated independently by each security site, and sent only to the user site; thus, no intruder, even with the complicity of an administrator, can impersonate the user site (except if he cracks the authentication algorithms).

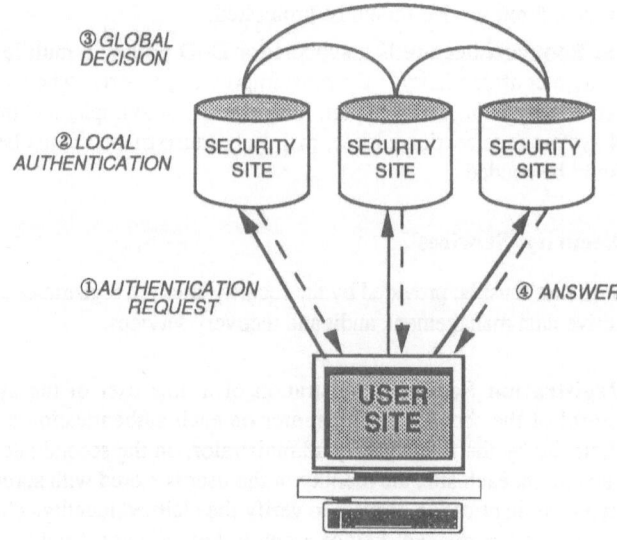

Fig. 11 - The Authentication Protocol

It would be difficult for a user to remember several independent passwords. A better solution would be to store secret keys on a personal smartcard, the owner of which has only to remember his PIN (Personal Identification Number). The current Delta-4 implementation uses Bull CP8 smartcards with shared keys, one smartcard for each user and one for each security site administrator. All the administrators of a given security site have identical master smartcards, except for the identification and the PIN of the administrator. On the user smartcard, there is a set of areas, one area for each security site, i.e., for each master smartcard. That means that when a user is registered by a security site, the local master smartcard generates a secret key that it writes within its own reserved area on the user smartcard. When the user has been registered by the N sites, his smartcard possesses N secret keys within N areas. In the authentication phase, each security site sends a challenge to the user. The user smartcard applies a one-way function to this challenge and the shared secret key, and then sends the result to the security site. The master smartcard makes the same operations to the same data and it compares the results. This protocol is performed by every security site.

13.3.2.3. Authorization Service. The role of this service is to check that the access to a secured service by a subject is authorized according to its access-rights. This service is made intrusion-tolerant by its implementation on the security sites, and by means of an authorization protocol between the user site and the security sites.

As a matter of fact, the authorization protocol is quite similar to the authentication protocol. Whatever the user request, a local decision is first taken by each security site according to the user access rights which are stored locally. This local decision is then broadcast to the other security sites. The authorization decisions received from the other sites are voted on locally (with the local decision), and, according to the result of the vote, the user request is locally executed (or not). This majority voting technique ensures that a legitimate request cannot be denied and that an illegitimate request cannot be granted, except if a majority of authorization data copies have been destroyed or altered.

The different phases of authorization are (figure 12):

- The subject asks the security servers for permission to access a secured service by sending its identifier (received in the authentication phase) (1).

- The access-rights stored on the security sites enable them to verify that the subject is authorized to access the requested service.

- The security sites vote to decide if the access is authorized using the same protocol as that defined for authentication (2).

- If the sites agree to permit access, they send a ticket to the subject and another ticket to the secured server (3).

- With the ticket, the subject can open a session with the server.

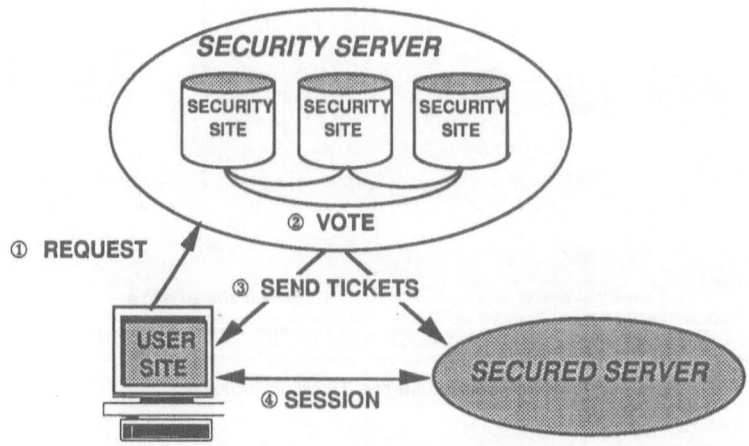

Fig. 12 - The Authorization Protocol

A subject may also want to access secured data stored on security sites such as access-right information or sensitive object descriptors. In this case, the access control protocol is the same as the one described above for phases 1 and 2 and then, if and when the security sites accept the access, it is performed on each site and the information or the affirmative response is sent to the user site.

13.3.2.4. Sensitive Data Management Service. The role of this service is to store, manage and retrieve the sensitive information on the security servers so that their protection

verifies the hypothesis made in section   13.3.1.2. This information consists of short data items
needed to achieve security services.

The data management service must enforce the three main security properties
(confidentiality, integrity and availability). The integrity property is provided by modification
detection mechanisms such as cryptographic signatures. According to the sensitivity of the
security data, it can be important to preserve both the confidentiality and availability of this data,
or only the availability. For this, two storage techniques can be applied: replication (for
availability) or threshold schemes (for confidentiality and availability). The algorithm used to
store data will depend on the functionality required (figure 13).

If a data item is replicated on N security sites, it is assumed:

- with respect to availability, that $N-1$ replicas can be lost (modified or destroyed),
- with respect to confidentiality, that one replica is sufficient to retrieve data.

If one data is shared on N security sites (in this case, fragments are called *shadows*) using a
threshold T, it is assumed:

- with respect to availability, that $N-T$ shadows can be lost,
- with respect to confidentiality, that T shadows are necessary and sufficient to
  retrieve data (less than T shadows gives *no information* about the data).

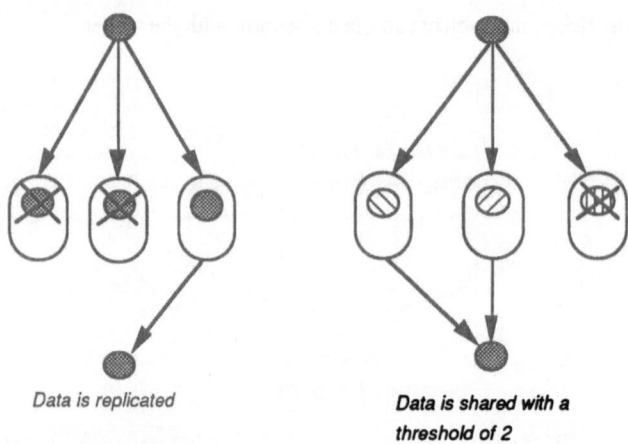

Data is replicated   Data is shared with a
                                                       threshold of 2

Fig. 13 - Replication and Threshold Scheme

13.3.2.5. Audit Service. The role of this service is to record all information related to
security. Such information is received from the services defined above. Two kinds of
information exist, authorized operations performed by authorized users (registration, access,
changes to existing access rights,...) and attempted or successful intrusions or misuses. It is
not the role of the services to identify intrusions or misuses by an authorized user. This is the
function of an audit trail analysis.

The audit information is sent not only by security sites but also by secured servers and user
sites. For the former, it will be access-requests, and for the latter it will be, for instance, information about correct or incorrect shared data sent by security sites (bad shadows received from
certain security sites).

The audit trails are stored on each security site. The information received on one site is not sufficient to compromise the security of the system.

The analysis of this audit information will be done off-line by security administrators. As one intrusion is masked, it is not necessary to detect intrusion on-line.

13.3.2.6. Recovery Service. This service acts as an error recovery mechanism to correct certain modified data (e.g., shadows of the threshold scheme). Other recovery functions can be performed manually by security administrators using audit trails.

13.4.   Selection and Implementation of an Authorization Policy

The security policy of a company strongly influences the design of the system intended to support applications for this enterprise. In general-purpose systems, the architecture should be flexible enough to adapt to any usual security policy. Nevertheless, in open systems such as Delta-4, mandatory policies cannot be securely implemented, except if it is specifically implemented within each site, and if all communications are under the control of some authorization mechanism implementing the same policy (which is contradictory with openness). This consideration has led to the choice of a discretionary authorization policy.

13.4.1.   Implementation of Discretionary Policy

Discretionary policy is an authorization policy where some authorized subjects have right-modify rights on an object. This means that such a subject has the right to modify the access-rights on the object. Discretionary policy means that subjects that have right-modify rights must evaluate the trustworthiness of other subjects for themselves. If a subject can read information, he can disclose this information to whoever he wants. Only some paths are trusted. Subjects can be identified as all processes acting on user sites. If this site can reassemble an archive, it can forward it to other user sites without any control. The unauthorized disclosure of information is only limited by the trust a subject puts in another subject. If integrity is ensured by security sites, confidentiality is partially the responsibility of the users.

The propagation of other rights (execute, write,...) can be limited by using hierarchical authorization (right-modify rights are given to the subjects at the top of hierarchy) or centralized authorization (right-modify right is given to one security administrator) [NCSC 1987a]. The centralized scheme can only be used in a rigid system where most subjects or objects are created during an initialization phase where changes have to be infrequent. This will certainly not be the case in Delta-4 applications. The hierarchical scheme is more flexible and simple to implement.

13.4.1.1. Access Lists and Access Tokens. The implementation of access-rights can be performed in two ways, either using *access control lists* or *capabilities* [Fabry 1974].

In this system, access control is done by security sites and then by archive sites. Security sites must authenticate subjects and authorize them to access archives. When they grant access, two solutions are possible. First, security sites could be the mandatory gateway between a user site and the archive sites during a session (figure 14). In this case performance and security decrease since:

- security sites become a fragment-transfer bottle-neck,
- security sites know information that they should not have to know (fragment, fragment name).

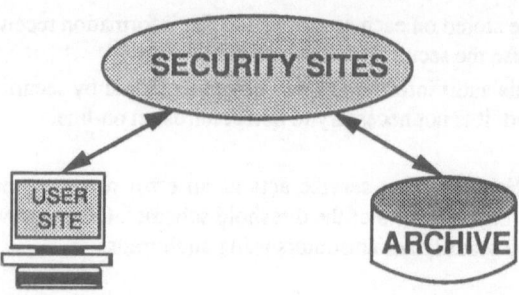

Fig. 14 - Security Sites act as a Gateway

The second solution is that security sites supply user sites and archive sites with elements that enable the creation of a direct session between them (figure 15). These elements form a ticket with cryptographic keys and access tokens (enciphered fragment name and access-rights). The access tokens will be used as "capabilities" and will be sent to archive sites to authorize users to access fragments.

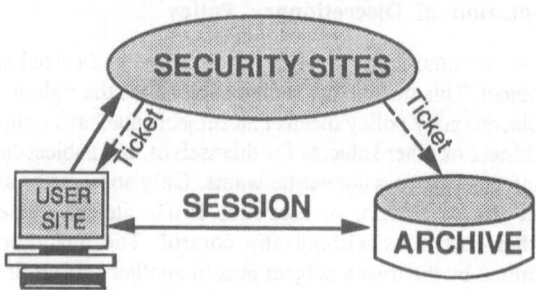

Fig. 15 - Opening Session with Tickets

The security sites have to verify if a subject is authorized to access an archive. The fastest way to check authorization is the use of an access control list that can be replicated on all sites.

13.4.1.2. Access Rights. To implement access control lists, it is possible to use a UNIX-like protection scheme. Under this scheme, the file authorization list is made up as follows:

- Access-right types are read, write and execute.
- A fixed list of three fields defines the rights granted to the owner, the members of the file group and the other users.

A fourth field, called dynamic access control list (ACL), is added. This access control list contains a set of groups or users with added or restricted rights (table 1). These groups or users have more or fewer rights than the ones defined in the standard fixed list. For instance, if the object group right is execute only, in order to give write right to a particular member of the group, an entry for this subject in the access control list with added write rights is created. This is the same thing for groups. Rights for some subjects can also be reduced. The advantage of this dynamic implementation is that it is close to complete access control list granularity without excessive data overhead.

Table 1 - Archive Access Control List

Fixed List	
Owner	rights
Group	rights
Others	rights
Dynamic ACL	
User or Group	Extended Rights
User or Group	Restricted Rights

13.4.2. Discretionary Multi-Categories Policy

Groups can be defined in a such a way that security categories can be built. Only the security administrators will be authorized to create and modify groups.

It has been shown that it is impracticable to implement a mandatory policy, but it is possible to make categories such as Bell-LaPadula categories [Bell and LaPadula 1974]. These categories represent interest classes. In military or commercial domains, it is easy to partition objects into categories. A subject can have access to more than one category whereas an object can only belong to one category (this is different from the Bell-LaPadula model). In the system presented here, categories are built with groups. These groups should be defined by security administrators when they implement the procedural part of the security policy. A group can also be added or subtracted when the system is in operation.

To implement the categories, UNIX group concepts are used with some differences. First, a subject can belong to several groups. This is already implemented in UNIX System V, however its use can be complicated. The file access control does not use the groups the user belongs to; instead, it uses the "current shell group". This is first defined in a login configuration (password file) and when a user wants to have other group permissions, he has to change his "current shell group" by a command (*newgrp*).

The idea is to verify group members during each access attempt. If a user belongs to several groups, he must have the access-rights of these groups all the time. These rights have to be defined using the object and subject creation rules.

13.4.2.1. Groups. Groups are subject sets and each group specifies one category. To partition categories, one category cannot include an object that belongs to another group, but a subject can belong to several groups (figure 16). A subject has all the rights of all the groups to which it belongs.

When the groups are specified by security administrators, and subjects are distributed over them, it is necessary to define the object creation rules that give rights to subjects.

The rules are:

- when a new object is created, the object owner is the object creator and has U-rights, the object group is the group of the owner and group members have G-rights, the other subjects have O-rights (often no rights).
- if a subject belongs to several groups, it may decide to which group the object is to belong.

These rules are modular enough to build different policies with different constraints. Security administrators have to choose the various U, G and O rights.

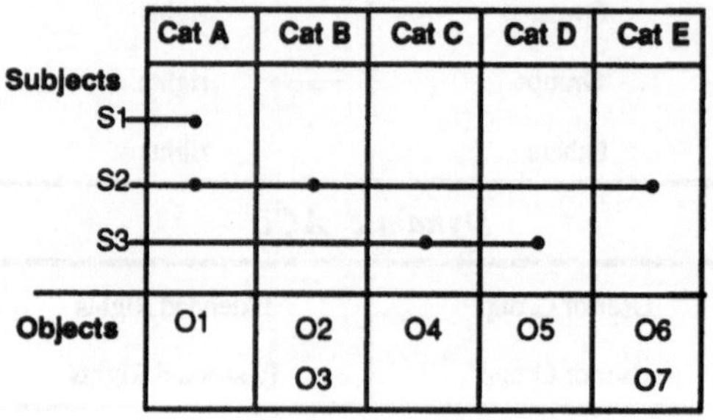

	Cat A	Cat B	Cat C	Cat D	Cat E
Subjects S1	•				
S2	•	•			•
S3			•	•	
Objects	O1	O2 O3	O4	O5	O6 O7

Fig. 16 - Example of Groups/Categories

In this example:

- S1 belongs to the category A and has G(A) rights on A-objects and O-rights on B, C, D and E-objects.
- S2 belongs to the categories A, B and E, and has G(A), G(B) and G(E) rights on the A, B and E-objects and O-rights on C and D-objects.
- S3 belongs to the categories C and D and has G(C) and G(D) rights on the C and D-objects and O-rights on the A, B and E-objects.

13.4.2.2. No Discretionary Control by Owner. The discretionary control on the objects is not carried out by the owner. The owner has particular rights but does not have the right-modify right on the objects. A user must not be able to give rights on an object in a given category to users in other categories.

The owner of the object will be the creator of the object, but the name of the owner cannot be changed by any user.

13.4.2.3. Category Managers. A superuser exists, called the category manager, for each category. He is a member of the corresponding group. This user has right-modify right on all objects that belong to the category. He can change the access-rights of all these objects. He does not have right-modify rights on objects in other categories. A category manager cannot change the owner or the group of an object. This can only be done by security administrators. This allows separation of duties. Users, category managers and security administrators all have different roles.

13.5. Future Extensions

In a large distributed system, a distributed security server would be responsible for the security management of only a subset of the whole system. That means that several security servers would have to cooperate for the global management of the distributed system. The Delta-4 intrusion-tolerance approach is compatible with this requirement. For instance, a user can be authenticated by a nearby distributed security server and this user may ask for access to an object managed by another security server. In such a case, the "nearby" security server can relay the request to the "more-remote" security server in the same way as in the "Red Book" approach.

Chapter 14

Software-Fault Tolerance

At the beginning of the Delta-4 project, a design assumption was made to the effect that only physical faults were to be taken into account when providing fault-tolerant mechanisms: the possibility of design faults could therefore be neglected. In the subsequent years, recognizing that software design faults are becoming a major source of system service disruption, it was decided to study how to provide the Delta-4 architecture with specific provisions to deal with this kind of faults.

Since the term software-fault tolerance may assume different meanings, let us also say that we intend here to deal with ways to tolerate design faults in software. The tolerance of hardware design faults will only be considered as a side issue. The general consideration applies that the effectiveness of fault-tolerance techniques is not usually limited to a precisely defined class of faults, and hardware design faults, software bugs and transient hardware faults often lead to similar behaviour, as discussed in section 6.4.2.

It should be noticed that design faults may be present in Delta-4 hardware, operating system software, and applications. Application-level software-fault tolerance can help against all three kinds of errors (assume an operating system error that causes messages to be delivered in the wrong order: an application will often be able to recognise this, based on the expected contents of the messages), but is mainly directed against errors resulting from faults in the application itself.

This chapter, after briefly recalling the main techniques presented in the literature to tolerate software design faults, focusses on the problem of applying some of these techniques in the Delta-4 architecture. Support mechanisms and structuring concepts are presented. It should be pointed out that the solutions shown below are still in the specification phase — no implementation has yet been carried out.

14.1.   A Brief History — the State of the Art

14.1.1. The Problem of Software Dependability

Software engineering has offered many improvements in the way software is produced, but no radical solution for producing dependable software is in sight [Brooks 1987]. Formal proofs of correctness are still unfeasible for many real-world products, and suffer from some inherent limitations (with respect to specification faults and/or ambiguities, for instance). Testing suffers from basic limitations, best stated as the fact that it can discover faults, but not prove absence of faults (sometimes not even upper bounds on the probabilities of failures caused by residual faults [Hamlet 1987, Miller 1986b]). Development methodologies and tools have improved, but none are known to avoid software faults.

From a management point of view, increasing the expense in fault avoidance methods beyond some threshold yields diminishing returns, because of technological limits, increased overhead and human communication problems.

Since creating perfect software components seems impossible anyway, to increase reliability, and in general dependability, of current, complex computing systems, we need strategies for coping with software design faults and techniques suitable to tolerate their effects.

According to [Gray 1986], for instance, once disc storage is duplicated, software faults become by far the most common cause of errors (except operator errors) in a transaction-oriented computer system. According to Giloth and Prantzen [Giloth and Prantzen 1983], about one fourth of the down time in the 4ESS electronic switching system was due to software problems. So was the January 1990 nationwide outage in the AT&T telecommunication network.

Software-fault tolerance, besides improving reliability, for a given cost, is thought to have the potential of: i) decreasing the cost of operational testing of the final product, through on-line error detection; ii) thus, shortening the teething problems of new products; iii) obtaining reliability levels that would be impossible by other means, regardless of the accepted cost, for applications where the cost of failure is very high.

Of course, software-fault tolerance techniques, given their cost, can only be cost-effective if they use reasonably reliable individual modules in their redundant configuration.

14.1.2. Software-Fault Tolerance Methods

Early proposals for software-fault tolerance began to appear in the mid 70s [Avizienis 1975, Horning et al. 1974, Kopetz 1974, Randell 1975, Shaw 1976, Yau and Cheung 1975]. By the late 70s, some methodological proposals had been defined. We will quote here:

1) *General concepts* about error detection and treatment (independent of the origin of errors): executable assertions, exception handling and such. These are proposals about language constructs, and corresponding program-structuring criteria, that would make it easier to: i) state conditions that the state of the computation must satisfy at given points in the execution, and ii) describe the actions that must be taken if an assertion is not satisfied [Cristian 1987].

2) Structured software-fault tolerance: *Specific proposals* for structuring the addition of redundant code in programs in a simple and coherent way:

    2a) *Recovery blocks* [Randell 1975], where a block[1] of code contains, besides the "primary" routine for the specified computation, i) an acceptance test to be performed on its results and ii) back-up ("alternate") routines, functionally equivalent to the primary, to be invoked, on the state of the data prior to the execution of the primary, if the results of the primary fail the acceptance test. So, the recovery block concept is based on error-detection through executable assertions plus backward error recovery; only if no alternate produces an acceptable result the block fails, and the resulting exception must be handled at a higher level.

    2b) *N-version programming* [Chen and Avizienis 1978], or "multiple version software": replicated, diverse, functionally equivalent software modules, referred to as "variants", are executed concurrently (e.g., on the replicated processors of an N-modular redundant system), and their results are compared

---

1 "Block" has the same sense here as in "block-structured languages".

and voted to mask errors and possibly correct the internal states of the variants themselves: it is a form of masking redundancy, with forward error recovery.

2c) *N-Self-Checking Programming* [Laprie et al. 1987]: N functionally equivalent self-checking components are executed concurrently; one is considered as being acting and delivers the results, the other self-checking components are considered as hot spares. A self-checking component can be either the association of an acceptance test to a variant or the association of two variants with a comparison algorithm.

Subsequent research has elaborated on these basic concepts; its topics may be divided as follows:

- *Architecture of fault-tolerant software*. A number of proposals have appeared, generally originating from one of the three concepts of recovery blocks, multiple-version software, and assertion checking, and often combining aspects of these techniques [Anderson 1986, Avizienis and Kelly 1984, Hecht 1976, Kim 1984, Kim and Ramamoorthy 1976, Strigini and Avizienis 1985] deal with how the component elements of those techniques can be subsumed in a more global vision of design possibilities. A series of contributions from the University of Newcastle (see [Randell 1987]) concentrates on the structuring of complex systems for fault-tolerance: nesting of recovery blocks, nesting of generic components and exception propagation, structuring fault-tolerance in distributed systems. An experimental supervisory system for running multiple-version software on Unix-like systems, called "Dedix" (Design Diversity Experiment), was developed at UCLA [Avizienis et al. 1985].

- *Ad hoc techniques for tolerating special classes of faults*. We shall quote the area of "robust data structures" [Taylor and Black 1985], dealing with techniques for protecting, through redundancy in their representation in storage, the "structural" information of data structures (e.g., the topology of a graph structure, regardless of the information located at the nodes).

- *Evaluation of software-fault tolerance strategies*. The effectiveness of software-fault tolerance techniques, as an improvement in the operational reliability or safety of software, has been studied along two main directions: experimental measurement and analytical estimation. Only very partial conclusions have been reached so far.

Documentation, and additional references, about most of the experimental work and industrial applications can be found in [Littlewood 1987, Strigini 1990, Voges 1988].

14.1.3. Software-Fault Tolerance in Practice

14.1.3.1. Fully Redundant Software. Fully redundant software is not yet widely used and is at present limited to highly critical applications. Nonetheless, there are a number of interesting applications. Most of these are documented in the book [Voges 1988].

These examples are in aerospace, railway and atomic energy applications. All these areas are influenced by the presence of regulatory agencies whose approval is needed for the operation of products. *Fail-safeness* is usually required (the computer system can stop, provided it leaves the controlled system in a safe state), sometimes with continuous fault-tolerant operation as a long-term objective, but in avionics, as computers take on flight-critical tasks, the basic requirement is becoming *continuous fault-tolerant* operation.

Nuclear applications have been only experimental so far.

Avionic multiple-version software systems have been operational for several years. We can quote the Airbus A-310 flap and slat control, the autopilot for the Boeing 737/300, the Airbus A-320 fly-by-wire flight control system, and the Space Shuttle Flight Control System.

There are examples of dual version (fail-safe) systems in railway applications in Sweden [Voges 1988] and in Austria [Theuretzbacher 1986].

14.1.3.2. Other Software-Fault Tolerance Techniques. Two outstanding examples of systems able to cope with software faults without massive (diverse) software redundancy are the Electronic Switching Systems of the Bell System, in its several versions, and the Tandem systems. Both systems were primarily designed to tolerate hardware faults by an error detection and backward recovery scheme; but including in programs means to detect erroneous behaviour proved to be quite effective to handle software faults as well.

A detailed description of the techniques used in the No.5 ESS system to detect software errors is given in [Haugk et al. 1985]. The software system in the No.5 ESS is structured in functional components that communicate by messages and operate with duplicated units in an active-standby configuration; at run time, relatively short lived processes are created to perform specific tasks. Error detection is based on in-line defensive checks to ensure quick error detection, audit programs to verify consistency of data using data redundancy, special processes to detect program loops, scheduling problems, unavailability of critical resources, time-out on watchdog timers and such. For each detection mechanism, a specific error recovery procedure is designed.

In the Tandem system, the main features that assist in tolerating software faults [Gray 1986] are:

- software modularity (the structuring of application software into processes that interact via messages reduces error propagation);

- fault containment through fail-fast software modules (processes are made "fail-fast" by defensive programming, i.e. they run many consistency or reasonableness checks on their inputs, intermediate results and outputs, and abort themselves if an error is detected);

- atomic transaction mechanisms to establish natural checkpoint-rollback points, masking the actions of the aborted process to the rest of the world;

- process pair mechanisms to allow easy repetition of failed computations (for processes that cannot use the atomic transaction mechanism, automatic checkpointing is provided).

14.1.4. Effectiveness

All experiments to date have shown that software-fault tolerance improves the reliability of software to which it was applied, but it is difficult to evaluate its effectiveness in quantitative terms.

Several analytical estimations of the effectiveness of structured software-fault tolerance techniques (and sometimes of performance) have been published, differing in the models and analytical tools used and in the assumptions made. Unfortunately, they all depend, in order to estimate reliability, on the probability of errors common to redundant components (replicated modules, or the producer of a result and its verifier). This information must be obtained experimentally, but we are still far from being able to determine it with any confidence. Much experimental work is aimed at selecting development rules that make that probability as small as possible, but the comparison between alternative rules is done mostly by qualitative observations on the anecdotal history of each development.

The experience in aerospace, railway and atomic energy applications cannot be used to measure the effectiveness of software-fault tolerance techniques: the experimental sample is small, little information is published, and the operational record of the software usually shows few or no serious errors. Such software is built to very high standards, and the desired reliability level is so high that it could hardly be measured experimentally with any reasonable degree of confidence. These techniques are used because of an *a priori* belief that they help, although in a non-quantifiable way, to reduce the probability of failures whose cost would be extremely high.

Even for the 5ESS system and Tandem no specific figures for the effectiveness of software-fault tolerance techniques are known; the high reliability and availability demonstrated in the operation of these systems is an indication of success of the mechanisms used.

14.2. Software-Fault Tolerance for Delta-4

14.2.1. Fields of Application

We must first recall that the intended product of the Delta-4 project is not a computing system that can be used in life-critical applications, but computing systems that will provide a significant increase in reliability and availability compared to those currently used.

There is a common misconception that structured software-fault tolerance is worth applying only for *life-critical* applications, where the supposedly very high costs of software-fault tolerance are acceptable. Actually, software-fault tolerance can be applied in different degrees depending on the applications, to match the cost of the redundancy to the expected benefits. Moreover, cases can be found where even expensive forms of software-fault tolerance are justified.

It is probably safe to assume that no life-critical safety functions will be programmed on Delta-4. Safety systems for industrial plants must usually be built independently of the main control systems, and they must be very simple and easy to validate. Hence, even in cases where they are implemented in software, this probably would not be run on a Delta-4 system.

However, in many applications that are not deemed life-critical, large amounts of money are staked on the dependability of the software. In some cases, the software can directly cause damage to valuable items (either information or physical objects); in others, the software may cause unavailability of a computer system, which in turn causes the financial loss.

Some examples of intended uses of Delta-4 systems and related risks are listed below.

- *High level planning and management functions* (in industrial applications). This means that faults may cause large inefficiencies in the allocation of resources and disruption in the production process. The distributed computer system is also bound to be involved in plant-wide emergency contingency plans, and so indirectly have life-critical functions, although the operation of individual machines in the plant is safeguarded by reliable safety systems.

- *Control of individual peripherals*. A chemical reactor can be induced to ruin a batch of an expensive or profitable product. Robots used in material handling or processing will be prevented from running into each other or rolling over people by safety interlocks, but they can still damage the objects they handle, if given wrong orders or information.

- *Electronic funds transfer and banking applications*. There have been several recent episodes of huge losses to banks and security brokers due to computer error or unavailability [Risks].

14.2.2. General Criteria

Software-fault tolerance support should be provided in Delta-4 according to the following criteria:

1) Consider the software component as the basic unit to which software-fault tolerance techniques may be applied (following a general Delta-4 rule).

2) Exploit basic Delta-4 fault tolerance mechanisms, e.g., atomic multicast, checkpointing, etc.

3) Offer a coherent set of support mechanisms, enabling the implementation of a wide spectrum of software-fault tolerance techniques. The user is left the choice of the technique best suited to the application, by trading cost against dependability improvement, as well as considering real-time requirements and other factors, e.g., error coverage. Simple techniques, featuring the notification of an internal error to abort the computation, up to structured solutions like N-Version-Programming and N-Self-Checking Programming can be implemented, as well as new paradigms, not bound to "classical" proposals.

   In fact, in some critical applications, there is a well-defined set of catastrophic failure modes that will invite the use of structured software-fault tolerance protections as an obvious and economical precaution. In others, audit and cross-checking procedures will play the some rôle. In some applications, the complexity of the system may make it desirable to apply a "blanket" approach, such as global replication, to reduce the residual failure probability.

4) Allow software-fault tolerance and hardware-fault tolerance to be configured independently. In the general case, however, an integral approach to fault tolerance has to be followed, to minimize the redundancy (and cost), to extend the coverage, and to facilitate fault diagnosis.

5) Obtain ease of use. An important feature for any reliability-increasing technique is the possibility of applying it without changing the representation of the system at some level of abstraction. This allows the application programmer to concentrate on the functional specifications of the product, and another designer to add redundancy without complicating the program.

   By contrast, many of the possible ways of providing software-fault tolerance require the application developer to think up tests based on the semantics of each individual module, to worry about whether the use of diverse modules might cause problems, etc.: in short, the application of redundancy complicates the development process.

   As this lack of transparency is a necessary consequence of the willingness to check for deviations from the specifications of the individual applications, as opposed to the specifications of the machine supporting the applications, it is desirable that at least the individual application programmers be shielded from the need to take care of software-fault tolerance at the same time as they program the functional parts of the application. The division of the software into work jobs assigned to different programmers should take into account the desirability of diversity, and configuration tools should make the task of assembling the redundant software parts relatively easy and error free. For instance, it must be possible for a configuration manager to vary the number of software variants in a redundant module without interfering with the individual variants.

In a similar way as for hardware-fault tolerance, strategies for tolerating software faults in Delta-4 can be developed for:

a) use of software components that are either self-checking or fail-uncontrolled; in the latter case error detection is accomplished through comparison with independently obtained results, or by separate components (consumers of results, auditors).

b) use of deterministic or non-deterministic replicas. Non-deterministic programs must be self-checking: non-deterministic non-self-checking programs can be made tolerant of software faults by ad hoc techniques that hardly suggest mechanisms of general use.

c) use of backward or forward recovery of faulty components.

The choice of how the basic support mechanisms are combined into a coherent strategy to cope with software faults is left to the application designer. The next section presents some typical combination schemes, that are thought to have wide applicability, and could be incorporated in tools for assisting the designer in this task.

In the presented solutions it is assumed that the hosts are *fail-uncontrolled*.

14.3. Support Mechanisms and Features

A number of specific mechanisms should be provided to support software-fault tolerance in Delta-4 in the several activities that it involves, namely error detection, error processing, recovery of damaged state in faulty variants; they are described in the following subsections.

14.3.1. Error Detection

Errors in a computation can be detected essentially in one of two ways: by checking that its results have some properties required by its specification (*acceptance test* or *executable assertion*), or by comparing them with independent results (here *independence* refers to the computation failure modes). These techniques are dual to self-checking and replicate-and-compare circuits, respectively, for hardware-fault tolerance (an acceptance test can be included in the same software component together with the code to be checked).

14.3.1.1. Self-Checking Software Components. Detection of internal faults during the execution of a program can be carried out by a variety of techniques, from simple data consistency checks (e.g., non-null pointers, range checks), computation of the inverse function with respect to the main program computation, up to, conceivably, execution of a different implementation of the specifications and comparison of results (inside the component).

Self-checking software components, when an error is detected that is not recoverable inside the component[2], are allowed to behave in two ways:

a) simply omit sending due messages (like fail-silent hardware) and abort;

b) explicitly notify the error occurrence to their rep_entities (cf. section 6.4.4).

The second alternative allows the system to react sooner, since there is no need to wait for a time-out to expire; it also offers some benefits for error processing, as shown later.

A self-checking software component is characterized by its use of error-detection mechanisms and by appropriate reactions to the detection of errors.

Error detection mechanisms include:

- tests explicitly programmed by the application programmer;

---

[2] Backward recovery *inside* a component can be done transparently to the rest of the system, but must satisfy the constraint of replica determinism and must not have side-effects outside the component.

- mechanisms existing in the virtual machine supporting the application component: hardware (divide-by-zero trap), operating system software (overflow of system data structures, illegal system calls), language support (array overflow checks).

The latter error detection mechanisms, in general-purpose computers, are typically configured to abort the application process, in a way transparent to the application programmer. However, many systems give the application programmer some limited control of the reactions to exceptions raised by the support virtual machine. Such control may be offered by the operating system (e.g., the UNIX *signal* system call) or as part of a language (programmed exception handling, as in Ada, for example).

For Delta-4, it is desirable that, to the largest degree allowed by the native LEX used, exceptions be made available to Deltase and/or the application program. Deltase must provide a default reaction, which must not only abort the erroneous component but also notify Delta-4 system administration.

The application programmer can instead customize the reaction to a detected error based on his knowledge of the context where it arose. Depending on the severity of the fault, two different actions can be performed:

a) If the software component cannot be (internally) recovered to an operational state, an exception-notifying message is generated; a subsequent system recovery and/or reconfiguration is required.

b) If a single output result is recognised as erroneous, but the internal state of the component still allows continued operation (e.g., an internal recovery action has been successful), the expected output message is generated, with the special format <ABSTENTION>. This allows the normal message flow to be maintained, avoiding the need for system recovery. Several <ABSTENTION> messages can be sent out by a software component during an execution.

14.3.1.2. Error Detection by Comparison.

In the Delta-4 fail-uncontrolled host paradigm, detection of hardware-induced errors is accomplished by comparing the results produced by two or more identical software components, running on separate hosts. This method relies on the assumption that faults in physically autonomous electronic circuits exhibit statistical independence: in fact, the assumption closely matches the actual behaviour, with limited exceptions (e.g., upon occurrence of large electromagnetic disturbances).

To detect errors caused by software bugs it is necessary to compare results generated by program modules, which are ideally free from common design faults. The assumption of statistically independent failures in redundant software modules has sometimes been used. Although convenient for mathematical analysis, this hypothesis represents neither reality, according to experiments, nor an ideal goal (an ideal goal would be zero probability of common error, i.e., negative correlation). It has been shown in the literature [Eckhard and Lee 1985, Littlewood and Miller 1987a, Littlewood and Miller 1987b] that programs developed "independently" (in the sense that they are extracted randomly and independently from a population of possible programs) do fail independently on individual inputs, but may exhibit failure correlation, on a population of inputs, if some inputs are more likely to induce failures than others. Anyway, there is evidence that forcing methodological diversity in development is generally better or no worse than not forcing it.

When error detection is based on comparison of results, the problem arises that diversely implemented replicas of a software component (variants) will often produce different correct results; such components will be referred to as *divergent variants*. For instance, diverse implementations of real-numbers arithmetic functions will usually produce results that differ in the least significant digits (specifications that allow multiple, very different correct solutions are not considered here).

For this reason, the possibility of inexact comparison is a requirement for practical systems. Otherwise, many redundant applications would always fail on a "no agreement" situation.

Since the result comparison is inherently connected to find out the "good" result from the set produced by several variants, the recommendations for its use in Delta-4 will be discussed below, together with the latter problem.

14.3.2. Error Processing

Several error processing strategies, allowing different levels of dependability, must be available.

The simplest way of processing an error condition, upon detection, is to abort the affected computation and discard the faulty modules. This solution is appropriate, of course, for applications having almost no requisite of dependability. The only support needed for such an action is the capability of self-aborting by application programs, which is ordinarily supplied by LEXes. However, such an occurrence has to be signalled to system administration; this can be implemented by intercepting the LEX's abort call.

As the next step in dependability levels, Delta-4 supports a range of applications that do not have stringent time constraints and allow the use of backward error recovery techniques.

To support applications with more demanding requirements in terms of real-time response, the voting mechanism (introduced in section 14.3.1.2 for the purpose of error detection and error masking) can be effectively exploited to determine a "correct" value out of N results obtained by diverse, concurrent, redundant variants.

While in the backward recovery technique both the correct result and a consistent computation state are obtained by the single recovery action, only the correct value can be obtained by voting. The state recovery of failed components is (optionally) obtained by ad–hoc means.

Backward error recovery and the problem of choosing a "good" result *(adjudication)* are discussed below.

14.3.2.1. Backward Recovery. Error-handling techniques based on backward error recovery can be reduced to a common scheme: the state of the computation is saved from time to time (checkpointing). In case of a fault detected by error detection mechanisms, the component execution is stopped, and a backward recovery action is taken, by restoring the state saved in the last checkpoint and the same computations are repeated (using the *same* code). For example, consider the case where an exception is raised by the operating system, due to some internal problem (say, overflow of the process table). The programmer can invoke the rollback primitive, since he may expect the problem to clear itself in short time. Notice that the implementation of the rollback primitive must i) be such as not to be likely to worsen the problems that may have caused the original exception (in this example, must not require new space in the overflowing table, but wait for the overflow to be resolved), and ii) allow a clean abort of the application component, with proper notification to system administration, if a rollback does not succeed.

This technique recovers only those faults (whether in hardware or in software), whose effects are expected not to last until the subsequent re-execution. To recover from solid hardware faults, the checkpoint data need to be stored on a redundant host (as in the Delta-4 passive replica model). There is a class of software faults that cause errors having a "transient" appearance: quoting from [Gray 1986], the residual bugs in good software are often "Heisenbugs" rather than "Bohrbugs" (see section 6.4.2).

Backward recovery techniques for software faults without *design diversity* are recommended on mature software only (where Heisenbugs only are likely to be left). The software components shall be equipped with extensive run-time error detection mechanisms, based on: defensive programming techniques; consistency and reasonableness checks on inputs, intermediate results and outputs; program structuring and run-time organization able to confine the effects of errors (such as processes communicating through messages only, object-oriented systems; e.g., Deltase programs).

These techniques are an extension of the passive replica model for hardware-fault tolerance (see section §6.6). However, unlike the passive replica model, the checkpoint data can be stored in the same node where the software component is running, if application software-caused errors are deemed to be the principal source of system problems.

Another way of implementing backward recovery in Delta-4 is to use a transactional model of computation. Transactions implement recoverable atomic actions at the application level. By forcing a transaction abort in case of error, the component itself, as well any other component involved in the transaction, is brought back to a consistent state.

The processing of software-originated errors by backward recovery, possibly through the structuring of applications in transactions, needs the same support as the processing of hardware-caused errors as shown in section 6.6, on passive replication. However, the process of rolling back to a previously saved checkpoint can be triggered by an autonomous action of the software component itself (which, of course, will act upon internal detection of an error). Furthermore, there is no need to clone the component on a different node, with savings on hardware redundancy requirements.

Handling also "Bohrbugs" requires the execution of diverse variants; this is accomplished by the recovery block structuring concept. No new architectural support, other than those well known in the literature, is needed in Delta-4 when using the active replica protocol for hardware-fault tolerance. The case of passive replicas on fail-silent hosts will be shortly discussed in the remarks concluding this chapter.

14.3.2.2. The Adjudication Problem. Structured software-fault tolerance techniques, such as N-Variant Programming and N-Self-Checking Programming, can be considered instances of a general model [Anderson 1986], based on the use of diverse variants, whose outputs are used by an *adjudicator* sub-component. The adjudicator [Di Giandomenico and Strigini 1990], or generalized decision function [Avizienis and Kelly 1984], or collator [Cooper 1984], must produce the most probably correct result.

Several adjudication functions have been presented in the literature. Some of them use only the normal outputs of the replicas (e.g., exact (bit-by-bit) majority voting; median adjudication function; mean adjudication function). Others use additional information, such as: results of acceptance tests, reasonableness tests, maximum distance between consecutive results, etc.

A comprehensive discussion can be found in [Di Giandomenico and Strigini 1990]. Summarizing, the adjudication problem, is based on algorithms much more complicated than exact (bit-by-bit) voting as provided by MCS. This implies that the definition of the adjudication function may often be application-specific, making it difficult to create a standard adjudicator.

As a consequence of the above considerations, a number of support mechanisms could be provided in Delta-4 to allow the application programmer to include appropriate adjudication procedures in standard Delta-4 components.

1) MCS-level adjudicators: used only when identical results are specified for the variants, and/or with self-checking variants capable of sending abstention messages;

2) Deltase-level support: the user is given means to embody his own voting procedures into Deltase run-time support code, by using templates available in system libraries;

3) Application-level support: a software component generation tool able to automatically expand a software component into its redundant modules plus the adjudicator module is provided. A library of adjudicators should then be set up, at least for the basic objects available in a given system. To simplify the process, a generic template of adjudicator objects could be designed; a specific adjudicator would be built by linking a user-supplied voting routine to the template.

Self-checking software components require some discussion. In a typical configuration, two such components run concurrently as a replicated Delta-4 component. In the absence of faults, the results of both of them are correct: it is sufficient to choose one of them, e.g., on the basis of a record of past errors. When a fault occurs, the affected component either aborts, or sends an *<ABSTENTION>* message. In either case, the task of choosing the (unique!) correct value is indeed trivial, disregarding the different time performance. However, when an integrated approach to hardware and software fault handling is taken, the adjudication task may present subtle problems, as will be shown by an example in section §14.3.2.6. Anyway, the decision function can be expressed in a simple table form, where the table contents depends only on the system hardware and software configuration; this allows the implementation of the adjudicator at the MCS level.

14.3.2.3. Application-Level Adjudicator. A possible structure for a software-fault tolerant application in Delta-4 is shown in figure 1.

All the modules shown are standard Deltase objects, whose corresponding software components are produced using the Deltase software component generation system, Deltase/GEN. Both client and server replicated objects are presented to the rest of the system as non-replicated (but software fault-free!) ones. This is obtained by hiding replicated objects behind adjudicator objects (CA.adj, CB.adj, S.adj in figure 1). The adjudicator objects should be protected from hardware faults by means of standard, transparent Delta-4 replication.

The ordinary client-server interaction between client *CA* and server *S* through, say, the RSR primitive, sketched as:

$$CA \xrightarrow{\text{RSR}} S$$

becomes:

$$CA.1 \xrightarrow{\text{RSR}} CA.adj$$
$$CA.2 \xrightarrow{\text{RSR}} CA.adj$$
$$CA.3 \xrightarrow{\text{RSR}} CA.adj$$

CA.adj adjudges requests

$$CA.adj \xrightarrow{\text{RSR}} S.adj,$$

$$\left\{ \begin{array}{l} S.adj \xrightarrow{\text{RSR}} S.1 \\ S.adj \xrightarrow{\text{RSR}} S.2 \end{array} \right\};$$

S.adj adjudges replies

S.adj replies to *CA.adj*

$$\left\{ \begin{array}{l} CA.adj \text{ replies to } CA.1 \\ CA.adj \text{ replies to } CA.2 \\ CA.adj \text{ replies to } CA.3 \end{array} \right\}$$

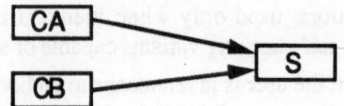

(i) A DELTASE CLIENT-SERVER CONFIGURATION

CLIENT A - 3 VARIANTS   SERVER - 2 VARIANTS

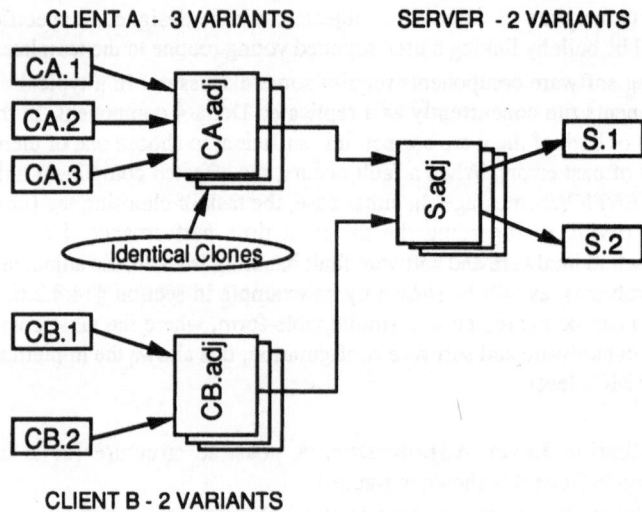

CLIENT B - 2 VARIANTS

(ii) REDUNDANT CLIENTS AND SERVERS

Fig. 1 - Separate Client and Server Adjudicators

Each server is accessed through a single adjudicator object (*S.adj* in figure 1), independent of the number of clients. The adjudicator should issue replicated service requests, avoiding the interleaving of requests coming from different clients. For this purpose there is no necessity of using atomic multicast; for better performance *S.adj* and *CA.adj* could use simple bi-point transmission to send messages to *S.1*, *S.2* and to *CA.1*, *CA.2*, *CA.3*.

In this solution, there is no need for special architectural support. However, a software component generation subsystem able to automatically expand a software component into its redundant modules plus the adjudicator could be very helpful to the application programmer. A library of adjudicator objects should be set up, at least for the basic objects available in a given system. Further, a generic template of adjudicator objects could be designed; a specific adjudicator would be built by linking a user-supplied adjudication function to the template.

No special restriction on software modules is assumed, i.e., divergent, non-self-checking components should be accommodated. Inexact voting procedures, which heavily depend on the specific application, can be provided.

The above configuration suffers some drawbacks:

1) The performance penalty imposed by the two added message levels between client and server.

2) The fairly large number of objects, considering that (as shown in the figure) the adjudicators need to be replicated to mask hardware faults.

3) The facilities already provided by the Delta-4 architecture, notably the atomic multicast communication protocol, are not exploited.

14.3.2.4. Adjudicator as a Unique Deltase Object. In this second solution, the adjudication of replicated client requests, as well as that of replies from replicated servers, are accomplished by a special Deltase object $S*$ (see figure 2). $S*$ is interposed between clients and servers by the Deltase subsystem for software component generation and installation (Deltase/GEN). A different adjudicator object $S*$ is required for each client-server pair.

Synchronization among replicas and error reporting are also implemented in this object.

$S*$ has a number of identical interfaces to the replicated clients; the number N of such interfaces, i.e., clients, actually linked to $S*$ can be dynamically changed. In a similar way, multiple interfaces to the replicated servers are provided.

For protection against hardware errors, $S*$ should be specified as a standard Delta-4 replicated object obeying the validate-before-propagate paradigm.

A template of a generic adjudicator object, including the communication interfaces, can be set up in a library, leaving to the application programmer the task of providing the procedures implementing specific adjudication algorithms. In the software component generation phase, the generic template should be linked with these procedures.

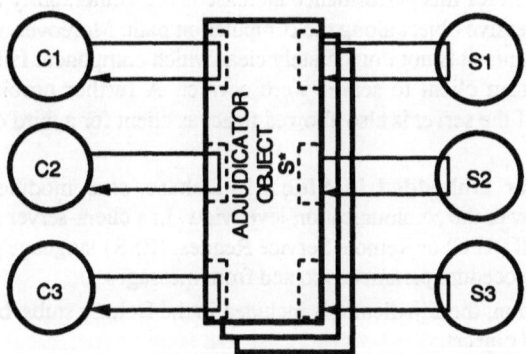

Fig. 2 - Combined Client-Server Adjudicator: Principle

The adjudicator implemented as an object has the advantage of allowing quite sophisticated fault-tolerance techniques.

An example is depicted in figure 3. Each client replica is associated to a specific server replica. Each service request is immediately forwarded to the associated server replica, without waiting for the other requests, therefore without going through the adjudication process. Adjudication is executed on the reply messages; the adjudged values are multicast back to clients.

On the request side, when all the replicated requests are available, $S*$ may compare them for error detection only.

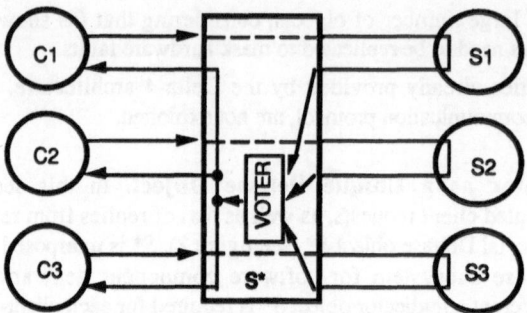

Fig. 3 - Combined Client-Server Adjudicator: Example

The advantages of the structure in this example are:

- It allows a high degree of diversity between the implementation of the clients (since their requests are not required to be compared). It is even conceivable that the formats of the RSR primitives are not the same for all the S_i-C_i pairs.

- The time to get through a chain of multiple clients and servers is approximately equal to the sum of the average execution and transmission times on the "request" path, instead of the sum of the worst times. Besides, the voting operation on requests can be executed concurrently with the main computation.

The price to be paid for this performance increase is the vulnerability to multiple software errors, namely in successive object along the computation path. Moreover, the recovery process is more complicated, since it is not immediately clear which component failed, and whether the interactions passed from client to server were correct. A further problem arises from the propagation of errors if the server is also allowed to act as client for a third object.

14.3.2.5. Adjudicator embedded in Stubs. In Deltase, *stub* modules are used to map the language-level view to the communication-level view. In a client-server interaction using the Remote Procedure Call (RPC) or Remote Service Request (RSR) language primitives, the stubs pack and unpack the procedure parameters to and from messages.

As an added function, the adjudicator is included in the Deltase stubs, both in the client and in the server replicated objects.

The adjudication procedures supplied by the user, according to a specific format, should be linked at component generation time to a modified Deltase stub module.

In figure 4, a subsystem composed of a 3-replicated client and a 3-replicated server is depicted. As all components are standard Deltase objects, communication among them takes place by means of RPCs or RSRs. The basic stub functions sketched out in [Powell 1988] should be complemented by: a) multicast of the message carrying the service request to all replicated servers; b) at the server input port, execution of the adjudication procedure on the data coming from the client variants. The same considerations apply for the handling of reply messages from the servers to the clients.

This configuration effectively distributes the adjudication service among the message-receiving modules. The consistency of servers, with regard to the sequence of input messages from different, replicated clients, is ensured by using the atomic multicast protocol. All the clients (even non-replicated ones) sending messages to a replicated server should use this facility.

Another source of inconsistency of servers can be the time-out mechanism, used to avoid hanging up while waiting for a message from a faulty client. In fact, whatever length of waiting time is chosen, it is possible that some replicas time-out whilst others do not. Since this problem is bound to the independence of decisions in the replicas, it can be avoided only by some form of agreement between replicas on the decision itself.

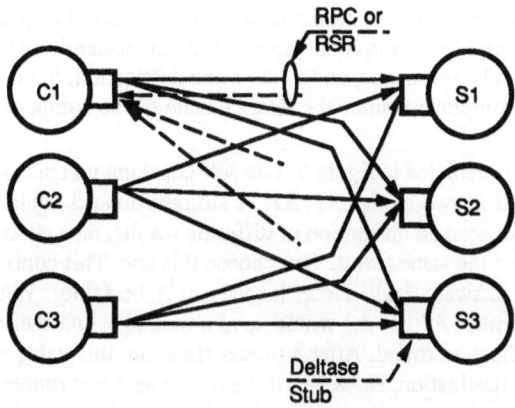

Fig. 4 - Adjudicator embedded in Stubs

If there are no specific safeguards, a slow replica can increasingly lag behind the other peers. It is conceivable to let faster replicas proceed, setting up a list of missing messages from the slower one(s) paired with the correct (adjudged) values. This allows safe discarding of late messages, as well as checking their correctness, for error detection purposes. This solution allows tolerance of temporary slowdown of a replica. After a given threshold in time or in the number of listed messages from an object, this one should be considered faulty, and the standard recovery procedure should be started (see also §6.5.2).

To avoid the possibility of erroneously declaring faulty a slow replica, some synchronization technique should be used. The cost is that the speed of progress of a replicated object over any given time interval is that of the replica that is slowest during that interval.

14.3.2.6. Communication-Level Adjudicator. If the application can be given specifications that reliably guarantee identical results, software-fault tolerance can be easily achieved using the N-Variant Programming model, as an extension of the N-modular redundancy already specified for hardware-fault tolerance. The MCS session-level, signature-based, IRp adjudicator can be used (cf. sections 6.5.2, 6.5.3 and 8.1.2.1). For example, a 1-fault-tolerant system would require 3 variants running on 3 hosts; to tolerate two simultaneous faults several configurations are available: i) 3 variants running on 9 nodes can mask 2 hardware faults, or 1 hardware plus 1 software fault; ii) the same result is obtained by using 4 variants on 7 nodes: more design effort traded against less run-time resources; iii) 5 variants on 5 hosts can also mask 2 software faults. The latter result stems obviously from the fact that each host equipped with a diverse variant is a fault-containment region (cf. chapter 4).

To allow wider use of software-fault tolerance, divergent variants are supported. Since application-dependent adjudication procedures, normally required in this case, cannot be included in the low-level communication software, the adopted solution, discussed in [Strigini 1988], is to use self-checking software components. The majority, signature-based, voting is of no use here; the adjudicator only has to choose among several correct results

The use of the <ABSTENTION> notification generated by failing self-checking components allows better performance, by i) reducing the use of time-outs in case of error, and ii) improving the adjudication process [Ciompi and Grandoni 1990]. As an example of the usefulness of this mechanism, consider the case where an application-level acceptance test shows that an individual output is erroneous. The programmer knows that this output is computed from the values of a data structure that is continuously updated with data from outside the software component. It is reasonable to send an abstention message instead of the output that was found to be erroneous, and to the same destination, but not to alter the normal execution flow of the component, since the internal state of the component may soon correct itself.

A simple example is depicted in figure 5: two self-checking variants, A and B, run on three hosts, with A replicated in two copies A1, A2. A straightforward adjudication policy is the following: i) if A1 or A2 send an abstention or differing results, then choose the result from B; ii) if A1 and A2 generate the same result, then choose this one. This configuration can tolerate one hardware or one software fault. Now, let variant A be faulty. Without the use of the abstention message, neither A1 nor A2 would send a message, and the value produced by B would trigger an adjudication round. After a proper time-out, this value would be considered valid and forwarded to destination. However, if the host where B is running fails by outputting an undue or "impromptu" message, this message would be considered valid and then erroneously forwarded. If abstention-messages are used instead, the adjudicator should expect to get, after the message from B, at least one message from A1 and/or A2. No more than one host can fail, and if the variant A has a bug, it sends an abstention notification message. This allows the discrimination of the above erroneous condition.

In summary, at the MCS level non-divergent variants with exact comparison, and self-checking variants, are supported. In both cases there is a need for new adjudication procedures (other than the present majority rule); in the second case the possibility of notifying an abstention is also required.

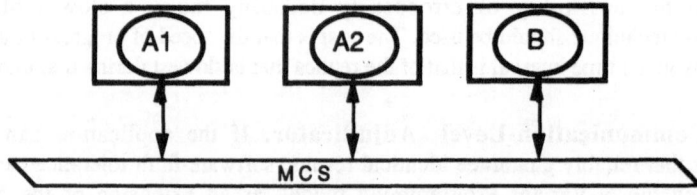

Fig. 5 - Communication-Level Adjudicator

The above illustrative examples of adjudication functions are based on fault hypotheses expressed in terms of number of faults. It is of course unrealistic to exclude the possibility of higher number of faults, and to treat all possible failures as though they had the same importance. Adjudicators that take into account the expected probabilities of all fault patterns can be designed as described in [Di Giandomenico and Strigini 1990].

14.3.3. The Problem of State Recovery of Variants

Once an error (or even a disagreement between correct variants) has been detected in some result, the first problem is how to mask the error and give the user (of the result itself) a correct result; once that is solved, a second problem is what to do with the failed (or disagreing) variant(s). The internal state of the variant is presumably corrupted. We may therefore expect subsequent results to be erroneous as well, and thus not only useless but dangerous.

A first choice is to kill the erroneous variant altogether. For short-lived components this seems advantageous. The component will be automatically "repaired" when its execution terminates and, at the next execution, a correct redundant module will be instantiated, from the same code, taking its internal state from an uncorrupted global state (data base, sensor readings, etc.).

For components with long execution life-spans, this is not appropriate: too many variants would be needed (the analogous hardware configuration is self-purging N-modular redundancy without repair). Then, there is a need to recover a variant that erred, so that it can catch up with the correct variants and carry on its work. Recovery brings with it an inherent problem that we discuss in the following.

Recovery may be:

1) programmed into the same variant that errs (estimation of approximate correct values, for instance, or reset to safe values). This is the simpler choice, but is not widely applicable; in particular, as noted earlier, perfectly healthy variants might drift apart from each other, with time, if no built-in way exists to recover them to consistent states.

or,

2) the recovery may use values produced by the other variants. The problems with this solution, i.e., correcting the internal state of a variant or create a state based on the states of the others, is that it requires either identical internal representations of those states (same internal variables laid out in the same way) or a way to translate between diverse representations.

For Delta-4, the latter solution is proposed. Specifically, a unique representation of internal state *for inter-variant exchange,* not necessarily for internal use, must be given in the object specification. Each variant must be equipped with special *Input_State, Output_State* routines; the latter should be able to be started, in pre-defined computation points, upon receiving a specific signal from System Administration (SA).

Assuming that the redundant configuration is made up by three variants $C1, C2, C3$, and that $C2$ has to be re-instantiated, the state cloning operation can be outlined as follows:

a) *SA* loads new $C2$ code;

b) *SA* activates $C2$ on the Input_State routine; $C2$ starts, waiting for the state information from *SA*;

c) *SA* sends $C1$ and $C3$ signals to enter the Output_State routine, which should be bound to *SA*;

d) At proper times, $C1$ and $C3$ output their state to *SA*, after executing some synchronization protocol, then they wait for a *Continue* signal from *SA*;

e) *SA* sends $C2$ the state received (options: send the first received; send all copies, with the number thereof as an additional parameter);

f) $C2$ sets internal state (option: checking validity if multiple copies received), then waits for the *Continue* signal (option: an acknowledgement signal to *SA* can be needed before beginning to wait);

g) *SA* sends all variants the Continue signal.

The state cloning operation describer above should be considered as a complement to the present cloning process.

14.4. Specifying Software Components for Software-fault Tolerance — Tradeoffs in an Object Environment

Any software-fault tolerance paradigm using redundant components needs to tradeoff between the conflicting needs: i) establish the minimum number of design constraints to allow diversity; the design manager can then force diversity among programmers, issuing individual constraints (e.g., development environment, language, etc.); ii) issue specifications that are tight enough to insure some pre-defined "homogeneity" among the diverse results.

As a simple example of the problem, consider the component A implemented as three variants $A.1$, $A.2$, $A.3$. If no specific restrictions are given to the variant designer, it can happen that, say, $A.1$ and $A.2$ are both programmed to import a service interface from the *same* component B, while $A.3$ uses services from another component C or, possibly, no external resources at all. The following problems arise:

a) A design fault in B can cause a failure in both $A.1$ and $A.2$; in this case, it is likely that the correct result computed by $A.3$ will be outvoted, leading to a catastrophic failure.

b) The variants exhibit different communication patterns. This makes it difficult to determine where and how to place adjudication mechanisms; for example, are the service calls to B from $A.1$ and $A.2$ to be adjudicated?

c) There may be consistency problems in shared components, like B in the example, which are quite unusual: in fact, $A.1$ and $A.2$ are semantically the same component, therefore for correct behaviour their operations on B should be made idempotent, in some way.

d) A problem related to the point c) above is that of recovery: if, say, variant $A.1$ fails after interacting with B, the recovery of $A.1$ mandates recovering B too, and then, in the classical domino effect, $A.2$ must be brought back to a consistent state. This operation, possibly triggered by a single fault, may require undoing preceding successful adjudications: a much more complex computational model would be required.

In the object model of computation, the conflict depicted above appears harder, because of the emphasis put on hiding the internal structure of the object, as well as in the methods used in implementing the offered services (in particular, which services from other objects are used).

On the other side of the coin, one can wonder if adjudicating results surfacing at the object boundary is enough: relatively complex, "taciturn" objects may suggest performing cross-checks among intermediate results, to prevent non-recoverable divergence among variants. However, this again contrasts the internal information hiding philosophy.

For the sake of simplicity, and as a first step in devising software-fault tolerance structures in object-oriented systems, like those based on the Delta-4 architecture, it is assumed here that adjudication is only performed at the level of object interfaces (as accomplished in the validate-before-propagate paradigm to mask hardware-caused errors). In other words, the scope of the attainable diversity is limited to the internal implementation of the object specification, since diverse use of external services is forbidden. Of course, the issue of how to organize diverse implementations across object boundaries is of great interest. In fact, some simple extensions can be easily given. For example, concurrent use of services exported by a memory-less object can be safely allowed to variants, from the point of view of information consistency. Such an object, is of course a reliability bottleneck. More general methods for wide-scope software-fault tolerance structures are under investigation.

14.5.   Concluding Remarks

Techniques and mechanisms to implement and support software-fault tolerance in systems built on Delta-4 machines have been described. According to the Delta-4 philosophy, a range of solutions is envisaged, differing in complexity and cost as well as in the attainable dependability.

Some tradeoffs presented here are likely to be superseded by better solutions in the near future. For example, the restriction that variant objects must exhibit identical interfaces may be relaxed, by introducing higher level structuring concepts that allow wider scope diversity, while ensuring data consistency and recoverability.

The proposed solutions are based on the assumption of fail-uncontrolled hosts. More restrictive assumptions may lead to simple solutions for software-fault tolerance.

Recovery blocks, for example, can be easily implemented in conjunction with the passive replication model for fail-silent hosts. In this model, the backups do not execute recovery blocks, just like any application code. However, the ordinary checkpoints are not sufficient to enable recovery from a software error occurring after a hardware fault. In this case, the next alternate in the recovery block must be executed *in the backup host*, after restoring the initial information, which was set at the *recovery point* in the primary host. The recovery point needs to be sent to the backup independently of the checkpointing mechanism: successive checkpoints along the execution of the recovery block supersede the previous one, while recovery point information must be held until the recovery block is exited.

A possible solution based on the leader-follower model could be the following. The follower does not maintain recovery points, and does not execute multiple alternates. On arrival at a recovery block, the followers suspend waiting for instructions from the leader. The leader runs through the recovery block and, on completion, informs the followers on which alternate to execute. The followers execute that alternate and the acceptance test. The replica determinism requirements of the leader/follower model would ensure that the alternate succeeds in the follower as it did in the leader in most cases. However, if a Heisenbug occurs in a follower, the acceptance test should fail; in that case, the classic recovery block paradigm would trigger the execution of the next alternate, starting a state divergence with the leader. The acceptance test failure should instead cause the abort and successive cloning of the follower. Therefore, a mechanism to trigger the abort of a follower from the component itself should be added to the standard leader/follower support.

Chapter 15

Validation

Users of the Delta-4 architecture must be able to have a *justified* confidence in its dependability. Consequently, such an architecture must undergo extensive *validation* both from the *verification* and *evaluation* viewpoints.

Verification is that part of the validation activity aimed at removal of design and implementation faults. Verification in Delta-4 is carried out at two levels:

- verification of the design of the communication protocols to discover and remove design faults,
- verification of the implementation by means of injection of hardware faults to verify the effectiveness of the architecture's self-checking and fault-tolerance mechanisms.

Dependability evaluation is that part of the global activity of validation that pertains to fault-forecasting, i.e., the estimation of the presence, the creation and the consequence of faults. Dependability evaluation is also carried out at two levels:

- evaluation of dependability measures of Delta-4 architecture configurations taking into account the nature of the different elements (e.g., fail-silent or fail-uncontrolled hosts, replication domain of the different components, replication techniques, reconfiguration possibilities, repair policies, ...)
- evaluation of software reliability through the application of reliability growth models.

The aim of each validation activity and the main results are summarised in the next section. Then, sections 14.2 - 14.5 provide further details on each activity.

15.1. Overview

Section 15.2 is devoted to protocol verification. The need to assure reliable communication among distributed sites has led to the design of specific protocols for the Delta-4 architecture, such as the Atomic Multicast protocol (AMp, cf. chapter 10) and the Inter Replica protocol (IRp, cf. section 8.1). To improve the quality of these protocols, formal methods have been used for their specification and verification.

The specifications of a protocol consist of the formal description of the distributed algorithm, the formalization of the assumptions about its execution environment and the formal definition of properties characterizing the service it should deliver. Formal specifications are useful on their own, since they force the specifier to formally explicit crucial features of the protocols. Furthermore, the verification of their consistency allows detection of possible errors early in the software development. We are concerned with *formal verification*, that means:

Given some description of an algorithm and given a description of its "service specifications", verify formally that the described algorithm delivers the specified service.

Verification is carried out by model checking techniques using the *Xesar* verification toolbox. This toolbox was developed to evaluate properties — given by formulas of temporal logic — on a model obtained from a program describing the algorithm to be verified (embedded in its execution environment). These methods have important advantages for discovering design faults since they are based on a complete search of the graph of all possible behaviours of the system to be verified.

The AMp and IRp families of protocols have been specified and verified. Inconsistencies of different nature have been detected such as incorrect initializations of local variables, or too weak conditions. Some of them are only detectable in some peculiar sequence of events that it would be unlikely to obtain by simulation, such as unspecified receptions, non termination of the monitor election phase or duplication of messages. Implementations have been derived from the formal specifications.

The section relative to formal verification overviews the method followed and contains the major results obtained for the critical protocols that have been studied. Annexes K and L complement this section by surveying the various models that can be used to represent reactive systems and techniques for specifying system behaviour.

Section 15.3 is dedicated to implementation validation by means of fault injection. Work on implementation validation is aimed at testing the basic building blocks that support the fault tolerance features of the Delta-4 architecture: the fail-silent assumption of the NAC components and the fault resiliency provided by AMp. The experimental validation carried out is based on the use of physical fault injection, i.e., physical faults are directly injected on the pins of the circuits that constitute the NAC.

The distributed testbed, including the fault injection tool *MESSALINE*, that has been developed to carry out this work provides an experimental environment that enables not only successive versions of the AMp software to be addressed[1], but also two distinct versions of the NAC hardware: one with restricted self-checking mechanisms (referred to as a *fail-uncontrolled* NAC) and another with improved self-checking (referred to as a *fail-silent* NAC).

The section relative to fault-injection validation contains an overview of the method, a brief description of the testbed and the major results obtained so far (essentially, those for the fail-uncontrolled NAC).

Section 15.4 deals with dependability modelling and evaluation. The objective of this activity is to provide the users with a quantified assessment of the amount of dependability that the architecture provides, i.e., the degree by which they can justifiably rely on the architecture.

Dependability modelling is based on Markov processes. Use of Markov process is first justified and general expressions giving the equivalent failure rate and the steady-state or asymptotic unavailability are presented.

Delta-4 is a modular and open architecture and all possible configurations have a common point: a communication support system that constitutes a hard core since its failure leads to loss of interactions between the hosts. In this section, emphasis is put on the selection of the "best" architecture among all possible ones.

Various communication topologies are considered (802.4 token bus, and 802.5 and FDDI token rings), for each of them a single and a double configuration architecture is modelled. Two dependability measures are evaluated: the equivalent failure rate and the asymptotic unavailability.

---

[1] Obtained in particular as a result of the corrections induced by the design errors detected during this validation phase.

It is shown that for the single media configurations the equivalent failure rates are limited by the failure rate of the medium and non-covered failures of the NACs and that, for the double media configurations, they are directly related to the failure rate of the non-covered failures only.

Comparison of these architectures showed that it is very difficult to classify them: the evaluated measures depend on several parameters and a tradeoff is needed to select the most suitable architecture. However it is shown that — whatever the architecture — the coverage factor of the NAC is of prime importance, it is thus worthwhile to put emphasis on this coverage (i.e., self-checking mechanisms) during development.

Finally, *section 15.5* is devoted to software reliability evaluation. Quantitative assessment of software reliability is usually carried out through the application of reliability growth models. These models enable prediction of either the number of failures to be activated for the next period of time or the mean time for the next failures or the software failure rate or some combinations of these measures.

Reliability growth models are generally parametric models and these parameters have to be estimated (i.e., model calibration) to carry out dependability predictions. Calibration of the models is fulfilled through failure data collected on the software either during development or in operation. Predictions are thus based on the observation of the software behaviour during a given period of time, calibration of the model using the observed failure data and application of the model to estimate dependability measures.

Data collection is a long process and is now being carried out on the developed software. Reliability growth models will be applied when sufficient data items have been collected.

Since insufficient data has been collected at the time of writing, this section is mainly devoted to data collection and the problems that this data collection induces.

15.2.   Protocol Validation

15.2.1.   Introduction

In the Delta-4 project, the need to assure reliable communication among distributed sites has led to the definition of specific protocols, such as the Atomic Multicast protocol (AMp) or the Inter Replica protocol (IRp). As the system architecture is based on these protocols, the quality of their development is crucial. One approach for improving quality is to carry out simultaneously the design and its verification by using formal methods and then to derive the implementation.

The usual informal specifications given in an implementation guide are not sufficient to achieve formal verification. The first task is to provide a structured formal description of the protocols and of the delivered services. These formalizations are useful on their own, as they enforce the specifier to explicit the assumptions on which the system is based. Furthermore, it is important to verify formally the consistency of the service definition with the protocol description to detect possible inconsistencies early in the software development activity.

15.2.1.1. Formal Verification Techniques. Formal verification of a design requires a description of an algorithm and a description of its "service specifications". The aim is then to verify formally that the described algorithm delivers the specified service.

The systems we are interested in interact with their *environment* and thus cannot be adequately described by a purely functional view, in terms of input/output. Typical examples are operating systems, communication protocols, distributed data bases, digital systems ... These systems, which are better specified in terms of their behaviours, are known as *reactive systems* [Pnueli 1986]. There is a large agreement on the fact that the algorithms involved in such systems are highly complex, and their development needs a large effort, especially concerning formalization and verification of their design. Furthermore, the verification of time bounds in the distributed algorithms used in communication protocols is crucial.

Two main classes of techniques have been proposed for formal verification:

- Program verification based on *deductive methods* [Misra and Chandy 1981, Owicki and Gries 1976]: a distributed system is described as an abstract program, and the service to be delivered is characterized by a set of properties, described for example by formulas of temporal logics [Pnueli 1977], which are transformed into assertions about the program. The proof of these assertions can be partially automated by using theorem provers, e.g., [Boyer and Moore 1979, Gordon et al. 1979]. These methods extend to concurrent programs the methods proposed by Floyd [Floyd 1967] and Hoare [Hoare 1972] for sequential programs.

- *Model checking* techniques consisting in: building a model from the description of the distributed system, and checking the service properties on this model by using appropriate algorithms. Usually these methods are restricted to finite state programs and are suitable for protocols. Many proposals based on this approach have been made, for example [Clarke et al. 1986, Fernandez et al. 1985, Holzmann 1984, West 1982, Zafiropulo et al. 1980]. These techniques are more adequate for automatic verification of large systems.

Both approaches require:

- A formal description of the algorithm under study, given, e.g., in the form of a labelled transition system or a set of equations, or a CCS [Milner 1980], CSP [Hoare 1985], Lustre [Caspi et al. 1987], Estelle [ISO 9074] or Lotos [ISO 8807] program. In the case of model checking, this description is translated into a *model* (see annexe K) representing the behaviour of the algorithm.

- Formal service specifications, given as a set of properties characterizing this service, or as an abstract algorithm describing it. In the case of model checking, each property is evaluated on the model.

- The assumptions about the *execution environment* of the algorithm.

For the result of the verification to be meaningful, some *soundness condition* must be satisfied. One possible soundness condition is that all properties that are formally verified must be true in reality. If one is more interested in error detection, the soundness condition may also be formulated the other way round, i.e., each error detected in the verification phase must correspond to a real error.

15.2.1.2. Verification of Timed Algorithms. Usual methods for reasoning about reactive systems abstract away from quantitative time, preserving only ordering properties, such as "whenever a process receives a message, it has been sent previously by another process". However, in the domain of distributed systems, especially fault-tolerant ones, the notion of time is important.

First, two remarks should be made:

- The notions of time in the algorithm and in the model need not to be the same.

- Formal verification is different from performance evaluation. Formal verification is performed on an abstraction of an implementation. Performance evaluation works on a particular implementation, for which intervals of execution times for all basic actions are supposed to be known, and thus an exact picture of the time behaviour may be obtained.

The notion of time needed for the verification depends obviously on the notion of time needed by the described algorithm. A classification of algorithms can be given [Cristian 1991], based essentially on two criteria: *synchrony* and presence or absence of a *global clock*:

- Asynchronous systems that work without taking time into account at all, in the sense that no upper bounds of execution times of actions need to be given for the system to work correctly; thus the given algorithm is supposed to work under any timing constraints.

- Synchronous systems in which all processes can be considered as working off a common clock, for example digital circuits or distributed systems in which the clocks local to each node are synchronized (see, for example, section 9.6.6 for the low-level clock synchronization proposed in XPA).

- Synchronous systems in which a global clock is not available, but an assumption is made about the upper bounds of the time needed for actions, where these upper bounds refer to local time.

The systems in the first category are also called *systems with unbounded delay*. Those in the last two categories are *systems with bounded delay*. The verified protocols of the Delta-4 architecture are in the third category. General verification methods must be adapted to this class of systems, the main problem being the introduction into the model of a suitable notion of time.

The remainder of this presents the actual verification work in the Delta-4 project and the main results that have been obtained. Two annexes provide a more detailed description of the formal verification method by model checking. Annexe K introduces different models and possible notions of time used for verification; annexe L presents the nature of the service specifications and gives some formalisms allowing us to describe the specifications concerning time.

15.2.2. Verification of Protocols in the Delta-4 Project

The work mainly concerns the formal specifications of the considered protocols, the clarification of the assumptions made on the behaviour of the environment and the formal verification, from which the new implementations have been structured and developed. The primitive material for the formal verification activity in Delta-4 was an existing communication stack software, including the critical protocols to be verified, that was partially and informally specified. The results obtained have required tight interaction between the designer-implementer and the specifier-verifier teams.

The verification work has been carried out in two steps: first, providing formal specifications, then verifying formally their consistency.

The specifications of each protocol have been structured into three formal descriptions:

- A description of the state machines implemented in the protocol.

- A description of the properties of the environment (e.g., duration of the message transmissions, limitation of the number of messages lost in sequence, ordering of the delivery of messages). These properties of the environment correspond to the concept of "operational envelope" used in section 5.1.2.

- A description of the expected service, by a set of logic formulas. A classification of the usual properties is discussed in annexe L.

Formal verification has enabled detection of inconsistencies in these formal specifications and thus led to removal of design faults. Model checking techniques have been applied, using the *Xesar* tool [Richier et al. 1987] developed at LGI. These methods were chosen because they are suitable for finite state machines such as protocols and can be automated to a large extent, even for complex systems (the *Xesar* tool allows verification of quite large models [Graf et al.

1989]). Furthermore, they provide considerable help for the detection of errors since they are based on a complete search of the graph of all possible behaviours of the system to be verified, using appropriate algorithms. Error states can not only be detected, but execution sequences leading to them can be displayed.

15.2.2.1. Verification using the *Xesar* Tool. The *Xesar* tool implements model checking techniques: it evaluates properties given by formulas of the branching time temporal logic CTL [Clarke et al. 1986] or by Büchi automata [Büchi 1962] on a model generated from a program written in Estelle/R, according to its formal semantics. More details concerning the model can be found in annexe L.

The specifications of the system to be verified are described by a finite set of communicating processes, forming a *closed system*; that is, for each possible communication, the complete description of both emission and reception must be provided. Therefore, the *protocol environment* must also be described explicitly by a (set of) process(es).

Each verification is carried out on a model obtained from a program describing a particular system configuration embedded in its environment, with a particular initialization; such a program is called in the sequel a *scenario*.

The protocol description language Estelle/R is a variant of Estelle. The main difference is that in Estelle/R communication is modelled by the *rendezvous* mechanism. Communication by rendezvous requires explicit representation of finite buffers by processes. The problem arising from "communication through unbounded buffers" of Estelle is that certain deadlock situations cannot be detected, since it is possible to carry on filling some message buffer forever.

The formal semantics of an Estelle/R model is based on the ATP algebra [Nicollin et al. 1990] where a "clocktick" event is used for the translation of the "delay" construction. The global model is obtained by using the semantics of ATP for the parallel composition, i.e., interleaving semantics with synchronization of all processes on "clocktick" as discussed in annexe K.

Formulas of the CTL temporal logics or automata describing the properties that characterize the service specification, are evaluated on the model; the complexity of the model checking algorithms is discussed in annexe L.

An error is detected if, during the traversal, a state not compatible with the property is encountered. In this case, the execution sequence leading to this state can be analysed. If the traversal can be completed without detecting an error, the model satisfies the property.

If a complete traversal is not possible in a reasonable time, the absence of detected errors does not allow one to deduce that the protocol is correct. Nevertheless, it increases the confidence we may have in it. In this case it is also possible to make several partial traversals with different criteria for the choice of the successor state in order to increase the coverage, as proposed in [Holzmann 1990].

15.2.2.2. Formal Verification of Protocols. Applying formal techniques to perform a brute force global verification of the innovative Delta-4 protocols is an unrealistic task since their execution environment involves the whole Delta-4 architecture. The two main problems to be solved first are: the characterization of the protocol execution to be verified and the establishment of a suitable model for bounded time protocols.

Formal verification using the *Xesar* tool allows automatic analysis of *specific complex configurations* of the protocol execution. In each configuration, exhaustive analysis allows detection of all design errors, even those occurring in complex execution sequences and which are very unlikely to be detected by simulation or test, as they might occur with very low probability.

To infer the validity of the protocol in *any configuration*, one needs some "inductive" method. Different solutions have been proposed, generally based on an invariant characterizing the behaviour of the protocol. This invariant must have a "good" structure, i.e., it must allow induction on the structure of the configuration. However, the protocols considered in this project are too complex since they are based on the (simultaneous and successive) use of different paradigms (two phase protocols, election, reliable multicast, use of time out to avoid livelocks, ...). The complexity of the interaction of these paradigms does not allow description of the complete behaviour of the protocol by a "suitable" invariant.

However, for any particular protocol, non-explicit reasonings, based on considerations about the overall structure (e.g., making symmetry considerations or exploiting the splitting of the protocol into independent phases) provide convincing arguments that the general validity can be deduced from the verification of a limited family of configurations. For example, it can be argued that in the case of AMp protocols, configurations with three machines are sufficient, and that we can further limit the possible actions in any of these machines.

Practically, a finite set of scenarios (judiciously selected configurations of a fixed number of networked protocol machines) is exhaustively verified.

A crucial point in the verification is the choice of an adequate model, suitable in the present case to time bounded systems, such that all properties formally verified on the model must be true in reality. This adequacy depends deeply on the abstraction level used for the protocol machine description and on the modelling of its environment. For example, if the environment is modelled by a process "chaos" (i.e., that can modify transmitted messages in any possible way), all detected errors may not correspond to real errors in the protocol.

In the present work, the environment consists of adjacent layers of the communication stack; the model must take into account the different features characterizing the behaviour of these layers, such as buffering, occurrence of faults, and timing constraints.

The first two features are explicitly represented in the model. Concerning time, two kinds of modelling have been used, depending on the nature of the system to be verified:

- In systems where all message transmissions are implicitly clocked, timeless models are used. For example, in the Turbo-AMp or token-based AMp protocol, (an extension of the IEEE 802.5 token ring MAC protocol, see section 10.8.1) all messages are implicitly clocked by the token circulation mechanism. Therefore, the verification of a service property is made under some fairness assumptions, stating that the property is true provided the token is effectively circulating.

- In other systems, e.g., the token-less AMp (see section 10.8.2) and IRp protocols, the passing of time is introduced explicitly in the model, by means of a specific clocktick action. Thus, time bounds can be verified without using fairness assumptions, and some statements concerning time limits can be made. However, all the obtained results can only be guaranteed to be valid for the particular timer values for which the verification has been carried out.

A survey on time modelling for the purpose of verification can be found in annexe K.

15.2.2.3. Results of the Verification Work.
Two families of protocols have been verified and consistent, structured specifications of both protocol machines and services have been provided:

- reliable group or multicast communication protocols (token-based and token-less implementations of AMp, see sections 10.8.1 and 10.8.2) — only the atomic quality of service (see section 10.4) has been verified;

- inter replica coordination protocols (IRp) managing the replicated aspects of the session service users, and supporting distributed fault tolerance by active replication.

For each protocol, formal specifications have been provided:

- The services have been characterized by a set of formulas, e.g., the AMp properties given in table 1 of chapter 10.

- All the environment assumptions have been specified. For instance, in the case of the token-less AMp protocol, this leads to the formal characterization of the abstract network given in table 3 of chapter 10. In the case of the IRp protocols, they consist of the atomic quality of xAMp service and additional properties stating that station failure indications are received in bounded time. Thus, network management does not need to be fully characterized.

- Only the abstraction of the protocol machine data structures relevant to verification have been defined. It is worth noting, however, that all transitions were in fact fully specified.

This task has been carried out in tight collaboration with the designer of the protocols because a detailed knowledge of the protocol architecture as well as of the verification method is needed. Some design errors were directly discovered during this collaboration.

The verification of the consistency of these specifications has been made as previously discussed. For example, in the case of token-less AMp, scenarios were grouped in different sets, covering each of the relevant steps of AMp execution, including: message transmission, monitor election, joining a group, leaving a group, and failure recovery. A justification for the chosen set of scenarios can be found in [Veríssimo and Marques 1990]. The "sub-services", i.e., the expected services for each of these steps, were also defined by sets of properties.

For instance, for the monitor election scenarios the following properties were defined:

- "There is one and only one winner of a monitor competition."

- "If the current monitor fails and there are still live members, another monitor competition takes place."

- "The monitor competition always terminates successfully unless all the group members fail."

Additional information is needed to define the scenarios associated with each step:

- The set of the possible initial states in which the steps can start. For example, for the monitor election step the initial states correspond to the states reached after the occurrence of an error in any other step: failure of the sender or of one of the receivers during a message transmission step; failure of one of the stations during a joining step; and so on. One or several scenarios are needed for each class of initial states

- The number of stations needed for the verification of each step. For example, for the monitor election step, it is possible to examine all possible error cases with three stations. The failed stations need not be represented.

Particular configurations with a fixed number (e.g., 3 for the token-less AMp) of interconnected protocol machines, covering critical cases derived from the structure of the protocol, have been verified with the *Xesar* tool. The generated models have reasonable size (on average 350,000 states for the token-less AMp). Non trivial inconsistencies, have been detected (and corrected) in the specifications.

Two types of inconsistencies have been identified: "superficial" ones that can be easily corrected, and "deep" ones that need more analysis to be corrected. Deep inconsistencies may show up inconsistencies in the application of the paradigms underlying the design of the protocol.

In both AMp and IRp protocols, a significant number of superficial inconsistencies were detected, for example: bad initializations of local variables; incorrect parameters; conditions in transitions that are too weak, etc.

A first example of a "deep" inconsistency was found in the token-less AMp: it corresponded to a livelock situation where, in some global states, a "monitor election" procedure can be restarted forever and will thus never terminate.

A second example is given by the IRp protocols.

Let us first describe shortly how these protocols work. The IRp protocols ensure that a family of replicated entities perform the same actions, and that an external user perceives all the replicas as a single one. Each correct replica has a complete knowledge of the state of the other correct replicas. To enforce this point, a replica uses the AMp protocol to inform the other replicas of any internal change.

When a message is sent outside the group of replicas, an election is carried out to select exactly one sender. Timers are used to control the relative speed of execution of the different replicas.

Three paradigms (atomic multicast, election, time-outs) are used together. As the number of possible behaviours is very large, conflicts occur. For example, two inconsistencies have been detected in the message send protocol:

- If a replica dies after its election as a sender, the other replicas are sometimes unable to elect another sender.

- If timers expire exactly at the moment a sender is elected, due to transmission delays, it can happen that the sender sends the message and a new election is performed, resulting in the message being sent twice.

Detailed results and the formal description of the services, of the protocol machines and of the environment assumptions can be found: for AMp in ([Baptista et al. 1990, Graf et al. 1989, Graf et al. 1990]) and for IRp in forthcoming reports.

To conclude, the benefit of this work is to force the definition of consistent formal specifications, that are of great importance when developing innovative protocols. Indeed, formal specifications do not only allow unambiguous characterizations of the protocol behaviour and facilitate reasoning about it, but some methodological points are to be stressed in the process of development for such systems:

- These verified formal specifications increase the confidence in a more correct design of protocols. As all possible behaviours are examined in a systematic way, the result is often more reliable than the result of mere simulation.

- The verification task and the definition of scenarios for verification require well-structured specifications; the software implementation can take advantage of this structuring of the specified distributed algorithm.

- The assumptions about the behaviour of the environment (for example the network in the case of AMp) are truly formalized.

- The service specifications are used for the implementation validation, for instance by fault injection (see next section).

- The automatic derivation from the formal specifications, of tests and assertions to be verified by the implementation can also be envisaged.

Note however, that in our experience, the automatic generation of executable code is not a realistic issue, in so far as the generation of *efficient* code depends highly on the characteristics of the environment in which the protocols are to execute. In the context of Delta-4, performance requirements do not allow automatic generation of code for significant parts of software, and many local optimizations are required by existing heterogeneous components of the architecture.

15.3. Fault Injection

15.3.1. Introduction

The fault tolerance features of the Delta-4 architecture are based on the multicast communication system (MCS) that provides generalized multicast services. The proper operation of MCS is further based on:

- the verification of a set of well defined properties that characterize the extended service provided by the atomic multicast protocol (AMp),

- the assumed fail-silent property of the underlying network attachment controllers (NACs) hardware modules that connect the stations to the Delta-4 network.

The validation described in this section is thus aimed at:

- estimating the coverage provided by the fault tolerance mechanisms, which incorporate two levels of coverage:

    - the local coverage achieved by the self-checking mechanisms which control the extraction of the NACs,

    - the distributed coverage corresponding to the fault tolerance provided by both the defensive characteristics of AMp and the NAC self-checking mechanisms,

- testing, in the presence of faults, the service provided by AMp.

This validation is also intended to address the successive versions of the AMp software (obtained in particular as a result of the corrections induced by the design errors detected during this validation phase) as well as two distinct versions of the NAC hardware featuring quite distinct levels of redundancy (fail-uncontrolled NAC and fail-silent NAC)[2]. Both NAC architectures are made up of two boards:

- a *main board* that ensures the interfacing with the host computer,

- a *specific board* that connects the main board to the physical medium.

15.3.2. Fault-Injection-Based Experimental Validation

Fault-injection is particularly attractive [Chillarege and Bowen 1989, Crouzet and Decouty 1982, Damm 1988, Gunneflo et al. 1989, Lala 1983, Segall et al. 1988] as a complement of other possible approaches such as proving or analytical modeling. By speeding up the occurrence of errors and failures, fault injection is in fact a method for *testing* the fault-tolerance mechanisms (FTMs) with respect to their own specific inputs: the *faults*.

Basically, fault injection has thus the same characteristics and limitations as any testing approach: its accuracy depends heavily on the representativeness of the inputs of the test and its actual impact is related to the number of significant events — erroneous behaviours of the FTMs — observed during the test sequence. Stated in other words, the coverage estimates derived from a fault injection test sequence are usually easier to extrapolate to the real operational domain when they are low. As an example, if an unacceptably low coverage figure is obtained in the case of a test sequence where only permanent faults have been injected, it can

---

2   A *fail-uncontrolled* NAC refers to a NAC characterized by restricted self-checking mechanisms, while a *fail-silent* NAC is provided with improved self-checking mechanisms.

be confidently assumed that this figure constitutes an upper bound for the "actual" coverage, i.e., when more demanding cases (e.g., transient faults) are also considered.

Two important contributions of fault injection concern the verification of the FTMs and the characterization of their behaviour, thus enabling any weakness in their design and/or implementation to be revealed. Also, a statistical analysis of the responses obtained during the fault injection experiments enables some relevant dependability parameters — coverage, fault dormancy, error latency, etc. — to be estimated.

Different forms of fault injection experiments (e.g., fault simulation [Choi et al. 1989], fault emulation [Gérardin 1986], error seeding [Mahmood et al. 1984], mutation testing [DeMillo et al. 1978], physical fault injection [Côrtes et al. 1987], etc.) can be considered depending on i) the complexity of the system to be validated (the *target system*), ii) the types of faults injected and iii) its level of application at various stages of the development process [Arlat et al. 1990]. The fault injection method used here is the *physical fault injection* method: in this case, the faults are directly injected on the pins of the integrated circuits (ICs) that implement the prototype of the target system. Although this methodology can only be applied at the final stages of the development process, its main advantages are that the tested prototype is close to the final system and that it enables a global validation of a complex system integrating both hardware and software features of the fault tolerance mechanisms.

Another practical limitation that is often opposed to pin-level fault injection is related to the representativeness of the injected faults with respect to the internal faults (in particular in the case of VLSI circuits). Such a limitation can be realistically — albeit partially — overcome by the application of multiple intermittent error patterns on the pins. The multiplicity and the values of the applied error patterns can be either randomly generated or possibly deduced from a fault simulation analysis of the IC considered.

From a general point of view, the experimental validation is based on the concept of a fault injection *test sequence*. More precisely, a fault injection test sequence is characterized by an *input* domain and an *output* domain. The *input* domain corresponds to a set of injected *faults F* and a set *A* that specifies the data used for the *activation* of the target system and thus, of the injected faults. The *output* domain corresponds to i) a set of *readouts R* that are collected to characterize the target system behaviour in presence of faults and ii) a set of *measures M* that are derived from the analysis and processing of the *FAR* sets. Together, the *FARM* sets constitute the major attributes that can be used to characterize fully a fault injection test sequence. In practice, the fault injection test sequence is made up of a series of *experiments*; each experiment specifies a point of the *{FxAxR}* space.

15.3.3. The Testbed

For the application of fault injection to validate the Delta-4 architecture a distributed testbed was built using the fault injection tool *MESSALINE* developed at LAAS. The hardware testbed configuration is shown on figure 1.

The *Target System* contains four stations interconnected by the *Target System Network*[3]. Each station is made up of a NAC containing the implementations of the AMp and Physical entities that are under test together with a *Target System Host (TSH)* which activates and observes the service offered by the AMp. The activation consists in the generation of traffic flow through the target system network and the observation consists in the collection of data obtained from each station.

---

3   All the experiments carried out to date concern the 802.5 token ring version of MCS.

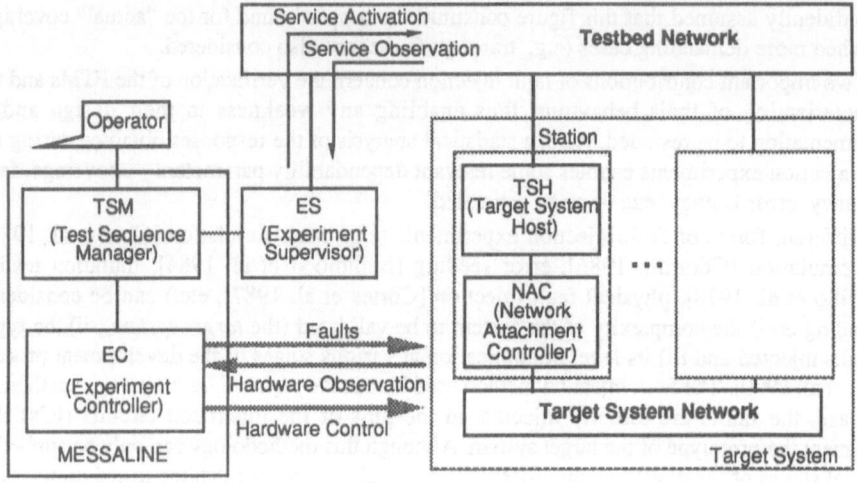

Fig. 1 - The Hardware Testbed Configuration

The purpose of the *Experiment Supervisor (ES)* is twofold: i) run-time control of the target system and ii) collection and analysis of the observed data. The *Testbed Network*, which is an Ethernet type LAN in the present implementation, ensures the communication between the ES and the TSHs. The ES is implemented on a Bull DPX2000 machine.

Most hardware and software necessary for the fault injection test sequence are part of *MESSALINE*. For sake of conciseness, only the two principal components are briefly presented here; a more detailed description of *MESSALINE* hardware architecture is given in [Arlat et al. 1990]

The *Experiment Controller (EC)* implements the injection of the elements of the F set and the collection of the hardware signals used to elaborate some elements of the R set. Faults are injected into the NAC component of a specific station (designated hereafter as the *injected station*) by the *forcing* technique[4]. A connection with the wiring concentrator of the target token ring network allows the states of the insertion relays of the stations connected on the ring to be read. These are used for the post-test analysis, as will be explained later. The only intervention of the operator is to position the probe on the circuit selected by the *Test Sequence Manager (TSM)*.

The TSM is implemented on a Macintosh II computer, connected to the EC and to the ES through serial lines. There are also physical connections between MESSALINE and all the stations (TSHs and NACs) to enable automatic hardware resets: the NACs are reset after each experiment whereas a host reset is requested only in case of a station crash.

In the reported experiments, the Target System is made up of four Bull SPS7 machines, running UNIX System V, as TSHs. The preliminary (fail-uncontrolled) NAC architecture tested so far contains only limited self-checking mechanisms, namely the mechanisms offered by the 802.5 token-ring standard (internal bus parity check, watch-dog timer, etc.). Improved self-checking (fail-silent) NACs, featuring duplicated processors and memory, that have been designed to interface a Ferranti Argus 2000 TSH are now being validated on the same testbed.

---

4   In the case of the *forcing* technique, voltage levels are directly applied by means of multi-pin probes on IC
    pins and associated equipotential lines.

15.3.4. Definition of the Test Sequence

This section presents successively the main parameters that specify the Fault, Activation, Readout and Measure *(FARM)* attributes of the test sequence.

15.3.4.1. The F Set. Faults are injected into a single NAC; the four stations were thus partitioned into two sets: the *injected station* (station S1) and the three *"correct"* (non-injected) *stations* (stations S2, S3 and S4).

For each IC, the injected pins, the nature and the timing characteristics of the injected faults were selected according to the order and form given below:

1) *Multiplicity (MX):* faults are injected on several (1, 2 or 3) pins of an IC, with a frequency of 50%, 30% and 20% respectively.

2) *Location:* selection of MX pins among the injectable pins, with a uniform distribution.

3) *Nature:* stuck-at-0 and stuck-at-1 faults, each with equal probability of occurrence.

4) *Timing characteristics:* the faults injected are essentially intermittent faults; their temporal parameters are composed of three values:

   • *lead time* (from start of experiment to first pulse): uniformly distributed between 1s and 40s,

   • *period:* logarithmically distributed between 10 µs and 30 ms,

   • *width:* uniformly distributed between 2 µs and 1 ms, with a duty cycle ≤ 50%.

In the lack of sound available data concerning actual IC failure modes, most of the parameters were selected according to a uniform distribution among range values selected quite arbitrarily. However, the limitation to a multiplicity order of 3 is to some extent substantiated by the results presented in [Gunneflo et al. 1989] for a microprocessor in the presence of single event upsets. These results show that more than 80% of the internal single-event upsets led to the occurrence of the first error pattern on 3 pins at most. The logarithmic distribution for the period is intended to obtain a significant number of short period intermittent faults while maintaining a wide selection range.

15.3.4.2. The A Set. The workload was varied to study its impact on the behaviour of the target system. The activation is characterized by the application of two types of traffic flows: *observed traffic* flow with respect to which AMp properties are tested[5] and *background traffic* flow to provide more realistic activation of the target system.

To provide a representative activation set, five activation modes have been considered for the stations; table 1 shows the transmitter and receiver allocations of the stations for each mode with respect to observed traffic and background traffic flows.

15.3.4.3. The R Set. Three types of readouts are collected for each experiment:

   • *binary* readouts: activation of the injected faults, status of the ring insertion relays for each NAC, AMp properties derived from the analysis of the messages,

   • *timing* readouts: time of activation of the injected fault, time of extraction of the stations,

   • *message* readouts: number of messages exchanged for both traffic flows and number of messages positively or negatively confirmed for the observed traffic.

---

5   The observed traffic consisted of 100 messages.

Table 1 - Transmitter and Receiver Allocation per Activation Mode

T: Transmitter present, R: Receiver present, —: No Traffic, X: Not inserted

Mode	Observed Traffic				Background Traffic			
	S1	S2	S3	S4	S1	S2	S3	S4
1	TR	TR	R	R	TR	TR	TR	TR
2	R	TR	TR	R	TR	TR	TR	TR
3	—	TR	TR	R	TR	TR	TR	TR
4	—	TR	TR	R	—	TR	TR	TR
5	X	TR	TR	R	X	TR	TR	TR

The information concerning whether or not a particular fault becomes activated during an experiment can be obtained from specific monitoring devices implemented in MESSALINE that sense current variations on the pin(s) where the fault is injected [Arlat et al. 1990]. The status of these devices can be read by software. Such information was used to eliminate from the statistics those experiments where the fault was not activated during the experiment and to perform a direct measurement of fault dormancy for faults that were observed to become activated.

The collection of occurrence and timing readouts enabled empirical distributions to be derived for the fault dormancy. Empirical distributions were also derived for the coverage achieved by the hardware error detection mechanisms of the NACs and by the AMp.

On the average, each fault injection experiment took about five minutes. This large value is mainly due i) to the experiment set-up and ii) to the possible execution of the automatic recovery and restart procedures, in case of failure of the testbed after a fault has been injected. More specifically, the watch-dog monitoring the useful duration of each experiment was set to 110s.

15.3.4.4. The M Set.
The measures considered for the analysis presented here consist of two types of measures: *predicates* and *time distributions*.

Let E designate the *error occurrence* predicate; i.e., E is true if the injected fault is activated on the faulted pin(s). Let I_i denote the status of the ring insertion relay of station S_i, i = 1, ..., 4; Ii is true if S_i is inserted into the ring.

The *local coverage* or *error detection predicate D* characterizes the efficiency of the NAC self-checking mechanisms:

$$D = E \cdot \overline{I_1}$$

D is true if the NAC of the injected station is extracted when the injected fault is activated (the expected behaviour in presence of faults)[6]. The notation: A • B is used to designate the conjunction between predicates A and B.

---

6 An opposite use of D has been made in the case of mode 5 to test if the station remained extracted when faults were injected; thus, in this case, a 100% coverage is assumed when the station remains extracted.

Let P designate the predicate corresponding to the conjunction of the subset[7] of the *AMp properties* considered (see table 1 of chapter 10 for a definition of the AMp properties). Let predicate C characterize the *confinement* of the fault/error (i.e., all the "correct" stations remain inserted in the ring). The *distributed coverage* or *fault tolerance* predicate T, that characterizes the defensive properties of the protocol at the MAC layer, can be expressed as:

$$T = E \cdot P \cdot C = E \cdot (Pa3 \cdot Pa6 \cdot Pa9) \cdot (I_2 \cdot I_3 \cdot I_4)$$

T is true whenever *all the AMp properties* are verified, the *confinement* of the fault/error is ensured when E is true. Although, it might *a priori* mask interesting characteristics, the grouping of the P and C predicates into one single predicate is substantiated by the two following remarks:

- the results obtained to date [Aguera et al. 1989, Arlat et al. 1989] have never led to an observation of P being false when C was true,

- the status of predicate P is of little interest when predicate C is false.

Thus, in subsequent analyses, the status of predicate T can be strongly related to the status of predicate C.

Two types of time distributions complement the analysis. The *fault dormancy* measures the time interval between the injection of a *fault* and its activation as an *error* at the point(s) of injection. If T_d denotes this random variable, and T_E and T_F the error and fault times respectively, then:

$$T_d = T_E - T_F$$

The *extraction latency* corresponds to the time interval between the injection of a *fault* and the *extraction* of the NAC of the injected station; if T_l denotes this random variable, and T_D the extraction time, then:

$$T_l = T_D - T_F$$

15.3.5. Major Results Obtained

15.3.5.1. Characterization of the Experimental Results.
For each circuit submitted to fault injection, 30 experiments were carried out for each of the 5 activation modes. Accordingly, the complete test of one IC consisted of a run of 150 experiments that was fully automated to enable the experiments to be carried out overnight. A total of 40 circuits out of the 101 that compose the NAC (with restricted self-checking) was submitted to fault injection. Even though this represents only a subset of the circuits, the use of the forcing technique allowed a high proportion of the actual equipotential lines to be faulted, resulting thus in a pin coverage of 84%.

Another level of uncertainty is attached to the practical restriction to 110s of the observation domain; in particular, it is clear that a distinction has to be made between:

- a fault which does not become active during an experiment,

- a fault that would never become active.

However, it has to be pointed out that such an uncertainty may lead [Arlat et al. 1990]:

- either to pessimistic estimates (e.g. in the case of the local coverage predicate D),

- or to optimistic estimates (e.g., in the case of the T predicate).

---

[7] The subset of AMp properties tested so far include *unanimity*, *non triviality* and *order*; testing properties such as *termination* and *causality* would require a global clock and global ordering of the observed interactions that were not implemented for the sake of simplicity.

Nevertheless, the analysis of the shape of the time distributions obtained provide an *a posteriori* means to support the acceptability of the observation domain. Furthermore, the exchange of several tens of messages (up to 100 messages) from the observed traffic flow during each experiment provides a sufficient activation domain enabling a reliable analysis of the P and C predicates to be carried out.

In this section, the focus is essentially on the presentation of summarized results concerning coverage estimations (local and distributed) and the evolution of three successive versions of the AMp. Detailed results and analysis of the influence of the activation modes, dormancy distribution, and others, can be found in [Aguera et al. 1989, Arlat et al. 1990, Arlat et al. 1989]. As the fifth activation mode is specific (faults are injected into a non-inserted NAC) and indicated a 100% coverage for the D and T predicates, the reported statistics concern the first four modes only.

15.3.5.2. Estimation of the Coverages. Figure 2 summarizes the major statistical results obtained concerning the estimation of the local and distributed coverages of the fail-uncontrolled NAC running the first version of the AMp software. The percents indicate the values of asymptotic coverage for the predicates E (error), D (detection at NAC level) and T (tolerance of AMp). The time measures indicate the means for fault dormancy and error latency distributions.

The minimum distributed coverage (NAC + AMp) is about 68 % (82 % x 83 %). The apparent fault coverage, i.e., the proportion of tolerated errors is about 85 % (17.5 % + 82 % x 83 %). The estimation of the actual coverage depends on the causes of the 17.5 % for the $\overline{D} \cdot T$ combination, which requires supplementary observations of the AMp.

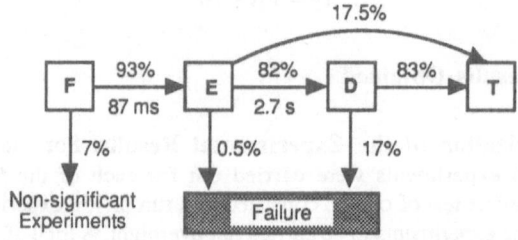

Fig. 2 - Summary of the Coverage Estimates

Further studies, including i) correlation tests and ii) the analysis of a supplementary set of 1600 experiments in which faults were injected directly onto the NAC relay control circuitry to ensure immediate ($T_1 = 0$) NAC extraction have been carried out. These studies made it possible to identify the causes of the 14 % [82 % x 17 %] of errors correctly detected but not tolerated ($D \cdot \overline{T}$). About 8 % of these failures can be attributed to an erroneous behaviour of AMp. The major proportion (92 %) is probably due to the excessive latency of NAC self-checking mechanisms (this hypothesis is substantiated by observed correlation between the $D \cdot \overline{T}$ occurrences and measured detected latency [Aguera et al. 1989]).

15.3.5.3. Impact of Fault Injection on the Development Process. Traces and memory dumps recorded for each experiment in which non-confinement occurred provided the protocol implementers with useful data for the fault removal task. As a consequence, two more versions of the AMp were submitted to the fault injection tests. For these tests, 8 of the circuits of the same NAC were selected among those that contributed most to the D• $\overline{\text{T}}$ proportion in the tests realized for the first version. The statistics presented were obtained in a set of 3600 experiments (8 x 150 x 3). The experiments showed that the distributed coverage was substantially enhanced, and that some types of errors were no longer obtained.

15.3.5.3.1. Evolution of the Distributed Coverage. Figure 3 illustrates the variation of the D• $\overline{\text{T}}$ proportion observed for three successive versions V1, V2 and V2.3.

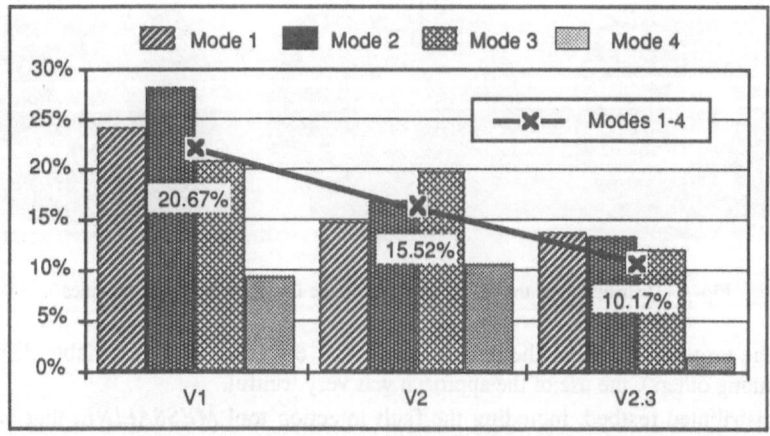

Fig. 3 - Variation of D• $\overline{\text{T}}$ Combination according to the Different Versions

It can be noted that passing from V1 to V2 had an impact on modes 1 and 2, although a certain degradation could be noted in mode 4. Instead, the passage from V2 to V2.3 caused an important decrease in the proportions of AMp failures, especially in modes 3 and 4. The percentages indicated on the histograms average the global variation for modes 1-4. The percentage of D• $\overline{\text{T}}$ is reduced by 50% from version V1 to V2.3, which demonstrates that there was an increase in AMp reliability.

15.3.5.3.2. Evolution of the Number of Errors per AMp Module. Figure 4 shows the distribution of the errors observed on the main software modules of the NAC for the three successive versions. For conciseness, only the modules in which errors were detected are shown. The *Monitor* operations enable the system to keep a coherent view of the multicast group in the network and to recover from station failures. The *Emitter* and *Receiver* entities perform the atomic data transfer operations. The *Driver* controls the exchange of information with the communication medium. A more detailed description can be found in [Ribot 1989].

15.3.6. Conclusion

In spite of the limitations of the physical fault injection approach (late application in the development process, for example) and the difficulties in applying it, especially in the case of distributed architectures (selection of faults to be injected; synchronization of the instant of fault injection with target system activity; parasitic mutations induced by the physical interference

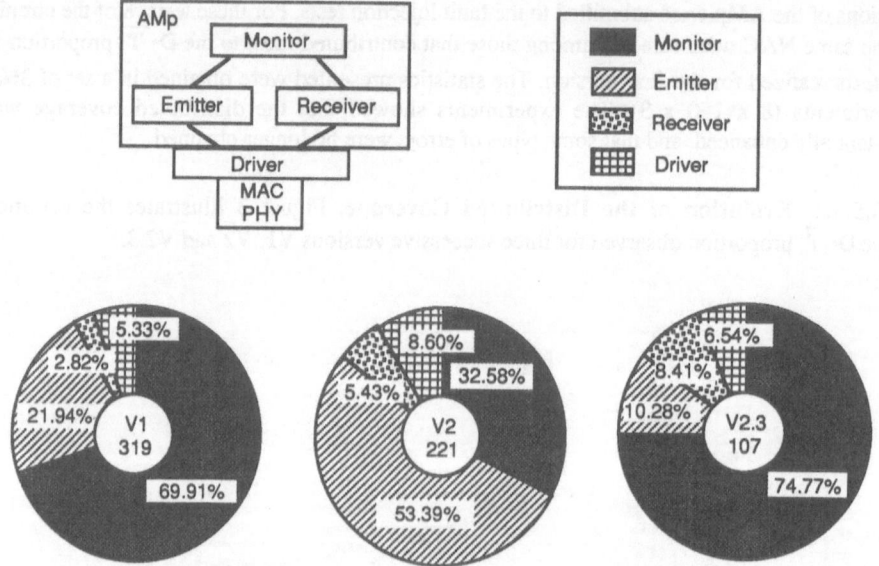

Fig. 4 - Distribution of the Errors observed in the Different Software Modules

between the target system and the experimental tool; effort to set up a reliable distributed testbed, among others), the use of the approach was very fruitful.

The distributed testbed, including the fault injection tool *MESSALINE*, that has been developed to carry out this work provides a fairly comprehensive experimental environment that enables:

a) a *global testing of the services* provided by AMp, which allows assessment of not only the interactions among several peer implementations but also those with the layers below,

b) the estimation of the effectiveness of the fault-tolerance features of the target system in the presence of *injected physical faults*,

c) the automated execution of a test sequence without operator intervention that was made possible only by the *integration of fault tolerance features in the testbed* itself. As an example, at the end of each experiment, should the injected station be found to have crashed, it is rebooted automatically.

For the sake of conciseness, only a fraction of the experimental results obtained has been included here. Other relevant contributions to mention are:

• the characterization of the behaviour of the system in the presence of fault, e.g., the impact of the activation modes; the influence of different types of faults (permanent, intermittent and transient faults); the impact of fault location (main versus specific board), among others;

• the identification of the limited performance of the self-checking mechanisms implemented on the tested NAC. Analysis based on a specific set of experiments showed that most AMp failure cases were caused by the excessive latency of the error detection mechanisms (especially for the main board). These results justify the need for the improved NAC architecture employing duplicated circuitry. The next step is thus the test of this NAC with enhanced self-checking (fail-silent NAC) to evaluate local and distributed coverages to help to decide whether the benefits obtained justifies the increased cost.

The most recent version of the AMp software is being used now for the comparison of the two NAC architectures for the tests mentioned in the above paragraph. For these experiments, the property relative to the reconfiguration (called *consistent group view*, see table 1 of chapter 10) is included in the analysis. The preliminary results of these experiments are reported in [Arlat et al. 1991].

Finally, it is worth noting that the coverage estimates presented in this section correspond essentially to conditional dependability measures for the fault tolerance mechanisms tested (i.e., conditioned by the occurrence of a fault or an error). In particular, they do not account at all for the fault occurrence process. Work is currently being carried out that is intended to refine these coverage estimates by integrating the experimental results with a model-based description of the fault occurrence process. Towards this end, two approaches are currently conducted which are aimed at:

- weighting the coverages obtained for the faults injected on one IC by the failure rate associated to that IC,

- implementing the link between experimental measures and Markov models to derive dependability measures.

15.4. Dependability Evaluation

15.4.1. Introduction

Dependability evaluation is viewed here as quantitative rather than qualitative. Its aim is to compare the various design solutions, to define some essential parameters and to study their effects on system dependability. A global dependability model will be defined that can be used to evaluate several measures of the dependability achievable by different configurations of the architecture. The dependability verification using fault-injection will provide assistance in the necessary quantification of the model parameters.

The results obtained from the different activities of this work should provide the users of the Delta-4 architecture with guidance in the decisions concerning configuration of their own system. The final assessment of the quality of the architecture should help the development process itself as well as maintenance planning.

It should be noted that this activity is centred on the OSA architecture although it is possible that the models could be easily extended to cover the XPA architecture.

15.4.2. First Analysis

The aim is to develop a global model of the dependability of the architecture including both hardware and software faults. It will then be necessary to estimate the parameters of this global model and finally evaluate the dependability measures to study the reliability, the safety, the availability, etc., achievable by various configurations of the architecture.

It is very difficult to establish directly a correct global model; so a progressive method will be used. It is necessary to establish a global evaluation strategy in terms of inter-connected sub-models. After the study of the sub-models, it is necessary to aggregate them and study the global model. This organization in sub-models should also give some early feedback about the design of the different components included in the sub-models.

For the purposes of the evaluation, three levels can be distinguished in the Delta-4 OSA architecture:

- the hardware level (networks, NACs, hosts ...),
- the system software level (local executive communication stack, administration system, Deltase/XEQ),
- the application software level.

These levels will give a solution for the design of interconnected sub-models and simplify the dependability study.

We have defined three main objectives:

- modelling and evaluation of the communication hardware (comparison of the various communication topologies)
- extension of the communication architecture model to include the host-resident management information base (this model will include the MIB and MIB-management taking into account the replication of these entities on several hosts, before the extension to the complete hardware and software architecture).
- establishment of the global model of a target application and the evaluation of its dependability measures (aggregation of the previous sub-models in order to provide a framework for quantifying the dependability offered by a particular configuration of the architecture).

The first objective has already been achieved and the results will be summarised hereafter (see §15.4.4). The next step is to include the host-resident management information base.

15.4.3. Evaluation Method

Several methods for evaluating dependability measures can be distinguished: reliability block diagrams, fault-tree (or event-tree) analysis and state diagrams [Laprie 1983]. The main advantages of the latter are:

- their ability to account for the stochastic dependencies which result for instance from maintenance and solicitation processes, or from simultaneous consideration of several classes of faults,
- various dependability measures can be derived from the same model.

A state diagram is a graph in which the nodes represent the states of the system and the edges the elementary events leading to system transition from one state to another. The system model may be viewed as a representation of (i) the modifications of the system structure resulting from the events likely to affect the system dependability (fault-error-failure process, maintenance actions) and of (ii) other events of interest (e.g., solicitation process corresponding to the user's requests).

15.4.3.1. Markov Chains. When the elementary events can be considered as exponentially distributed (constant failure rates) the state diagram corresponds to a time-homogeneous Markov chain. Markov modelling is well adapted to dependability evaluation, it is well-suited for comparing different possible structures at the design phase (or during operational life if the architecture of the system allows this possibility) in order to select the "best" one.

Evaluating system dependability measures using Markov chain may be viewed as being composed of two tasks:

- model construction: derivation of the system behaviour model from the elementary stochastic processes,
- model processing: derivation of dependability measures from the system behaviour model.

The model may be very large for complex systems, in which case we need formal methods to construct it and program packages to handle it.

The current approaches aimed at formalizing Markov chain construction when accounting for stochastic dependencies are either algebraic approaches or graphic approaches.

Graphic approaches are based on stochastic Petri nets [Béounes and Laprie 1985, Beyaert et al. 1981, Molloy 1982] for which the basic idea is very simple, and thus attractive: when the transitions of the Petri net are weighted by hazard rates, the reachability graph may be interpreted as a Markov transition graph.

Advantage is taken of the SURF-2 dependability evaluation tool that is currently being developed by LAAS (independently of the Delta-4 project). This tool is based on Markov process evaluation techniques and model description can be carried out either directly using a Markov chain or using stochastic Petri nets.

15.4.3.2. Constant and Non-Constant Hazard Rates. The constancy of failure rates is a widely-recognised, and widely-used, assumption for element failures due to physical faults. Concerning the maintenance rate, it has been shown [Laprie 1975, Laprie et al. 1981] that considering corrective maintenance (repair) rates as being constant is, although not physically realistic, a perfectly satisfactory hypothesis for reliability evaluation. It is less satisfactory for availability evaluation since asymptotic unavailability may vary by a factor 1 to 2 when considering a constant repair rate or a constant repair time.

More generally, it can be considered that, under the assumption of exponentially distributed times to failures, assuming a constant hazard rate for the other processes is a satisfactory hypothesis as long as the mean values are small when compared to the mean times to failures.

Concerning software failure rates, due to the corrections introduced during the software life cycle, the failure rate generally decreases during the development and even in operational life. However, if no more modifications are performed or if the modifications still performed do not significantly affect the failure rate, consideration of a constant failure rate constitutes a good assumption. Usually this situation corresponds to an advanced phase of the operational life. Experimental results confirm this assumption [Kanoun and Sabourin 1987, Nagel and Skrivan 1982]. This failure rate (denoted residual failure rate) can be derived by applying a reliability growth model to data collected on the software in operation: the hyperexponential model developed at LAAS and used for several projects [Kanoun et al. 1988, Laprie 1984, Laprie et al. 1990] is the only model able to predict this measure.

What are the alternatives in terms of modelling techniques? There are three ways for handling non-constant hazard rates:

- time-varying Markov processes [Howard 1971],
- semi-Markov processes [Howard 1971],
- transformation of a non-Markov process to a Markov process by adding either (i) supplementary variables [Cox and Miller 1968] or (ii) fictitious states (the method of stages [Cox and Miller 1968, Singh et al. 1977].

Our recommendation (which has been put into practice for several years in the dependability group at LAAS) is the following [Costes et al. 1981, Laprie 1975, Laprie et al. 1990]:

- consider all the hazard rates as constant and derive a time-homogeneous Markov chain,
- perform sensitivity studies using the device of stages for those rates that are considered to be non-constant, starting with one fictitious state for each non constant rate, and stopping when addition of more states is of non-perceptible influence.

This recommendation stems from the following facts:

- a model is always an approximation of the real word, and this approximation has to be globally consistent,
- when modelling phenomena stochastically, the first moment generally determines the order of magnitude, the further moments bringing in refinements; an exponential distribution can be seen as the distribution corresponding to the knowledge of the first moment only.

15.4.3.3. Equivalent Failure Rate. When the non-absorbing states (non failed states) constitute an irreducible set (i.e., the graph associated with the non absorbing states is strongly connected), it can be shown that the absorption process is asymptotically a homogeneous Poisson process, whose failure rate (denoted equivalent failure rate) is given by:

$$\lambda_{eq} = \sum_{\substack{\text{paths from} \\ \text{initial state (I)} \\ \text{to the failed state}}} \frac{\prod \text{transition rates of the considered path}}{\prod_{\substack{\text{states in path} \\ \text{(except I)}}} \sum \text{output rates of the considered state}}$$

The reliability is then given by:

$$R(t) = \exp(-\lambda_{eq} t)$$

and the asymptotic unavailability is equal to:

$$\overline{A}(t) = \frac{\lambda_{eq}}{\mu}$$

where μ is the repair rate from the failed state.

Since the Delta-4 system is repairable, the associated graphs are generally strongly connected, and this approach will be adopted in the following: the different sub-systems will be evaluated through their equivalent failure rates.

15.4.3.4. Parameters of the Sub-Models. The parameters needed to establish the sub-models are (i) the failure rates of the different components of the architecture, (ii) the repair rates as well as the repair policy and (iii) the coverage factors that quantify the effectiveness of error-detection and fault-tolerance mechanisms. The estimation of these parameters will entail the use of failure rate data banks when such banks exist as well as the use of the results obtained from hardware fault-injection. In some cases, extrapolation of failure data obtained on similar projects will be of great help. Concerning the maintenance process, one has to assume some realistic repair policies (see subsection 15.4.4.1).

At the moment, reasonable figures have been guessed and the values of these parameters will be used before the end of the project to predict system dependability from a modular and parametric model. When the "real" values are evaluated from the definite project the corresponding parameters will be replaced in the model.

15.4.4. Modelling and Dependability Evaluation of the Communication System

The various hardware communication architectures are the 802.4 token bus, and 802.5 and FDDI token rings. In each architecture, every host possesses a NAC that interfaces the host and the underlying media. The communications software and a part of the administration software

are executed on the NACs. The couple host-NAC form a node or a station, and the set of the NACs with the underlying media constitute the communication system.

There are thus two essential aspects to be taken into account in the models: the communication topology and the nature of the NACs: fail-silent (with extended self-checking mechanisms) or fail-uncontrolled (with limited self-checking mechanisms).

15.4.4.1. Model Assumptions. We assume that a non-covered failure of one element leads to a total system failure and we will consider the following maintenance policy:

- a covered failure (of the NAC or of the medium) does not affect service delivery, moreover the repair of such a failure does not need service interruption,

- after a non-covered failure (of the NAC or of the medium), service delivery is interrupted, repair of all the failed elements is carried out before service is resumed,

- in case of one or several covered NAC failures, followed by a covered failure of the medium, repair priority is given to the medium,

- for the double ring, the wiring concentrators have the higher repair priority.

15.4.4.2. Notation and Numerical Values of the Parameters. The two types of NACs are modelled in the same manner. They differ by the numerical values of the parameters: for the fail-silent NAC, error detection coverage should be higher, the failure rate is also higher due to the greater amount of hardware necessary to enhance the self-checking. The equivalent failure rate of the communication system is evaluated as a ratio of the failure rate of the NAC. The double FDDI ring has the same model as the double 802.5 ring.

The main parameters of the models are:

- λ_N, the failure rate of the NAC — a value of 10^{-4} / h has been taken as a reference and corresponds to 1 failure per year,

- λ_{WC}, the failure rate of a wiring concentrator in the double ring, it should be about the same as the failure rate of a NAC (it has been taken in fact equal to λ_N in this study),

- n, the number of stations — fixed (arbitrarily) at 15,

- N, the number of wiring concentrators in the double ring,

- λ_B, the failure rate of the bus — a value of 2 10^{-5} / h has been taken; this corresponds to 1 failure per 5 years,

- λ_R, the failure rate of the ring — it has been taken equal to λ_B,

- μ, the repair rate, a repair duration of 2 hours has been adopted.

The coverage factors for the different elements are denoted: p_N, for the NAC, p_B, for the bus, p_l, for a link in the ring, p_{WC}, for the wiring concentrator and p'_{WC}, for the correct reconfiguration of the double ring after a non-covered failure of a wiring concentrator.

Let $\overline{p}_X$ be defined as: $\overline{p}_X = 1-p_X$ where $X \in \{N,B,l,WC\}$.

15.4.4.3. Summary of the Results. The expressions for the equivalent failure rates of the different architectures are very complex for the double media. However they can be simplified using the fact that: $\dfrac{\lambda_B}{\mu} \ll 1$.

The expressions of these failure rates considering only the first order terms are:

Single ring : $\lambda_{eq} \approx \lambda_R + n\,\overline{p}_N\,\lambda_N + n\,p_N\dfrac{\lambda_R}{\mu}\,\lambda_N$

Double ring : $\lambda_{eq} \approx 2\,\overline{p}_l\,\lambda_R + n\,\overline{p}_N\,\lambda_N + N\,\overline{p}_{WC}\,p'_{WC}\,\lambda_{WC}$

Single bus : $\lambda_{eq} \approx \lambda_B + n\,\overline{p}_N\,\lambda_N + n\,p_N\dfrac{\lambda_B}{\mu}\,\lambda_N$

Double bus : $\lambda_{eq} \approx 2\,\overline{p}_B\,\lambda_B + n\,\overline{p}_N\,\lambda_N + 2\dfrac{\lambda_B}{\mu}\,(n\,p_N\,\overline{p}_B\,\lambda_N + p_B\,\lambda_B)$

It can be noted that, for the single media the equivalent failure rates are limited by the failure rate of the medium and the non-covered failures of the NACs and that, for the double media, they are directly related to the failure rate of the non-covered failures.

The coverage factors are thus of prime importance (this has important consequences on the dependability of a system containing both Delta-4 NACs and non-Delta-4 NACs, see annexe M). However, due to the numerical values of the different failure rates the coverage of the NACs, p_N has more influence than p_B and p_l.

For the ring, duplication is worthwhile only for $\lambda_R > 5\ 10^{-5}\,/\,h$ even with a perfect coverage of the NACs, $p_N = 1$, (figure 5, for which $p_l = 0.95$ and $p_{WC} = p'_{WC} = 0.9$). This is due to the introduction of the wiring concentrators whose failure rates are of the same order of magnitude than the NACs. For instance, considering $p_N = 1$, duplication of the ring acts as follows:

- for $\lambda_B = 10^{-5}\,/\,h$, it increases the unavailability from 11 min. to 48 min. / year,

- for $\lambda_B = 10^{-4}\,/\,h$, it decreases the unavailability from 1 h 45 min. to 58 min. / year.

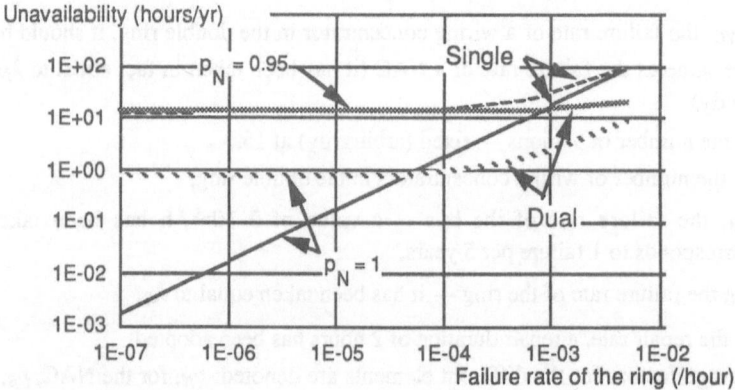

Fig. 5 - Communication System Unavailability (in hours per year) for the Single and the Double Ring, with $p_N = 0.95$ and 1.

With respect to dependability improvement due to the duplication of the bus, the unavailability of the communication system with a single and a double bus, versus the failure rate of the bus (λ_B) and for $p_N = 0.95$ and 1, is given in figure 6.

When the coverage factors are less than 1, duplication can lead to dependability deterioration depending on the value of the bus failure rate: for instance for $p_N = p_l = 0.95$, improvement is effective only for $\lambda_B \geq 2 \cdot 10^{-5}$ / h. This is because when λ_B is low, dependability is conditioned by the NAC failures.

When the coverage factor of the NAC is equal to 1, duplication is worthwhile, for example:

- for $\lambda_B = 10^{-5}$ / h, duplication of the bus decreases the unavailability from 0.2 (11 min. / year) to 1 min. / year,

- for $\lambda_B = 10^{-4}$ / h, unavailability is decreased from 1 h 45 min. to 11 min. / year,

which is a significant improvement.

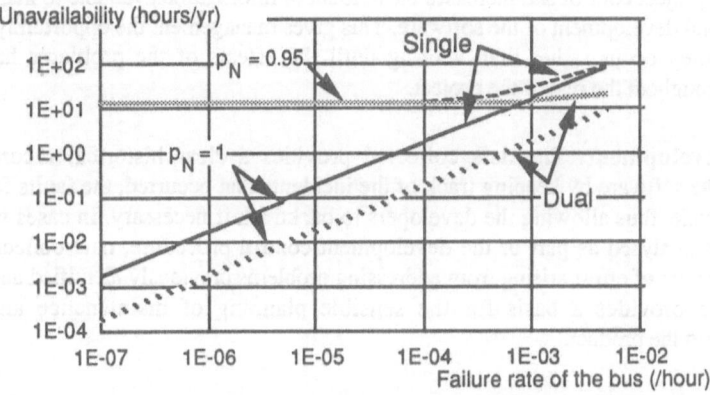

Fig. 6 - Communication System Unavailability (in hours per year) for the Single and the Double Bus, with $p_N = 0.95$ and 1.

Concerning comparison of these architectures, it is very difficult to classify them. Assuming the same failure rate for the bus and for the ring leads to the same expression of the equivalent failure rate and for the unavailability of the single medium. Figures 5 and 6 show that for reasonable failure rates ($\lambda_B \approx \lambda_R < \lambda_N$) dependability measures are independent from this failure rate. Which means that the single bus and the single ring are equivalent.

Duplication of the medium can even deteriorate the dependability measures depending in the parameter values. The results enable the different architectures to be compared according to the various parameters in order to make a tradeoff and to select the more suitable architecture. For instance, for the considered values, the double bus seems more interesting than the double ring for $\lambda_B < 4 \cdot 10^{-3}$ / h, however, the value of the failure rate of the wiring concentrator is of prime importance: a lower value of λ_{wc} acts in favour of the double ring.

15.5. Software Reliability

15.5.1. Introduction

The idea of using the collection of data to carry out a reliability evaluation is not a new one. The techniques have been applied in several areas, including hardware reliability evaluations, over several years. Data collection for software has also been successfully implemented in several projects and was introduced within Delta-4 as the basis for both quantified and qualitative evaluation. This section provides a general overview of the reasons for data collection, its relationship to software testing and software reliability prediction and a discussion of the problems of data collection. It then goes on to describe the data collection and reliability modelling activities undertaken within Delta-4 and to discuss some project-specific topics.

15.5.2. Reasons for Data Collection

There are several reasons for collecting data. These can be broken down into three separate levels of interest as below:

15.5.2.1. Managerial. At the highest level, data collection allows managers to have a clear picture of the progress of the development of the software and the problems encountered. It aids in overall project control and increases the amount of information available to management during the actual development of the software. This gives management the opportunity to solve problems as they occur rather than waiting until the effects of the problems have been propagated throughout the rest of the project.

15.5.2.2. Development. The data collected provides a clear historical record of the evolution of the software by keeping track of the incidents that occurred, the faults found and the changes made, thus allowing the developers to backtrack if necessary. In cases where the information is analysed as part of the development control procedure, data collection may prevent duplication of effort arising from addressing problems previously identified and solved. The data also provides a basis for the sensible planning of maintenance and future enhancements to the product.

15.5.2.3. Research. The data collected can provide the information necessary for further research into means of achieving and predicting software reliability using static analysis tools and modelling techniques. Significant research has already been done on the modelling side but work on the use of static analysis tools and their effect on reliability has so far been limited to a very small section of the software community. This research in the long term should benefit both management and software developers by enabling managers to locate problems earlier in the project and developers to have the opportunity to gauge how well the module has been constructed, before any testing has taken place. Frequently, extensive maintenance can prove to be detrimental to the structure and complexity of the code. The application of static analysis tools during routine maintenance and enhancements should allow the effects of the changes made after initial development to be assessed.

15.5.3. Planning and Implementation of Data Collection

To ensure that the data collected is of high quality, the planning and organization of the collection activity should take into account the need for adequate resources and skilled staff. It is also desirable to automate the data collection as much as possible to improve consistency and validity and to establish a project contact within each participating company. This contact's duties should include the management of data and the data collection activity to ensure the completeness of the data and data set, with an emphasis on accuracy and quality of data. The contact will also be able to act as an intermediary between those storing and analysing the data and those collecting it.

Some projects have taken the approach of collecting as much data as possible, as it may be of potential use in the future, although at the time it would serve no purpose. Given endless resources, time and man-effort, this may be a very useful thing to do. However, collecting data incurs cost and the financial constraints placed upon the project will dictate the depth and range of the data collection activity. With this in mind, some thought should be given towards the most cost effective means of collection.

Some data can be accumulated for little cost, e.g., by use of static and dynamic analysis, whereas collection of other data can be more expensive. Static analysis tools provide valuable data in their own right but, the very fact that they are static analysers means that the code is not executed, and hence they cannot provide dynamic data, e.g., test coverage, time between failures.

The data collected should be "multi-purpose" to enable the use of the original data in a variety of areas. For some applications, minor amendments may be necessary but it is important to ensure that these amendments do not violate the integrity of the data.

On the research level, the data collected can be used both in a reliability evaluation and to assess the effectiveness and correctness of other techniques. Perhaps the best way to illustrate this point is by an example — several sets of data are input to a new software engineering tool that predicts very high failure rates. If we already know that the data has come from code that has been highly reliable in operational use over a considerable period of time, then confidence in the tool will be significantly reduced.

The data can also be used as a yardstick for project planning and projections on a managerial level and as a basis for future work on similar projects on the development level. It provides developers with the opportunity of learning from their mistakes and achievements by applying the data to the identification of their main problem areas or strengths, hence potentially improving the quality of their work.

15.5.4. Data Collection and Software Developers

To gain the best results from any data collection procedures, management should be aware of the reasons for data collection so that they can appreciate both the need for data collection and the need to motivate those people actually collecting the data. The collection mechanism should be made as simple as possible to ensure that the data collected is of an acceptable quality. Once the data collection methods are produced and agreed, they should be disseminated in the most appropriate manner.

Feedback of the results obtained on the data should be given at regular intervals in a clear and concise manner. Individuals are generally keen to improve the standard and quality of their work and are therefore interested in receiving feedback as long as it is done constructively. If the data collection activity is perceived in any way to be a means of performance measurement then the developers will be demotivated and the data collection activity will fail [Littlewood 1987]. It is important that this feedback is provided as quickly as possible to improve co-operation, even more so in cases where the staff involved in the software design and development are also responsible for testing.

The analysis of the data collected may allow software developers to discover answers to some very important questions:

- *Am I always making the same sort of mistakes in the same areas?*

  Records will indicate whether or not there are constant sources of error affecting the quality of particular products.

- *How effective are my testing procedures?*

  Analysis of the records may show that the testing itself, rather than the items tested, tends to be at fault. This can lead to a more effective testing practice.

- *How do I know what to test?*

The results of any walkthroughs that have been done allow developers to pinpoint any areas of weakness inherent in the code. Static analysers provide the developers with an internal picture of the data or control flow within the modules. This allows the developer to produce test data that will address the areas of weakness and increase the structural coverage of the code.

- *How do I know when to start testing?*

By using walkthroughs as well as static analysers it is possible to significantly reduce the time spent in finding those faults that can be identified before the dynamic testing activity begins. Dynamic testing should begin only once there is a significant decrease in the cost-effectiveness of other techniques and the developers are satisfied that errors already identified have been removed.

- *How do I know when to stop testing?*

The data collected provides an overview of the number of faults found and allows these faults to be classified. It cannot provide a hard and fast rule on when the testing should be stopped but can give an indication of the effectiveness and coverage of the testing already completed.

It should however be reiterated that while it is of the utmost importance to encourage the data collectors to proceed in a diligent and professional manner, so that the best results possible are attained, their own management needs to be persuaded about the importance of what they are doing. Without the approval of management, the data collection procedure will not be given the priority it requires.

15.5.5. Data Collection and Project Managers

Managers of all software projects need to have access to information on the development of a particular piece of software. They need to be able to collect and review operational information in order to have the fullest possible picture of the current project. The data collected as part of a reliability evaluation provides additional information on 'good' and 'bad' trends in performance, suspected or actual deviations from experience of previous projects, as well as providing input to future projects to help support:

- feasibility studies,
- project planning,
- project costing,
- resource allocation,
- choice of development methods, testing methodologies, process control and support facilities.

This enables managers to do their job more efficiently and make cost-effective use of the resources available.

Data collection must be purposeful. It should avoid obstructing or duplicating other activities, and if possible be built in to the development process, in order to make it acceptable to project staff. In many cases, those involved with the project may not fully appreciate the potential future benefits from regular data collection and may be tempted to give it a low priority at times when it should have high priority, e.g., during testing. In addition, the short-term benefits of data collection and analysis may not be immediately apparent as it takes some time to gather a suitable subset to make any reasonable statistical comparisons. However, the benefits of some data collection may be immediately available if the data required for such a subset is obtainable for other sources or there are commonly accepted precedences, e.g., McCabe's complexity measure [McCabe 1976].

For the best possible results, it is necessary to start data collection as early as possible in the project but this will vary from project to project. Data collected at a very early stage during testing should provide information on the progress being made by developers of individual modules, allowing managers to monitor closely those modules that are critical to the development of the system. Thus, a greater managerial control can be exercised over the entire software life-cycle.

15.5.6. Data Collection and Testing

Before the testing of any software product, it is very important to have produced a test plan defining the testing work in detail. This plan should provide a baseline for the work to be done and will help to provide effective status reporting. Data collection should be seen as an integral part of the testing activity and provides an ideal medium for the recording of the test results.

If data has already been collected before the start of the testing activity the results obtained can be used to indicate those modules that are more likely to contain faults. It is then possible to test these "faulty" modules first so that any modifications required can be made while the rest of the system is being tested. Using data previously collected from similar projects it may be possible to predict roughly the number and type of faults inherent in the software. Combining this prediction with the number of errors actually detected can provide an indication of the success of the testing activity.

The approved ANSI standard [IEEE 829] for software test documentation describes a test log and a test incident report and concentrates on providing a framework for collecting data during dynamic testing. The ANSI test log describes a chronological record of relevant details about each test performed, and the subsequent results (error messages generated, aborts, requests for operator action, etc.). Discrepancies, including what happened before and after any unexpected events, are described.

The ANSI test incident report documents any event that requires investigation or correction, summarising the discrepancies noted and refers to the appropriate test specification and test log entry. It includes the expected and actual results, the environment, anomalies, attempts to repeat, and the names of the testers and observers. Related activities and observations that may help to isolate and correct the problem are also included [Hetzel 1984].

15.5.7. Data Collection Problems

All data collected must be accurate and of the necessary level of detail so both developers and analysts can understand, and in some way measure, the software development process. Often the inconsistency of data definitions and a lack of commonality of terminology impose barriers to the general usage and effective interpretation of the available information.

The ANSI standard on Software Test Documentation described in the previous section goes some way to eliminating the problems encountered with the software data collection process, by providing collection standards. However, for data collection in general, there is still a need for commonly accepted terminology and data definitions. This would provide a solution to some of the common data collection problems listed below:

1. Inconsistent Definitions:

   • Inconsistent phase definitions.
   • Using terms without actually defining them.
   • Inconsistent use of terms which have standard definitions.
   • Standard definitions that do exist may not be freely available.
   • Accepted definitions that are contradictory.

2. *Observational Bias*. The problem with any subjective measure is that for a single given condition people will supply different answers. Some people are fundamentally very generous or critical when making subjective judgements. In other cases, ratings are conditioned by the individual's desire for the project to succeed, or at least appear to be successful.

3. *Local vs. Global Frame of Reference*. Consider a small department that consists exclusively of highly experienced and talented personnel who have been using a given development method for several years. If they are asked to rate the level of rigour with which they have applied that development method on a particular project, they may rate it as medium rigour with respect to the rest of the department's projects. However, with respect to other departments within the same organization that have limited experience of the development method, they should be rated as applying a high level of rigour.

4. *Averaging and Side Effects*. When project or subsystem assessments establish ratings of complexity, required reliability, timing constraints, etc., they tend to provide the rating for the most highly-stressed portion rather than the average rating across the project or subsystem. Also, people will intuitively give large software projects higher complexity ratings than small projects. This can be avoided by using tools such as COCOMO (COnstructive COst MOdel) [Boehm 1981] and several of the static analysers to calculate intrinsic complexity

5. *Double Counting*. In cases where a particular piece of code is being developed for use in two separate subsystems the information on the software may be included twice.

During the data collection process, problems will usually arise concerned with the accurate recording of software metrics. The values collected ought to be:

- *Repeatable* — two independent data collectors would obtain the same value if they were to measure the same item.

- *Comparable* — the metric values obtained from different items have been obtained using the same procedures or an equivalent translation process has been produced.

- *Verifiable* — the values can be checked for clerical errors and inconsistencies.

To satisfy these requirements, it is necessary that data collection procedures be implemented and adhered to, since the way in which data is recorded, verified and analysed will influence the success of any data collection activities. The following suggestions will however ensure that the data collection is as successful as possible:

- data collection should be integrated into the development process

- data collection should be automated wherever feasible

- data should be treated as a company resource and facilities should be available to keep historical records of projects, as well as monitoring current projects

- data that cannot be collected automatically should be collected at the time, not based on the recollection of past events, and verified immediately

- software engineers should be motivated enough to keep accurate records, and to participate in the data collection process

- the timescales between data collection and analysis should be minimized

A data collection activity following some, if not all, of the suggestions above will benefit the project as a whole and provide important information for use during reliability evaluations.

15.5.8. Data Collection and Delta-4

Within Delta-4, consideration was given to the method of data collection, its use within the project and the definition, storage and analysis of the data. This helped to remove ambiguities in the results of the collection caused by poorly specified requests for information and therefore improved the validity of and confidence in the data. The data collection manuals used attempt to either define the terms used or provide examples in cases that are open to misinterpretation.

As it was also important to reduce the costs of data collection to a minimum, the most cost-efficient techniques were implemented wherever possible. However, because the reliability growth models being used for the Delta-4 reliability evaluation require inter-failure times, it was necessary to accept a slightly larger overhead in order to be able to provide the data to drive the models.

Unfortunately it was not possible to introduce data collection until most of the code had been written and, in most cases, some initial testing had already been undertaken. It was however envisaged that sufficient data would be collected to allow a detailed reliability evaluation to take place during the project.

The original data collection activities that were produced ranged from extremely detailed collection activities to a set of simplified data collection forms. The detailed collection activity that was to be carried out at Ferranti on the development of Deltase for Ada consisted of a comprehensive set of collection forms as outlined below:

A) An operating environment form containing information on the hardware and software resources in use at each development and/or test site.

B) Several forms that only need to be filled in once for each product (e.g., Deltase for Ada). These forms provide background information and, although they are not directly used in the reliability evaluation, deliver a basic description of the structure of the software product and process. They are:

- Product form — contains information on the components and tasks associated with the product.

- Task structure form — contains information on the relationships between tasks

- System/subsystem structure form — contains information on the relationships between the different software components and how they are integrated.

- Textual system/subsystem structure form — contains information on the relationships between the textual components of the system.

C) Two forms that need to be completed for each task:

- Task definition form — contains information on the type of task being performed and dependencies between tasks.

- Task resource form — contains information on task duration and effort expended. This has been separated from the task definition form to ensure data confidentiality.

D) A form that needs to be completed for each component. The form required differs depending on whether the component is a software component or a textual component:

- Software component form — contains basic metrics and cross-references to the system/subsystem form and the task form.

- Textual component form — as for the software component form.

E) A final set of forms that provide the data required to drive the reliability growth
   models described later. This data is collected from several phases of the
   development process — in the early phases from design and code walkthroughs, in
   the later phases from dynamic testing and finally during maintenance. These forms
   are all cross referenced to allow the data to be processed into the correct format for
   the models.

- Test history — contains information on the behaviour of the software under
  test.
- Incident form — completed for each incident. Contains information on the
  cause and severity of the incident.
- Fault form — completed for each fault. Contains information on how the fault
  was discovered, when it was actually introduced (i.e., during which phase of
  development) and its cause.
- Change form — completed for each change. Contains information on how the
  need for change was determined and the purpose of the change (e.g., planned
  enhancement, fault correction).

The use of several forms at different levels of detail was designed to provide a
comprehensive picture of the software being developed and to ensure the completeness of the
data set over time. However, due to a change of direction in the project out of the control of the
data collectors, this detailed data collection was not implemented.

A simplified data collection manual has been provided to other Delta-4 partners who were
willing to collect data but did not have the time and effort available for an extensive data
collection activity. This manual consists of only three forms — a software log, an incident
report and a change report. The software log provides a detailed history of the execution of the
software over time and includes details of the test phase and test type for data collected during
testing. An example of the form is shown in figure 7.

This form is, however, not restricted to the testing phase but could be used throughout
software development. The other forms provide more detailed information on the incidents and
changes referenced in the software log and are very similar to the incident and change forms
used in the more extensive data collection activity. Using this simplified approach minimizes the
time spent on data collection but still allows sufficient data to be collected to be able to apply the
reliability growth models to the software.

The forms in the simplified data collection manual are compliant with the ANSI Standard
for Software Test Documentation described earlier. However certain areas of the standard are
not necessary for the Delta-4 data collection activity. Where there is a mapping between the
needs of Delta-4 and the standard, the two have been kept as consistent as possible.

Within Delta-4, to complement the work done on data collection, the mechanisms for data
storage, transfer and analysis were set up as shown:

- *Data storage.* All data collected has been stored in a central analysis database and
  consideration has been given to the content and structure of this database. Plans
  were made to validate and verify the data before entry, incorporating the lessons
  learnt from previous data collections. A consistency checker is available to ensure
  that the data sets within the database are as complete as possible.
- *Data transfer.* Applications have been developed to execute some static analysis tools
  in a consistent manner and output the results to standard flat files. This standard flat
  file format allows the data to be fed directly into the central database.
- *Data analysis.* The data in the database can be output in the format required by a
  generic statistical package. The analysis process has been automated so that output

from the static analysis tool can be fed into the statistical package and analysed. A list of the code metrics is then automatically produced that indicates which metrics are outliers.

SOFTWARE LOG

DATA COLLECTOR : _____ COMPONENT ID : _____ VERSION ID : _____

DATE	START TIME	END TIME	TEST PHASE	TEST TYPE	INCIDENT ID	CHANGE ID

Fig. 7 - The Delta-4 Software Log

Although data collection could not be introduced right at the start of the project, every effort has been made to ensure the effectiveness and success of the data collection activity by providing automated versions of the forms and keeping the collection as simple as possible. The implemented mechanisms allowed prompt and accurate feedback directly to the relevant project members.

15.5.9. Data Collection and Predictions

Continuous data collection means that it is easier to statistically identify trends as and when they occur. The summary statistics provided as a result of statistical analysis should include information on the most frequent error causes, the most common means of detection and the most error prone modules. Detailed statistical analyses should also be provided for any trends that give cause for concern.

The reliability of software can be qualitatively and quantitatively assessed during its development, provided the use of the software is restricted to its specified purpose. Data collection can lead to an improved understanding of the software development process and this improved understanding extends to the reliability assessment of the software product itself. There is currently no single agreed method of producing reliable software and therefore, the collection, storage and analysis of data can aid in the evaluation of the links between the development method(s) and the reliability of the software.

Software reliability models use data on the times at which software failures are observed to occur, although the data can be processed for input to models that require data about the number of failures within a given timescale. For use in driving the software reliability models, the failure data will need to be processed, as several failures may result from one error.

All the reliability growth models rely on inter-failure times and require that a minimum number of errors occur before the data can be fitted to the model. Once the data has been processed, the use of a general linear model may help to identify problem areas in the software. When enough error data has been generated, the data is run through the reliability growth models. Unlike hardware reliability modelling, where large quantities of a particular piece of equipment can be tested, software reliability modelling involves just one unique software product. The profile of the software changes when each error is encountered and repaired, with the result that it is impossible to determine in advance which model will best fit the data. Not only is the piece of software itself unique but so is each operational environment in which it is to be used. Therefore an assessment of a piece of software will contain restrictions as to the applicability of the reliability attached to the software.

A considerable number of reliability growth models have been published over the past twenty years and there has gradually developed an underlying unity in the approach of the models to the problem of software reliability [Mellor 1987], in that they adopt the following assumptions:

- the system contains a set of errors on delivery,
- each error causes failure independently of all the others,
- each error causes failure at its own particular rate,
- failures due to the manifestation of a single error occur as a Poisson process[8].

Some of the models also assume that on manifestation, an error is immediately and perfectly removed.

A software suite of nine models is currently produced by Reliability Consultants Ltd. and can be used to fit data to the models and to enable an analysis of the predictive accuracy of the models for each data set.

To use these models, the following procedure is implemented:

- assume that there has been a sequence of failures during continuous use of the software
- collect at least the minimum number of processed failures required to run the model and the times between failures
- collate the inter-failure times of the software, $t_0,......t_n$
- verify that the chosen model fits the data reasonably well
- take the chosen model and fit parameters using statistical estimation techniques
- use the fitted model to estimate quantities of interest, e.g., current failure rate, time to next failure, number of remaining errors, etc.

The technique used to analyse the models ([Rook 1990], chapter 6) is that of partitioning the processed failure data so that the observed data on the first failures $t_0......t_{j-1}$ are fitted to the models and the predictions made by the models when the data have been fitted to them are compared with the subsequent failure data $(t_j,....t_n)$ thereby giving a measure of the predictive accuracy of the models. The accuracy of the predictions is analysed using standard statistical techniques, such as u-plots and prequential likelihood, which will determine bias and noisiness in the dataset.

The testing scenario should match, as closely as possible, the environment in which the software will be put to use. To this end, records of all data input during testing should be kept, so that comparisons of testing and the operational profile can be made. It is very important that

---

8   If a number of incidents are recorded within a set of time periods, and the following three conditions are
     satisfied, then the process is a Poisson process:
     a.   Events occurring in two disjointed time intervals are independent
     b.   The number of events occurring in a time interval (t_1, t_2) is only dependent on the length of the
          interval $t=t_2-t_1$ and *not* on t_1
     c.   The probability that more than one event will occur at the same time, and the probability that an
          infinite number of events will occur in some finite time interval are both zero.

the data input during testing is carefully selected to optimize the test coverage of the paths through the software that will be exercised in operational use. Figure 8 shows one possible relationship between the set of inputs leading to failure and the set of inputs covered by testing.

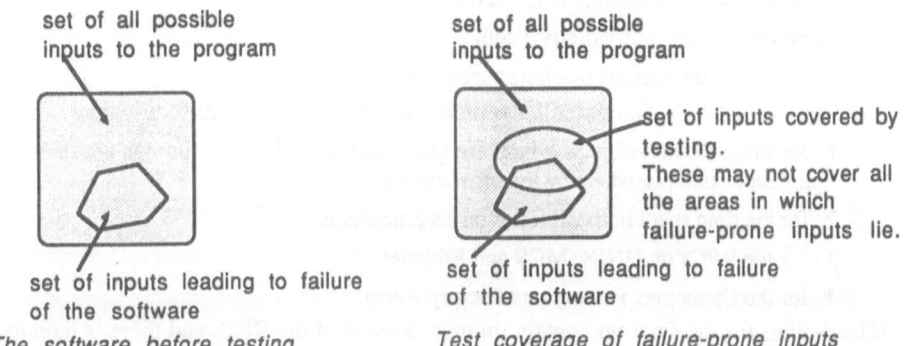

Fig. 8 - The Software Before and After Testing

There is an important potential problem in the simulation of the operational profile. A relatively small change in the characteristics of the profile can often lead to the execution of a previously rarely used piece of code, resulting in a whole new cluster of errors being found. Conversely, a larger change in the profile can have little or no effect on reliability. This is counter intuitive, but a corollary of this property is that a small error in determining the operational profile can cause a massive error in determining the reliability ([Rook 1990], chapter 1).

15.5.10. Predictions in Delta-4

The suite of nine models discussed above is being used to evaluate the reliability of the software. However, since Deltase is a support environment, this has raised several points of interest:

- The support environment does not execute continuously and has been designed to supply many services by multiple and diverse mechanisms in such a way as to minimize overhead. As a result, the measurement of execution time is difficult to achieve and an instrumented version of the code has been produced to provide execution time measurements. It is probable that the presence of the instrumentation may affect the results obtained since the process of measuring will affect the quality of the measurements.

- The distributed architecture means that ports on multiple different machines are neither fully independent nor fully interdependent. This has implications for the timing measurements, as different machines have different timing capabilities.

- The replication of software components raises problems in the processing of the data for the models. If the times of their parallel executions are summed, this will give an inaccurate measure of the reliability, since it is the overall reliability produced by the system architecture that interests us, not the duplicated reliability measurements of the replicas.

The instrumentation inserted into the code has entailed a prolonged effort to analyse the data produced by the instrumentation. Times for the remote procedure calls (RPCs) have had to be extrapolated from cumulative figures for:

- the number of RPCs sent/received,
- the number of characters sent/received,
- timings on RPCs of various length,
- timings on the various machines used on the network.

The instrumentation has produced (after processing) four different sets of timings:

1. for three Deltase objects which are quite distinct and whose timings are therefore not duplicated elsewhere by the instrumentation,
2. for the time spent in the libraries on each machine,
3. for the RPCs on UNIX, MCS and Ethernet,
4. for the Catalogue, Factory and Factory Agent.

The timings for the final set contain timings of some of the RPCs and there is therefore duplication of timings that is impossible to calculate from the measurements.

To enable a meaningful analysis, the software problem reports have had to be cross-referenced to the incident report forms generated by the instrumentation and the faults relating to each of the subsets of the timings extracted.

It is intended that a quantitative assessment will be available by the end of the project since there is now enough failure data to drive the models but the results are not available at the time of writing of this document.

ANNEXES

ANNEXES

Annex A
Safety-Critical and Safety-Related Systems

A *safety-critical system* can be defined as any system for which there exists a possibility for catastrophic failure, i.e., a failure for which the consequences on the system's environment are incommensurably greater than the benefit provided by service delivery in the absence of failure (see chapter 4). (The corollary definition of a "non-safety-critical system" would then be a system for which failures can be considered as "benign", i.e., for which the consequences are of the same order of magnitude — generally in terms of cost — as the benefit provided by service delivery in the absence of failure).

Defining recursively a *system* as a set of components and a *component* as another system, consider the case of a (global) safety-critical system (i.e., an entire nuclear power station). Now, the failure of any single component, when combinations with other component failures are considered, could *possibly* lead to a catastrophic failure of the global system. Should therefore all the components of such a safety-critical system be considered as safety-critical (sub-)systems or are there *degrees* of safety-criticality?

In the aviation community, this notion of *degrees of criticality* is embodied in the following definitions for the *functions* of a system [RTCA 178A]:

- *critical:* functions for which the occurrence of any failure would prevent the continued safe flight and landing of the aircraft;

- *essential:* functions for which the occurrence of any failure would reduce the capability of the aircraft or the ability of the crew to cope with adverse operating conditions,

- *non-essential:* functions for which the failure could not significantly degrade aircraft capability or crew ability.

From this viewpoint then, a *safety-critical* component would be any component of a safety-critical system carrying out critical functions, i.e., a component whose failure can *lead directly* to, or *significantly increase the risk* of, a catastrophic failure of the global system.

Similarly, a *safety-related* component could be defined as any component of a safety-critical system carrying out essential (or possibly non-essential) functions, i.e., a component whose failure could *lead indirectly* to, or only *slightly increase the risk* of, a catastrophic failure of the global system. However, the imprecision of such adverbs as "significantly" or "slightly" introduces a fuzziness in the distinction between safety-critical and safety-related which should ideally be avoided.

When stressing that *the Delta-4 architecture does not address safety*, it is meant that the architecture does not address systems or system components for which the primary measure of dependability is safety, i.e., the probability that no catastrophic failure will occur. This implies that the *architecture is not suitable for implementing safety-critical (sub-)systems* in the sense defined above. The design of such a safety-critical computing system should be based on the most stringent fault prevention and fault tolerance techniques, including, for instance, formal proof of correctness, diverse design, etc., which would be far too expensive to consider in more mundane application areas of Delta-4 such as office automation and computer-integrated manufacturing.

However, *the Delta-4 architecture does address reliability and availability*, i.e., systems or system components for which the primary measures of dependability are the probability of continuous proper service or the probability of proper service at a given instant. This means that the Delta-4 architecture is dependable in the sense that one can be confident that the mean times to failure or between failures will be far greater than a non-fault-tolerant architecture. This, together with the flexibility advantages of the open approach to fault-tolerance adopted in Delta-4, means that it would seem reasonable to envisage it being usefully applied to *safety-related* (sub-)systems. Indeed, increased reliability of such (sub-)systems can only increase the safety of the global system, e.g., by reducing the frequency of shutdowns due to false alarms and thereby reducing wear-out of the primary, safety-critical protection system. However, the word of caution in distinguishing safety-*critical* systems from safety-*related* ones on a yes or no basis must again be re-iterated; a thorough risk analysis of the *global* safety-critical system is essential.

Annex B

Deterministic UNIX

With the increasing popularity of the "shrink-wrap" approach to software retailing in mind, a study has been initiated on the feasibility of constructing a "deterministic UNIX" (referred to, for short, as D_UNIX in the text below). Initial results are encouraging. The function of D_UNIX is to enforce scheduling and control the interactions of the UNIX processes in a multi-process package so that the package as a whole behaves as a state machine and hence exhibits replica-determinism. The basic requirements are twofold; to replace all non-deterministic interactions between UNIX processes and the UNIX environment with replica-deterministic alternatives, and to enforce deterministic scheduling between members of a set of associated UNIX processes called a *family*. The definition of what constitutes a family is: that set of processes for which it is desired to enforce mutual determinism locally rather than use distributed replica group determinism mechanisms. By arranging a single cyclically-ordered execution queue for each family, under the discipline that fault-free replicas run until they suspend or terminate, with no concurrent activity by any other family member permitted, deterministic scheduling is assured, through all the following process state transitions: Activate, Suspend, Terminate, Run, Block, Unsuspend.

Generally, if the head of the queue is suspended, the entire family is blocked; however, a further state transition, Shift (to end of queue), is needed to deal with the situation that a process at the head of the queue must be unsuspended by another family member. This is a controversial requirement; if a family member is suspended awaiting an event, it must be possible for D_UNIX to determine whether this is an internal-to-family event or an external event such as those involving disc or clock drivers. Consider, for example, use of D_UNIX message queues. A particular queue must either be used intra-family and invoke the shift mechanism, or must be managed by D_UNIX using Delta-4 replica message ordering and delivery mechanisms. Such mechanisms carry performance costs. At one limit, all activity of a node could execute deterministically and all blocking for external events (input messages, disc transfers, etc.) leads to processor idling. At the other limit, all interprocess communications are via Delta-4 and involve LAN delays. As well as making use of such Delta-4 mechanisms as ordered message delivery and replica message validation, we make the assumption that the start condition is with a single replicated process; members of the family arise through descent and in an ordered manner via Fork/Exec action. The study has yet to determine the most appropriate mechanisms for the cloning of families; this will be the subject of the continuation of this work. All remaining non-deterministic interactions with UNIX (reading the time, for example) must be replaced by a replica-deterministic version, using a replica negotiation protocol based on actions initiated by D_UNIX itself on the application's behalf. Note that, as with all Delta-4 replica interaction protocols, this must be executed in a fail-silent environment and hence requires a suitable MCS service to be defined; the replicas can, however, execute in a fail-uncontrolled regime. Interactions with UNIX could also be rendered deterministic through use of the leader-follower notification protocol carried out by D_UNIX on the application's behalf.

The result would have performance advantages over the more complex alternative negotiation protocol, but in this case all code including the application software must of course execute under a fail-silent regime.

Annex C

Statistical Effects with Schedules of many Activities

Over a large number of activations, the execution times for an activity form a distribution characteristic of that activity. A schedule consists of an ordered sequence of activities. Provided that the activities are independent and their execution times do not depend on the time at which they are executed, the execution time for each activity in a given schedule can be modelled by randomly selecting one of the execution times from the activity's distribution function. For any one schedule a (possibly very large) number of model schedules can be constructed.

Let:

m_i be the mean execution time of activity i

σ_i be the standard deviation of the distribution function for the execution time for activity i

m be the mean schedule execution time

σ be the standard deviation of m

N be the number of activities in the schedule.

The total time required to complete each model schedule is the sum of the execution times for all activities in the model.

$$m = \sum_i m_i$$

Over all possible models the variance in the expected schedule completion time is the sum of the variances in the completion times for each activity.

$$\sigma^2 = \sum_i \sigma_i^2$$

Hence, for schedules with more than one activity having a non-zero variance, the standard deviation in the expected completion time is less than the sum of the standard deviations in the completion times for each activity.

$$\sigma < \sum_i \sigma_i$$

Hence, the fractional standard deviation (standard deviation divided by the mean) in the schedule execution time is less than the sum of the fractional standard deviations of the execution times of the activities in the schedule.

$$\frac{\sigma}{m} < \frac{1}{m} \sum_i \sigma_i < \sum_i \frac{\sigma_i}{m_i}$$

As an example: for the special case where a schedule consists of N activities each having identical execution time distribution functions with mean m and standard deviation σ:

- the expected time to execute the schedule is N.m
- the standard deviation of the schedules execution time is $\sqrt{N}.\sigma$
- the fractional standard deviation of the schedule's execution time is $\dfrac{\sigma}{m.\sqrt{N}}$, which is

   less than $\dfrac{\sigma}{m}$, the fractional standard deviation for each activity.

If both the execution times of individual activities and the mean execution times for each activity have Gaussian distributions, then the fractional standard deviation in the schedules execution time is less than the mean fractional standard deviation of the individual activities.

PROOF

We wish to prove the fractional standard deviation in the schedule execution time is less than the mean fractional standard deviation in the individual activities' execution times. (Note: for clarity the summation index is dropped)

On squaring
$$\frac{\sum \sigma_i^2}{(\sum m_i)^2} < \left(\frac{1}{N} \cdot \sum \frac{\sigma_i}{m_i}\right)^2 \tag{1}$$

Assume the Cauchy inequality:

$$(\Sigma a.b)^2 \le \sum a^2 \sum b^2$$

Identify σ_i with $a.b$ and m_i with a.

Then
$$\frac{(\sum \sigma_i)^2}{\sum m_i} \le \sum \left(\frac{\sigma_i}{m_i}\right)^2 \tag{2}$$

Consider a set of positive random variables $x_1, x_2,..., x_N$, and let δ_i be the deviation of x_i from the mean of the set.

Then
$$\sum x_i^2 = \frac{1}{N} \cdot \left(\sum x_i\right)^2 + \sum \delta_i^2$$

Let k_x be defined such that:

$$\sum \delta_i^2 = k_x \cdot \frac{1}{N} \cdot \left(\sum x_i\right)^2$$

Then
$$\sum x_i^2 = (1+k_x) \cdot \frac{1}{N} \cdot \left(\sum x_i\right)^2$$

Using this result for δ_i , m_i and δ_i / m_i in 2 above gives:

$$\frac{\sum \sigma_i^2}{(\sum m_i)^2} \le \frac{(1+k_{\sigma/m}) \cdot (1+k_\sigma) \cdot (1+k_m)}{N} \cdot \left(\frac{1}{N} \cdot \sum \frac{\sigma_i}{m_i}\right)^2 \tag{3}$$

For a Gaussian distribution the mean value must be much larger than the standard deviation to make the probability of negative times (or standard deviations) very small. This implies that the k values are ≈ 0 and so equation 1 follows from equation 3.

For distributions in which the variance is defined, equation 1 follows from equation 3 for all N greater than some value (any activity that forms a distribution having an undefined variance will create problems on a real-time system and so we can assume that such activities will not be used). For many commonly encountered distributions (e.g., rectangular and Poisson) the mean is at least as large as the standard deviation and so $0 \leq k \leq 1$ For these distributions, equation 1 follows from equation 3 for even small values of N (greater than 8).

For real-time systems executing a large number of activities the Central Limit Theorem is normally applicable (especially for the large numbers of activities per schedule that occur in, for example, process control systems). The Central Limit Theorem, as applied to an execution schedule, states that the time to execute the schedule will approximate a Gaussian distribution function when the number of activities comprising the schedule is large.

This Gaussian approximation is good for times near to the mean execution time (i.e., within a few standard deviations of the mean execution time).

The Gaussian approximation is less accurate in the tails of the distribution (at least for realistic numbers of activities) and so can not be used as basis for assessing the very low probabilities required to assert that some combinatorial eventuality is unlikely during the lifetime of the system. For such predictions an accurate knowledge of the behavior of each individual activity is needed.

Some model schedules may require more time to complete than is available. In such cases only a subset of the activities will have completed by the end of the available time. Where the conditions above for the whole schedule can be applied to such subsets of the schedule, then by the same arguments, the fractional standard deviation in the completion time for the subset is less than the mean fractional standard deviations of the activities comprising the subset. It follows then that in the allocated time T, the fractional standard deviation in the expected number of activities to have been executed grows at a slower rate than T as T is increased.

As a practical example, the distributions of many individual activity execution times found in real-time systems resemble exponential decay functions, though many cases have a maximum execution time and a peak for some small non-zero execution time, which for simplicity can be ignored. Examples of activities exhibiting such exponential decay functions are handling messages in a message-switching system, or logging reports, or updating disc based data. The probability density function for such a process is:

$$\lambda.e^{\lambda t} \quad \text{for } t > 0$$

In general, each process will have a different value of λ. In such cases for large N the schedule is a non-homogeneous Poisson process. The probability of executing K activities in a time T is:

$$\frac{1}{K!} \cdot \left(\int_0^T \lambda(t).dt \right)^K \cdot e^{-\int_0^T \lambda(t).dt}$$

For such a process the expected number of activities, N, executed in time T is:

$$\int_0^T \lambda(t).dt$$

This value is also the variance in N. In the available time T, the fractional standard deviation in N is $\frac{1}{\sqrt{N}}$.

Annex D

Assigning Precedence Parameters

This annexe offers some suggestions as to how a system designer might assign precedence parameters to the various services in the system in order to achieve required timeliness properties. In practice, the system designer would also select the number of priority levels and the precise scheduling algorithm to be used at each priority level; see chapter 9.

Priority scheduling is important in overload situations, when it ensures that only those timing constraints are violated which will occasion minimal cost. So the system designer should assign the highest priority to those components whose untimeliness would occasion maximum cost, or lose maximum benefit.

On the other hand, targetline scheduling is important in normal situations, when it finds a feasible schedule so that each component can deliver its service at or before the selected targetline. The system designer should therefore choose the targetline as the optimum time of service delivery. (Components that are ready to deliver service before the targetline can always suspend until they receive an event at the targetline.)

There are several heuristics that can assist a system designer to model the required timeliness of each component and select suitable scheduling parameters for it. The first option is to use benefit/delivery-time graphs, (see figures at the start of chapter 5). Priority = 1 could be assigned to those hard real-time components whose graphs show considerable costs. Priority = 2 could then be assigned to soft real-time components whose graphs show smaller costs; Priority = 3 to non-real-time components that show positive benefit unless never delivered.

The targetline could be assigned as a time at which the graph shows maximum benefit. However, if there is a range of such times, as in figure 1 of chapter 5, it is generally better to choose a "just in time" targetline towards the end of the range; this may release resources to some other component that has a more urgent need for them.

Figure 2 of chapter 5 shows the benefit/cost of delivering a soft real-time service at various times. Using the above guidelines, the priority is set equal to 2, and the targetline has been chosen as the time of maximum benefit.

Figure 3 of chapter 5 shows the corresponding graph for a component that fires an anti-aircraft gun. The aircraft can only be hit between the liveline and deadline, which are very close together. The priority is set equal to 2, since this is soft real-time. The targetline is equated to the estimated liveline, and when it is reached, the component receives a signal ordering it to fire.

Figure 4 of chapter 5 shows a non-real-time service that has positive value at any finite time, so it is given the lowest priority. Its output is normally collected at a certain time, which is chosen as the targetline. If it is delivered after the targetline, increasing inconvenience is caused, so it should be preferred to other non-real-time components with a later targetline.

However, other system designers may find it more natural to use a quite different "urgency/time" graph, i.e., one that plots the dynamic urgency of the component against time.

The urgency/time graph corresponding to figure 1 of chapter 5 would rise between the liveline and the deadline, because it is getting increasingly urgent to schedule the component. If the urgency is estimated so as to reflect the cost of missing the deadline, the priority of a component could be equated to its maximum urgency. If two components have the same priority, we want to prefer the one that reaches maximum urgency first, so the targetline could be equated to the time of maximum urgency.

Yet another type of graph, which reflects the XPA scheduling philosophy more closely, is shown in figure 1 below. The graph shows XPA priority against time for the hard real-time points switching component and the soft real-time anti-aircraft gun controller. Both priorities are static between the start time and the targetline, but the targetline event raises the priority of the gun controller, since the gun must then be fired in a short window of opportunity. The graph helps one to visualize the hard real-time component preempting the soft at one time and returning the processor to it at another.

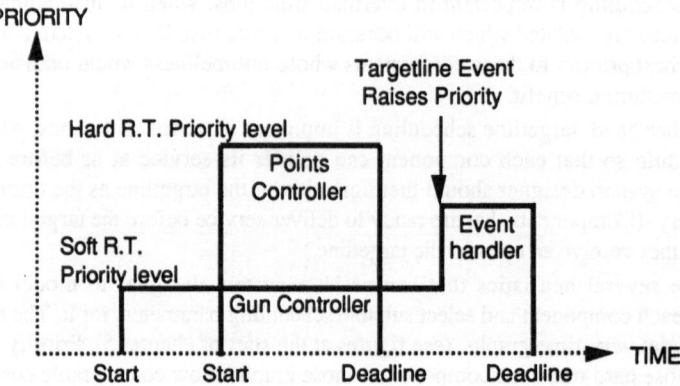

Fig. 1 - Priority of Two Components against Time

Annex E

Formal Failure Mode Assumptions

This annexe gives some formal definitions of the failure modes of message-passing[1] components such as nodes, hosts and network attachment controllers that were described briefly in chapter 6. The definitions are based on a notation defined in [Powell 1991].

E.1. Perfect Component

The correct behaviour of a message-passing component can be formally modelled as a (possibly-infinite) sequence of messages m_i, i=1, 2,... each characterized by a pair $<vm_i, tm_i>$ where vm_i is the *value* (information content) of the message m_i and tm_i is the *time* or instant of delivery (by the component) of the message m_i. A given message m_i is defined to be *correct* iff:

$$\{[vm_i \in SV_i] \wedge [tm_i \in ST_i]\}$$

where SV_i is the specified set of values for message m_i (often a single value) and ST_i is the specified set of delivery instants for message m_i. Of course, SV_i and ST_i depend on the prescribed activity of the component that, in turn, may depend on the passage of time and/or inputs to the component from other components. The various SV_i and ST_i are an abstract specification of correct component behaviour; thus a hypothetical perfect component (i.e., one that never fails) is characterized by the following assertion:

$$A[PF] := \{\forall i=1,2,... [vm_i \in SV_i] \wedge [tm_i \in ST_i]\} \tag{1}$$

Note that the formal definition given above also includes constraints on the timing behaviour of a non-failed component; this is essential in order for it to be possible to achieve a distributed consensus in the presence of faults [Fischer et al. 1985].

E.2. Fail-Silent Component

A fail-silent component is defined here to be one that, viewed from *outside* the component, either operates in conformance with its specification or remains silent [Powell et al. 1988]. In particular, any message sent by a fail-silent component is a message that is correct in both value and time. In terms of the above model of component behaviour, a component that is assumed to be fail-silent is characterized by the assertion:

$$A[FS] := \{\forall i=1,2,... [vm_i \in SV_i]$$
$$\wedge [(tm_i \in ST_i) \vee (\forall j \geq i, (tm_j = \infty))]\} \tag{2}$$

---

[1] The notion of a "message" should be taken broadly in that it covers not only strings of bits sent over a communication network but also other discrete "items of service" such as blocks of data or handshake signals exchanged over a parallel bus.

This assertion states that until message j, all messages that the component should have delivered are delivered within the specified time-bounds and with correct content and that all messages after j are "never" sent. "Never" is taken to mean "not until after the component is repaired and re-initialized".

E.3. Fail-Uncontrolled Component

A fail-uncontrolled component is defined here to represent the opposite extreme of the failure mode spectrum, i.e., a component that may fail in a quite arbitrary fashion. The corresponding assertion for the assumed failure mode is:

$$A[FU] := \left\{ \forall i=1,2,... \left[(vm_i \in SV_i) \vee (vm_i \notin SV_i) \right] \right.$$
$$\left. \wedge \left[(tm_i \in ST_i) \vee (tm_i \notin ST_i) \right] \right\} \tag{3}$$

which trivially states that every output message is either correct or incorrect (in value and/or time); a proposition that is of course always true. Note that this definition includes the following possibilities for faulty component behaviour:

a) for any i, $(ST_i \neq \emptyset) \wedge (tm_i > \max(ST_i))$: late messages or missing messages $(tm_i = \infty)$,

b) for any i, $(ST_i \neq \emptyset) \wedge (tm_i < \min(ST_i))$: early messages,

c) for any i, $(SV_i \neq \emptyset) \wedge (vm_i \notin SV_i)$: messages with erroneous content,

d) for any i, $(ST_i = \emptyset) \wedge (SV_i = \emptyset)$: extra or "impromptu" messages, i.e., messages for which no time and value domain specification exists and which should therefore never occur.

Remark that in case (d) above, then *a fortiori:*, for an impromptu message m_i:

$$(vm_i \notin SV_i) \wedge (tm_i \notin ST_i)$$

i.e., an impromptu message is erroneous in both value and time.

From a practical viewpoint, an impromptu message can be detected as being erroneous in *either* the value domain *or* the timing domain, depending on the particular error detection scheme that is used. A real observer (i.e., some error detection mechanism) would declare an impromptu message to be erroneous in value if it had deduced a value domain SV_j and a time domain ST_j for an expected message m_j and the impromptu message m_i falls within ST_j but with $vm_i \notin SV_j$. Similarly, an impromptu message would be declared as a being untimely if the impromptu message m_i was delivered outside any time window in which a message was expected.

E.4. Fail-Silent Component with Coded Messages

The definition given earlier in expression 2 states a very strong assertion on component behaviour. A slightly weaker assertion is of practical significance in a message-passing system in which messages are coded by some checksum mechanism that allows corrupted messages to be detected. If a fail-silent processor sends coded messages over a transmission channel that may corrupt messages, then the corresponding assertion on the behaviour of the composite processor/channel component is:

$$A[FSC] := \left\{ \forall i=1,2,... \left[(vm_i \in SV_i) \vee (vm_i \notin CM) \right] \right.$$
$$\left. \wedge \left[(tm_i \in ST_i) \vee (\forall j \geq i, (tm_j = \infty)) \right] \right\}$$

where CM defines an error-detecting code such that $\forall i$, $sm_i \neq \emptyset$, $SV_i \in CM$.

This assertion states that until message j, all messages that the component should have delivered are delivered within the specified time-bounds and with correct, or detectably erroneous, content and that all messages after j are never sent.

E.5. "Weak" Fail-Silent Component with Coded Messages

A further relaxation of assertion A[FSC] is of practical useful significance. Consider now that the processors' messages are responses to (coded) request messages received over the transmission channel. Clearly, if a request message is corrupted, the processor can choose to ignore it (over a shared channel, this is the only reasonable action following a corrupted request message since the processor will not know to whom a response should be sent). Therefore, the composite processor/channel component can be seen to omit certain (response) messages. In this common situation, a requesting processor is faced with the predicament of deciding whether its own request message was corrupted (in which case it may choose to repeat it) or the remote processor has failed (in which case it is no use repeating the request). An issue to this predicament is to assume that a "fair" channel will not corrupt messages indefinitely, i.e., that the number of consecutive omissions is bounded to a degree k [Veríssimo 1988]. The corresponding assertion on what can be termed a "weak" fail-silent component, is given by:

$$A[WFSC] := \Big\{ \forall i=1,2,\dots \big[(vm_i \in SV_i) \vee (vm_i \notin CM)\big]$$
$$\wedge \big[(ts_i \in TS_i) \vee \big([(ts_i = \infty) \wedge (\exists j \in [i+1,i+k], ts_j \in TS_j)]$$
$$\vee [\forall j \geq i, (ts_j = \infty)]\big)\big]\Big\}$$

This assertion states that until message j, all messages that the component delivers within the specified time-bounds have correct, or detectably erroneous, content. The component may however omit to deliver up to k consecutive messages; if more than k consecutive messages are omitted, then the component remains silent for ever.

Annex F

Assumption Coverage

In a fault-tolerant system, the coverage of an activated fault, defined as:

$$Pr\{correct\ error\ processing\ |\ component\ failed\}$$

can be decomposed into the product of two terms:

$$Pr\{correct\ error\ processing\ |\ X{=}true\} \times Pr\{X{=}true\ |\ component\ failed\}$$

where X is an assumption made during the design of the error-processing mechanisms regarding the failure mode of the component [Powell 1991].

The first term defines the coverage of a system's error processing mechanisms, conditioned on the validity of the assumption X and the second term defines the *coverage of the assumption* X, i.e., the conditional probability of the assumption X being true given that a component has failed. Since the overall coverage is the product of two coverages, either coverage sets bounds the overall coverage from above.

The only assumption for which the coverage can be set identically equal to 1 is the fail-uncontrolled assumption since, being the worst case imaginable, the assertion is trivially true. In all other definitions of more well-behaved failure modes, the assumption coverage *must* be less than 1. In the extreme case of a "perfect" component (cf. annexe E, assertion 1), the assumption coverage is 0 since, if such a perfect component does in fact fail, the assumption of perfect behaviour can only be violated.

Often, components for which a fail-controlled behaviour has been assumed are designed internally in such a way as to substantiate the fail-controlled assumption. For instance, hosts or network attachment controllers that are assumed to fail-silent should be designed with built-in self-checking. The coverage of the fail-silent assumption is then equivalent to the coverage of these internal self-checking mechanisms. Two different "fail-silent" components may thus have different coverages according to the efficacy of their self-checking mechanisms.

It is important to point out then that the assumption of a fail-controlled behaviour for a component, while allowing simplification of the system fault-tolerance techniques, is accompanied by an assumption coverage $p < 1$. Whenever a component fails, since the system fault-tolerance techniques were designed under the assumption of a fail-controlled behaviour, then the system will[1] fail with a probability of at least $1{-}p$.

The importance of this less-than-unity coverage for fail-controlled components can be illustrated by a very simple example. Suppose that N nodes are interconnected by a perfect communication network and that a distributed fault-tolerance strategy is implemented based on the fail-silent assumption for nodes. Suppose also that the system is capable of delivering a

---

[1] Adopting a worst-case approach that precludes the "chance" tolerance of faults that result in a less-controlled behaviour than that defined by the assumption.

correct service if at least one node is functioning correctly (and all failures of other nodes have been successfully tolerated).

Let λ be the node failure rate and μ the node repair rate. Assuming that only one node can be repaired at a time and that $\dfrac{N\lambda}{\mu} \ll 1$ the asymptotic unavailability UA is given by:

$$UA = \frac{N\lambda}{\mu} \cdot \left[(1-p) \cdot \sum_{i=1}^{N-1} \frac{(N-1)!}{(N-i)!} \cdot \left(\frac{p\lambda}{\mu}\right)^{i-1} + (N-1)! \left(\frac{p\lambda}{\mu}\right)^{N-1} \right] \qquad (0)$$

The important point to note from expression (0) is that when p=1 (perfect coverage), the first term in brackets is zero and the expression for unavailability becomes a strictly decreasing function of N (since $\dfrac{\lambda}{\mu} < 1$), i.e., as redundancy is added, availability *increases* (as would normally be expected). However, when p<1, the first term in brackets can add to unavailability and, according to the value of p, adding redundancy can actually decrease the system availability. Figure 1 illustrates this phenomenon for systems whose nodes have an individual mean time between failures (MTBF) of 1 year and which, on the average, require an intervention of 3 hours to effect a repair (MTTR: mean time to repair). Although when coverage is only 90%, annual unavailability is decreased from 180 to 38 minutes by adding a redundant node, any further increase in N results in an *increase* in unavailability. Thus, to attain, say an annual unavailability of less than 1 minute, the coverage must be between 99% and 99.9%. Only when coverage is greater than about 99,99% does one get any improvement in availability by adding a third unit.

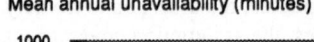

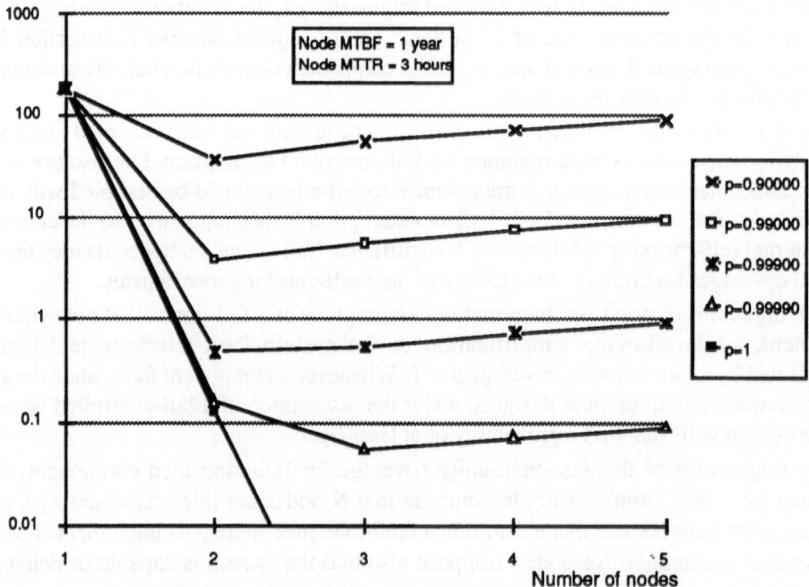

Fig. 1 - Example illustrating the Importance of Assumption Coverage on System Dependability

Annex G

Propagate-before-Validate Error Processing Technique

A potential disadvantage of the validate-before-propagate techniques described in chapter 6 is that there is a systematic time overhead — even if the first replica to produce a message produces an error-free message, propagation must wait until at least t slower replicas agree. The aim of the propagate-before-validate technique is, on the contrary, to allow computation to proceed at the rate of the fastest replica as long as no fault is activated. The comparison of the propagated message is carried out afterwards and is thus pipelined with the useful computational process. (In fact, if message production times are randomly distributed then, in the fault-free situation, this technique is capable of better burst performance than a none fault-tolerant system. On average, however, all replicas must be globally synchronized and thus proceed at the same average rate).

On the contrary to all the active replication techniques for fail-silent or fail-uncontrolled hosts described in chapter 6, the propagate-before-validate technique only allows for error-*detection*. At some point in computation, result propagation must be suspended until all previous computation has been validated — thus providing error *confinement*.

One way of achieving this is by creating global checkpoints so as to allow computation to be carried out tentatively and to be undone if any message used in part of that computation is subsequently found to be erroneous. A transactional model of computation (cf. §6.3.2) provides a suitable framework for such a global checkpointing facility. Before committing a transaction (i.e., before making results of computation permanent and visible outside the transaction), all messages sent by replicas of components involved in the transaction must have been validated, i.e., checked against at least t other replicas. Furthermore, to ensure that no faults remain dormant, an attempt must be made to compare all messages from all $2t+1$ replicas. If a minority of slower replicas are subsequently found to disagree with the messages that were propagated over the network, then, although no computation need be undone, the administration system must be notified of the nodes that are suspected so that fault-treatment can be carried out.

Figure 1 illustrates a burst of error-free executions of the propagate-before-validate mode compared to the competitive and round-robin validate-before-propagate modes (for fail-uncontrolled active replicas). A client-server style of interaction between two triplicated software components is assumed. For simplicity, it is assumed that:

- all replicas are executed by different hosts,
- the slowest replica at each step has time to complete before the next step is initiated,
- that communication times are negligible compared to execution times,

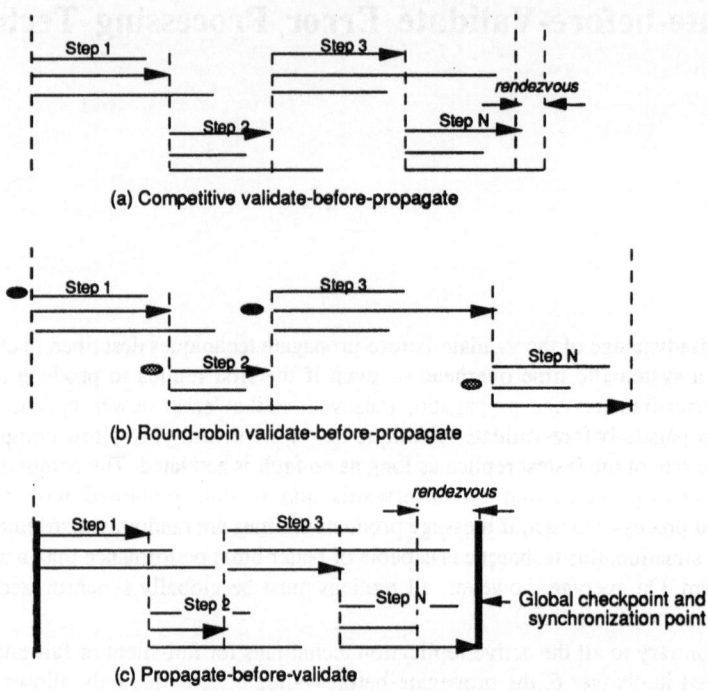

(a) Competitive validate-before-propagate

(b) Round-robin validate-before-propagate

(c) Propagate-before-validate

Fig. 1 - Active Replication Burst Execution Times: No Error

Figure 2 illustrates the same execution sequence with a single value error activated during step 2. For clarity, it is further assumed that:

- the error was caused by a soft fault (cf. chapter 4),
- the replicas do not have any persistent data (state re-initialization is not necessary before re-execution of a failed replica)

Figure 1.a shows for the competitive validate-before-propagate mode that, in the error-free case, computation of step $i+1$ can proceed as soon as two replicas have completed computation of step i. Periodically, an extra waiting period must occur in order to rendezvous with the slowest replica. In the single-error case (figure 2.a), an extra time overhead is incurred while waiting for the slowest replica during step 2.

In figures 1.b and 2.b for the round-robin mode, the order of token rotation is assumed to be from top to bottom of each replica group; the token possessor at the beginning of each step is indicated. The figures show that, in the error-free case, computation of step $i+1$ can proceed as soon as the token-possessor and its successor in the virtual ring have completed computation in step i; this corresponds thus to either the second or third replica to complete. In the single-error case, the extra time overhead is again that due to waiting for the slowest replica during step 2.

It can be observed from figure 1.c that validation of each step is effectively pipelined with the execution of the following steps. A synchronization time overhead is incurred only once every N steps. However, since in figure 2.c, a value error during step 2 affects the fastest replica, the following computation must be undone — this rollback to the previous global checkpoint thus results in an important time overhead.

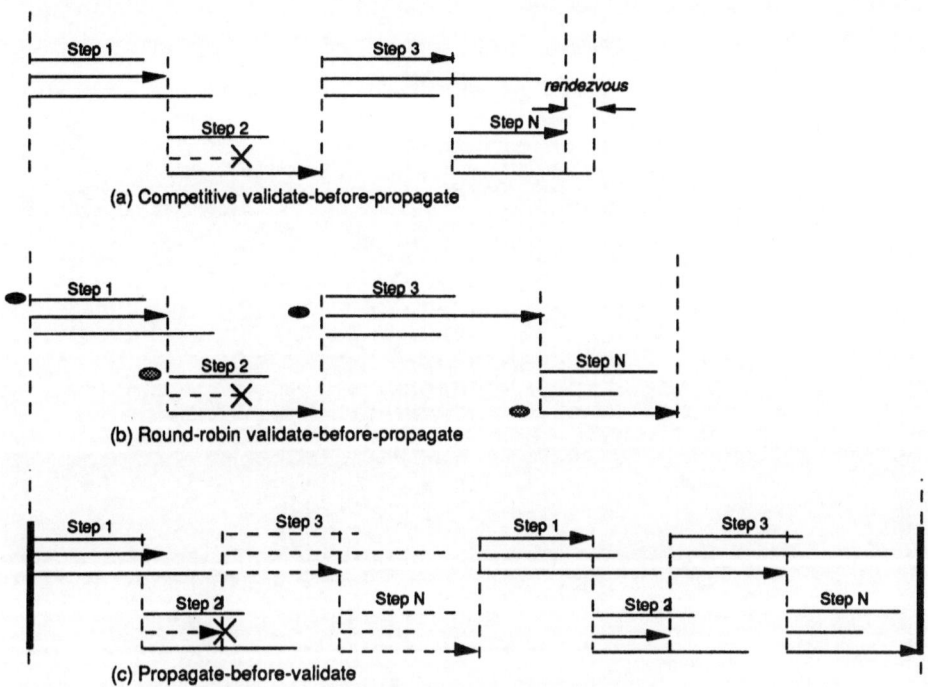

Fig. 2 - Active Replication Burst Execution Times: Single Error

The propagate-before-validate technique has not been implemented by the project. It has however been extensively simulated and shown to be a feasible alternative to the validate-before-propagate techniques.

Annex H

Interface between XPA and OSA

H.1. Portability of Applications

Portability of applications between OSA and XPA worlds is achieved at the level of Deltase Objects, which can be regenerated with Deltase envelopes appropriate to either an XPA or OSA environment, and to a particular model of replication. Not all applications are likely to be ported successfully from XPA to OSA. The constraint is that applications that must achieve timeliness and high performance requirements on XPA may fail to meet these requirements in OSA.

Binary Deltase Software Components are not portable, because the XPA Communications Library contains performance enhancements that bypass the UNIX device driver interface currently used by implementations of OSA.

Some software packages are represented in the Deltase environment by transformers, for example the Dependable Database (see chapter 7). These can run in either an OSA or an XPA environment, provided their transformers are linked to the appropriate communications library.

H.2. Co-existence and Inter-working

If XPA and OSA co-existed on the same physical LAN segment, OSA could compromise the performance and real-time behaviour of XPA. It is therefore preferred that XPA and OSA should exist on separate LAN segments, maybe with a gateway between them. They can then inter-work with the limitation that an XPA Client that invokes an OSA Server cannot expect normal XPA performance.

The node with the gateway would therefore have two NACs connecting it to two LANs; since a single NAC cannot support both the XPA and OSA communication protocols. The gateway may be replicated, e.g., on two fail-silent nodes.

A similar gateway between MCS and an 8802.3 network was already designed and implemented, as part of the process of extending Delta-4 to foreign Workstations with 8802.3 connections. Here the gateway is a simple message-switching application, consisting of two UNIX processes which copy messages transparently in the two directions MCS-to-8802.3 and 8802.3-to-MCS. For 8802.3 to MCS transfers, the gateway must sometimes create an endpoint on the destination association.

The gateway only decodes the message far enough to determine its ultimate destination, and does not require an interface that mimics that of all possible ultimate destinations. The same principle can be used between any two LANs, whether or not they are using the same subset of the Delta-4 Architecture.

H.3. Common Infrastructure

Although the XPA emphasis on optimal performance leads to the use of the Collapsed Layer
Communication System (CLCS, see chapter 9) rather than the MCS, there are several areas in
which OSA and XPA are able to make use of common infrastructure, notably the following:

- Common hardware for the Network Attachment Controller (NAC). The appropriate
  software can be downline-loaded to the NAC without powering down the node; thus
  an OSA node can be transformed "on-line" into an XPA node or vice versa.

- A common extended Atomic Multicast Protocol (xAMp) is used in both OSA and
  XPA, although OSA does not at present make use of some of the extra features
  added for XPA. This should help to reduce protocol validation and software
  maintenance costs.

- The semi-active replication protocol, developed first in XPA, will be ported onto
  OSA. In fact, the version of this protocol used to support Dependable Databases is
  being prototyped on OSA.

- Most of the objects and mechanisms supporting Deltase (chapter 7), e.g., the
  catalogues and factories that support interface trading and the cloning of new
  replicas, are common between OSA and XPA. Differences arise only where XPA
  contains extra mechanisms to ensure high performance and predictable timeliness
  properties.

Annex I

Timing Characteristics of Preemption Points

As explained in section §9.4.3.2, many "C" compilers permit the insertion of in-line assembler code at every preemption point. The preemption point then appears as follows, using assembler code for the Motorola MC68030:

```
ADDQ.L &1, Preempt_point_no; /* add 1 to preempt pt. no. */
TST.B Preempt_soon_flag; /* is flag set? */
BNE Preempt_code; /* if so, preempt thread */
```

In the normal case, where no preemption is imminent, the branch "BNE" will not be taken, and the current thread will continue. Assuming the MC68030 data and instruction caches are not enabled, and a 1/2 wait state is required on each access, the three instructions then take 26 clock cycles, or 1.3 microseconds on a 20 MHz CPU. If no wait states are required, the figures are reduced to 22 clock cycles or 1.1 microseconds.

A capsule with an average interval between preemption points of 100 microseconds therefore suffers only a 1.3% overhead executing the preemption points. The overhead would be lower on the hosts used in the XPA prototype, where caching is used.

Formal Presentation of AMp

In this annexe, we give a formal presentation of AMp and outline a correctness demonstration. The protocol is presented in figures 1 and 2; the state machines are locally instantiated for each group, to process relevant transmissions or receptions. The fundamental assumption presented in section §10.6 is recalled here:

- A1 — **In each group, there is at most one message transmission pending from each node, at any time.**

Protocol progress is straightforward in absence of faults. When faults occur, one may argue about the best way to tolerate them. Reliability of pure diffusion protocols is based on two fundamental conditions: synchronism and error masking. Thus, network delivery delay must be bounded — this is guaranteed by the network (Pn6, Pn7) — and there must be redundancy. The n-redundant broadcast channels of [Babaoglu, 1985, Streets of Byzantium] or the redundant point-to-point graphs of [Cristian et al. 1985] are an example of space redundancy. Given that standard, non-replicated networks are used in Delta-4, time redundancy must be used. In this case, further to bounding the number of faulty components, one should bound the extent (number or duration) of the system errors they produce. These assumptions were made in §10.3.2 and lead to properties Pn3 and Pn6.

The protocol is sketched out in figure 1. Three protocol machines are identified (for each participant): a sender machine, a recipient machine and a timer machine. Two classes of subsidiary functions are used which will be described first:

- basic transmission/reception functions;
- the group monitor function.

J.1. Basic Transmission/Reception Functions

Given that the anticipated channel omission error rate is low, instead of masking errors, by systematic $(k+1)$ transmissions (k being the allowed omission degree), which causes a fixed overhead, they are detected through an acknowledgment mechanism. $TxwResp(Gdest,Psour,data,id; nrResp,nrTries)$, is a transmission-with-response function, sketched in figure 2, used by an AMp sender identified by its individual physical address $Psour$, to transmit a frame containing information $data$ with an identifier id, to a group address $Gdest$ (line 52); after transmission, a number of responses $nrResp$ is awaited for, from group members (line 53); the waiting period is given by a timer, $timerWaitResponses$, hidden by the $waitResponsesPutIn$ function, which returns when either $nrResp$ responses arrive or $timerWaitResponses$ expires; responses are put in a bag $Resp$, and a number of tries $nrTries$ are made, until a full bag is obtained — the bag is emptied before each try — despite the occurrence of errors ($nrTries=k+1$) (line 51). The bag is returned to the invoker. A simpler,

```
/* Sender E execution: */
00 status(e) := Ok; i := 0; plase(e) := halted;
...

01 do phase(e) = halted ∧ status(e)= Ok ∧ send(e,m) → /*dissemination*/
02          phase(e) := dissemination;        i := i+1;
03          Resp(e) := TxwResp(e, E,m,i; n,k+1)
04              if result(Resp(e)) ≠ full → status(e) := probableRfailure fi
05          phase(e) := decision
06 ❑ phase(e) = decision Æ /*decision*/
07              if ∀ r ∈ Resp(e): r = Ok  → d := accept
08          q ∃ r ∈ Resp(e): r = notOk       → d := reject fi
09          Resp(e) := TxwResp(e,E,d,i; n,K+1)
10          if result(Resp(e)) ≠ full status(e) := probableRfailure fi
11          phase(e) := halted
12 ❑ phase(e) = halted ∧ status(e)= probableRfailure → /*monitor*/
13          status(e) := GroupMonitor(status(e);e;Resp(e))
14 od
/* Any Recipient R execution: */
20 status(E) := Ok; state(E) := null
...

21 do state(E) = null ∧ status(E)= Ok ∧ RxwResp(e,E,m,it) → /*reception*/
22          state(E) := i;   start(timerWaitDecision(E,i))
23          if canAccept(e,m) = true r(E,i) := Ok
24              ❑ canAccept(e,m) = false r(E,i) := notOk fi
25          ptpTx(E,R,r(E,i),it)
26 ❑ state(E) = i ∧ status(E) = Ok ∧ RxwResp (e,E,m,it) → /*repetition*/
27          start(timerWaitDecision(E,i));    discardAnypending(E,it-1);
28          ptpTx(E,R,r(E,i),it)
29 ❑ state(E) = i ∧ status(E)= Ok ∧ RxwResp(e,E,d,it) → /*decision*/
30          stop(timerWaitDecision(E,i));    ptpTx(E,Rack,it);
31              if d= accept → deliver m(E,i) to user(e)
32          ❑ d= reject → discardAnypending(E,it) fi
33          state(E) := null;      status(E) := Ok
34 ❑ state(E) = i ∧ status(E) = probableEfailure → /*monitor*/
35          state(E) := null; status E := GroupMonitor(status(E);e,E;i)
36 ❑ anyFrameReceived(E) ∧ status(E) ≠ Ok → /*discard*/
37          discardJustreceived
38 od
/* Timer Actions: */
40 do expired(timerWaitDecision(E,i)) → /*timeout*/
41 status(E) := probableEfailure od
```

Fig. 1 - Two-Phase Accept Protocol — Main Body

```
            Outline of TxwResp(Gdest,Psour,data,id;nrResp,nrTries)
50 tires := 0;         Resp := empty
51 do tries < nrTries ∧ Resp ≠ full →
52                Resp := empty;  Tx(Gdest,Psour,data,idtries);
53 waitResponsesPutIn(Resp,nrResp);   tries := tries +1    od
54 return Resp
```

Fig. 2 - Two-Phase Accept Protocol — Outline of TxwResp.

point-to-point, transmission function is the *ptpTx(Pdest, Psour, data, id)*, a datagram from sender *Psour* to a recipient *Pdest*, both individual addresses (e.g., line 25). *RxwResp*, at a recipient (e.g., line 21), is the indication to AMp of an incoming *TxwResp* reception. Lemma 1 illustrates the properties of *TxwResp*:

> **Lemma 1** — If *TxwResp(e,E,m,i;n,k+1)* is invoked by *E*, there is at least one try, where a reply to reception of frame *m,i* is provided by all those among the *n* intended recipients of logical destination *e*, which remain correct, if *E* and the network remain correct.

The proof, based on Pn3 (see section §10.5), is left to the reader.

J.2. Group Monitor

Given the clock-less structure of the protocol, its actions rely on a consistent *view* of each group, i.e., the stations where gates of the group are open. For this reason, all actions that are bound to modify this view are performed through a specialized *GroupMonitor* function. Normal joins and leaves are triggered by calls from the user interface; the sender may raise a suspicion about the state of recipients (*probableRfailure*); a recipient may raise a suspicion about a sender (*probableEfailure*) and, in this latter case, it may be necessary to terminate the execution started by a failed sender. There is no permanent monitor activity however — the GroupMonitor function is only invoked when needed.

The information required to perform a monitor action on a group is supplied by the recipient or sender machines of that group. Because the protocol is clock-less and based on acknowledgements, participant decisions must be established by agreement, rather than locally; the GroupMonitor acts thus by consensus, established atomically. The invoker of *GroupMonitor(reason; target; information)* starts by stating the reason, identifying the target of the action (a group or an individual station), and supplying the initial information, when required (e.g., identification of failed station, or pending message). Once activated, the action of the monitor is structured recursively in two-phased transmissions, like the normal atomic multicast. This solves the problem of monitor failure recovery, introduced by centralizing monitoring functions: if an active monitor fails, it is replaced by another GroupMonitor, invoked, in the same way as for a normal transmission, by a recipient that detects the failure. The cooperating process between local machines then starts with an investigation phase, where information about the local contexts of group members is gathered, and ends with a decision, disseminated to those members. The decision contains the new group view, after insertion of new members, or elimination of members leaving or having failed. The function then returns *Ok* to the invoker.

J.3. Execution in Absence of Permanent Faults

The sender, recipient and timer machines of the protocol are sketched in figure 1. Declarations and some obvious initializations are omitted, for space reasons. Group *e* is considered formed and has *n* members; the allowed omission degree is *k*. Delivery of message *m* to group *e* is requested (line 1) by a primitive *send(e,m)*, which triggers execution of the protocol at the node with individual address *E*. The (/*dissemination*/) action is thus started: message *m*, together with an appended sequence number *i*, is sent to *e*, by a *TxwResp* (line 3).

At the recipient, there is initially no context related with sender *E* (*state(E) = null*). Upon receiving the indication of reception (line 21), it sets *state(E)* to the pending message identification *i* (a timer is also started: this is discussed later in §J.4). The recipient then invokes the function *canAccept*, which verifies the accessibility of the recipient for the message (lines 23,24). The resulting response *r* is placed in a point-to-point frame addressed to *E* (line 25).

Note also that the pair E,i uniquely identifies all recipient variables concerned with a given protocol execution; i is indexed by the sender with the try number (figure 2, line 52), so that frames are individually related to a given try. In case of repetitions (line 26), the response process is the same, but this allows the recipients to suppress duplicates, by eliminating the older frames — *discardAnypending(E,i_{t-1})* gets rid of all frames of execution *(E,i)* with indexes up to and including *t-1* (line 27).

If no permanent failures occur, all responses are received by the sender. The *result* function, applied by the sender, on *Resp*, decides whether it is *full* (line 4). The sender then proceeds to the (/*decision*/) action, analysing the content of *Resp* (lines 7,8) and putting its decision d in a *TxwResp* frame that it sends to the group (line 9). The decision is *accept*, if all responded *Ok*. At this time, the atomic transmission is over for the sender (line 11). The recipients, upon reception of the decision (line 29), finally deliver the message to the user, or discard any frames related to it (lines 31,32), according to the decision sent. The reception is then considered ended (line 33).

Using transmission-with-response has the additional advantages of embedded permanent failure detection, discussed below.

J.4. Permanent Faults and Recovery

It was seen how to process temporary omission errors. It remains to be discussed how to deal with permanent recipient failures, and how to deal with sender failure, during protocol execution.

If the omission degree k is exceeded, there is no longer a temporary failure: if the bag *Resp* is not full after *TxwResp*, there is a probable permanent failure of one or more recipients. This is detected by the *result* function — bag not full (lines 4,10). AMp accomplishes the service for the remaining correct participants and after the protocol is ended (line 12), the (/*monitor*/) action is executed. The sender executes the (/*monitor*/) action invoking the GroupMonitor function, for reason *probableRfailure*, target e, information supplied: the non-empty bag. The GroupMonitor restores the coherence of the group view accordingly, and returns *Ok* (line 13).

Sender failure, on the other hand, is detected by recipients. Let us go back to the (/*reception*/) action, line 22: a *timerWaitDecision* is started for this E,i execution, restarted by (/*repetition*/) actions (line 27), and obviously stopped by the (/*decision*/) action (line 30). These timers, set to an expected maximum delay to get a decision, perform surveillance of the sender. If the sender fails, the timers expire (/*timeout*/, line 40), raising a suspicion about the sender's state, by asserting *state(E)* to *probableEfailure* (line 41), prompting for a (/*monitor*/) action that starts as soon as the previous action is over. The recipient terminates the reception and invokes the GroupMonitor (line 35), supplying the sender identity and the message reference. The GroupMonitor will restore the coherence of the group view, and terminate the pending transmission on behalf of the sender.

J.5. Discussion of Protocol Correctness

In this section, we outline the demonstration of how the two-phase accept protocol fulfils *atomic* AMp properties, taking into account that AMp validation is complemented by the methods discussed in chapter 15.

Unanimity: Given A1, $\exists$ 1 m pending from any E; given lemma 1 (L1) it is the same and is in every queue. If there has been a decision, sent by a *TxwResp*, then all have decided on the same message.

Non-triviality: Given Fs and L1, any *m,E,i* pending to be delivered in any recipient, is *m,i* sent by E.

Accessibility: If a participant is not correct, its gate is closed, so no message is delivered. We go by contradiction now. If a message is delivered by a recipient *R_j* to a participant, inaccessible for it, *R_j* has to have remained correct up to the end of that protocol execution. Given Fs, it must have responded *r_j = notOk* to the message *TxwResp*, in (/*reception*/) action. Given L1, *r_i ∈ Resp, ∀R_i*. Given Fs, the sender has to make a correct decision in (/*decision*/) action, which is reject, if there is any response *notOk* in *Resp*. Given L1, if *R_j* remained correct, it must have received that reject decision, sent by a *TxwResp*, and never delivered the message, which is a contradiction with our original assumption.

Delivery: We analyse the causes for a message not to be delivered, in the protocol. Observe that both in (/*dissemination*/) and (/*decision*/) actions of the sender, recipients may be considered permanently failed, if *result full*, however, the protocol proceeds. If the sender does not fail, and *canAccept accessible*, then a message is only undelivered in case of inaccessibility of at least one participant. If the sender fails during a transmission of *m*, the AMp entity that becomes group monitor acquires the state of *m* in all recipients, using *TxwResp*: if any participant accepted, then the decision to all (using *TxwResp*) is *accept*, otherwise, it is *reject*, regardless of whether all recipients are accessible.

Order: We just discuss order consistency, the interested reader may find the proof of causality in [Veríssimo et al. 1989]. Messages to be delivered, are delivered in the order they arrive from the abstract network, in the sense of Pn5. Given L1 and namely, *TxwResp* and the recipient (/*repetition*/) action in figure 1, transmissions where omission errors occur are discarded, so a message that remains in a queue, exists in all queues, and comes from a single transmission. Given this fact and Pn5, any messages from competing senders in the network have to exist in the same order in all participants' queues. Given Pa3, any two messages delivered anywhere, are delivered to all, and in consequence, in the same order.

Synchronism means existence of a known bound for protocol execution time. Synchronism is discussed section 10.1.3.

Group View: the *GroupMonitor (GM)* controls (knows about) every group reconfiguration; if it uses atomic transmissions (consistently ordered) to disseminate changes, then they will be ordered consistently with the remainder of the information flow.

Annex K

Models for the Representation of Reactive Systems

For the verification of a reactive system, a *model* representing all its possible *behaviours* is needed, on which the specifications are verified. Many such models have been proposed, e.g., transition systems [Plotkin 1981], refusal models [Hoare 1985], event structures [Winskel 1984], etc. We concentrate here on transition systems that can be considered as the basis for most of the proposed models. A transition system consists of a set of states and transitions relating them.

A suitable notation for transition systems is provided by process algebras, which are basic languages with operators such as "action", "choice", "parallel composition" or "abstraction of actions". Transition systems can be represented by terms of a process algebra.

To be verified, reactive systems must be described in some formalism for which a *formal semantics* is defined that allow its model to be obtained automatically, e.g., a transition system. Process algebras are the most primitive of these formalisms. Examples are CCS [Milner 1980], CSP [Hoare 1985], ACP [Bergstra and Klop 1985], ATP [Nicollin et al. 1990], Esterel [Berry and Cosserat 1985] and Lotos [ISO 8807]. In process algebras, semantics can be defined by an *equivalence relation* that may be given by a set of equalities on terms. This induces an equivalence on transition systems, e.g., as for Theoretical CSP [Brookes et al. 1984] or ATP.

Many other formalisms for the description of reactive systems have been proposed, for example programming languages like Estelle, or graphical formalisms such as Petri nets or Statecharts [Harel et al. 1987, Maraninchi 1990]. Statecharts are graphical formalisms allowing design by stepwise refinement; their formal semantics is defined in term of a process algebra.

The choice of one of the above formalisms is motivated by the underlying semantics. Between formalisms with comparable semantics, the choice depends on the commodity with which a given problem can be described and on the availability of appropriate verification tools. Below, we discuss mainly the different semantics that can be associated with the operator "parallel composition".

K.1. The Transition System Model

In order to associate transition systems with terms of a process algebra, we associate an operation on transition systems with any operator of the algebra. Here, we focus on the parallel operator as it is the most important one, and it can be interpreted in different ways.

We therefore consider reactive systems to be given by *sets of communicating processes*, where each process is modelled by a *transition system* [Plotkin 1981]. *States* represent the values of variables and *transitions* correspond to atomic steps (*actions*). These actions may be either *internal* actions (that may occur independently of the other processes) or *synchronizations* (e.g., *communications*) with other processes.

The transition system associated with the whole system of communicating processes (*global model*) is obtained as a *"product"* (corresponding to the parallel operator) of the transition systems of the individual processes. The state space of the global model is a subset of the Cartesian product of the state spaces of the individual processes. The set of global transitions may be obtained in different ways corresponding to different families of semantics of the parallel operator:

- In the case where each global transition corresponds to a product of atomic steps of each process (lockstep progress), we say that we have a *synchronous* semantics (this corresponds to the tensor product, and has been used for example to define the parallel operator of SCCS [Milner 1983]).

- In the case where the set of transitions is obtained by pure interleaving of all transitions of the individual processes, we have *pure interleaving* semantics. This corresponds to the Cartesian product.

- It is possible to have a mixed form, also called in the sequel interleaving semantics. In this model all the global transitions correspond either to an internal action of one individual process, or of the common execution of corresponding synchronization actions in two or more processes. If communication is carried out by means of synchronization, we have communication by *rendezvous*. Actions that are executed synchronously are defined by a *synchronization product* [Winskel 1980] on the set of actions.

Pure interleaving semantics is only interesting in two cases, either if no communication takes place between processes (but then the global model is not needed for the verification of the system) or if communication takes place by shared memory or unbounded buffers, since in this model no other possibility of communication exists. In the other cases, communication may take place either by shared memory or by rendezvous. Most process algebras have a parallel operator whose formal semantics corresponds to the third kind of product. This form is the most general one since it allows many variants to be defined depending on the synchronization product determining the actions that have to be executed synchronously. For example, pure interleaving semantics is obtained if no synchronization is allowed and synchronous semantics is obtained if all actions are obliged to synchronize.

Most equivalences defining the semantics of process algebras (and therefore inducing an equivalence on transition systems) are based on different criteria of *observability*. These equivalences generally distinguish two types of actions: *externally observable* ones and *unobservable* ones, which are considered as irrelevant for the observable behaviour. Different *equivalence relations* can be defined allowing abstraction from these unobservable actions.

Observational equivalence identifies all states with the same observable behaviour, where the structure of the model is taken into account. *Testing equivalence* identifies all processes that cannot be distinguished by a given "set of tests" (which may be infinite). This gives raise to a whole family of equivalences since the equivalence between two processes depends on the chosen set of tests. The coarsest equivalence that is usually defined is *trace equivalence* that distinguishes only transition systems with different traces of externally observable events. All these equivalences express different notions of observability.

Among the equivalence relations that are compatible with the soundness condition mentioned at the end of section   15.2.1.1, one should choose one that allows many abstractions and for which a minimization procedure (see section   K.3) with low complexity exists.

For example, the semantics of CCS is usually defined by *observational equivalence* [Milner 1980] and the semantics of Lotos can be defined by *testing equivalence* [ISO 8807].

K.2. Hypotheses on the Environment

An algorithm describing a reactive system is generally based on a set of hypotheses constraining the possible behaviours of an environment. These hypotheses can be taken into account in different ways:

- One possibility is not to take into account the environment for the construction of the model, but only for the verification of the specifications. For example, in [Josko 1987] the specifications are given as a pair (environment hypotheses, specifications) meaning that the specifications must be verified on all execution sequences satisfying the environment hypotheses. For this method, model checking is exponential in the size of the description of the environment.

- Another possibility is to describe a process representing the constraints of the environment and to construct the global model by adding this environment process to the global system. As the result of the verification should be valid for any environment satisfying the given environment hypotheses, the environment process is often highly non-deterministic, whereas the processes described by an algorithm are deterministic.

  For example, consider a communication protocol where stations exchange messages through a medium (= environment) that may lose messages from time to time. A solution is to describe the medium as a process that non-deterministically loses or transmits messages. A difficulty arises from the fact that transition systems are not expressive enough to take into account some usual hypotheses on the environment, such as "an arbitrary amount of messages may be lost, but not all of them". Therefore the resulting global behaviour cannot be represented by a transition system.

  This raises the problem of fairness, which is discussed in more detail in annexe L.

K.3. The State Explosion Problem

The modelling of a reactive system by a global model, as required by model checking techniques, poses the well-known problem of *state explosion* wherein the state space of the global model grows exponentially with the number of processes in the system. Thus, if we are not careful enough, the models may become too large to be processed by automated tools in reasonable time.

Deductive methods mentioned in section 15.2.1.1 do not require the generation of such a global model, but they pose another problem: the verification algorithms themselves are very time-consuming, or even worse, their termination cannot always be guaranteed.

Different methods have been proposed, trying to avoid the state explosion problem, but obviously none of them allows more than a heuristic gain in complexity without quite deep insight into the system under study. In all cases, good system design structure is of great importance for verification. Without discussing them much further, we just enumerate some of the more important results. Unfortunately, some of the solutions mentioned below cannot be adapted easily to existing tools, as for example the tools described in [Clarke et al. 1986, Cleaveland et al. 1989, Fernandez et al. 1985, Har'El and Kurshan 1990].

- One possibility to obtain a smaller global model is to represent it in the form of an event structure, a partial order representing causality between events and a symmetric relation representing possible conflicts between events. Unfortunately, due to the model checking algorithms for these models, based on the generation of a minimal representative set of execution sequences, the expected gain of time complexity is lost. Even the memory requirement, which is in most cases smaller, is in the worst case the same as for model checking on transition systems.

- As already discussed in section §K.1, semantics is often defined by an equivalence relation on models. The practical interest of such an equivalence relation is the existence of a minimization procedure allowing the construction of the smallest representative of each equivalence class. Whereas the size of the model obtained by composing directly all processes of a system grows exponentially with the number of processes, the size of its minimal representative grows in some examples in a linear way. The *Aldebaran* tool [Fernandez 1990] implements such minimization procedures, e.g., for observational equivalence.

  A more difficult task is to construct this smallest representative *directly* without first constructing the large model. To achieve this goal for observational equivalence, methods based on symbolic computation [Bouajjani et al. 1990] or methods based on some insight in the problem [Graf and Steffen 1990] have been developed.

- Attempts have been made to verify systems *compositionally*, that means, to deduce the specifications of a composed system from specifications of its components. Unfortunately, these methods are adequate only in the case of pure interleaving semantics, otherwise most interesting properties of systems cannot be derived compositionally in presence of communication without taking into account hypotheses on the environment of each process. Thus, compositional methods based on *guesses for the environment* of each component have been developed, e.g., in [Shurek and Grumberg 1990]. The correctness of these guesses must be proved in a separate step. Also, the method in [Josko 1987] is intended for compositional verification, but the complexity of its decision procedure lessens its interest.

- It has also been proposed to evaluate specifications *"on the fly"*. Instead of constructing first the global model and then checking the formulas on it, the idea is here to traverse the model in a single generation and checking phase without storing it completely. A method for the verification on the fly of properties expressed by automata is given in [Jard and Jeron 1989], and a method for the verification on the fly of bisimulation is given in [Fernandez and Mounier 1990]. All properties cannot be verified on the fly. The expected gain is to detect errors and sometimes even to terminate the check before the model is completely traversed. However, the main advantage of this method is that less memory is required; unfortunately, in many cases, not only the memory but also execution time is a problem (see section §L.1.1).

- A slightly different problem stems from the fact that an algorithm does not generally describe a single system, but a possibly infinite set of systems with different numbers of instantiations of some processes [Stadler and Grumberg 1989] (e.g., networks with an arbitrary number of nodes). Inductive methods have been proposed to verify such sets of systems. Unfortunately, these have either a very restricted field of application or require some *invariant* to be guessed (see, e.g., the methods given in [Stadler and Grumberg 1989, Wolper and Lovinfosse 1989]) and have therefore mainly been applied to rather simple examples.

We are thus often obliged to verify only systems with a small number of components, and then try to extend the result obtained to systems with any number of components by a different, unfortunately often informal, argumentation. The general problem is undecidable. Such verification is nevertheless important since it allows errors to be eliminated in a systematic way.

K.4. Introducing Time into the Models

Until now, *time* has not been taken into account explicitly in our models. Time may be introduced in transition systems in the following ways:

- For synchronous semantics of the parallel operator, an obvious solution is the following: each unit of time allows one transition to occur (examples may be found in [Berry and Cosserat 1985, Caspi et al. 1987]).

- For interleaving semantics, time must be introduced explicitly. A first possibility is to associate a duration with each action, e.g., as in timed Petri nets: the duration of a global transition is a function of all local transitions participating in it; the duration of a global transition corresponding to an internal transition of a single process is the duration of this local transition.

  In this case, time corresponds simply to a strictly growing global variable. This notion of time is obviously problematic since we want to construct a finite model. It should also be mentioned that the execution times obtained in this way correspond to the execution times of the sequentializations of a system. These execution times do not make sense in a real parallel execution.

- Another way to introduce time, in both synchronous and interleaving semantics, is to introduce a particular *"clocktick"* action meaning that "one unit of time has elapsed". If such an action is *executed simultaneously by all processes*, all processes are synchronized regarding the progress of time. Many formalisms have been proposed introducing this notion of time, e.g., [Groote 1990, Hennessy and Regan 1990, Nicollin et al. 1990, Quemada et al. 1989].

  Clearly, all actions different from "clocktick" have to be considered as instantaneous — all actions between ticks N and $N+1$ are considered to happen at time N. Therefore, infinite execution sequences without occurrence of "clocktick" must be avoided. In each process, it must be indicated explicitly where the action "clocktick" is possible. The progress of time is generally under the responsibility of the user or of the description language. This means that the progress of time must be stated in the service specifications, except if the description language allows only constructions enforcing progress of time (e.g., as in Esterel).

- It is also possible to introduce time in transition systems by changing the nature of the model. One considers a time domain, e.g., the domains of natural or real numbers; a model is the set of all possible functions from the time domain into an execution structure, as it has been done in [Alur et al. 1990].

In the first and the third cases, only discrete time can be modelled, whereas the second and the fourth cases also allow *dense* time to be expressed, e.g., as in [Baeten and Bergstra 1990] or [Alur et al. 1990]. The advantage of dense time is to allow exact execution times to be defined, whereas discrete time cannot be more precise than to express "some event occurs between clockticks n and $n+1$". However, at least in the domain of protocol verification, this is not really useful since it is never possible to give any execution time more exactly than in terms of some interval.

The equivalences defined on transition systems without explicit notions of time can easily be adapted to the above models with a notion of explicit time.

K.5. Modelling Time for Verification

The appropriate notion of time to be introduced in the model and the properties that can sensibly be specified at the design level depend on the type of the algorithm (according to the classification above) and on the complexity of the problem under study.

It is not always necessary to reflect all details of the algorithm in the model. It is sufficient to take into account those needed to decide the validity of the design specifications. In particular, it is generally not necessary to model the granularity of time used in the algorithm in order to decide whether the specification is satisfied. Therefore, the distinctions between external and internal time, e.g., introduced in [Kopetz and Schwabl 1989], are not relevant for design verification, except perhaps to verify a clock synchronization algorithm. This will become clear in the example given in the third point below.

We examine the three classes of systems introduced in section   15.2.1.2.

K.5.1. Asynchronous Systems

The introduction of time is useless, especially as the proposed algorithm is supposed to work under any timing constraints. Obviously, properties specifying time bounds make no sense in this context.

Thus, this kind of system is best verified by using interleaving semantics since it allows any ordering of independent events. This semantics introduces causality between events exclusively by explicit synchronization.

Unfortunately, as shown in [Fischer et al. 1985], this kind of algorithm is of little use in distributed systems, since there exists no algorithm of this type allowing safe communication between partners, even in presence of a single fault.

K.5.2. Systems with a Global Clock

In systems depending on a global clock, it is essential to include time in the design since the ordering of observable events may depend on relative execution times. Such systems may be modelled by using the synchronous semantics of the parallel composition, and by defining the transitions in the individual process such that they take exactly one unit of time. This corresponds to the first alternative mentioned in section K.4. This is only possible if the number of actions executed between two clockticks of each process can be determined statically.

A description in which an execution time multiple of some time unit is associated with each transition of an individual process (as in timed CSP [Davies and Schneider 1989]) can be transformed into this kind of synchronous model.

If the number of transitions between two clockticks cannot be determined statically, the third alternative for the modelling of time given in section K.4 can be used: a particular "clocktick" action is introduced that must occur simultaneously in all processes whereas all other actions interleave or communicate.

The same effect can be obtained by partitioning the set of actions into those which can only be started by a clocktick and those which can occur between clockticks.

In general, the assumption is made that the system is stable when a new clocktick occurs; therefore, it is possible to enable the event "clocktick" only when the system is stable, i.e., when no other action remains possible. The process algebra ATP given in [Nicollin et al. 1990] takes into account this feature.

K.5.3. Systems without a Global Clock

Synchronous systems without a global clock are the most difficult ones to model. The synchronous semantics is certainly not appropriate for this kind of system. In interleaving semantics, a notion of local clock cannot be used to synchronize processes as local clocks are supposed to be independent and therefore all possible interleavings of clockticks of local clocks would be possible in the global model. This has the same effect as not introducing time at all.

One possibility to model local clocks is by using a global clock, and by defining local clocks in terms of the global clock, as it has been proposed, for example, in [Lamport 1978]. This restricts the generality of the verification result since particular local clocks must be chosen to obtain a global model. The verification results can therefore only be guaranteed for this particular definition of the relationship between the local clocks.

However, in this kind of system, the notion of time effectively used by the algorithm is often rather weak. For example, in many communication protocols, time is only needed to decide when a message has to be considered as lost (a time-out expires). Furthermore, the correctness of the system should not be endangered by the fact that a time-out expires from time to time even if the message is not lost (early time-out).

Two modelling approaches are possible:

1) A timeless modelling approach as for asynchronous systems can be used: time-outs become simply events occurring non-deterministically. This allows verification of all time-independent safety properties. Liveness properties can be verified under fairness conditions on the time-outs and the environment.

   The correctness of such a model, with respect to its time independent properties, can be interpreted as follows: for any timing constraints and hypotheses on the environment, provided they fulfil the fairness conditions used for the verification of the liveness properties, the system works correctly.

   Consequently, the correctness result is very general and highly implementation independent. However, nothing about time bounds can be said.

2) If, for the same type of system, assertions about time bounds are part of the design specifications, all the assumptions under which these time bounds are supposed to be respected must be given. These assumptions concern essentially the environment and the local execution times.

   Now, we again need a model with a global clock. We probably have to choose the third solution given in section K.4, where all processes are synchronized on the "clocktick" and all other events interleave. Naturally, the granularity of the time scale needed depends on the granularity of the time scale defined by the assertions of time bounds and the assumptions about the environment made in the design. As described in section 15.2.2.3 this approach has been used for the verification of token-less AMp.

   In fact, it would be interesting to obtain execution time limits as function of parameters such as timer values but to the knowledge of the authors, no general methods yielding such results have been proposed.

As a conclusion, there are mainly two classes of techniques allowing time to be modelled explicitly in transition system models, that both use a notion of global time. This is independent of the fact that the modelled algorithm uses local or global clocks. One possibility is to use the synchronous product and make sure that each transition in each process takes exactly one unit of time. The other possibility is to introduce a new type of action, called "clocktick", that is executed simultaneously in all processes, whereas interleaving semantics is used for all other actions.

Annex L

Service Specifications

Service specifications define the expected *observable behaviour* of a system. There are essentially two alternatives for describing service specifications:

- The first is to describe the service in the same formalism as the algorithm, in some sense as a more abstract version of the same algorithm (this is typically the case in process algebras). In this case, formal verification means comparing models obtained from the different descriptions by using a partial order relation, e.g., the specification-implementation relation defined in [Walker 1988], or an equivalence relation, e.g., one of those mentioned in section K.1.

  In the case that a normal form exists for an equivalence relation, its minimization algorithm can be used for the comparison of models.

- The second alternative is to describe the service specifications by a set of properties given, for example, by formulas of a logic or by automata accepting infinite languages of actions. In this case, formal verification means: given a *satisfaction relation* between models and properties, check if a pair (model, property) belongs to this relation.

  The set of properties expressible in a formalism defines another equivalence relation on models: two models are equivalent if and only if they satisfy the same set of properties.

In both approaches the equivalence defining the underlying semantics on the model must be compatible with the equivalence induced by the specification formalism in the same sense as indicated by one of the soundness conditions mentioned in section 15.2.1.1.

We have already seen that there is a correspondence between properties and sets of models: a property can be identified with the set of models satisfying it. Therefore, logical operations on properties can be translated into operations on sets of models. Also a term of an algebra can be considered as a property since it defines a set of (equivalent) models. The relationship between process algebras and logics has been widely studied, e.g., in [Brookes and Rounds 1983, Graf and Sifakis 1986, Hennessy and Milner 1985]. In the sequel, we only treat the case where the service specifications are given by a set of properties.

L.1. The Nature of the Service Specifications to be Verified

The properties expressing the service of reactive systems can be divided following different criteria. The choice of the formalism used for the expression of service depends on the classes of properties to be expressed, and on the complexity of the verification algorithms associated with the formalism.

L.1.1. Safety and Liveness Properties

Two important classes of properties are *safety* and *liveness* properties [Lamport 1977]. In fact, it has been shown in [Alpern and Schneider 1987] and [Bouajjani et al. 1991] that any property (in most of the interesting specification formalisms) can be expressed as the intersection of a safety and a liveness property, where as above, a property is characterized by a set of models. This distinction between safety and liveness properties is very useful since much more efficient model checking algorithms exist for the first class of properties and much more powerful equivalence relations for reduction may be applied without losing soundness. For example, "safety equivalence" [Rodriguez 1988] allows us to eliminate all unobservable actions of a model.

Safety. Intuitively, safety properties express *invariants*, remaining always true. For example, the following properties are all safety properties:

- "Deadlock freedom."
- "Whenever a process p_2 receives a message, it has been sent previously by process p_1."
- "There is never more than one token in the ring."
- "The order of the received messages is the same as the order of the sent messages."

Safety properties can be formally defined as the sets of models which are prefix and limit closed with respect to some preorder, see [Abadi and Lamport 1988, Bouajjani et al. 1991, Manna and Pnueli 1989].

Intuitively, if one considers partial executions of a system (from the initial state), "prefix closed" means that if such a property is false in an execution, it will be false in any "larger" execution; the property cannot "become true" by extending the partial execution.

Intuitively "limit closed" means that such a property can be verified by considering only finite executions. Transition systems can only describe limit closed behaviours.

Even if they are often considered as the most important part of the specifications, safety properties alone are unable to ensure the correct operation of a system. For example, an idle system doing nothing useful at all, may satisfy all safety properties. Properties expressed by process algebras are safety properties.

Liveness. Liveness properties are needed in order to guarantee some progress in a system. Intuitively, liveness properties state that "eventually something good will happen" [Lamport 1977]. Typical liveness properties are for example:

- "The process will terminate."   (1)
- "A message that has been sent, will *eventually* reach its destination."   (2)

Formally, liveness properties can be defined as the sets of models whose limit closure is the set of all models, i.e., the dense sets for the same topology for which safety properties are the closed sets.

The example of section L.2.2 emphasises the difference between safety and liveness.

When describing a process by a finite transition system, only limit closed behaviours can be described, whereas neither liveness properties nor the hypotheses on the environment are in general of this nature. The evaluation of liveness properties on finite transition systems poses the problem of *fairness*. This problem has been widely studied, e.g., in [Francez 1986, Lehmann et al. 1981, Park 1980, Queille and Sifakis 1983].

Take the example of property (2) to be verified on the buffer losing messages defined in section K.2 with the following hypothesis on the environment:

- "The buffer may lose an arbitrary amount of messages, but it must however deliver a message from time to time." (3)

The modelling we have proposed in section K.2, where arbitrary loss of messages is described by a non-deterministic choice, allows an execution sequence in which the buffer loses all messages. That means that the obtained global model does not satisfy the requirement described by property (3). Notice that this requirement does not describe a limit closed set of sequences, and cannot therefore be defined by a transition system.

This modelling of a buffer does not pose any problem for the verification of safety properties since safety properties cannot distinguish between a model in which the sequences with infinite loss are allowed and one in which they are not.

One possible solution for this problem is to use this inexact model nevertheless, and to verify, instead of the property (2), the following property containing a *fairness condition*:

- "Under the condition that the communication channel does not lose messages forever from some moment on, a message that has been sent will eventually be received."

The verification algorithms for logic formulas with fairness conditions take into account only the fair executions of the model.

Another way to deal with this problem is to use more sophisticated models, describing only the fair execution sequences. This can be done as in [Har'El and Kurshan 1990] where processes and models are represented by infinite automata, instead of transition systems. These automata define particular acceptance conditions for infinite sequences and are discussed in section L.2.

L.1.2. Branching versus Linear Properties

Another point of view is to divide properties into those of "linear" and "branching" type. The first sort depend only on the execution sequences of a model, e.g.

- "In all execution sequences an action a will inevitably be followed by an action b." (4)

The second sort of properties depend also on the structure of the model, e.g.

- "From any state of any execution sequence it is possible to reach a state in which an action b has been executed." (5)

Property (4) is of linear type, for its satisfaction depends only on the set of sequences of a transition system. Property (5) is of branching type: it may be satisfied in some transition system, but not in another with the same set of sequences. For example in figure 1, the two systems $S1$ and $S2$ have the same set of execution sequences: Property (4) is false on both systems as the execution sequence consisting only of infinitely many a-actions, denoted a^ω, does not satisfy it. Whereas property (5) is true for system $S1$ and false for system $S2$ since in $S2$ there exists a state from which it is impossible to reach a state allowing execution of an a-action.

Most service specifications are of linear type. However, as we will see later, for some branching time temporal logics, which allow most of the important linear properties to be expressed, there exists a linear verification algorithm whereas general linear liveness properties require an exponential verification algorithm.

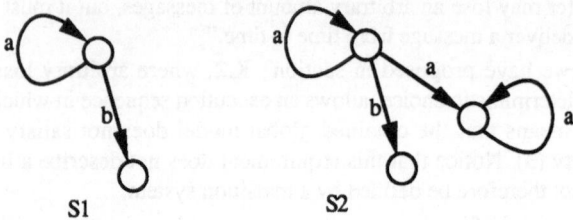

Fig. 1 - Illustration of Branching versus Linear Properties

L.1.3. Properties expressing Time Bounds

Until now, we have not mentioned properties concerning time explicitly. All the above mentioned examples of properties are only concerned with *ordering* of events. The following are typical examples of properties expressing time requirements:

- "The motor must be turned off after at most 3 ms after some anomaly has occurred."
- "If a token is lost, it must be regenerated within at most 3 units of time."
- "The light is switched on, exactly 5 seconds after the bell rings."
- "The time between the emission of a message and its reception must be in at least 90% of the cases less than 5ms, and in at most 1% of the cases more than 10ms."

The first two properties are of the same nature, only the time units are defined in a different way. The third property is probably never true in any real system as we have already mentioned at the end of section K.4.

The last property expresses in fact a requirement in the domain of performance evaluation and not of design specification, and therefore we do not consider this kind of properties here.

Notice also that in a model in which the progress of time is guaranteed, the first two properties mentioned above are safety properties. It is easy to see that in the case where we specify time bounds for all *eventualities*, they become safety properties and the fairness conditions are no longer needed.

Time must also be introduced in the model and the environment hypotheses must be refined. In order to verify stronger specifications, a more detailed model and stronger hypotheses on the environment are needed.

As an example, consider again the channel losing messages introduced in section K.2. To verify that a message reaches its destination in at most n time units, the hypothesis that the medium transmits a message "from time to time" is no longer sufficient, but a concrete assumption about the number of messages it is allowed to lose is necessary.

The detail in which time must be specified depends to a great extent on the type of application under study.

L.2. Expression of Service Specifications

Many formalisms have been proposed for the description of the service specifications of reactive systems. We mention here the three main families of formalisms that are generally used for the expression of the service specifications of reactive systems in the context of model checking.

L.2.1. Formalisms

As already mentioned, all process algebras are specification formalisms. In this case, terms defining the service specifications of a system generally have no occurrences of the parallel and the abstraction operator. Therefore, these terms can be interpreted as a transition system or an automaton accepting finite or limit closed sequences. Thus, terms of process algebras are suitable for the expression of safety properties.

Temporal logics [Pnueli 1977] are often used for the description of properties. They are extensions of the propositional calculus by addition of so-called *modalities*, allowing expression of, for example:

- "it is *inevitable* to reach a state in which predicate *p* holds"   or
- "predicate *p* is *always* true".

Like the properties themselves, temporal logics can be divided in two classes, *linear time* temporal logics, expressing only linear properties and *branching time* temporal logics, expressing branching properties. However, the formalisms of the second class also allow most of the interesting linear properties to be expressed. LTL [Pnueli 1977] is a representative of the first class, and CTL [Clarke et al. 1986] of the second. CTL* [Emerson and Halpern 1986] contains the formulas of both CTL and LTL. The μ-calculus [Kozen 1982], which is a very powerful formalism and has the form of a branching time logic, is more expressive than CTL* and allows one to describe all ω-regular languages[1] and sets of trees[2].

All temporal logics allow one to express safety as well as liveness properties.

Automata are also an interesting formalism for the expression of properties of sequences or trees. The expressiveness of automata recognizing infinite languages can be compared with linear time logics, and that of automata recognizing infinite trees with branching time logics. For example, Büchi automata [Büchi 1962], which recognise all infinite sequences having infinitely many occurrences of so called "repetitive" states, allow one to describe any ω-regular language.

L.2.2. Example

In order to give an idea of the nature of these different formalisms, consider the property:

- "In all execution sequences, after the occurrence of an action *a* it is inevitable to have an action *b*, and without having any occurrence of an action *a* in between." (6)

This property of linear type contains a liveness part that can be expressed in LTL by:

$$AG\,(\,\text{'after}(a)\text{'} \Rightarrow (\neg\text{'after}(a)\text{'}\ U\ \text{'after}(b)\text{')}\,)$$

where "after(*a*)" is a predicate defined on the states of the model, meaning that this state has been reached by executing an *a*-action. *A*, *G* and *U* are modal operators (taken from the syntax of CTL*) meaning respectively:

- "the argument is true in all sequences starting in the current state",
- "the argument is true in all subsequences of the considered sequence",
- "in the considered sequence, the first argument is true until the second becomes true, and the second argument becomes eventually true".

---

[1] ω-regular languages are extensions of regular languages to infinite sequences.

[2] Trees are often used to represent behaviours. We consider finitely branching infinite trees, with arcs labelled by actions. They can be obtained by unfolding transition systems.

L.2.1. Formalisms

As already mentioned, all process algebras are specification formalisms. In this case, terms defining the service specifications of a system generally have no occurrences of the parallel and the abstraction operator. Therefore, these terms can be interpreted as a transition system or an automaton accepting finite or limit closed sequences. Thus, terms of process algebras are suitable for the expression of safety properties.

Temporal logics [Pnueli 1977] are often used for the description of properties. They are extensions of the propositional calculus by addition of so-called *modalities*, allowing expression of, for example:

- "it is *inevitable* to reach a state in which predicate p holds"   or

- "predicate p is *always* true"

Like the properties themselves, temporal logics can be divided in two classes, *linear time* temporal logics, expressing only linear properties and *branching time* temporal logics, expressing branching properties. However, the formalisms of the second class also allow most of the interesting linear properties to be expressed. LTL [Pnueli 1977] is a representative of the first class, and CTL [Clarke et al. 1986] of the second. CTL* [Emerson and Halpern 1986] contains the formulas of both CTL and LTL. The μ-calculus [Kozen 1982], which is a very powerful formalism and has the form of a branching time logic, is more expressive than CTL* and allows one to describe all ω-regular languages[1] and sets of trees[2].

All temporal logics allow one to express safety as well as liveness properties.

Automata are also an interesting formalism for the expression of properties of sequences or trees. The expressiveness of automata recognizing infinite languages can be compared with linear time logics, and that of automata recognizing infinite trees with branching time logics. For example, Büchi automata [Büchi 1962], which recognise all infinite sequences having infinitely many occurrences of so called "repetitive" states, allow one to describe any ω-regular language.

L.2.2. Example

In order to give an idea of the nature of these different formalisms, consider the property:

- "In all execution sequences, after the occurrence of an action a it is inevitable to have an action b, and without having any occurrence of an action a in between."   (6)

This property of linear type contains a liveness part that can be expressed in LTL by:

$$AG \, (\, \text{'after}(a)\text{'} \Rightarrow (\neg\text{'after}(a)\text{'} \; U \; \text{'after}(b)\text{'}) \,)$$

where "after(a)" is a predicate defined on the states of the model, meaning that this state has been reached by executing an a-action. A, G and U are modal operators (taken from the syntax of CTL*) meaning respectively:

- "the argument is true in all sequences starting in the current state",

- "the argument is true in all subsequences of the considered sequence",

- "in the considered sequence, the first argument is true until the second becomes true, and the second argument becomes eventually true".

---

[1] ω-regular languages are extensions of regular languages to infinite sequences.

[2] Trees are often used to represent behaviours. We consider finitely branching infinite trees, with arcs labelled by actions. They can be obtained by unfolding transition systems.

Formulas of LTL have exactly one occurrence of A as the outermost operator, since they express

- "all sequences starting in the initial state have some linear property".

Property (6) can be expressed in CTL (using the same syntax) as:

$$AG\,(\,\text{'after}(a)\text{'} \Rightarrow A(\neg\text{'after}(a)\text{'}\ U\ \text{'after}(b)\text{'})\,)$$

Since in CTL all formulas are evaluated on states, only combinations of operators such as AG or $A(...U...)$ are allowed, expressing respectively:

- "the argument is true in all states reachable from the current state",
- "in all execution sequences starting from the current state, as long as the first argument is true in all states, the second is also true".

A deterministic Büchi automaton expressing the property is shown in figure 2 where a square state is a "repetitive" one. An infinite sequence is accepted only if it contains infinitely many repetitive states.

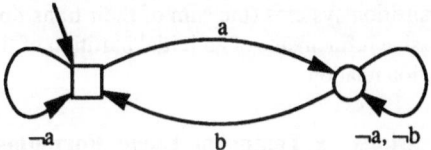

Fig. 2 - Büchi Automaton illustrating Property (6)

Since this property contains a liveness part, it cannot be expressed by a term of a process algebra.

Consider now a slightly different property that is a safety property:

- "In all execution sequences, after any occurrence of an action a, a new occurrence of action a is only possible after an occurrence of action b."   (7)

This property can be expressed in LTL by:

$$AG\,(\,\text{'after}(a)\text{'} \Rightarrow (\neg\text{'after}(a)\text{'}\ WU\ \text{'after}(b)\text{'})\,)$$

where WU is the "weak until" operator, $f\,WU\,g$ means that

- "either f is true until g is true (as for $f\,U\,g$), or f remains true forever".

In CTL, property (7) can be expressed by:

$$AG\,(\,\text{'after}(a)\text{'} \Rightarrow A(\neg\text{'after}(a)\text{'}\ WU\ \text{'after}(b)\text{'})\,)$$

A deterministic Büchi automaton expressing this property is shown in figure 3.

Notice that in this automaton, both states are repetitive. The difference between this Büchi automaton and the previous one is that the second automaton accepts the sequence a^*c^ω, whereas the first automaton rejects it.

This figure can also be viewed as a transition system defining the same property.

L.3. Model Checking Algorithms and Complexity

For all the above-mentioned specification formalisms, model checking algorithms exist allowing one to evaluate their satisfaction on a transition system. We give here the principle of some of these algorithms and their complexities. In all algorithms below, it is supposed that the representation of the model under study can be completely stored in memory, and all the complexity measures are based on the hypothesis that its states and transitions can be accessed in constant time.

In all cases, it is possible to produce the sequences leading to error states.

L.3.1. Service Specifications as Terms of Process Algebra

Model checking means to compare two transition systems (the model and the one obtained from the term representing the service specifications) by means of an equivalence or a preorder relation.

For example, for comparison by observational equivalence, an algorithm exists which is cubic in the size of both transition systems (the sum of their transitions plus the sum of their states). It is based on successive refinements of an initial partition of the states until it becomes "compatible" with the transition relation.

L.3.2. Service Specifications as Temporal Logic Formulas

Models for temporal logics are Kripke structures. A Kripke structure is obtained from a transition system by associating predicates with states.

There exists for CTL an algorithm with a complexity linear in the product of the size of the model and the formula.

Unfortunately, the complexity of model checking for LTL is linear in the size of the model, but exponential in the size of the formula. The same is true for CTL* even if it is more expressive than both CTL and LTL.

This is the reason why it may be advantageous to use CTL for the expression of the specifications even if they can be expressed (sometimes more easily) in LTL.

L.3.3. Service Specifications as Büchi Automata

The evaluation of a (linear) property on a model reduces to testing whether the product automaton of the model and the automaton recognizing the negation of the given property is empty or not (this is known as the *emptiness problem* for an automaton). For Büchi automata, the emptiness problem is of linear complexity. The size of this automaton may be exponential in the size of the LTL formula expressing the same property; thus the complexity of evaluation of LTL formulas cannot be improved in this way.

If the property is initially given in the form of a deterministic Büchi automaton, its complementation is linear in time and size and the whole evaluation remains of linear complexity.

In the special case of deterministic Büchi automata, in which all states are repetitive (i.e., recognizing a safety property), there exists a different algorithm. It is still linear in the size of the model and of the automaton but the complementation is not necessary. It is based on the traversal of the product of the model with the automaton. An error is detected whenever a transition is possible in the model but not in the automaton.

These methods have a lower space complexity than the ones for temporal logics since it is not necessary to store the whole product automaton, but only the states that have already been reached (this is an important difference). Thus, the model need not to be built first, but is traversed "on the fly". If all states cannot be stored the time complexity is increased.

Annex M

Co-existence with non-Delta-4 Stations on the same LAN

The Delta-4 communications systems, either the Multipoint Communication System for OSA, or the XPA communications sub-system, use standard communication support in the lowest layers of the architecture. Physical layers are fully compatible with standards (ISO 8802.4 for Token Bus, ISO 8802.5 for Token Ring, ISO 9314 for FDDI). The software-implemented Atomic Multicast Protocol relies on an abstraction of services provided by commercially available VLSI circuits implementing standard MACs; at this level, the Delta-4 logical addressing is mapped onto functional addressing that is offered by all these standards. Even the "Turbo AMp" for Token Ring implements additional services to the Token Ring MAC in a way that allows co-existence on the same medium with other nodes using ISO 8802.5 in a fully conforming way.

There is therefore no apparent problem that would prevent co-existence on the same LANs of Delta-4 systems and non-Delta-4 systems, with the assumption that if these non-Delta-4 systems also use functional addressing at the MAC level, the addresses do not conflict with the ones selected by Delta-4.

It is thus conceivable that Delta-4 applications and non-Delta-4 applications co-exist on the same nodes interconnected by the same standard LANs, using either the Delta-4 communication stack or the standard ISO stack (or other stack). The communication stacks used by Delta-4 and non Delta-4 applications can be implemented on different NACs or even on the same NACs if sufficient resources are available on these NACs.

However, it is necessary to consider what would be the effects of this co-existence on the properties of the Delta-4 system. It is obvious that the co-existence on a same network of a Delta-4 XPA system and a non Delta-4 system cannot be envisaged, since the performance and real-time aspects inherent to XPA could be severely compromised by such a co-existence. However, it is necessary to raise a word of caution regarding dependability properties of a local area network in which there are both fail-silent network attachment controllers (i.e., Delta-4 NACs) and non-fail-silent network attachment controllers (e.g., off-the-shelf non-Delta-4 NACs). The following is a tentative of formalizing these effects:

Consider a single LAN that interconnects m Delta-4 stations and n foreign (i.e., non-Delta-4) stations and let:

- R be the reliability of a network attachment controller (assumed for the purposes of this demonstration to be the same for both Delta-4 and foreign NACs)
- p_d be the self-check coverage of a Delta-4 NAC
- p_f be the self-check coverage of a foreign NAC

then, assuming the (optimistic) case where failures of the physical transmission links can be ignored, the reliability (R_p) of the path between a given pair of Delta-4 hosts is given by:

$$R_p = R^2 . (R + p_d.(1-R))^{m-2} . (R + p_f.(1-R))^2$$

since:

- R^2 is the reliability of the pair of NACs that are in the direct path between the considered pair of hosts,
- $R+p.(1-R)$ is the probability for another NAC of coverage p (=p_d or p_f) on the LAN to be either correctly operational or failed, but correctly eliminated from the network.

The *unreliability* of the path between the two stations can thus be approximated by (ignoring all terms in (1-R) of order greater than or equal to two):

$$Q_p = 1 - R_p \approx (1-R).(2 + (m-2).(1-p_d) +n.(1-p_f)) \qquad (1)$$

Now, since Delta-4 NACs are equipped with self-checking mechanisms to support the fail-silence assumption, i.e., $p_d \approx 1$, the second term in (1) can be neglected, whence:

$$Q_p \approx 2.(1-R) + n.(1-p_f).(1-R) \qquad (2)$$

The first term in (2) above is unavoidable since it is due to the pair of NACs of the considered pair of stations. The second term is due to the lack of coverage of the foreign NACs and is linearly proportional to n. For large values of n and moderately-low values of p_f, this term could dominate and be very detrimental to path reliability. More precisely, the second term is only negligible if:

$$n \ll 2/(1-p_f)$$

or, equivalently:

$$p_f \gg 1-2/n \qquad (3)$$

Moreover, when calculating the unreliability of a fault-tolerant Delta-4 application, i.e., a *redundant* application, the effect of the first term would become second-order or higher whereas the second term would always remain first-order in (1-R) (a demonstration of this is left as an exercise to the reader).

By way of examples, table 1 gives the constraint on p_f obtained from expression 3 for different numbers of foreign hosts.

Table 1 - Examples of Constraint on Foreign NAC Coverage

n	$p_f \gg$
100	0.98
20	0.9
10	0.8

Glossary of Abbreviations

ACL ....Access Control List

ACSE ....Association Control Service Element

AE ....Application Entity

AMp..............Atomic Multicast protocol

ANSA............Advanced Networked Systems Architecture

ANSI.............American National Standards Institute

ASN.1 ....Abstract Syntax Notation One

AT ....Administrative Task

CDM ....Communication Domain Manager

CIDL.............."C" Interface Definition Language

CIM ....Client (Service User) Interface Module

CIME.............Computer Integrated Manufacturing and Engineering

CIMG............"C" Interface Module Generator

clabel ....Causal Ordering Label

CLC..............Collapsed-Layered Communication stack

CLCS ....Collapsed-Layered Communication System

CMIP.............Common Management Information Protocol

CMISE...........Common Management Information Service Element

CNMA ....Communications Network for Manufacturing Applications (ESPRIT Project 2617)

CRC..............Cyclic Redundancy Checksum

DCS ....Dependable Communication System

DDBM ....Distributed Data Base Management

Delta-4 ....Definition and Design of an open Dependable Distributed systems architecture (ESPRIT Project 2252)

Deltase ....Delta-4 Application Support Environment

Deltase/GEN .....Deltase Generation Support

Deltase/XEQ......Deltase Execution Support

DM ....Domain Manager

DMA ....Direct Memory Access

DMI ....Domain Manager Interface

DMI-AM..........Domain Manager Interface — Accounting Management

DMI-CM..........Domain Manager Interface — Configuration Management

DMI-FM..........Domain Manager Interface — Fault Management

DMI-PM..........Domain Manager Interface — Performance Management

DMI-SM..........Domain Manager Interface — Security Management

EC................Experiment Controller

ECMA............European Computer Manufacturers' Association

EMA..............Enterprise Management Architecture

ES ....Experiment Supervisor

FDDI ....Fiber Distributed Data Interface

FM................File Manager

FM-AE...........File Manager Application Entity

FM-DM..........File Manager Domain Manager

FM-OME ....File Manager Object Manager Entity

FSM..............Finite State Machine

FTAM............File Transfer and Access Method

FTM..............Fault-Tolerance Mechanism

GM ....Group Manager

IDL...............Interface Definition Language

IIP................Integrated Information Processing

IPC...............Inter-Process Communication

IRp ....Inter-Replica protocol

ISA ....Integrated Systems Architecture for ODP (ESPRIT Project 2267)

ISO...............International Standards Organization

LAN..............Local Area Network

LEX..............Local Execution Environment

LLC ....Logical Link Control (sub-layer)

LME..............Layer Management Entity

LMI..............Layer Management Interface

LPC ....Local Procedure Call

MAC ....Medium Access Control (sub-layer)

MAP..............Manufacturing Automation Protocol

MCM.............Multipoint Communication Management

M-CMIP..........Multipoint Common Management Information Protocol

M-CMISE........Multipoint Common Management Information Service Element

MCP..............Multipoint Communication Protocol stack

MCS..............Multipoint Communication System

MDL..............Management Description Language

MDN.............Multicast Data Number

MGS ....Multicast Group of Stations

MIB ....Management Information Base

MIB-M...........Management Information Base — Manager

MMS.............Manufacturing Message Specification

M-SAP...........Multicast Service Access Point

NAC..............Network Attachment Controller

NI.................Network Interface

NTCB............Network Trusted Computing Base

ODL..............Object Description Language

ODP..............Open Distributed Processing

OME..............Object Manager Entity

OMI ....Object Manager Interface

OSA ....Open System Architecture

OSI................Open Systems Interconnection

PDCS ....Predictably Dependable Computing Systems (ESPRIT Project 3092)

PDU..............Protocol Data Unit

PHY..............Physical (layer)

PIN...............Personal Identification Number

PLC ....Programmable Logic Controller

QOS ....Quality of Service

RDM ....Replication Domain Manager

RM-ODP.........Reference Model for Open Distributed Processing

RNG ....Random Number Generator

RPC..............Remote Procedure Call

RSR..............Remote Service Request

RSW ....Remote Service Wait

RT-UNIX ....Real-Time UNIX

RTM...............Real-Time Monitor

SA.................System Administration

SAP...............Service Access Point

SE-ODP..........Support Environment for Open Distributed processing

SIM...............Server (Service Provider) Interface Module

SMAP............Systems Management Application Process

SQL...............Structured Query Language

TB.................Token Bus

TCB...............Trusted Computing Base

TIU...............Trusted Interface Unit

TR.................Token Ring

TSH...............Target System Host

TSM...............Test Sequence Manager

WAN..............Wide Area Network

xAMp............Extended Atomic Multicast protocol

XPA...............Extra Performance Architecture

References

Abadi, M. and Lamport, L. (1988) *The Existence of Refinement Mappings*, DEC Research Center, Technical Report N°SRC-29.

Adams, N. (1984), "Optimizing Preventive Service of Software Products", *IBM Journal of Research and Development*, 28 (1), pp. 2-14.

Adrion, W.R., Branstad, M.A. and Cherniavsky, J.C. (1982), "Validation, Verification and Testing of Computer Software", *Computing Surveys*, 14 (2), pp. 159-192.

Aguera, M., Arlat, J., Crouzet, Y., Fabre, J.-C., Martins, E. and Powell, D. (1989) *Results of Fault-injection into an MCS Network Attachment Controller with Limited Self-Checking*, LAAS-CNRS, Report N°89.071 (Delta-4 Document N°R89.022/I2/P) (available from LAAS-CNRS, 7 Ave. Colonel Roche, 31077 Toulouse, France).

Almes, G.T., Black, A.P., Lazowska, E.D. and Noe, J.D. (1985), "The Eden System: a Technical Review", *IEEE Transactions on Software Engineering*, SE-11 (1), pp. 43-59.

Alpern, B. and Schneider, F.B. (1987), "Recognizing Safety and Liveness", *Distributed Computing*, 2, pp. 117–126.

Alur, R. and Henzinger, T.A. (1989) "A Really Temporal Logic", in *Proc. 30th. Symposium on Foundations of Computer Science (FOCS 89)*, pp. 164-169 (IEEE).

Alur, R., Courcoubetis, C. and Dill, D. (1990) "Model-Checking for Real-time Systems", in *Proc. 5th. Symposium on Logic in Computer Science (LICS 90)*, pp. 414-425 (IEEE).

Anderson, D.A. and Metze, G. (1972) "Design of Totally Self-Checking Check Circuits for M-out-of-N Codes", in *Proc. 2nd. Int. Symp. on Fault Tolerant Computing (FTCS-2)*, Newton, MA, USA, pp. 30-35 (IEEE).

Anderson, T. (1986) "A Structured Decision Mechanism for Diverse Software", in *Proc. 5th. Symp. on Reliability in Distributed Software and Data Base Systems*, Los Angeles, CA, USA, pp. 125-129.

Anderson, T.A. and Lee, P.A. (1981), *Fault Tolerance — Principles and Practice*, (Prentice-Hall).

ANSA (1987) *The ANSA Reference Manual*, Advanced Network System Architecture (Release 00.03) (available from Architecture Project Management Limited, 24 Hills Road, Cambridge CB2 1JP, UK).

ANSA (1989) *The ANSA Reference Manual*, Advanced Network System Architecture (Release 01.00) (available from Architecture Project Management Limited, 24 Hills Road, Cambridge CB2 1JP, UK).

Arlat, J., Aguera, M., Crouzet, Y., Fabre, J., Martins, E. and Powell, D. (1990), "Experimental Evaluation of the Fault Tolerance of an Atomic Multicast Protocol", *IEEE Transactions on Reliability*, 39 (4), pp. 455-467 (Special Issue on Experimental Evaluation of Computer Reliability).

Arlat, J., Aguera, M., Crouzet, Y., Fabre, J.-C., Martins, E. and Powell, D. (1989) *Dependability Testing Report LA1*, LAAS-CNRS, Report N°89.410 (Delta-4 Document N° R89.139/I1/P) (available from LAAS-CNRS, 7 Ave. Colonel Roche, 31077 Toulouse, France).

Arlat, J., Aguerra, M., Amat, L., Crouzet, Y., Fabre, J.-C., Laprie, J.-C., Martins, E. and Powell, D. (1990), "Fault-Injection for Dependability Validation — A Methodology and Some Applications", *IEEE Transactions on Software Engineering*, 16, pp. 166-182.

Arlat, J., Crouzet, Y. and Laprie, J.-C. (1989) "Fault Injection for Dependability Validation of Fault-Tolerant Computing Systems", in *Proc. 19th. Int. Symp. on Fault Tolerant Computing (FTCS-19)*, Chicago, MI, USA, pp. 348-355 (IEEE).

Arlat, J., Crouzet, Y., Martins, E. and Powell, D. (1991) *Dependability Testing Report LA2 - Fault-Injection on the Fail-Silent NAC: Preliminary Results*, LAAS-CNRS, Report N°91.043 (Delta-4 Document N° R90.206/I1/P) (available from LAAS-CNRS, 7 Ave. Colonel Roche, 31077 Toulouse, France).

Arlat, J., Kanoun, K. and Laprie, J.-C. (1988) "Dependability Evaluation of Software Fault-Tolerance", in *18th. Int. Symp. on Fault Tolerant Computing (FTCS-18)*, Tokyo, Japan, pp. 142-147 (IEEE).

Arnold, T.F. (1973), "The Concept of Coverage and its Effect on the Reliability Model of Repairable Systems", *IEEE Transactions on Computers*, C-22 (3), pp. 251-254.

Avizienis, A. (1975), "Fault-Tolerance and Fault-Intolerance: Complementary Approaches to Reliable Computing", *ACM SIGPLAN Notices*, 10 (6), pp. 458-464.

Avizienis, A. (1978), "Fault Tolerance, the Survival Attribute of Digital Systems", *Proceedings of the IEEE*, 66 (10), pp. 1109-1125.

Avizienis, A. and Kelly, J.P.J. (1984), "Fault-Tolerance by Design Diversity: Concepts and Experiments", *Computer*, 17 (8), pp. 67-80.

Avizienis, A. and Laprie, J.-C. (1986), "Dependable Computing: from Concepts to Design Diversity", *Proceedings of the IEEE*, 74 (5), pp. 629-638.

Avizienis, A., Gunningberg, P., Kelly, J.P.J., Strigini, L., Traverse, P.J., Tso, K.S. and Voges, U. (1985) "The UCLA DeDiX (Design Diversity Experiment) SYSTEM: A Distributed Testbed for Multiple-version Software", in *Proc. 15th. Int. Symp. on Fault-Tolerant Computing (FTCS-15)*, Ann Arbor, MI, USA, pp. 126-134 (IEEE).

Babaoglu, O. and Drummond, R. (1985), "Streets of Byzantium: Network Architectures for Fast Reliable Broadcasts", *IEEE Transactions on Software Engineering*, SE-11 (6), pp. 546-554.

Babaoglu, O. and Drummond, R. (1987) "(Almost) No Cost Clock Synchronization", in *Proc. 17th. Int. Symp. on Fault-Tolerant Computing (FTCS-17)*, Pittsburgh, PA, USA, pp. 42-47 (IEEE).

Baeten, J.C.M. and Bergstra, J.A. (1990) *Real Time Process Algebra*, University of Amsterdam.

Banâtre, J.P., Banâtre, M. and Muller, G. (1988) "Main Aspects of the GOTHIC Distributed System", in *European Teleinformatics Conf. (EUTECO'88): Research into Networks and Distributed Applications*, Vienna, Austria (R. Speth, Eds.), pp. 747-760 (Elsevier Science Publishers B.V. (North-Holland)).

Banâtre, J.P., Banâtre, M., Lapalme, G. and Ployette, F. (1986), "The Design and Building of ENCHERE, a Distributed Electronic Marketing System", *Communications of the ACM*, 29 (1), pp. 19-29.

Baptista, M., Graf, S., Richier, J.-L., Rodrigues, L., Rodriguez, C., Veríssimo, P. and Voiron, J. (1990) "Formal Specification and Verification of a Network Independent Atomic Multicast Protocol", in *3rd. Int. Conf. on Formal Description Techniques (FORTE'90)* (J. Quemada, J. Mañas and E. Vazquez, Eds.) (North-Holland).

Barrett, P.A., Hilborne, A.M., Bond, P.G., Seaton, D.T., Veríssimo, P., Rodrigues, L. and Speirs, N.A. (1990) "The Delta-4 XPA Extra Performance Architecture", in *Proc. 20th. Int. Symp. on Fault-Tolerant Computing Systems (FTCS-20)*, Newcastle upon Tyne, UK, pp. 481-488 (IEEE).

Bell, D.E. and LaPadula, L.J. (1974) *Secure Computer Systems: Unified Exposition and Multics Interpretation*, The MITRE Corporation N°MTR-2997 (AD/A-020-445).

Béounes, C. and Laprie, J.-C. (1985) "Dependability Evaluation of Complex Computer Systems: Stochastic Petri Net Modeling", in *Proc. 15th. Int. Symp. Fault Tolerant Computing (FTCS-15)*, Ann Arbor, MI, USA, pp. 364-369 (IEEE).

Bergstra, J.A. and Klop, J.W. (1985), "Algebra of Communicating Processes with Abstraction", *Theoretical Computer Science*, 37, pp. 77–121.

Bernstein, P.A. (1988), "Sequoia: A Fault Tolerant Tightly Coupled Multiprocessor for Transaction Processing", *IEEE Computer*, 21 (2), pp. 37-45.

Bernstein, P.A., Hadzilacos, V. and Goodman, N. (1987), *Concurrency Control and Recovery in Database Systems*, (Addison-Wesley).

Berry, G. and Cosserat, L. (1985) "The ESTEREL Synchronous Programming Language and its Mathematical Semantics", in *Proc. CMU Seminar on Concurrency*, Lecture Notes in Computer Science, 197, pp. 389-448 (Springer-Verlag).

Beyaert, B., Florin, G., Lonc, P. and Natkin, S. (1981) "Evaluation of Computer Systems Dependability using Stochastic Petri Nets", in *Proc. 11th. Int. Symp. Fault-Tolerant Computing (FTCS-11)*, Portland, Maine, USA, pp. 79-81 (IEEE).

Birman, K. and Joseph, T. (1987), "Reliable Communication in the Presence of Failures", *ACM Transactions on Computer Systems*, 5 (1).

Birman, K.P. (1985), "Replication and Fault-Tolerance in the ISIS System", *ACM Operating Systems Review*, 19 (5), pp. 79-86 (Proc. 10th. ACM Symp. on Operating System Principles, Orcas Island, WA, USA, December 1985).

Birman, K.P. and Joseph, T.A. (1987), "Exploiting Virtual Synchrony in Distributed Systems", *ACM Operating Systems Review*, 21 (5), pp. 123-128 (Proc. 11th. ACM Symp. on Operating System Principles, Austin, TX, USA, November 1987).

Birrell, A. and Nelson, B. (1984), "Implementing Remote Procedure Calls", *ACM Transactions on Computer Systems*, 2 (1), pp. 39-59.

Bishop, P.G. (1988), "The PODS Diversity Experiment", in *Software Diversity in Computerized Control Systems*, (U. Voges, Eds.), pp. 51-84 (Springer-Verlag).

Black, A., Hitchinson, N., Jul, E., Levy, H. and Carter, L. (1987), "Distribution and Abstract Types in Emerald", *IEEE Transactions on Software Engineering*, SE-13 (1), pp. 65-75.

Boehm, B.W. (1979) "Guidelines for Verifying and Validation Software Requirements and Design Specifications", in *Proc. EURO IFIP'79*, London, UK, pp. 711-719 (IFIP).

Boehm, B.W. (1981), *Software Engineering Economics*, (Prentice Hall).

Boehm, B.W. (1988), "A Spiral Model of Software Development and Enhancement", *IEEE Computer*, 21 (5), pp. 61-72.

Bond, P.G. (1987) *Real-Time Extensions to UNIX*, Delta-4 Project Consortium, Delta-4 Document (Specification S5, Phase 1).

Bond, P.G., Drackley, S.D., Powell, D. and Seaton, D.T. (1987) *The Impact of Real-Time and Dependability on Open, Distributed Systems - Interim Analysis of Requirements: Input to ECMA TC32/TG2*, Delta-4 Project Consortium, LAAS Report N°87.415 (available from LAAS-CNRS, 7 Ave. Colonel Roche, 31077 Toulouse, France).

Borg, A., Baumbach, J. and Glazer, S. (1983) "A Message System supporting Fault Tolerance", in *Proc. 9th. Symp. on Operating System Principles*, pp. 90-99 (ACM).

Bouajjani, A., Fernandez, J.-C. and Halbwachs, N. (1990) "Minimal Model Generation", in *Proc. Conf. on Automatic Verification (CAV 90)*, Rutgers, NJ, USA, Lecture Notes on Computer Science, 531 (Springer Verlag).

Bouajjani, A., Fernandez, J.-C., Graf, S., Rodriguez, C. and Sifakis, J. (1991) "Safety for Branching Semantics", in *Proc. 18th. Int. Conf. on Automata, Languages and Programming (ICALP 91)*, Lecture Notes in Computer Science, 510, pp. 72-92 (Springer-Verlag).

Bouricius, W.G., Carter, W.C. and Schneider, P.R. (1969) "Reliability Modeling Techniques for Self-Repairing Computer Systems", in *Proc. 24th. National Conference*, pp. 295-309 (ACM).

Boyer, R.S. and Moore, J.S. (1979), *A Computational Logic*, (Academic Press).

Brookes, S.D. and Rounds, S.D. (1983) "Behavioral Equivalence Relations induced by Programming Logics", in *Proc. 10th. Int. Conf. on Automata, Languages and Programming (ICALP 83)*, Lecture Notes in Computer Science, 154, pp. 97-108 (Springer Verlag).

Brookes, S.D., Hoare, C.A.R. and Roscoe, A.W. (1984), "A Theory of Communicating Sequential Processes", *Journal of the ACM*, 31 (3), pp. 560-599.

Brooks, F.P. (1987), "No Silver Bullett - Essence and Accidents of Software Engineering", *IEEE Computer*, 20 (4), pp. 10-19.

Büchi, J.R. (1962) "On a Decision Method in Restricted Second Order Arithmetic", in *Proc. Int. Congress on Logic, Method and Philosophy of Science*, pp. 1-11 (Stanford University Press).

Burns, A. (1990a) "Distributed Hard Real-time Systems: what Restrictions are Necessary?", in *Proc. 1989 Real-Time Symp.* (H. Zedan, Eds.) (Elsevier Scientific).

Burns, A. (1990b) *Scheduling Hard Real-Time Systems: A Review*, Department of Computing, University of Bradford, UK, Research Report.

Cart, M., Ferrie, J. and Mardyanto, S. (1987) "Atomic Broadcast Protocol, Preserving Concurrency for an Unreliable Broadcast Network", in *Proc. IFIP Conf. on Local Communication Systems: LAN and PBX* (J. Cabanel, G. Pujole and A. Danthine, Eds.) (North-Holland).

Carter, W.C. (1979) "Fault Detection and Recovery Algorithms for Fault-Tolerant Systems", in *Proc. EURO IFIP'79*, London, UK, pp. 725-734 (IFIP).

Carter, W.C. (1982) "A Time for Reflection", in *Proc. 12th. Int. Symp. on Fault Tolerant Computing (FTCS-12)*, Santa Monica, CA, USA, pp. 41 (IEEE).

Carter, W.C. and Schneider, P.R. (1968) "Design of Dynamically Checked Computers", in *Proc. IFIP'68 Cong.*, Amsterdam, The Netherlands, pp. 878-883 (IFIP).

Carter, W.C., Joyner, W.H., Brand, D., Ellozy, H.A. and Wolf, J.L. (1978) "An Improved System to Verify Assembled Programs", in *Proc. 8th. Int. Symp. on Fault Tolerant Computing (FTCS-8)*, Toulouse, France, pp. 165-170 (IEEE).

Caspi, P., Halbwachs, N., Pilaud, N. and Plaice, J. (1987) "LUSTRE, a Declarative Language for Programming Synchronous Systems", in *Proc. 14th. Symp. on Principles of Programming Languages (POPL 87)*, Munich, Germany, pp. 178-188 (ACM).

Castillo, X. and Siewiorek, D.P. (1981) "Workload, Performance and Reliability of Digital Computing Systems", in *Proc. 11th. Int. Symp. on Fault Tolerant Computing (FTCS-11)*, Portland, Maine, USA, pp. 84-89 (IEEE).

Chang, J. and Maxemchuck, N. (1984), "Reliable Broadcast Protocols", *ACM Transactions on Computer Systems*, 2 (3).

Cheheyl, M.H., Gasser, M., Huff, G.A. and Miller, J.K. (1981), "Verifying Security", *ACM Computing Surveys*, 13 (3), pp. 279-339.

Chen, L. and Avizienis, A. (1978) "N-Version-Programming: A Fault-Tolerance Approach to Reliability of Software Operation", in *Proc. 8th. Int. Symp. on Fault-Tolerant Computing (FTCS-8)*, Toulouse, France, pp. 3-9 (IEEE).

Cheriton, D. and Zwaenepoel, W. (1985), "Distributed Process Groups in the V-Kernel", *ACM Transactions on Computer Systems*, 3 (2).

Chesson, G. (1988) "XTP/PE Overview", in *Proc. 13th. Local Computer Network Conference*, Minneapolis-USA (IEEE).

Chillarege, R. and Bowen, N.S. (1989) "Understanding Large System Failures — A Fault Injection Experiment", in *Proc. 19th. Int. Symp. on Fault-Tolerant Computing (FTCS-19)*, Chicago, MI, USA, pp. 356-363 (IEEE).

Choi, G.S., Iyer, R.K. and Carreno, V. (1989) "A Fault Behavior Model for an Avionic Microprocessor: a Case Study", in *Proc. 1st Int. Working Conf. on Dependable Computing for Critical Applications*, Santa Barbara, CA, USA (A. Avizienis and J.-C. Laprie, Eds.), pp. 171-195 (Springer-Verlag).

Ciompi, P. and Grandoni, F. (1990) *SWFT in Delta-4: Selection of Techniques*, Delta-4 Project Consortium, Delta-4 Document N°R89.146/I1/P (available from Ferranti International, Wythenshawe, Manchester M22 5LA, UK).

Clark, D.D. and Wilson, D.R. (1987) "A Comparison of Commercial and Military Computer Security Policies", in *Proc. Symp. on Security and Privacy*, Oakland, CA, USA, pp. 184-194 (IEEE).

Clarke, E.M., Emerson, E.A. and Sistla, E. (1986), "Automatic Verification of Finite State Concurrent Systems using Temporal Logic Specifications: A Practical Approach", *ACM Transactions on Programming Languages and Systems*, 8 (2), pp. 244-263.

Cleaveland, R., Parrow, J.G. and Steffen, B. (1989) "The Concurrency Workbench", in *Proc. Workshop on Automatic Verification Methods for Finite State Systems*, Grenoble, Lecture Notes in Computer Science, 407, pp. 24-37 (Springer Verlag).

CNMA (1990) *Network Management*, ESPRIT Project 2617: Communications Network for Manufacturing Applications (CNMA).

Cooper, E.C. (1984) "Circus: A Replicated Procedure Call Facility", in *Proc. 4th. Symp. on Reliability in Distributed Software and Database Systems*, Maryland, USA, pp. 11-24 (Silver Spring).

Côrtes, M.L., Millman, S.D., Goosen, H.A. and McCluskey, E.J. (1987) *Techniques for Injecting Non Stuck-at Faults*, Center for Reliable Computing (CRC), Technical Report N°87-21.

Costes, A., Doucet, J.-E., Landrault, C. and Laprie, J.-C. (1981) "SURF: a Program for Dependability Evaluation of Complex Fault-Tolerant Computing Systems", in *Proc. 11th. Int. Symp. Fault-Tolerant Computing (FTCS-11)*, Portland, Maine, USA, pp. 72-78 (IEEE).

Cox, D.R. and Miller, H.D. (1968), *The Theory of Stochastic Processes*, (Methuen).

Craigen, D. (1987) "Strengths and Weaknesses of Program Verification Systems", in *Proc. 1st European Software Engineering Conf.*, Strasbourg, France, pp. 421-429.

Cristian, F. (1980) "Exception Handling and Software Fault Tolerance", in *Proc. 10th. Int. Symp. on Fault Tolerant Computing (FTCS-10)*, Kyoto, Japan, pp. 97-103 (IEEE) (also in *IEEE Transactions on Computers*, C-31(6), pp. 531-540, June 1982).

Cristian, F. (1987) *Exception Handling*, IBM Almaden Research Center, RJ5724 (57703).

Cristian, F. (1988) "Agreeing on Who is Present and Who is Absent in a Synchronous Distributed System", in *Proc. 18th. Int. Symp. on Fault-Tolerant Computing (FTCS-18)*, Tokyo, Japan, pp. 206-211 (IEEE).

Cristian, F. (1989), "Probabilistic Clock Synchronization", *Distributed Computing*, 3 (3), pp. 146-158.

Cristian, F. (1991), "Understanding Fault-Tolerant Distributed Systems", *Communications of the ACM*, 34 (2), pp. 56-78.

Cristian, F., Aghali, H., Strong, R. and Dolev, D. (1985) "Atomic Broadcast: From Simple Message Diffusion to Byzantine Agreement", in *Proc. 15th. Int. Symp. on Fault-Tolerant Computing (FTCS-15)*, Ann Arbor, MI, USA, pp. 200-206 (IEEE).

Cristian, F., Dancey, B. and Dehn, J. (1990) "Fault-Tolerance in the Advanced Automation System", in *Proc. 20th. Int. Symp. on Fault-Tolerant Computing (FTCS-20)*, Newcastle upon Tyne, UK, pp. 6-17 (IEEE).

Cristian, F., H., A. and Strong, R. (1986) "Clock Synchronization in the Presence of Omission and Performance Faults", in *Proc.16th. Int. Symp. on Fault-Tolerant Computing (FTCS-16)*, Vienna, Austria (IEEE).

Crouzet, Y. and Decouty, B. (1982) "Measurements of Fault Detection Mechanisms Efficiency: Results", in *Proc. 12th. Int. Symp. Fault-Tolerant Computing (FTCS-12)*, Santa Monica, CA, USA, pp. 373-376 (IEEE).

Damm, A. (1988) *Experimental Evaluation of Error-detection and Self-Checking Coverage of Components of a Distributed Real-time System*, Doctoral Dissertation, Technical University, Vienna, Austria.

David, R. (1986), "Signature Analysis for Multiple Output Circuits", *IEEE Transactions on Computers*, C-35 (9), pp. 830-837.

David, R. and Thévenod-Fosse, P. (1981), "Random Testing of Integrated Circuits", *IEEE Transactions on Instrumentation and Measurement*, IM-30 (1), pp. 20-25.

Davies, J. and Schneider, S. (1989) *An Introduction to Timed CSP*, Oxford University Computing Laboratory, Technical Monograph N°PRG-75.

Delta-4 (1990) *Delta-4 Application Support Environment*, Implementation Guide IG2, Delta-4 Project Consortium, Delta-4 Document (available from Bull SA, 1 Rue de Provence, 38423 Echirolles, France).

Delta-4 (1991) *Implementation Guide IG2*, Delta-4 Project Consortium, Delta-4 Document (available from Bull SA, 1 Rue de Provence, 38423 Echirolles, France).

DeMillo, R.A., Lipton, R.J. and Sayward, F.G. (1978), "Hints on Test Data Selection: Help for the Practicing Programmer", *Computer*, 11 (4), pp. 34-41.

Denning, D.E. (1982), *Cryptography and Data Security*, (Addison-Wesley).

Denning, D.E. (1986) "An Intrusion-Detection model", in *Proc. Symp. on Security and Privacy*, Oakland, CA, USA, pp. 118-132 (IEEE).

Dertouzos, M.L. and Mok, A.K. (1989), "Multiprocessor On-Line Scheduling of Hard Real-Time Tasks", *IEEE Transaction on Software Engineering*, 15 (12), pp. 1497-1506.

Deswarte, Y., Blain, L. and Fabre, J.-C. (1991) "Intrusion Tolerance in Distributed Systems", in *Proc. Symp. on Research in Security and Privacy*, Oakland, CA, USA, pp. 110-121 (IEEE).

Di Giandomenico, F. and Strigini, L. (1990) "Adjudicators for Diverse-redundant Components", in *Proc. 9th. Symp. on Reliable Distributed Systems (SRDS-9)*, Huntsville, AL, USA.

Diaz, M. (1982), "Modeling and Analysis of Communication and Cooperation Protocols using Petri Net Based Models", *Computer Networks*, 6 (6), pp. 419-441.

DoD 5200.28 (1985) *Trusted Computer System Evaluation Criteria*, US Department of Defense, STD N°5200.28.

Duane, J.T. (1964), "Learning Curve Approach to Reliability Monitoring", *IEEE Transactions on Aerospace*, 2, pp. 563-566.

Dugan, J.B. and Trivedi, K.S. (1989), "Coverage Modeling for Dependability Analysis of Fault-Tolerant Systems", *IEEE Transactions on Computers*, 38 (6), pp. 775-787.

Duran, J.W. and Ntafos, S.C. (1984), "An Evaluation of Random Testing", *IEEE Transactions on Software Engineering*, SE-10 (4), pp. 438-444.

Eckhard, D.E. and Lee, L.D. (1985), "A Theoretical Basis for the Analysis of Multiversion Software subject to Coincident Errors", *IEEE Transactions on Software Engineering*, SE-11 (12), pp. 1511-1517.

ECMA 127 (1990) *Remote Procedure Call using OSI (RPC)*, ECMA N°127.

ECMA TR/49 (1990) *Support Environment for Open Distributed Processing (SE-ODP)*, ECMA, Technical Report N°49.

Eich, M.H. (1988), "Graph Directed Locking", *IEEE Transactions on Software Engineering*, 14 (2), pp. 133-140.

Elmendorf, W.R. (1972) "Fault-Tolerant Programming", in *Proc. 2nd. Int. Symp. on Fault Tolerant Computing (FTCS-2)*, Newton, MA, USA, pp. 79-83 (IEEE).

Emerson, E.A. and Halpern, J.Y. (1986), "'Sometimes' and 'Not Never' Revisited: On Branching versus Linear Time Temporal Logic", *Journal of the ACM*, 33 (1), pp. 151-178.

Emerson, E.A., Mok, A.K., Sistla, A.P. and Srinivasan, J. (1989) "Quantitative Temporal Reasoning", in *Proc. Workshop on Automatic Verification Methods for Finite State Systems*, Grenoble, France, pp. 15 (Participants' version of Proceedings).

Ezhilchelvan, P.D. and Shrivastava, S.K. (1986) "A Characterization of Faults in Systems", in *Proc. 5th. Symp. on Reliability in Distributed Software and Database Systems*, Los Angeles, CA, USA, pp. 215-222 (IEEE).

Ezhilchelvan, P.D. and Shrivastava, S.K. (1991) "A Distributed Systems Architecture supporting Availability and Reliability", in *Proc. 2nd. Conf. on Dependable Computing for Critical Applications*, Tucson, AZ, USA (IFIP).

Fabry, R.S. (1974), "Capability-Based Addressing", *Communications of the ACM*, 17 (7), pp. 403-412.

Fernandez, J.-C. (1990), "An Implementation of an Efficient Algorithm for Bisimulation Equivalences", *Science of Computer Programming*, 13 (2-3), pp. 219-236.

Fernandez, J.-C. and Mounier, L. (1990) "Verification Bisimulations on the Fly", in *Proc. 3th. Int. Conf. on Formal Description Techniques (FORTE'90)* (J. Quemada, J. Mañas and E. Vazquez, Eds.) (North-Holland).

Fernandez, J.-C., Richier, J.-L. and Voiron, J. (1985) "Verification of Protocol Specifications using the Xesar System", in *Proc IFIP WG6.1 5th. International Conference on Protocol, Specification, Testing and Verification*, pp. 71-90 (North-Holland).

Fischer, M.J. (1983) "The Consensus Problem in Unreliable Distributed Systems (A Brief Survey)", in *Proc. Int. Conf. on Foundations of Computations Theory*, Borgholm, Sweden, pp. 127-140.

Fischer, M.J., Lynch, N.A. and Paterson, M.S. (1985), "Impossibility of Distributed Consensus with One Faulty Process", *Journal of the ACM*, 32 (2), pp. 374-382.

Floyd, R.W. (1967) "Assigning Meaning to Programs", in *Proc. Symposium in Applied Maths*, Providence, RI, USA, vol. XIX , pp. 19–32 (AMS).

Fraga, J. (1985) *Data Security by Intrusion-Tolerance*, Doctoral Dissertation, Institut National Polytechnique, Toulouse, France (LAAS Report N°85.133, in French).

Fraga, J. and Powell, D. (1985) "A Fault and Intrusion-Tolerant File System", in *Proc. 3rd. IFIP Int. Cong. on Computer Security*, Dublin, Ireland, pp. 203-218 (North-Holland).

Francez, N. (1986), *Fairness*, Monographs in Computer Science, (Springer Verlag).

Fray, J.M., Deswarte, Y. and Powell, D. (1986) "Intrusion-Tolerance using Fine-Grain Fragmentation-Scattering", in *Proc. Symp. on Security and Privacy*, Oakland, CA, USA, pp. 194-201 (IEEE).

Frison, S.G. and Wensley, J.H. (1982) "Interactive Consistency and its Impact on the Design of TMR Systems", in *Proc. 12th. Int. Symp. on Fault Tolerant Computing (FTCS-12)*, Santa Monica, CA, USA, pp. 228-233 (IEEE).

FTCS12 (1982) *Fundamental Concepts of Fault-Tolerance*, Proc. 12th. Int. Symp. on Fault-Tolerant Computing (FTCS-12), IEEE.

Garcia-Molina, H., Kogan, B. and Lynch, N. (1988) "Reliable Broadcast in Networks with Nonprogrammable Servers", in *Proc. 8th. Int. Conf. on Distributed Computing Systems (ICDCS-8)*, San Jose, CA, USA, pp. 428-437 (IEEE).

Gasser, M. (1988), *Building a Secure Computer System*, (Van Nostran Reinhold).

Gérardin, J.P. (1986), "Design Aid to Reliable and Safe Systems: The DEFI", *Electronique Industrielle* 116, pp. 58-63 (in French).

Giloth, F.K. and Prantzen, K.D. (1983) "Can the Reliability of Digital Telecommunication Switching Systems be Predicted and Measured?", in *Proc. 13th. Int. Symp. on Fault-Tolerant Computing (FTCS-13)*, Milano, Italy, pp. 392-397 (IEEE).

Goel, A.L. and Okumoto, K. (1979), "Time-Dependent Error-Detection Rate Model for Software and other Performance Measures", *IEEE Transactions on Reliability*, R-28 (3), pp. 465-484.

Golberg, A. and Robson, D. (1981), *Smalltalk-80: The Language and its Implementation*, (Addison-Wesley).

Goldberg, J. (1982) "A Time for Integration", in *Proc. 12th. Int. Symp. on Fault Tolerant Computing (FTCS-12)*, Santa Monica, CA, USA, pp. 42 (IEEE).

Goodenough, J.B. and Gerhart, S.L. (1975), "Toward a Theory of Test Data Selection", *IEEE Transactions on Software Engineering*, SE-1 (2), pp. 156-173.

Gordon, M., Milner, R. and Wadsworth, C. (1979), *Edinburgh LCF*, (Springer Verlag).

Graf, S. and Sifakis, J. (1986), "A Logic for the Description of Non-Deterministic Programs and their Properties", *Information and Control*, 68 (1-3), pp. 125-145.

Graf, S. and Steffen, B. (1990) "Compositional Minimization of Finite State Systems", in *Proc. Conf. on Automatic Verification (CAV 90)*, Rutgers, NJ, USA, Lecture Notes in Computer Science, 531 (Springer Verlag).

Graf, S., Richier, J.-L., Rodriguez, C. and Voiron, J. (1989) *Protocol Verification Report*, Delta-4 Project Consortium, Delta-4 Document N°R89.147/l1/P (available from Ferranti International, Wythenshawe, Manchester M22 5LA, UK).

Graf, S., Richier, J.-L., Rodriguez, C. and Voiron, J. (1990) *Protocol Verification Report*, Delta-4 Project Consortium, Delta-4 Document N°R90.225 (available from Ferranti International, Wythenshawe, Manchester M22 5LA, UK).

Graf, S., Richier, J.L., Rodriguez, C. and Voiron, J. (1989) "What are the Limits of Model Checking for the Verification of Real Life Protocols", in *Proc. Workshop on Automatic Verification Methods for Finite State Systems*, Grenoble, France, Lecture Notes on Computer Science, 407, pp. 275-285 (Springer-Verlag).

Gray, J. (1986) "Why do Computers Stop and What can be done about it?", in *Proc. 5th. Symp. on Reliability in Distributed Software and Database Systems*, Los Angeles, CA, USA, pp. 3-12 (IEEE).

Groote, G.F. (1990) *Specification and Verification of Real Systems in ACP*, CWI, Amsterdam, Technical Report N°CS-R9015.

Guérin, C., Raison, H. and Martin, P. (1985) *Procedure for Dependable Message Broadcasting in a Ring and Mechanism for Implementing the Procedure*, French Patent N°85.2.2 (in French).

Gunneflo, U., Karlsson, J. and Torin, J. (1989) "Evaluation of Error Detection Schemes using Fault Injection by Heavy-ion Radiation", in *Proc. 19th. Int. Symp. Fault-Tolerant Computing (FTCS-19)*, Chicago, MI, USA, pp. 340-347 (IEEE).

Gunningberg, P. (1983) "Voting and Redundancy Management Implemented by Protocols in Distributed Systems", in *Proc. 13th. Int. Symp. on Fault-Tolerance Computing (FTCS-13)*, Milano, Italy, pp. 182-185 (IEEE).

Habermann, A.N. (1969), "Prevention of System Deadlocks", *Communications of the ACM*, 12 (7), pp. 373-385.

Halang, W.A. (1986), "Implications on Suitable Multiplrocessor Structures and Virtual Storage Management when applying a Feasible Scheduling Algorithm in Hard Real-time Environments", *Software Practice and Experience*, 16 (8), pp. 761-769.

Halpern, J.Y., Simons, B., Strong, H.R. and Dolev, D. (1984) "Fault-Tolerant Clock Synchronisation", in *Proc. 3rd. Symp. on the Principles of Distributed Computing*, Vancouver, Canada, pp. 89-102 (ACM).

Hamlet, R. (1987) "Testing for Trustworthiness", in *Proc. DIAC 87 Symp. on Directions and Implications of Advanced Computing*, Seattle, WA, USA, pp. 87-93.

Har'El, Z. and Kurshan, R.P. (1990), "Software for Analytical Development of Communication Protocols", *AT&T Technical Journal*, 60 (1), pp. 45-58.

Harel, D., Pnueli, A., Schmidt, J.P. and Sherman, R. (1987) "On the Formal Semantics of State Charts", in *Proc. 2nd Symp. on Logic in Computer Science (LICS 87)*, pp. 54-64 (IEEE).

Harel, E., Lichtenstein, O. and Pnueli, A. (1990) "Explicit Clock Temporal Logic", in *Proc. 5th. Symp. on Logic in Computer Science (LICS 90)*, pp. 402-413 (IEEE).

Haugk, G., Lax, F.M., Rover, R.D. and Williams, J.R. (1985), "The 5 ESS Switching System: Maintenance Capabilities", *AT&T Technical Journal*, 64 (6), pp. 1385-1416.

Hecht, H. (1976), "Fault-Tolerant Software for Real-time Applications", *ACM Computing Surveys*, 8 (4), pp. 391-407.

Hecht, H. and Fiorentino, E. (1987) "Reliability Assessment of Spacecraft Electronics", in *Proc. Annual Reliability and Maintainability Symp.*

Hennessy, M. and Milner, R. (1985), "Algebraic Laws for Nondeterminism and Concurrency", *Journal of the ACM*, 32 (1), pp. 137-161.

Hennessy, M. and Regan, T. (1990) *A Temporal Process Algebra*, University of Sussex, Computer Science, Report N°2/90.

Hetzel, W. (1984), *The Complete Guide to Software Testing*, (QED Information Science).

Hoare, C.A.R. (1969), "An Axiomatic Basis for Computer Programming", *Communications of the ACM*, 12 (10), pp. 576-583.

Hoare, C.A.R. (1972), "Proof of Correctness of Data Representations", *Acta Informatica*, 1, pp. 271–281.

Hoare, C.A.R. (1985), *Communicating Sequential Processes*, (Prentice Hall International).

Holzmann, G.J. (1984), "The Pandora System: an Interactive System for the Design of Data Communication Protocols", *Computer Networks*, 8 (2), pp. 71–81.

Holzmann, G.J. (1990), "Algorithms for Automated Protocol Validation", *AT&T Technical Journal*, 60 (1), pp. 32–44.

Horning, J.J., Lauer, H.C., Melliar-Smith, P.M. and Randell, B. (1974), "A Program Structure for Error Detection and Recovery", in *Operating Systems*, (G. Goos and J. Hartmanis, Eds.), Lectures Notes in Computer Science, 16, pp. 172-187 (Springer-Verlag).

Hosford, J.E. (1960), "Measures of Dependability", *Operations Research*, 8 (1), pp. 204-206.

Howard, R.A. (1971), *Dynamic Probabilistic Systems*, (John Wiley)..

Howden, W.E. (1987), *Functional Program Testing and Analysis*, (McGraw-Hill).

Huang, K.K. and Abraham, J.A. (1982) "Low Cost Schemes for Fault Tolerance in Matrix Operations with Processor Arrays", in *Proc. 12th. Int. Symp. on Fault Tolerant Computing (FTCS-12)*, Santa Monica, CA, USA, pp. 330-337 (IEEE).

IEC 191 (1985) *Reliability, Maintainability and Quality of Service*, International Electrotechnical Vocabulary, Chapter 191, International Electrotechnical Commission, Document N°1-IEV-191-Central Office-1243 and 56-IEV-191-Central Office-119 (Geneva).

IEEE 729 (1982) *Standard Glossary of Software Engineering Terminology*, IEEE N°729.

IEEE 829 (1983) *Standard for Software Test Documentation*, IEEE N°829.

IEEE P1003 (1989) *Realtime Extension for Portable Operating Systems*, IEEE N°P1003/P9.

ISO 7498-1 *Basic Reference Model*, Information Processing Systems: Open Systems Interconnection, IS N°7489-1 (1984).

ISO 7498-4 *Basic Reference Model, Part 4: OSI Management Framework*, Information Processing Systems: Open Systems Interconnection, ISO, IS N°7498-4 (E) (1989).

ISO 8571 *File Transfer and Access Method (FTAM)*, Information Processing Systems: Open Systems Interconnection, ISO, IS N°8571.

ISO 8802-4 *Token-passing Bus Access Method and Physical Layer Specifications*, ISO, DIS N°8802-4 (1984).

ISO 8802-5 *Token Ring Access Method and Physical Layer Specifications*, ISO, DP N°8802-5 (1985).

ISO 8807 *LOTOS: a Formal Description Technique based on the Temporal Ordering of Observational Behaviour*, ISO, IS N°8807 (1989).

ISO 9074 *ESTELLE: A Formal Description Technique based on an Extended State Transition Model*, ISO, IS N°9074 (1989).

ISO 9314 *Fiber Distributed Data Interface (FDDI)*, ISO, DP N°9314 (ANSI Standard X3.139, 1987).

ISO 9506 *Manufacturing Message Specification (MMS)*, Information Processing Systems: Open Systems Interconnection, ISO, IS N°9506.

ISO 9595 *Common Management Information Service Definition*, Information Processing Systems: Open Systems Interconnection, ISO/IEC N°9595 (E) (1991).

ISO 10040 *Systems Management Overview*, Information Processing Systems: Open Systems Interconnection, ISO, DIS N°10040 (E) (1990).

ISO 10165-1 *Management Information Services - Structure of Management Information, Part 1: Management Information Model*, Information Processing Systems: Open Systems Interconnection, ISO/IEC, DIS N°10165-1 (E) (1990).

ISO 10165-4 *Management Information Services - Structure of Management Information, Part 4: Guidelines for the Definition of Managed Objects (GDMO)*, Information Processing Systems: Open Systems Interconnection, ISO/IEC, DIS N°10165-4 (E) (1990).

ISO N2031 *Working Draft addendum to ISO 7498-1 on Multipeer Data Transmission*, ISO N°TC97/SC21 WG1 N2031 (1987).

Iyer, R.K., Butner, S.E. and McCluskey, E.J. (1982), "A Statistical Failure/Load Relationship: Results of a Multi-Computer Study", *IEEE Transactions on Computers*, C-31, pp. 697-706.

Jard, C. and Jeron, T. (1989) "On-Line Model Checking for Finite Linear Temporal Logic Specifications", in *Proc. Workshop on Automatic Verification Methods for Finite State Systems*, Grenoble, France, Lecture Notes on Computer Science, 407, pp. 189-196 (Springer Verlag).

Jensen, E.D. (1991) *Alpha: a Non-Proprietary Experimental Operating System for Distributed Mission-Critical Real-Time Applications — An Overview of its Objectives and Kernel Abstractions*, Concurrent Computer Corporation, Draft (available from Concurrent Computer Corporation, One Technology Way, Westford, MA 01886, USA).

Jensen, E.D. and Northcutt, J.D. (1990) "Alpha: a Non-Proprietary Operating System for Mission-Critical Real-Time Distributed Systems", in *Proc. Workshop on Experimental Distributed Systems*, Huntsville, AL, USA (IEEE).

Jensen, E.D., Locke, C.D. and Tokuda, H. (1985) "A Time-Value Driven Scheduling Model for Real-Time Operating Systems", in *Proc. Symp. on Real-Time Systems* (IEEE).

Jessep, D.C. (1977) *Fault-Tolerant Computing, Definition of Terms*, IEEE Computer Society N°P 610/DI.

Johnson, H.H. and Madison, M.S. (1974) "Deadline Scheduling for a Real-Time Multiprocessor", in *Proc. Eurocomp Conference*.

Joseph, M.K. and Avizienis, A. (1988) "A Fault Tolerance Approach to Computer Viruses", in *Proc. 1988 Symp. on Security and Privacy*, Oakland, CA, USA, pp. 52-58 (IEEE).

Josko, B. (1987) "MCTL - An Extension of CTL for Modular Verification of Concurrent Systems", in *Proc. Workshop on Temporal Logic in Specification*, Manchester, UK, Lecture Notes on Computer Science, 398, pp. 165-187 (Springer Verlag).

Kanoun, K. and Powell, D. (1991) "Dependability Evaluation of Bus and Ring Communication Topologies for the Delta-4 Distributed Fault-Tolerant Architecture", in *Proc. 10th. Symp. on Reliable Distributed Systems (SRDS-10)*, Pisa, Italy (IEEE).

Kanoun, K. and Sabourin, T. (1987) "Software Dependability of a Telephone Switching System", in *Proc. 17th. Int. Symp. on Fault Tolerant Computing (FTCS-17)*, Pittsburgh, PA, USA, pp. 236-241 (IEEE).

Kanoun, K., Bastos, M.R. and Moreira de Souza, J. (1988) *A Method for Software Reliability Analysis and Prediction: Application to the TROPICO-R Switching System*, LAAS-CNRS, Report N°88.337.

Kieckhafer, R.M., Walter, C.J., Finn, A.M. and Thambidurai, P.M. (1988), "The MAFT Architecture for Distributed Fault Tolerance", *IEEE Transactions on Computers*, 37 (4), pp. 398-405.

Kim, K.H. (1984) "Distributed Execution of Recovery Blocks: an Approach to Uniform Treatment of Hardware and Software Faults", in *Proc. 4th. Int. Conf. on Distributed Computing Systems*, pp. 526-532.

Kim, K.H. and Ramamoorthy, C.V. (1976) "Failure-Tolerant Parallel Programming and its Supporting System Architecture", in *Proc. 4th. Int. Conf. Distributed Computing System (ICDCS-4)*, San Francisco, CA, USA, pp. 413-423 (IEEE).

Knuth, D.E. (1973), *The Art of Computer Programming*, (Addison-Wesley).

Koga, Y., Fukushima, E. and Yoshihara, K. (1982) "Error Recoverable and Securable Data Communication for Computer Network", in *Proc. 12th. Int. Symp. on Fault-Tolerant Computing (FTCS-12)*, Santa Monica, CA, USA, pp. 183-186 (IEEE).

Kopetz, H. (1974) "Software Redundancy in Real-time Systems", in *Proc. IFIP Conf. Information Processing*, Stockholm, Sweden, pp. 182-186 (North-Holland).

Kopetz, H. (1986) "Scheduling in Distributed Real-Time Systems", in *Proc. Advanced Seminar on R/T Lans*, Bandol, France (INRIA).

Kopetz, H. and Merker, W. (1985) "The Architecture of MARS", in *Proc. 15th. Int. Symp. on Fault-Tolerant Computing (FTCS-15)*, Ann Arbor, MI, USA, pp. 274-279 (IEEE).

Kopetz, H. and Ochsenreiter, W. (1987), "Clock Synchronization in Distributed Real-Time Systems", *IEEE Transactions on Computers*, C-36 (8), pp. 933-940.

Kopetz, H. and Schwabl, W. (1989) *Global Time in Distributed Real-Time Systems*, Technische Universität Wien, Research Report N°15/89.

Kopetz, H., Damm, A., Koza, C., Mulazzani, M., Schwabl, W., Senft, C. and Zainlinger, R. (1988), "Distributed Fault-Tolerant Real-Time Systems: The MARS Approach", *IEEE Micro*, 9 (1), pp. 25-40.

Kopetz, H., Kantz, H., Grünsteidl, G., Puscher, P. and Reisinger, J. (1990) "Tolerating Transient Faults in MARS", in *Proc. 20th. Int. Symp. on Fault-Tolerant Computing (FTCS-20)*, Newcastle upon Tyne, UK, pp. 466-473 (IEEE).

Kozen, D. (1982) "Results on the Propositional μ-calculus", in *Proc. 9th. Int. Conf. on Automata, Languages and Programming (ICALP 82)*, Lecture Notes on Computer Science, 140, pp. 348-359 (Springer Verlag).

Kuipers, B. (1985), "Commonsense Reasoning about Causality: Deriving Behavior from Structure", in *Qualitative Reasoning about Physical Systems*, (D. G. Bobrow, Eds.), pp. 169-203 (MIT Press).

Lala, J.H. (1983) "Fault Detection, Isolation and Reconfiguration in FTMP: Methods and Experimental Results", in *Proc. Digitals Avionics Systems Conf.*, pp. 21.3.1-21.3.9 (AIAA/IEEE).

Lala, J.H. (1986) "A Byzantine Resilient Fault-Tolerant Computer for Nuclear Power Plant Applications", in *Proc. 16th. Int. Symp. on Fault Tolerant Computing (FTCS-16)*, Vienna, Austria, pp. 338-343 (IEEE).

Lamport, L. (1977), "Proving the Correctness of Multiprocess Programs", *IEEE Transactions on Software Engineering*, SE-3 (2), pp. 125-143.

Lamport, L. (1978), "Time, Clocks and the Ordering of Events in a Distributed System", *Communication of the ACM*, 21 (7), pp. 558-565.

Lamport, L. (1981), "Password Authentication with Insecure Communications", *Communications of the ACM*, 24 (11), pp. 770-772.

Lamport, L. (1984), "Using Time instead of Timeout for Fault-Tolerant Distributed Systems", *ACM Transactions on Programming Languages and Systems*, 6 (2), pp. 254-280.

Lamport, L. and Melliar-Smith, P.M. (1985), "Synchronizing Clocks in the Presence of Faults", *Journal of the ACM*, 32 (1), pp. 52-78.

Lamport, L. and Schneider, F.B. (1985), "Formal Foundation for Specification and Verification", in *Distributed Systems, Methods and Tools for Specification*, Lecture Notes in Computer Science, 5 (Springer-Verlag).

Lamport, L., Shostak, R. and Pease, M. (1982), "The Byzantine Generals Problem", *ACM Transactions on Programming Languages and Systems*, 4 (3), pp. 382-401.

Lampson, B.W. (1981), "Atomic Transactions", in *Distributed Systems — Architecture and Implementation*, Lecture Notes in Computer Science, 11 (Springer-Verlag).

Laprie, J.-C. (1975) "Reliability and Availability of Repairable Systems", in *Proc. 5th. Int. Symp. on Fault Tolerant Computing (FTCS-5)*, Paris, France, pp. 87-92 (IEEE).

Laprie, J.-C. (1983) "A Trustable Evaluation of Computer Systems Dependability", in *Proc. Int. Workshop on Applied Mathematics and Performance/Reliability Models of Computer/Communication Systems*, Pisa, Italy (IEEE).

Laprie, J.-C. (1984), "Dependability Evaluation of Software Systems in Operation", *IEEE Transactions on Software Engineering*, SE-10 (6), pp. 701-714.

Laprie, J.-C. (1985) "Dependable Computing and Fault Tolerance: Concepts and Terminology", in *Proc. 15th. Int. Symp. on Fault Tolerant Computing (FTCS-15)*, Ann Arbor, MI, USA, pp. 2-11 (IEEE).

Laprie, J.-C. (1989), "Dependability: A Unifying Concept for Reliable Computing and Fault Tolerance", in *Dependability of Resilient Systems*, (T. Anderson, Eds.), pp. 1-28 (Blackwell Scientific Publications).

Laprie, J.-C., Arlat, J., Béounes, C., Kanoun, K. and Hourtolle, C. (1987) "Hardware and Software Fault-Tolerance: Definition and Analysis of Architectural Solutions", in *Proc. 17th. Int. Symp. on Fault-Tolerant Computing (FTCS-17)*, Pittsburgh, PA, USA, pp. 116-121 (IEEE).

Laprie, J.-C., Béounes, C., Kaâniche, M. and Kanoun, K. (1990) "The Transformation Approach to the Modeling and Evaluation of the Reliability and Availability Growth of Systems in Operation", in *Proc. 20th. Int. Symp. on Fault Tolerant Computing (FTCS-20)*, Newcastle upon Tyne, UK, pp. 364-371 (IEEE).

Laprie, J.-C., Costes, A. and Landrault, C. (1981), "Parametric Analysis of 2-unit Redundant Computer Systems with Corrective and Preventive Maintenance", *IEEE Transactions on Reliability*, R-30 (2), pp. 139-144.

Le Lann, G. (1989) "Critical Issues in Distributed Real-time Computing", in *Proc. of Workshop on Comunication Networks and Distributed Operating Systems within the Space Environment* (ESTEC).

Lehmann, D., Pnueli, A. and Stavi, J. (1981) "Impartiality, Justice and Fairness: the Ethics of Concurrent Termination", in *Proc. 8th. Int. Conf. on Automata, Languages and Programming (ICALP 81)*, Lecture Notes on Computer Science, 115, pp. 264-277 (Springer Verlag).

Levendel, Y. (1986), "Fault Simulation", in *Fault-Tolerant Computing, Theory and Techniques*, (D. K. Pradhan, Eds.), pp. 184-264 (Englewood Cliffs: Prentice Hall).

Lewis, H. (1990) "A Logic of Concrete Time Intervals", in *Proc. 5th. Symp. on Logic in Computer Science (LICS 90)*, pp. 380-389 (IEEE).

Littlewood, B. (1979), "A Software Reliability Model for Modular Program Structure", *IEEE Transactions on Reliability*, R-28, pp. 241-246.

Littlewood, B. (1987), *Software Reliability Achievement and Assessment*, (Blackwell Scientific Publications).

Littlewood, B. (1988), "Forecasting Software Reliability", in *Software Reliability Modeling and Identification*, (S. Bittanti, Eds.), pp. 140-209 (Springer-Verlag).

Littlewood, B. and Miller, D.R. (1987a) "A Conceptual Model of Multi-Version Software", in *Proc. 17th. Int. Symp. on Fault-Tolerant Computing (FTCS-17)*, Pittsburgh, PA, USA, pp. 150-155 (IEEE).

Littlewood, B. and Miller, D.R. (1987b) "A Conceptual Model of the Effect of Diverse Methodologies on Coincident Failures in Multi-Version Software", in *Proc. 3rd. Int Fault-Tolerant Computer Systems Conf.*, Bremerhaven, Germany (also Centre for Software Reliability Technical Report, May 1987).

Liu, C.L. and Layland, J.W. (1973), "Scheduling Algorithms for Multiprogramming in a Hard Real-Time Environment", *Journal of the ACM*, 20 (1), pp. 46-61.

Loques, O.G. and Kramer, J. (1986), "Flexible Fault Tolerance for Distributed Computer Systems", *Proceedings of the IEE*, E-133 (6), pp. 319-332.

Mahmood, A., Andrews, D.M. and McCluskey, E.J. (1984) "Executable Assertions and Flight Software", in *Proc. 6th. Digital Avionics Systems Conf.*, Baltimore, Maryland, USA, pp. 346-351 (AIAA/IEEE).

Mancini, L.V. and Shrivastava, S.K. (1989) "Replication within Atomic Actions and Conversations: a Case Study in Fault Tolerance Duality", in *Proc. 19th. Int. Symposium on Fault-Tolerant Computing Systems (FTCS-19)*, Chicago, MI, USA, pp. 454-461 (IEEE).

Manna, Z. and Pnueli, A. (1989), "The Anchored Version of the Temporal Framework", in *Linear Time, Branching Time, and Partial Order in Logics and Models for Concurrency*, (J. W. De Bakker, W. P. De Roover and G. Rozenberg, Eds.), Lecture Notes on Computer Science, 354, pp. 201-284 (Springer Verlag).

MAP *Network Management Requirements Specification*, MAP/TOP 3.0, Chapter C11 (1987).

Maraninchi, F. (1990) *ARGOS, a Graphical Language for the Design, Description and Validation of Reactive Systems*, Doctoral Dissertation, Université J. Fourier, Grenoble, France (in French).

McCabe, T.J. (1976), "A Complexity Measure", *IEEE Transactions on Software Engineering*, SE-2 (4), pp. 308-320.

McCluskey, E.J. (1986), "Design for Testability", in *Fault-Tolerant Computing, Theory and Techniques*, (D. K. Pradhan, Eds.), pp. 95-183 (Prentice Hall).

Melliar-Smith, P.M. (1987) "Extending Interval Logic to Real Time Systems", in *Proc. Workshop on Temporal Logic in Specification*, Manchester, UK, Lecture Notes in Computer Science, 398, pp. 224-242 (Springer-Verlag).

Melliar-Smith, P.M. and Randell, B. (1977), "Software Reliability: The Role of Programmed Exception Handling", *ACM SIGPLAN Notices*, 12 (3), pp. 95-100 (also in *Reliable Computer Systems*, S.K. Shrivastava (Ed.), Springer-Verlag, 1985, pp.143-153).

Melliar-Smith, P.M. and Schwartz, R.L. (1982), "Formal Specification and Mechanical Verification of SIFT: A Fault-Tolerance Flight Control System", *IEEE Transactions on Computers*, C-31 (7), pp. 616-630.

Mellor, P. (1987), "Software Reliability Modelling: the State of the Art", *Information and Software Technology*, 29 (2), pp. 81-98 (Butterworth Scientific Press).

Meyer, J.F. (1978) "On Evaluating the Performability of Degradable Computing Systems", in *Proc. 8th. Int. Symp. on Fault Tolerant Computing (FTCS-8)*, Toulouse, France, pp. 44-49 (IEEE).

Miller, D.R. (1986a), "Exponential Order Statistic Models of Software Reliability Growth", *IEEE Transactions on Software Engineering*, SE-12 (1), pp. 12-24.

Miller, D.R. (1986b), "Making Statistical Inferences About Software Reliability", *Software Reliability and Metrics Newsletter*, 4, pp. 3 (Center for Software Reliability - Alvey Club).

Milner, R. (1980), *A Calculus for Communicating Processes*, Lecture Notes in Computer Science, 92 (Springer Verlag).

Milner, R. (1983), "Calculii for Synchrony and Asynchrony", *Theoretical Computer Science*, 25 (3), pp. 267-310.

Minet, P. and Sedillot, S. (1987), "Integration of Real-Time and Consistency Constraints in Distributed Databases: The SIGMA Approach", *Computer Standards & Interfaces*, 6 (1), pp. 97-105.

Mishra, S., Peterson, L.L. and Schlichting, R.D. (1989) "Implementing Fault-Tolerant Replicated Objects using Psync", in *Proc. 8th. Symp. on Reliable Distributed Systems (SRDS-8)*, Seattle, Washington, USA, pp. 42-52 (IEEE).

Misra, J. and Chandy, M. (1981), "Proofs of Networks Processes", *IEEE Transactions on Software Engineering*, 7 (4), pp. 417-426.

Molloy, M. (1982), "Performance Analysis using Stochastic Petri Nets", *IEEE Transactions on Computers*, 39 (9), pp. 913-917.

Moreira de Souza, J. and Peixoto Paz, E. (1975), "Fault-Tolerant Clocking System", *Electronic Letters*, 11 (19), pp. 433-434.

Moreira de Souza, J., Peixoto Paz, E. and Landrault, C. (1976) "A Research Oriented Microcomputer with Built-In Auto-Diagnostics", in *Proc. 6th. Fault-Tolerant Computing Symposium (FTCS-6)*, Pittsburgh, PA, USA, pp. 3-8 (IEEE).

Moss, J.E.B. (1981) *Nested Transactions: an Approach to Reliable Distributed Computing*, Massachussetts Institute of Technology.

MP MMS *Specification of a Multipoint MMS*, Delta-4 Implementation Guide, Section 1, Chapter 3 (1991) (available from Bull SA, 1 Rue de Provence, 38423 Echirolles, France).

Myers, G.J. (1979), *The Art of Software Testing*, (John Wiley & Sons).

Nagel, P.M. and Skrivan, J.A. (1982) *Software Reliability: Repetitive Run Experimentation and Modeling*, NASA, Report N°CR-165836.

Navaratnam, S., Chanson, S. and Neufeld, G. (1988) "Reliable Group Communication in Distributed Systems", in *Proc. 8th. Int. Conf. on Distributed Computing Systems (ICDCS-8)*, San Jose - USA, pp. 428-437 (IEEE).

NCSC (1987a) *A Guide to Understanding Discretionary Access Control in Trusted Systems*, National Computer Security Center N°NCSC-TG-003 (Version-1).

NCSC (1987b) *Trusted Network Interpretation of the Trusted Computer System Evaluation Criteria*, National Computer Security Center N°NCSC-TG-005 (Version 1).

Needham, R.M. and Schroeder, M.D. (1978), "Using Encryption for Authentication in Large Networks of Computers", *Communications of the ACM*, 21 (12), pp. 993-999.

Nicolaidis, M., Noraz, S. and Courtois, B. (1989) "A Generalized Theory of Fail-Safe Systems", in *Proc. 19th. Int. Symp. on Fault Tolerant Computing (FTCS-19)*, Chicago, MI, USA, pp. 398-406 (IEEE).

Nicollin, X., Richier, J.-L., Sifakis, J. and Voiron, J. (1990) "ATP, an Algebra for Timed Processes", in *Proc. IFIP WG2.2/2.3 Working Conference on 'Programming Concepts and Methods'*, Sea of Galilee, Israel, pp. 415-442.

Norman, D.A. (1983), "Design Rules Based on Analyses of Human Error", *Communications of the ACM*, 26 (4), pp. 254-258.

Ntafos, S.C. (1988), "A Comparison of Some Structural Testing Strategies", *IEEE Transactions on Software Engineering*, SE-14 (6), pp. 868-874.

Osterweil, L.J. and Fodsick, L.D. (1976), "DAVE — A Validation Error Detection and Documentation System for Fortran Programs", *Software Practice and Experience*, 6 (4), pp. 473-486.

Owicki, S. and Gries, D. (1976), "An Axiomatic Proof Technique for Parallel Programs", *Acta Informatica*, 6, pp. 319–340.

Park, D. (1980) "On the Semantics of Fair Parallelism", in *Proc. Abstract Software Specification*, Lecture Notes in Computer Science, 86, pp. 504-524 (Springer Verlag).

PDCS (1990) *Timeliness*, Specification and Design for Dependability, Esprit Project N°3092 (PDCS: Predictably Dependable Computing Systems), PDCS First Year Report.

Perry, K.J. and Toueg, S. (1986), "Distributed Agreement in the Presence of Processor and Communication Faults", *IEEE Transactions on Software Engineering*, SE-12 (3), pp. 477-482.

Peterson, L.L., Buchholdz, N.C. and Schlichting, R.D. (1989), "Preserving and Using Context Information in Interprocess Communication", *ACM Transactions on Computer Systems*, 7 (3).

Peterson, W.W. and Weldon, E.J. (1972), *Error-Correcting Codes*, (MIT Press).

Pignal, P.I. (1988), "An Analysis of Hardware and Software Availability Exemplified on the IBM 3725 Communication Controler", *IBM Journal of Research and Development*, 32 (2), pp. 268-278.

Plotkin, G.D. (1981) *A Structural Approach to Operational Semantics*, DAIMI, Aarhus, Technical Report N°FN-19.

Pnueli, A. (1977) "The Temporal Logic of Programs", in *Proc. 18th. Symp. on Foundations of Computer Science (FOCS 77)* (IEEE) (Revised version published in *Theoretical Computer Science*, 13:45–60, 1981).

Pnueli, A. (1986) "Specification and Development of Reactive Systems", in *Proc. 10th. IFIP Int. World Computer Congress (Information Processing 86)*, Dublin, Ireland, pp. 845-858 (North-Holland).

Powell, D. (1991) *Fault Assumptions and Assumption Coverage*, LAAS-CNRS, Report N°90.074 (in PDCS Second Year Report, vol.1) (available from LAAS-CNRS, 7 Ave. Colonel Roche, 31077 Toulouse, France).

Powell, D. (Ed.) (1988) *Delta-4 Overall System Specification (OSS)*, (Delta-4 Project Consortium) (Delta-4 Document S88.040/I2/P).

Powell, D., Bonn, G., Seaton, D., Veríssimo, P. and Waeselynck, F. (1988) "The Delta-4 Approach to Dependability in Open Distributed Computing Systems", in *Proc. 18th. Int. Symp. on Fault-Tolerant Computing Systems (FTCS-18)*, Tokyo, Japan, pp. 246-251 (IEEE).

Pradhan, D.K. (1986), "Fault-Tolerant Multiprocessor and VLSI-Based System Communication Architectures", in *Fault-Tolerant Computing, Theory and Techniques*, pp. 467-576 (Prentice Hall).

Puscher, P. and Koza, C. (1989), "Calculating the Maximum Execution Time of Real-Time Programs", *Real-Time Systems*, 1 (2), pp. 159-176.

Queille, J.P. and Sifakis, J. (1983), "Fairness and Related Properties in Transition Systems: a Temporal Logic to deal with Fairness", *Acta Informatica*, 19 (3), pp. 195-220.

Quemada, J., Aczorra, A. and Frutos, D. (1989) "A Timed Calculus for Lotos", in *Proc. 2nd. Int. Conf. on Formal Description Techniques (FORTE 89)* (North-Holland).

Rabin, M.O. (1989), "Efficient Dispersal of Information for Security, Load Balacing and Fault Tolerance", *Journal of the ACM*, 36 (2), pp. 335-348.

Ramamoorthy, C.V., Prakash, A., Tsai, W.T. and Usuda, Y. (1984), "Software Engineering: Problems and Perspectives", *IEEE Computer*, 17 (10), pp. 191-209.

Randell, B. (1975), "System Structure for Software Fault Tolerance", *IEEE Transactions on Software Engineering*, SE-1 (2), pp. 220-232.

Randell, B. (1987), "Design Fault Tolerance", in *The Evolution of Fault-Tolerant Computing*, (A. Avizienis, H. Kopetz and J.-C. Laprie, Eds.), vol. 1 pp. 251-270 (Springer-Verlag).

Randell, B. and Dobson, J.E. (1986) "Reliability and Security Issues in Distributed Computing Systems", in *Proc. 5th. Symp. on Reliability in Distributed Software and Database Systems*, Los Angeles, CA, USA, pp. 113-118 (IEEE).

Randell, B., Lee, P.A. and Treleaven, P.C. (1978), "Reliability Issues in Computer System Design", *ACM Computing Surveys*, 10 (2), pp. 123-165.

Rapps, S. and Weyuker, E.J. (1985), "Selecting Software Test Data using Data Flow Information", *IEEE Transactions on Software Engineering*, SE-11 (4), pp. 367-375.

Rennels, D.A. (1986) "On Implementing Fault-Tolerance in Binary Hypercubes", in *Proc. 16th. Int. Symp. on Fault Tolerant Computing (FTCS-16)*, Vienna, Austria, pp. 344-349 (IEEE).

Ribot, R. (1989) *Specifications of AMp Protocol for the Token Ring*, Delta-4 Document N°W89.125 (available from Bull SA, 1 Rue de Provence, 38423 Echirolles, France).

Richier, J.-L., Rodriguez, C., Sifakis, J. and Voiron, J. (1987) *XESAR: a Tool for Protocol Validation - User Manual*, Laboratoire de Génie Informatique (Edition 1.2).

Risks *Forum on Risks to the Public in Computer Systems*, ACM Committee on Computers and Public Policy, Computer-Distributed Newsletter.

Rivest, R., Shamir, A. and Adleman, L. (1978), "A Method of obtaining Digital Signatures and Public-key Cryptosystems", *Communications of the ACM*, 21 (2), pp. 120-126.

Robinson, D.C. (1988) *Domains - A Uniform Approach to Distributed Systems Management*, Imperial College of Science & Technology, London, UK, Research Report N°88/9.

Rodrigues, L. and Verissimo, P. (1991) *A Posteriori Agreement for Clock Synchronization on Boradcast Networks*, INESC, Technical Report (available from INESC, Rua Alves Redol, 9, 1000 Lisboa, Portugal).

Rodriguez, C. (1988) *System Specification and Validation in XESAR*, Doctoral Dissertation, Institut National Polytechnique, Grenoble, France (in French).

Rook, P. (1990), *Software Reliability Handbook* , (Elsevier Applied Science).

Roth, J.P., Bourricius, W.G. and Schneider, P.R. (1967), "Programmed Algorithms to Compute Tests to Detect and Distinguish between Failures in Logic Circuits", *IEEE Transactions on Electronic Computers*, EC-16 (October), pp. 567-579.

RTCA 178A (1985) *Software Considerations in Airborne Systems and Equipment Certification*, Radio Technical Commission for Aeronautics, Document N°RTCA/DO-178A.

Rudin, H. (1985), "An Informal Overview of Formal Protocol Specification", *IEEE Communications*, 23 (3), pp. 46-52.

Rushby, J.M. and Randell, B. (1983), "A Distributed Secure System", *IEEE Computer*, 16 (7), pp. 55-67.

Rutledge, L.S. (1987) *A Spatial Encoding Mechanism for Network Security*, PhD Thesis, Institute for Information Science and Technology, Washington .

Saltzer, J., Reed, D. and Clark, D. (1984), "End-to-End Arguments in System Design", *ACM Transactions on Computer Systems*, 2 (4).

Schlichting, R.D. and Schneider, F.B. (1983), "Fail-Stop Processors: An Approach to Designing Fault-Tolerant Computing Systems", *ACM Transactions on Computer Systems*, 1 (3), pp. 222-238.

Schneider, F.B. (1984), "Byzantine Generals in Action: Implementing Fail-Stop Processors", *ACM Transactions on Computing Systems*, 2 (2), pp. 145-154.

Schneider, F.B. (1986) "A Paradigm for Reliable Clock Synchronization", in *Proc. Advanced Seminar on Real-Time Local Area Networks*, Bandol, France, pp. 85-104 (INRIA).

Schneider, F.B. (1990), "Implementing Fault Tolerant Services using the State Machine Approach: a Tutorial", *ACM Computing Surveys*, 22 (4), pp. 229-319.

Segall, Z., Vrsalovic, D., Siewiorek, D., Yaskin, D., Kownacki, J., Barton, J., Rancey, D., Robinson, A. and Lin, T. (1988) "FIAT — Fault Injection based Automated Testing Environment", in *Proc. 18th. Int. Symp. on Fault-Tolerant Computing (FTCS-18)*, Tokyo, Japan, pp. 102-107 (IEEE).

Sha, L., Lehoczky, J.P. and Rajkumar, R. (1988) *Meeting the Timing Requirements of Distributed Real-time Systems*, Carnegie Mellon University.

Sha, L., Rajkumar, R. and Lehoczky, J.P. (1987) *Priority Inheritance Protocols: An Approach to Real-Time Synchronization*, Carnegie-Mellon University.

Shamir, A. (1979), "How to Share a Secret", *Communications of the ACM*, 22 (11), pp. 612-613.

Shaw, D.E. (1976), "Managing a Software Emergency", *Datamation*, 22 (11), pp. 48-50.

Sheridan, C.T. (1978), "Space Shuttle Software", *Datamation*, 24 (7), pp. 128-131.

Shrivastava, S.K., Dixon, G.D., Hedayati, F., Parrington, G.D. and Wheater, S.M. (1988) *A Technical Overview of Arjuna: a System for Reliable Distributed Computing*, University of Newcastle upon Tyne, Technical Report Series N°262.

Shrivastava, S.K., Ezhilchelvan, P.D., Speirs, N.A., Tao, S. and Tully, A. (1991) *Principal Features of the VOLTAN Family of Reliable Node Architectures for Distributed Systems*, University of Newcastle up Tyne, Technical Report.

Shurek, G. and Grumberg, O. (1990) "The Modular Framework of Computer-Aided Verification: Motivation, Solutions and Evaluation Criteria", in *Proc. Conf. on Automatic Verification (CAV 90)*, Rutgers, NJ, USA, Lecture Notes in Computer Science, 531 (Springer Verlag).

Siewiorek, D.P. and Johnson, D. (1982), "A Design Methodology for High Reliability Systems: The Intel 432", in *The Theory and Practice of Reliable System Design*, (D. P. Siewiorek and R. S. Swarz, Eds.), pp. 621-636 (Digital Press).

Singh, C., Billington, R. and Lee, S.Y. (1977), "The Method of Stages for Non-Markov Models", *IEEE Transactions on Reliability*, R-26 (2), pp. 135-137.

Sloman, M. (1987) *Distributed Systems Management*, Imperial College, Research Report N°87/6.

Sloman, M. (1989) "Domain Management for Distributed Systems", in *Proc. IFIP TC6/WG6.6 Symp. on Integrated Network Management*, Boston, MA, USA (B. Meandzija and J. Wescott, Eds.) (Elsevier Science Publishers BV (North Holland)).

Smith, R.M., Trivedi, K.S. and Ramesh, A.V. (1988), "Performability Analysis: Measures, an Algorithm, and a Case Study", *IEEE Transactions on Computers*, 37 (4), pp. 406-417.

Speirs, N.A. and Barrett, P.A. (1989) "Using Passive Replicates in Delta-4 to provide Dependable Distributed Computing", in *Proc. 19th. Int. Symp. on Fault-Tolerant Computing Systems (FTCS-19)*, Chigago, MI, U.S.A, pp. 184-190 (IEEE).

Srikanth, T.K. and Toueg, S. (1987), "Optimal Clock Synchronization", *Journal of the Association for Computing Machinery*, 34 (3).

Stadler, Z. and Grumberg, O. (1989) "Network Grammars, Communication Behaviours and Automatic Verification", in *Proc. Workshop on Automatic Verification Methods for Finite State Systems*, Grenoble, France, Lecture Notes in Computer Science, 407, pp. 151-165 (Springer Verlag).

Steiner, J.G., Neumann, C. and Schiller, J.I. (1988) "Kerberos: an authentication service for open network systems", in *Proc. USENIX Winter Conference*, Dallas, TX, USA, pp. 191-202.

Strigini, L. (1988) *Support of Design Fault-Tolerance in Delta-4 Software: Proposals*, Delta-4 Project Consortium, IEI Report N°B4-54.

Strigini, L. (1990) *Software Fault Tolerance*, Esprit Project N°3092 (PDCS: Predictably Dependable Computing Systems), PDCS First Year Report.

Strigini, L. and Avizienis, A. (1985) "Software Fault-Tolerance and Design Diversity: Past Experience and Future Evolution", in *Proc. 4th. IFAC Workshop SAFECOMP'85*, Como, Italy, pp. 167-172.

SVC200 *User's Guide to Ferranti Real-Time System V (SVC200)*, Ferranti International, Document N°U20410 (Release 1, Issue 1a) (available from Ferranti International, Wythenshawe, Manchester M22 5LA, UK).

Taylor, D.J. and Black, J.P. (1985) "Guidelines for Storage Structure Error Correction", in *Proc. 15th. Int. Symp. on Fault-Tolerant Computing (FTCS-15)*, Ann Arbor, MI, USA, pp. 20-22 (IEEE).

Taylor, D.J., Morgan, D.E. and Black, J.P. (1980), "Redundancy in Data Structures: Improving Software Fault Tolerance", *IEEE Transactions on Software Engineering*, SE-6 (6), pp. 383-394.

Thatte, S.M. and Abraham, J.A. (1978) "A Methodology for Functional Level Testing of Microprocessors", in *Proc. 8th. Int. Symp. on Fault Tolerant Computing (FTCS-8)*, Toulouse, France, pp. 90-95 (IEEE).

Theuretzbacher, N. (1986) "Using AI-methods to improve Software Safety", in *Proc. 5th. IFAC Workshop SAFECOMP'86*, Sarlat, France (W. J. Quirk, Eds.), pp. 99-105 (Pergamon Press).

Tohma, Y., Tokunaga, K., Nagase, S. and Murata, Y. (1989), "Structual Approach to the Estimation of the Number of Residual Faults Based on the Hyper-Geometric Distribution", *IEEE Transactions on Software Engineering*, SE-15 (3), pp. 345-355.

Trivedi, K.S. (1984), "Reliability Evaluation for Fault-Tolerant Systems", in *Mathematical Computer Performance and Reliability*, (G. Iazeolla, P. J. Courtois and A. Hordijk, Eds.), pp. 403-414 (North-Holland).

Tully, A. and Shrivastava, S.K. (1990) "Preventing State Divergence in Replicated Distributed Programs", in *Proc. 9th. Symp. on Reliable Distributed Systems (SRDS-9)*, Huntsville, AL, USA, pp. 104-113 (IEEE).

Veríssimo, P. (1988) "Redundant Media Mechanisms for Dependable Communication in Token-Bus LANs", in *Proc. 13th. Local Computer Network Conf.*, Minneapolis, USA, pp. 453-462 (IEEE).

Veríssimo, P. (1990) "Real-Time Data Management with Clockless Reliable Broadcast Protocols", in *Proc. Workshop on the Management of Replicated Data*, Houston, Texas, USA (IEEE).

Veríssimo, P. and Marques, J.A. (1990) "Reliable Broadcast for Fault-Tolerance on Local Computer Networks", in *Proc. 9th. Symp. on Reliable Distributed Systems*, Huntsville, AL, USA, pp. 54-63 (IEEE).

Veríssimo, P. and Rodrigues, L. (1990) "Reliable Multicasting in High-Speed LANs", in *NATO Advanced Research Workshop on Architecture and Performance issues of High-Capacity LANs and MANs*, INRIA, Sophia Antipolis, France.

Veríssimo, P., Rodrigues, L. and Baptista, M. (1989) "AMp: A Highly Parallel Atomic Multicast Protocol", in *Proc. SIGCOM'89 Symp.*, Austin, TX, USA, pp. 83-93 (ACM).

Veríssimo, P., Rodrigues, L. and Marques, J. (1987) "Atomic Multicast Extensions for 802.4 Token-Bus", in *Proc. FOC/LAN 87 Conference*, Anaheim, USA.

Voges, U. (1988), *Software Diversity in Computerized Control Systems*, 2 (Springer-Verlag).

Wakerly, J. (1978), *Error Detecting Codes, Self-Checking Circuits and Applications*, (Elsevier North-Holland).

Walker, D.J. (1988) "Bisimulation and Divergence in CCS", in *Proc. 3rd. Symp. on Logic in Computer Science (LICS 88)*, pp. 186-192 (IEEE).

Walter, C.J., Kieckhafer, R.M. and Finn, A.M. (1985) "MAFT: A Multicomputer Architecture for Fault-Tolerance in Real-Time Control Systems", in *Proc. 6th. Real-Time Systems Symp.*, pp. 133-140 (IEEE).

Wensley, J.H., Lamport, L., Goldberg, J., Green, M.W., Levitt, K.N., Melliar-Smith, P.M., Shostack, R.E. and Weinstock, C.B. (1978), "SIFT: The Design and Analysis of a Fault-Tolerant Computer for Aircraft Control", *Proc. IEEE*, 66 (10), pp. 1240-1255.

West, C. (1982) "Applications and Limitations of Automatic Protocol Validation", in *Proc. IFIP WG6.1 2nd. Int. Conf. on Protocol Specification, Testing and Verification*, pp. 361-371.

Weyuker, E.J. (1982), "On Testing Non-Testable Programs", *Computer Journal*, 25 (4), pp. 465-470.

Williams, T.W. (1983), "Design for Testability — A Survey", *Proceedings of the IEEE*, 71 (1), pp. 98-112.

Winskel, G. (1980) *Events in Computation*, PhD Thesis, University of Edinburgh .

Winskel, G. (1984), "Synchronization Trees", *Theoretical Computer Science*, 34 (1-2), pp. 33-82.

Wolper, P. and Lovinfosse, V. (1989) "Verifying Properties of Large sets of Processes with Network Invariants", in *Proc. Workshop on Automatic Verification Methods for Finite State Systems*, Grenoble, France, Lecture Notes in Computer Science, 407, pp. 68-80 (Springer Verlag).

Xu, J. and Parnas, D.L. (1990), "Scheduling Processes with Release Times, Deadlines, Precedence and Exclusion Relations", *IEEE Transactions on Software Engineering*, SE-16 (3).

Yamada, S. and Osaki, S. (1985), "Software Reliability Growth Modeling: Models and Applications", *IEEE Transactions on Software Engineering*, SE-11 (12), pp. 1431-1437.

Yau, S.S. and Cheung, R.C. (1975) "Design of Self-Checking Software", in *Proc. 1st. Int. Conf. on Reliable Software*, Los Angeles, CA, USA, pp. 450-457.

Zafiropulo, P., West, C., Rudin, H., Cowan, D. and Brand, D. (1980), "Towards Analysing and Synthesizing Protocols", *IEEE Transactions on Communication*, Comm-28 (4), pp. 651-661.

Zhao, W., Ramamirithan, K. and Stankovic, J.A. (1987), "Scheduling Tasks with Resource Requirements in Hard Real-Time Systems", *IEEE Transactions on Software Engineering*, SE-13 (5), pp. 564-577.

Ziegler, B.P. (1976), *Theory of Modeling and Simulation*, (John Wiley).

Author Affiliations

Jean Arlat	LAAS-CNRS
Peter Barrett*	MARI
David Benson	Ferranti International plc
Laurent Blain	LAAS-CNRS
Peter Bond	Ferranti International plc
Gottfried Bonn	IITB-Fraunhofer
Ulrich Bügel	IITB-Fraunhofer
Lynda Burrill**	SRD, AEA Technology
Marc Chérèque	Bull S.A.
Paolo Ciompi	IEI-CNR
Yves Crouzet	LAAS-CNRS
Yves Deswarte	LAAS-CNRS & INRIA
David Drackley	Ferranti International plc
Jean-Charles Fabre	LAAS-CNRS & INRIA
Brian Gilmore	Ferranti International plc
Susanne Graf	LGI-IMAG
Fabrizio Grandoni	IEI-CNR
Andrew Hilborne*	MARI
Nigel Howard	Ferranti International plc
Fritz Kaiser	IITB-Fraunhofer
Karama Kanoun	LAAS-CNRS
Jean-Claude Laprie	LAAS-CNRS
Marion Le Louarn	SEMA Group
Eliane Martins	LAAS-CNRS
Anne MacInnes	SRD, AEA Technology
Jean-Michel Pons	LAAS-CNRS
David Powell	LAAS-CNRS
Jean-Luc Richier	LGI-IMAG
Luís Rodrigues	INESC
José Rufino	INESC
Douglas Seaton	Ferranti International plc
Gérard Ségarra	Renault
Santosh Shrivastava	University of Newcastle upon Tyne
Neil Speirs	University of Newcastle upon Tyne
Lorenzo Strigini	IEI-CNR
Thomas Usländer	IITB-Fraunhofer
Paulo Veríssimo	INESC
Jacques Voiron	LGI-IMAG
John Waddington	Ferranti International plc

---

*   Now with the University of Newcastle upon Tyne
** Now with Longman Cartermill

The Delta-4 Consortium

BULL	Bull SA, B.S.I Tour Bull BSI (FRTB0812) 92309 Puteaux Cedex 74 PARIS-LA-DEFENSE France	Contact: Mr. P.Sommet Tel: +33/ 1.46.96.79.36 Telex: 616254 Fax: +33/ 1/46/96.88.98 e-mail: p_sommet@eurokom.ie
	Bull SA 1 rue de Provence B.P. 208 F-38432 ECHIROLLES France	Contacts: Mr. M.Chérèque Tel: +33/ 76.39.76.64 Telex: 980648 Fax: +33/ 76.39.77.02 e-mail: Marc.Chereque@ec.bull.fr
Crédit Agricole	Unibanque Groupe Crédit Agricole 40 rue d'Oradour-sur-Seine F-75015 PARIS France	Contact: Mr. G. Pellerin Tel: +33/ 1.43.23.61.63 Telex: 270566 Fax: +33/ 1.43.23.61.34
Ferranti International plc	Ferranti International plc Commercial and Industrial Systems Division Concord Business Park Wythenshawe MANCHESTER M22 5LA United Kingdom	Contacts: Mr. S.D.Drackley Tel: +44/ 61.499.9900 Telex: 668084 Fax: +44/ 61.499.6000
IEI-CNR	IEI-CNR Via S. Maria 46 I-56100 PISA Italy	Contact: Dr. F.Grandoni Tel: +39/ 50.500159 Telex: 590305 Fax: +39/ 50.554342 e-mail: procis@icnucevm.cnuce.cnr.it
IITB-Fraunhofer	Fraunhofer Institut FhG-IITB Fraunhoferstraβe, 1 D-7500 KARLSRUHE 1 Germany	Contact: Mr. G.Bonn Tel: +49/ 721.6091.301 Telex: 7825931 Fax: +49/ 721.6091.413 e-mail: u_buegel@eurokom.ie
INESC	INESC Rua Alves Redol, 9 P-1000 LISBOA Portugal	Contact: Dr. P.Veríssimo Tel: +351/ 1.52.62.27 Telex: 15696 Fax: +351/ 1.52.58.43 e-mail: paulov@inesc.inesc.pt
LAAS-CNRS	LAAS-CNRS 7 avenue du Colonel Roche F-31077 TOULOUSE France	Contact: Dr. D.Powell Tel: +33/ 61.33.62.87 Telex: 520930 Fax: +33/ 61.55.64.11 e-mail: dpowell@laas.fr

LGI-IMAG	LGI-IMAG B.P.53X F-38041 GRENOBLE France	Contact: Dr. J.Voiron Tel: +33/ 76.51.48.32 Telex: 980134 Fax: +33/ 76.54.66.15 e-mail: voiron@imag.fr
MARI	The MARI Group MARI House Old Town Hall Gateshead TYNE AND WEAR NE8 1HE United Kingdom	Contact: Mr. J.Clarke Tel: +44/ 91.490.1515 Telex: 537038 Fax: +44/ 91.490.0013 e-mail: jsc@mari.co.uk
SRD	SRD, AEA Technology Wigshaw Lane Culcheth WARRINGTON WA3 4NE United Kingdom	Contact: Ms. A.MacInnes Tel: +44/ 925.254482 Telex: 629301 Fax: +44/ 925.254539
RENAULT	Renault Automobile — SCE 0450 34 Quai le Gallo F-92109 BOULOGNE BILLANCOURT France	Contact: Mr. G. Ségarra Tel: +33/ 1.46.09.62.19 Telex: 204000 Fax: +33/ 1.46.09.63.30 e-mail: g_segarra@eurokom.ie
SEMA Group	SEMA Group 56 rue Roger Salengro F-94126 FONTENAY-SOUS-BOIS France	Contact: Mr. T.Vassigh Tel: +33/ 1.43.94.57.10 Telex: 264709 Fax: +33/ 1.48.77.72.02 e-mail: t_vassigh@eurokom.ie
UNIVERSITY of NEWCASTLE upon TYNE	University of Newcastle upon Tyne Computing Laboratory Claremont Tower Claremont Road NEWCASTLE-UPON-TYNE NE1 7RU United Kingdom	Contact: Prof. S.K.Shrivastava Tel: +44/ 91.222.8038 Telex: 53654 Fax: +44/ 91.222.8232 e-mail: Santosh.Shrivastava@newcastle.ac.uk